Handbook on Digital Twin and Artificial Intelligence Techniques for Rail Applications

With contributions from experts from around the world, this handbook aims to systemize the existing experience and knowledge that can be used to the development of more efficient and controlled railway systems. As a result, this handbook showcases the modern methods, methodologies and frameworks for the development of DT and AI architectures and apparatus in the area of the existing railway systems and transport engineering tasks. The chapters cover such varied and specialized topics as the processes related to the transformation of a physical twin into a digital twin; the application of data-driven and physics-based simulation approaches in the development of digital twins; asset management application tasks with the implementation of DT and AI; and the experimental and field applications of the DT and AI concepts and technologies in railway transport system design and predictive maintenance tasks.

Handbook on Digital Twin and Artificial Intelligence Techniques for Rail Applications is essential reading for engineers, practitioners and researchers involved in the development of railway transport and transit systems.

Handbook on Digital Twin and Artificial Intelligence Techniques for Rail Applications

Edited by

Maksym Spiryagin, Stefano Bruni, and Colin Cole

CRC Press
Taylor & Francis Group
Boca Raton London New York

CRC Press is an imprint of the
Taylor & Francis Group, an **informa** business

Designed cover image: Shutterstock

First edition published 2025
by CRC Press
2385 NW Executive Center Drive, Suite 320, Boca Raton FL 33431

and by CRC Press
4 Park Square, Milton Park, Abingdon, Oxon, OX14 4RN

CRC Press is an imprint of Taylor & Francis Group, LLC

ISBN: 9781032776026 (hbk)
ISBN: 9781032794631 (pbk)
ISBN: 9781003492146 (ebk)

DOI: 10.1201/9781003492146

Typeset in Times
by codeMantra

Contents

Preface

Digital twin (DT) and artificial intelligence (AI) technologies have been developing exponentially during recent decades, this being dictated by the demand to develop increasingly complex systems or, in other words, the ultimate *system of systems* based on the concepts of Industry 4.0, IoT, and big data, which can easily provide answers to the existing multidisciplinary tasks related with the optimization and improvement of design and operational processes and predictive maintenance. The general railway is itself a *system of systems* represented by multiple amounts of moving trains with their own subsystems and operating on infrastructure with other subsystems such as tracks, bridges, tunnels, stations, building structures, power supply, signaling, and communications, each of which consists of other multiple subsystems, all interconnected with each other using new computerized technologies establishing different subsystems in a virtual metaverse. In addition, there is always the presence of humans that can be classified as additional subsystems like pedestrians, customers, and railway personnel such as drivers and maintenance staff. Addressing the obvious rail domain challenges clearly requires quite specific knowledge, not only in DT and AI technologies but also calls for tailoring the knowledge to the railway world specificities. Considering that the developments and deployments of novel technologies for railway systems are comparatively slow to be accepted in the rail domain due to the existing conservatism and safety-first guiding rules, this means that technologies should be heavily tested prior to introduction. This scenario also raises a high potential for a DT that is commonly defined as an accurate virtual duplicate (replica) of a real physical system to be considered not only at the operational stage but also at the design stage, which also leads to the need to resolve some challenges with the transition of DT technologies from existing to non-existing physical system designs. When we talk generally about design in the rail domain, we assume that applications should be based on a combination of modeling and data, and this represents a high potential for analysis and prediction techniques to be built using integrated DT and AI technologies and their basic principles in rail domain applications.

The aim of this handbook is to systemize the existing experience and knowledge that can be used in the development of more efficient and controlled railway systems. As a result, the handbook aims to show the modern methods, methodologies and frameworks for the development of DT and AI architectures and apparatus in the area of the existing railway systems and transport engineering tasks.

However, considering that the area is still in the earliest stages of development and limited developments in the existing application cases, the editors should emphasize that this handbook can provide only some limited directions to assemble the existing DT and AI theories and technologies in one entire system of systems in the rail domain. More standardization and integration work should be performed in both focused areas of this handbook because, for railways, both DT and AI are novel research areas, and the system architectures and developmental frameworks require more investigations and improvements based on the design and deployment scenarios. In this regard, this handbook aims at both summarizing the present status of research and identifying emerging trends toward the full implementation of DT and AI in railways.

The editors of this handbook hope that readers find the presented materials useful and systemized. The theoretical background, concepts, design and deployment cases and applications for railway transport and transit systems should not only lead to safe and reliable operations from the existing realities but also provide engineers and researchers working in this field the directions to future significant challenges in the development of the tools and techniques to build more modern railway systems.

Maksym Spiryagin, Stefano Bruni, and Colin Cole
November 2024

Acknowledgments

The editors would like to thank their colleagues and friends for their assistance with the preparation of this book. In addition, the editors would like to thank their families for their significant support during the work on this handbook.

The contributors are also recognized with special thanks by the editors for their efforts to deliver the best content that can be produced in their areas of expertise, as well as their support and significant contribution to the development of the content of each chapter.

Our sincere thanks also go to the team at CRC Press who worked on this book. Their acceptance of our proposal followed by ongoing support and helpful instructions during the writing process made the project much more systemized.

The editors would like to sincerely thank Mr. Tim McSweeney, who is an Adjunct Research Fellow at the Centre for Railway Engineering of Central Queensland University, for the professional proofreading of all materials presented in this handbook.

Our humble apologies if we have forgotten to mention anyone. Doubtless, there are also others we are not aware of behind the scenes who have helped with our project.

Finally, our huge thanks go to all those involved.

Declaration

We would like to acknowledge the usage of the special artificial intelligence tools in this book.

The contributors of Chapters 1, 4, and 19 acknowledge the use of the Gemini software platform (Gemini 1.5 Flash) in helping them to review and grammatically and linguistically improve their writing at the final stage of preparing the chapters. The contributors of Chapters 1, 4, and 19 reviewed the feedback generated by the software critically and, based on this, revised the writing and presented it in their final forms.

The contributors of Chapter 3 acknowledge the use of Grammarly (version 8.912.0) in helping them to grammatically and linguistically improve their writing at the final stage of preparing the chapter. The contributors of Chapter 3 reviewed the feedback generated by the software critically and, based on this, revised the writing and presented it in its final form.

The contributor of Chapter 18 acknowledges the use of the Anthropic software platform (Claude 3.5 Sonnet), in helping him to review and grammatically and linguistically improve their writing at the final stage of preparing the chapter. The contributor of Chapter 18 reviewed the feedback generated by the software critically and, based on this, revised the writing and presented it in its final form.

Editors

Maksym Spiryagin is Professor of Engineering and the Deputy Director of the Centre for Railway Engineering at Central Queensland University. Prof. Maksym Spiryagin's current research work includes almost all stages of the development and realization of an engineering project from the definition of its aims to the development of design concepts and, if required, advanced computer simulation approaches and physics-based digital twin techniques including their verification by experimental programs and the development of the final solutions. Prof. Spiryagin is Associate Editor of the *Journal of Rail and Rapid Transit*, the *Journal of Vehicle System Dynamics* and the *Journal of Railway Engineering Science*. Prof. Spiryagin has published five authored books (including *Design and Simulation of Heavy Haul Locomotives and Trains* in 2017 and *Rail Vehicle Mechatronics* in 2021, both CRC Press) and two edited books (including *Handbook of Railway Vehicle Dynamics*, Second Edition in 2020, CRC Press), and he has more than 330 other scientific publications. Prof. Spiryagin is a Chartered Professional Engineer, RPEQ and RPEV in Australia and a Chartered Engineer in the UK.

Stefano Bruni is Full Professor at the Department of Mechanical Engineering, Politecnico di Milano, Italy, where he teaches applied mechanics and dynamics. He is the leader of the Railway Dynamics research group, carrying out research on rail vehicles and their interaction with the infrastructure. Prof. Bruni authored over 330 scientific papers, mostly related to rail vehicle dynamics, train–track interaction, wheel–rail contact forces, damage and wear of wheels and rails, active control and condition monitoring of rail vehicles and pantograph–catenary interaction. He is/has been the lead scientist for several research projects funded by the railway industry and by the European Commission. He is past Vice-President of the International Association for Vehicle System Dynamics (IAVSD) and was chairman of the IAVSD'05 International conference held in Milano in 2005. He is Editorial Board member for some international journals in the field of Railway Engineering.

Colin Cole is Professor of Engineering and Director of the Centre for Railway Engineering at Central Queensland University, Rockhampton, Australia. He has worked in the Australian rail industry since 1984, starting with six years in mechanized track maintenance for Queensland Railways. Since then, he has focused on a research and consulting career involving work on track maintenance, train and wagon dynamics, train control technologies and the development of onboard devices. He has been extensively engaged with industry via the past nationally funded rail CRC programs and the Australasian Centre for Rail Innovation. His PhD was in longitudinal train dynamics modeling. He has authored and/or co-authored over 300 technical papers, three books, numerous commercial research and consulting reports and has developed two patents relating to in-cabin locomotive technologies. He has continued to develop railway and road-specific modeling and software tools with a focus on traction, vehicle stability and energy consumption. Prof. Cole is a Chartered Professional Engineer and RPEQ in Australia.

List of Contributors

Sanjar Ahmad is a senior research officer at Central Queensland University (CQU), Australia, with more than a decade of experience in railway vehicle dynamics modeling and multibody simulations. Graduated in mechanical engineering discipline, he obtained his master's and Ph.D. degrees from CQUniversity, Australia. He has been involved in more than 20 research and commercial consultancy activities in the field of railway engineering since 2018. His primary research focus is on railway vehicle system dynamics, physics-based digital twins, multibody modeling systems, rail wear, and railway track systems.

Esteban Bernal Arango is a research fellow at the Centre for Railway Engineering, Central Queensland University, Australia. During his doctoral dissertation, he studied innovative on-board condition monitoring technologies for railway vehicles. He has participated in research and consultancy projects involving railway digital twins, artificial intelligence, vehicle dynamics simulations, condition monitoring systems, and wheel-rail wear and damage studies.

Wolfgang Birk is a CTO at Predge AB and professor of automatic control. He holds an M.Sc. degree in electrical engineering from the University of Saarland, Germany (1997); a Ph.D. degree in automatic control from Luleå University of Technology (2002); and a Professor of Automatic Control (2015). Since 2021, he has been a CTO at Predge. He has a background in the development of condition monitoring systems, process control systems, and active safety systems. His research work has led to innovations for control and monitoring, increasing resource efficiency, utilization, and availability in energy systems, process industrial systems, and railway systems. In the railway sector, his main interest and expertise is the use of on-board, way-side, and track monitoring systems for condition monitoring, condition-based maintenance, and predictive maintenance using fully automated analytics and artificial intelligence approaches. It includes the development and provisioning of predictive decision support and automated decision-making.

Chris Bosomworth has been a software engineer and an analyst at Insyte Solutions since 2007. During this time, he collaborated extensively with the team at the Centre for Railway Engineering, CQUniversity on a range of hardware and software products in the Heavy Haul rolling stock and physics simulation space. This collaboration also included a number of authored research journal articles and book chapters. Since 2015, advanced train control, with a particular focus on explicitly solved in-train force control, unlimited remote group distributed power control, and regenerative dynamic brake control, has been a key R&D project and passion for the team at Insyte Solutions. Other areas of R&D include electronically controlled pneumatic brake systems and renewable locomotives.

Nicola Bosso is a full professor at the Department of Mechanical and Aerospace Engineering at Politecnico of Torino. He leads the railway engineering laboratory at Politecnico di Torino. Currently, he works in the Design Simulation and Experimentation of Industrial and Railway Systems. He also works in the field of longitudinal train dynamics, machine learning, digital twins, numerical simulation of brake problems, and railway maintenance.

Marco Carnevale, Ph.D. in mechanical systems engineering with Doctor Europaeus certification, is an associate professor of applied mechanics at Università di Pavia (Italy). He is the author of more than 75 scientific papers in journals and conference proceedings (Scopus-indexed) and the author of the national patent "Damping device for railway pantographs, and railway vehicle using this device". He is the associate editor of *Proceedings of the Institution of Mechanical Engineers, Part F: Journal of Rail and Rapid Transit* (SAGE Publications) and a member of the editorial board of the journal *Designs* (MDPI). His research interests include the modeling of railway vehicles and infrastructure, as well as their condition monitoring. He also focused on applied robotics, immersive simulation systems, and human-robot interaction. Since July 2019, he has been a member of the board of directors of A.T.B. Servizi S.p.a., the public transportation company serving the city of Bergamo, Italy, and its surrounding areas.

Carlos Casanueva received a Ph.D. in mechanical engineering from the University of Navarra in Donostia-San Sebastian, Spain. He then moved to KTH Royal Institute of Technology in Stockholm as a postdoctoral fellow, where he is currently an associate professor in the engineering mechanics department and part of the leadership team at the Centre for ECO2 Vehicle Design. He teaches about rail vehicles at bachelor and master levels, and his main research interests are rail vehicle dynamics, wheel-rail interaction and damages, and railway traction, with a special focus on subsystem interaction modeling. He is the director of the 5-year program in aeronautical and vehicle engineering at KTH and leads educational development, from developing individual course modules to entire program organization and setup, with a special interest in the integration of environmental and social sustainability in engineering disciplines.

Luigi Castagna is a Ph.D. candidate in the Department of Mechanical Engineering at Politecnico di Milano, Italy. He earned his master's degree in mobility engineering in 2021. His research interests focus on public transport systems simulation and econometric models to benchmark transportation systems efficiency. He has contributed to various transportation projects in Italy and participated in the task force Urban Rail Benchmarking led by the World Conference on Transportation Research Society (WCTRS).

Gabriele Cazzulani received his Ph.D. in mechanical engineering at Politecnico di Milano in 2012 and is currently an associate professor of applied mechanics in the Department of Mechanical Engineering at Politecnico di Milano. He is the author of more than 100 publications in international journals and conferences. In recent years, his research activities have focused on structural health monitoring techniques applied to both mechanical systems and civil structures. Other research activities involve smart structures and metamaterials for wave propagation control and for wave-based structural health monitoring. He is the co-founder of the innovative startup Displaid s.r.l., which aims at introducing innovative monitoring strategies on the market, mainly based on the development of new wireless sensors, together with AI-based data processing.

Yit Hong Choo (Kelvin) received his Ph.D. in optimization and operations research and bachelor's in civil engineering (Honours) and is currently a research fellow in operations analytics at the Institute for Intelligent Systems Research and Innovation (IISRI) at Deakin University in Australia, with sponsorship from the Rail Manufacturing Cooperative Research Centre (RMCRC). His research focuses on designing and developing advanced multi-objective optimization algorithms, specifically for addressing complex maintenance scheduling problems for rolling stock. He currently holds the position of Research Fellow at IISRI, where he collaborates with industrial partners in the transportation sector to create AI-based tools and solutions for video and image analytics.

Pierluigi Coppola, Ph.D. in road infrastructure and transportation systems, is a full professor of transportation planning at the Politecnico di Milano (Italy) and lecturer of the "Transportation Networks and Smart Mobility: Methods and Solutions" summer course at the Massachusetts Institute of Technology (MIT). He is the author of more than 130 scientific papers in journals and conference proceedings and a member of the editorial boards of transportation research journals such as *Case Studies on Transport Policy* (Elsevier). His research interests include land-use and transports interactions, sustainable urban mobility planning, travel demand behavior modeling, and forecasting. He has been the Chair of the Board of Directors of the Association for European Transport (AET) and is currently a Steering Committee member of the World Conference of Transportation Research Society (WCTRs). He has been involved in several transportation projects in Italy, Netherlands, Portugal, and Spain. He is an expert advisor of the Italian Ministry of Infrastructure and Transport and Coordinator of the Pillar 2 (Transport sub-group) of the European Strategy of the Adriatic-Ionian Region (EUSAIR).

Douglas Creighton is a Deakin distinguished professor, with Ph.D. in industrial engineering, Honours in systems engineering, and bachelor's in Physics. He is a professor of systems engineering and the director of Deakin University's Institute of Intelligent Systems, Research, and Innovation (IISRI), leading a team of 100 researchers at the forefront of innovation and technology in defense, transport, and advanced manufacturing. He is a research leader in transport and mobility. In 2020, he was elected to the ARAs National Rail Industry Group Executive Committee. His research focuses on the theory and application of modeling, simulation, and systems thinking to further understanding, derive insight, and support decision-making in complex engineered systems. He has completed numerous projects over the last 20 years with impact in automotive, aerospace, and rail, delivering productivity, sustainability, accessibility, and capability benefits.

Egidio Di Gialleonardo graduated in mechanical engineering in 2007 and obtained his Ph.D. in mechanical systems engineering in 2011. During his Ph.D., he was a visiting Ph.D. student at the Technical University of Denmark (DTU). From 2015 to 2021, he was a researcher at the Politecnico di Milano, Department of Mechanical Engineering, and since November 2022, he has been an associate professor. His research interests include dynamics, vibrations, and control of mechanical systems, with applications in railway vehicle dynamics, train-track interaction, and diagnostics of vehicles and railway infrastructure. In these fields, he is the author of over 90 scientific publications, 35 of which are in peer-reviewed international journals. He has received several awards for his publications.

Johannes Edelmann received his Ph.D. in 2008 and his venia docendi for technical dynamics in 2011 from TU Wien, Austria. From 2018 to 2021, he was an associate professor at the Institute of Mechanics and Mechatronics, TU Wien. Since 2021, he has been a professor of technical dynamics at the Institute of Mechanics and Mechatronics. His research interests include applied multi-body dynamics, stability of dynamical systems, and vehicle dynamics modeling and control.

Francesco Flammini received his master's (2003), cum laude, and doctoral (2006) degrees in computer engineering from the University of Naples Federico II, Italy. He is currently a full professor of computer science at the University of Florence (Italy), and the head of Trustworthy Autonomous Systems (TAS) Group of Dalle Molle Institute for Artificial Intelligence (IDSIA USI-SUPSI) at the University of Applied Sciences and Arts of Southern Switzerland. He has been a professor of computer science with a focus on cyber-physical systems at Mälardalen University (Sweden) and the chair of the cyber-physical systems (CPS) environment at Linnaeus University (Sweden). He has worked for 15 years in private and public companies, including Ansaldo STS (now Hitachi Rail) and IPZS (Italian State Mint and Polygraphic Institute), on large international projects addressing intelligent transportation systems, critical infrastructure protection, and cybersecurity, as a technical leader and unit head. He is an IEEE senior member, an IEEE Computer Society distinguished visitor, an ACM distinguished speaker, the chair of the IEEE SMC Technical Committee on Homeland Security, and a member of the BoG of the IEEE SMC Society. He has (co)authored more than scientific publications, and he has served as a chair, invited speaker, steering/program committee member, and editor for many international conferences, books, and journals. For his research, he has won several prizes, including the latest TRA Visions Senior Researcher Competition Award 2024 and the Dalle Molle Award for the Quality-of-Life Label 2024.

Xiaohua Ge (Senior Member, IEEE) received his Ph.D. in computer engineering from Central Queensland University, Rockhampton, QLD, Australia, in 2014. From 2015 to 2017, he was a research fellow with the Griffith School of Engineering, Griffith University, Gold Coast, QLD, Australia. He is currently a senior lecturer with the School of Science, Computing and Engineering Technologies, Swinburne University of Technology, Melbourne, VIC, Australia. His research interests include networked, secure, and intelligent control and estimation theories, and their applications in mobile robots, electric vehicles, autonomous road and rail vehicles, connected vehicles, intelligent transportation systems, and so on. He is currently an associate editor of *IEEE Transactions on Industrial Informatics*; *IEEE Transactions on Systems, Man, and Cybernetics: Systems*; *IEEE Transactions on Intelligent Vehicles*; and *IEEE/CAA Journal of Automatica Sinica*. He was an associate editor of *IEEE Transactions on Circuits and Systems-II: Express Briefs* and a guest editor of several prestigious journals such as *Vehicle System Dynamics* and *Control Engineering Practice*.

Florian Klinger received his Ph.D. in 2018 from the Faculty of Mechanical and Industrial Engineering at TU Wien, Austria. Since 2018, he has been a senior lecturer at the Institute of Mechanics and Mechatronics at TU Wien. His research interests include vehicle dynamics modeling and control with a focus on two-wheeled vehicles and on vertical dynamics, and strategies for digital twin modeling in vehicle dynamics applications.

Ivano La Paglia received his Ph.D. in mechanical engineering at Politecnico di Milano in 2022. Since February 2023, he has been a researcher in the Department of Mechanical Engineering at Politecnico di Milano, where he is involved in the field of condition monitoring of railway infrastructure. His research focuses on the development of monitoring techniques based on the adoption of sensors installed on-board the train, aimed at the predictive maintenance of the infrastructure. Other research activities include railway signaling systems, rail vehicle dynamics, rolling tire dynamics, and vehicle NVH. He is the author of several scientific publications in journals and proceedings of international congresses.

Jonathan Leung is an assistant professor at KTH Royal Institute of Technology in Stockholm. He holds an M.Sc. in mechanical engineering and a Ph.D. in material science and engineering from the Georgia Institute of Technology, Atlanta, USA. He has a background in understanding and predicting fatigue and damage behavior of engineering materials under extreme conditions and harsh environments. By developing custom computational tools that capture multi-physics mechanisms for damage formation and integrating them into system-level fatigue prediction tools, I increase the ability to predict the influence of various loading, material, and operational factors at multiple length scales. He is involved in projects on vehicle dynamics, wheel-rail interface wear fatigue and damage prediction, and structural fatigue assessments for vehicle reliability.

Giovanni Lugaresi is an assistant professor in the Department of Mechanical Engineering at KU Leuven. In 2018, he co-founded the The Factory laboratory at Politecnico di Milano, an innovative space based on the construction of lab-scale models of manufacturing systems using easy-to-access materials such as LEGO® and Arduino. His research focuses on production planning and control, smart manufacturing systems, digital twins, robust optimization for the design, development, and operation of innovative and sustainable production systems.

Marco Macchi is a full professor at the Department of Management, Economics and Industrial Engineering at the Politecnico di Milano. His teaching and research activities regard industrial technologies, asset lifecycle management, operations and maintenance management, and smart manufacturing. Serving the scientific community, he is currently the chair of the technical committee IFAC TC 5.1 Manufacturing Plant Control; he is a member of the IFIP WG 5.7 Advances in Production Management Systems; he is the co-leader of the Special Interest Group (SIG) on "Product and Asset

Lifecycle Management" at the IFIP TC5 WG 5.7 Advances in Production Management Systems; he is a fellow of the International Society of Engineering Asset Management (ISEAM). Concerning his role in scientific journals, he is an associate editor of the *Journal of Intelligent Manufacturing* and a member of the International Editorial Board of Production Planning & Control: the Management of Operations and of the International Editorial Board of Reliability Engineering & System Safety.

Matteo Magelli is a researcher at the Department of Mechanical and Aerospace Engineering at Politecnico di Torino, Italy. He achieved his Ph.D. in 2023 with a thesis on numerical and experimental tools for the simulation of braking operations. His research activity concerns the numerical simulation of several phenomena related to railway vehicle dynamics (wear, wheel-shoe thermal behavior in tread braking operations, running safety and stability, etc.) with longitudinal train dynamics and multibody codes as well as the experimental investigation of wheel-rail adhesion phenomena with laboratory test rigs. Recent activities are dealing with the integration of longitudinal train dynamics within multibody codes by means of digital twins. He is a co-author of more than 20 scientific works.

Travis March is a principal design manager within WSP's Major Projects Team with over 24 years in the engineering industry and is a chartered professional engineer and RPEQ. He has worked on a wide variety of major transport infrastructure projects across Australia and New Zealand in both road and rail sectors. His background as a civil designer allows him to understand the fundamental design requirements and how they must be applied and where they can be safely challenged. Throughout his career, Travis has been on the digital engineering frontline and pushing the use of digital models and attributes to help successfully deliver both brownfield and greenfield construction projects, and from this experience, he has developed a key understanding of data management and digital workflows throughout a project life cycle.

Andrea Matta is a full professor of manufacturing and production systems in the Department of Mechanical Engineering at Politecnico di Milano and a guest professor at Shanghai Jiao Tong University. He graduated in industrial engineering at Politecnico di Milano where he developed his teaching and research activities since 1998. He was a distinguished professor at the School of Mechanical Engineering at Shanghai Jiao Tong University from 2014 to 2016. He has been a visiting professor at Ecole Centrale Paris (France), University of California at Berkeley (USA), and Tongji University (China). He is the scientist responsible for the Research Area Design and Management of Manufacturing Systems at MUSP (Laboratory for Machine Tools and Production Systems). His research area includes analysis, design and management of manufacturing and healthcare systems. He has published 130+ scientific papers in international and national journals/conference proceedings. He has been the editor in chief of *Flexible Services and Manufacturing Journal* since 2017; he is an editorial board member of *OR Spectrum* and *IEEE Robotics and Automation Letters*. He is the co-chair of the technical committee of IEEE RAS Sustainable Production Automation. He is a member of the scientific committee at several international conferences. He was awarded the Shanghai One Thousand Talent and Eastern Scholar in 2013. He has been cited among the leading scholars in production research by the *International Journal of Production Research*.

Jacek Mocki is a qualified and highly experienced chartered transport engineer confirmed with a master's and a Ph.D. degrees. He is a good analytical and strategic thinker and is clear in the outcomes he expects from the activities he is undertaking. With over 20 years of professional and scientific experience in transport and transport systems at all levels (consulting, engineering, product supply, manufacturing, construction, maintenance, operation, governance, and software development), he provides specialized services to ensure the outcome is worth the customer journey. He has extensive systems knowledge, experience, and ability to deliver outstanding results.

Thomas Mond graduated with honours from the Louvain School of Engineering (Belgium) in 2016, where he obtained his master's electromechanical engineering with a professional focus on energy and automotive engineering, under the supervision of Prof. Paul Fisette. He has since relocated to Australia where he is now leading a team of 10 engineers for Sogeclair Rail Simulation (formerly Oktal-Sydac) in Adelaide and coordinating the technical strategies to enhance simulation modeling in Pune (India) and Toulouse (France). He is a member of Engineers Australia and the RTSA Technical Society. His fields of expertise include simulation modeling of locomotive control, power electronics, and complex pneumatic systems for freight rail applications. He is currently directing the development of Sogeclair's simulation products destined to transform freight train driver training and operation analysis.

Elena Napoletano received her master's degree, cum laude, in management engineering (2015) and her doctoral degree in information technology and electrical engineering (2018) from the University of Naples Federico II, Italy. She is currently a research fellow at the University of Naples Federico II. She participated in national and international research projects addressing the design, modeling, validation of railway-embedded systems, intelligent transportation systems, and smart mobility. She is the co-founder and CEO of the academic spin-off company *DAS2 s.r.l.* (Distributed Automation System Solutions), approved by the University of Naples Federico II, which aims at promoting the emerging paradigms of Mobility As A Service (MAAS), Smart and Sustainable Mobility, and Smart Roads.

Anjum Naweed is a professor, an applied researcher, and an Australian Research Council Future Fellow based at the Appleton Institute, the Adelaide-based campus of Central Queensland University. He is internationally regarded for his applied and theoretical advancements in human factors, particularly in safety-critical systems and transport domains. His research interests include complex decision-making and knowledge representation in skilled work, job design, creative and cultural applications of simulation, and systems thinking. His expertise is in understanding the relationship between people, technologies, tools, environments, and systems, and practically translating this across complex domains. He has co-edited books, written over 150 peer-reviewed articles, and secured over AU$7.5 million in research funding. At the Appleton Institute, Anjum leads the Human Factors and Operational Readiness research cluster. He currently serves as a scientific editor for *Applied Ergonomics*, an international journal aimed at all those interested in applying ergonomics and human factors in the design, planning, and management of technical and social systems. He has been a scientific convenor for the Australasian Simulation Congress on five consecutive occasions and for the Annual Conference of the Human Factors and Ergonomics Society of Australia twice. His research has received numerous accolades. In 2012, he was awarded the Body of Knowledge prize from Simulation Australasia and the John Lane Award from the Human Factors and Ergonomics Society of Australia (HFESA). He has twice been a recipient of the Vice-Chancellor's Award for Outstanding Researcher of the Year (2013; 2018), recognizing his excellence in research activity, publication, and engagement. In 2022, he received the Simulation Achievement Award for clear and unambiguous, outstanding contributions to science, technology, and education, and Simulation Australia's Lifetime Membership Award.

Jake Pisano graduated from Griffth University with a bachelor's with Honours degree in electrical and electronic engineering. He completed his thesis with an industry partner on developing an asset condition assessment tool to be used in rail. He now works as a consultant primarily focusing on heavy asset industries and optimizing operations.

Adalberto Polenghi is an assistant professor in the Manufacturing Group at the Department of Management, Economics and Industrial Engineering at Politecnico di Milano. He earned his Ph.D. in industrial engineering from the same university with a dissertation focused on improving maintenance practice through ontological models for advanced information management and integration. His current research interests include operations management, particularly maintenance and industrial asset management, with expertise in data modeling, ontology engineering, reliability engineering, and data analytics. He is actively involved in international and national research projects, as well as industrial collaborations, in various sectors, like manufacturing, railway, and utilities. He teaches and manages undergraduate and graduate courses at Politecnico di Milano, specializing in the design and management of production systems. He is part of the Core Faculty of the POLIMI GSOM (Politecnico di Milano Graduate School of Management), for which he leads various courses on maintenance and industrial asset management. Additionally, he is part of the IFAC TC5.1 (Manufacturing Plant Control), AMEST WG (Advanced Maintenance Engineering, Services and Technologies Working Group), and a member of the UNI/CT 153 – Industrial Automation Committee (Italian Standards Body).

Leron Postelnik is a senior modeling engineer at Sogeclair Rail Simulation. He received his Ph.D. in control systems in 2016 from the University of Auckland. He has over six years of industry experience in the modeling and simulation of rail vehicle systems for driver training simulators.

MD Mamunur Rashid, Ph.D. (Monash), M.Eng. (Kyung Hee Uni), B.Sc. Eng. (KUET), MACS, MIEEE, is a senior lecturer in ICT at CQUniversity. He received his Ph.D. in computer science from the Faculty of Information Technology (IT), Monash University, Australia. His research interests include knowledge discovery and data mining, Internet of Things (IoT), cybersecurity, and big data analytics. His research works have been published in prestigious journals in the computer science discipline including *IEEE Transactions on Parallel and Distributed Processing, IEEE Transactions on Computers*, and mainstream IEEE conferences. He has served the professional community in the capacity of reviewers in international conferences and journals (e.g., *Future Generation Computer Systems, Journal of Network and Computer Applications (JNCA), IEEE Transactions on Industrial Electronics*, etc.). He is an experienced IT professional with more than 15 years of experience in academics and research who earned a Ph.D. in computer science from Monash University, Australia in October 2016. He has a strong commitment to academic achievements and is keen to learn new topics to teach to the students. He is a passionate learner of new technologies from IT and cutting-edge research and a dedicated team player with a knowledge-sharing attitude.

Stefano Ricci received his master's degree, cum laude, in civil engineering – transport section (1988) and his doctoral degree in transport engineering (1993) from Sapienza University of Rome. He is currently a full professor of transport engineering and director of the Postmaster Course in Railway Infrastructure and Systems Engineering at Sapienza University of Rome. He has worked for 8 years in the international consulting company SOTECNI, specializing in transport engineering, and 28 years in transport engineering education and research, coordinating academic programs and participating in 23 international research projects, mainly EU-funded, and 62 national research projects addressing railway engineering, maritime transport and transport economy. He supervised 19 Ph.D. thesis and 294 B.Sc. and M.Sc. theses. He is a founding member of the European Rail Research Network of Excellence (EURNEX) and a member of the European Railway Research Advisory Council (ERRAC). He has authored or co-authored 234 scientific publications and 19 educational publications. He is the editor in chief of *IF – Ingegneria Ferroviaria*. He has served as a chair, invited speaker, steering/program committee member, and editor for many international conferences, books, and journals. He is the founding member of the academic spinoff company Development and Innovations in Transport Systems (DITS), participated by Sapienza University and Hitachi Rail.

Irene Roda is Associate Professor in the Department of Management, Economics and Industrial Engineering at the Politecnico di Milano, and she holds a Ph.D. in industrial engineering. Her main research and teaching areas are related to industrial asset lifecycle management, operations and maintenance management, and smart manufacturing. She is a member of the IFIP TC5 WG 5.7 Advances in Production Management Systems and a co-leader of the Special Interest Group on "Product and Asset Lifecycle Management" at the IFIP TC5 WG 5.7. She is a member of the AMEST Working Group (Advanced Maintenance Engineering, Service and Technology) of the IFAC TC 5.1. on Manufacturing Plant Control. She is Community Engagement editor of the international journal *Production Planning & Control*: The Management of Operations, Taylor & Francis. She is the author of several scientific publications in national and international journals.

Fariza Sabrina, Ph.D. (UNSW), MER (USyd), GCETT (Feduni), B.Sc. Eng (BUET), HNC (Bus&Fin) (Coverntry Uni, UK), is a passionate ICT professional with a career spanning across universities, industries, and government research agencies. She received her Ph.D. in computer science and engineering from the University of New South Wales (UNSW), Master of Engineering (Research) from The University of Sydney (USYD), and Bachelor of Science in Engineering (Hons.) degree in electrical and electronics engineering from Bangladesh University of Engineering and Technology (BUET), Bangladesh. After completing her Ph.D., Fariza worked as a postdoctoral research fellow and then as a research scientist in the Networking Technologies Lab, Commonwealth Scientific and Industrial Research Organisation (CSIRO) ICT Centre, Sydney and worked on a wide variety of R&D and consulting projects. Currently, she is a senior lecturer-ICT and the discipline lead for networking and information security in the School of Engineering and Technology at CQUniversity Sydney. Prior to joining CQUniversity, Fariza worked as a senior lecturer and academic coordinator at the School of IT and Engineering for the Sydney Campus at Melbourne Institute of Technology. Her past industry experience also includes working as an IT consultant on large projects. Currently, her research interests include networking and information security, artificial intelligence, machine learning, blockchain, Internet of Things, and cybersecurity.

Hong Shen, Ph.D., PLic, M.Sc., B.Sc., joined the Central Queensland University (CQU) in November 2023 as a professor of information technology in the School of Engineering and Technology. With main research interests in parallel and distributed computing, network optimization, privacy and security, data mining, and machine learning, Prof. Shen has received over $20m research funding and led eight projects supported by major national competitive grants and numerous other projects in Australia, Japan, and China. He has published more than 500 research papers including over 200 papers in a variety of IEEE/ACM transactions and other major international journals. With more than 7,100 citations (Google Scholar), Prof. Shen has an H-index 40+ and i10-index 180+ in the hardcore fields of systems and algorithms (parallel and distributed computing). He received his B.Sc. (Hons) degree from the Beijing University of Science and Technology, M.Sc. degree from the University of Science and Technology of China, and PLic and Ph.D. degrees from Abo Akademi University, Finland. Prior to joining CQU, Prof. Shen held professorial appointments including Professor under the title of "China National Endowed Expert" at Sun Yat-sen University (#89 ARWU21); Professor of Computer Science at the University of Adelaide, Australia; and Professor and Head of Computer Networks Laboratory at Japan Advanced Institute of Science and Technology (Japanese graduate-program only national university).

Fulvio Silvestri is a research fellow in the Department of Mechanical Engineering at Politecnico di Milano (Italy). He earned his Ph.D. from the University of Rome "Tor Vergata" (Italy) in 2022. His research focuses on transport planning, with a specific emphasis on travel behavior modeling and demand forecasting for innovative mobility solutions. He is the author of numerous scientific publications indexed in Scopus and Web of Science and serves as a reviewer for the journals *European Transport Research Review* (Springer) and *Case Studies on Transport Policy* (Elsevier). In 2019, he has been appointed as Ambassador to Italy for the Association for European Transport (AET). He participates in the Young Researchers' and Practitioners' Forum of the European Transport Conference (ETC).

Sebastian Stichel is a professor at KTH Royal Institute of Technology in Stockholm and vice dean of the School of Engineering Sciences. He holds a B.S. and a M.S. in vehicle engineering from TU Berlin. He also holds a Ph.D. in vehicle dynamics from TU Berlin. From 2011 to 2024, he was the director of the KTH Railway Group. From 2013 to 2023, he was the vice head and head of the Department of Aeronautical and Vehicle Engineering and the Department of Engineering Mechanics, respectively. He was employed at Bombardier Transportation in Sweden between 2000 and 2011 where he, part of the time, headed a vehicle dynamics department. He was the Chairman of the Shift2Rail Scientific Committee from 2016 to 2018 and Vice-Chair Academia of the European Rail Research Advisory Council (ERRAC) between 2017 and 2022. He has been a board member of SWEDTRAIN – the Swedish Association of Railway Industries – since 2019. His primary research interest is in dynamic vehicle-track interaction. The main concerns are improved ride comfort and reduced wheel and track damage. He is also involved in research on active suspension and the dynamic interaction between pantograph and catenary.

George Stojkovski is a proven executive leader with over 20 years of senior-level leadership experience across the private and public sectors in complex, highly dynamic, customer-facing environments. He has an impeccable track record in strategic planning and cultural transformation, resulting in a significant shift change in improving customer service, safety, and financial results. He completed an 18-year career with Sydney Trains, finishing as Executive Director Operations, with full executive responsibility for the operations function. He moved to the Operations role following a lengthy career in customer service, having served as Executive Director Customer Service and earlier General Manager Customer Service for several Sydney Trains (then RailCorp) regions. He is currently a senior transport strategic consultant, working with global and local mobility, logistics, and construction organizations.

Shuai Su is the Vice Dean and Deputy Secretary of the School of Automation and Intelligence at Beijing Jiaotong University, Deputy Director of the Frontier Science Center for Intelligent High speed Railway Systems, Chief Professor of the State Key Laboratory of Advanced Rail Autonomous Operation. He has been engaged in teaching and research work in the field of rail train operation and control for a long time. During the last decades, Professor Su focused on the academic research and have hosted more than 40 key, general, and national research and development projects funded by the National Natural Science Foundation of China, and the Ministry of Science and Technology. His research has resulted in over 100 high-level publications, primarily in the Transportation Research series journals. His contributions have been recognized with numerous prestigious awards, including the first prize of Natural Science from the Ministry of Education, the first prize of Science and Technology Progress from the China Railway Society, the first prize of Science and Technology Progress from the China Urban Rail Transit Association, the first prize of Science and Technology Progress from the China Instrumentation Society.

Yan Quan Sun completed his bachelor's and master's degrees in mechanical engineering at North-Eastern University, China, in 1983 and 1985, respectively. He then joined the University of Science and Technology Beijing, where he served as a lecturer in 1986 and was promoted to associate professor in 1995, a position he held until 1998 when he relocated to Australia. In 2002, he completed his Ph.D. research in mechanical engineering, specializing in rail vehicle-track interaction dynamics, at the Centre for Railway Engineering, Central Queensland University, earning his Ph.D. in 2003. He has published over 100 scientific papers and authored two books. His research interests include rail vehicle-track performance and safety analysis using system modeling and dynamics, train crashworthiness, rail vehicle-track performance monitoring via vehicle mass measurements, rail track condition assessment through vehicle performance, longitudinal train dynamics, locomotive traction, energy analysis, and rail vehicle and suspension design based on vehicle dynamics.

Farzaneh Tahmoorian is a senior lecturer at Central Queensland University (CQU), Australia, with more than a decade of experience in construction and infrastructure engineering. She holds an M.Sc. from the University of Technology Sydney (UTS) and a Ph.D. from Western Sydney University. She has extensive experience in both academia and industry. Throughout her career, Dr. Tahmoorian has been involved in diverse projects, including road construction, landfill development, and infrastructure management. Her research primarily focuses on ground improvement techniques for roads and railways, pavement engineering, sustainability, life cycle analysis, and pavement designs.

Qing Wu (Member, IEEE) received the B.Eng. degree in vehicle engineering (mechanical) from Southwest Jiaotong University, Chengdu, China, in 2010; the M.Eng. degree in vehicle engineering (mechanical) from the State Key Laboratory of Traction Power, Chengdu, in 2012; and the Ph.D. degree in engineering (multiobjective optimization and parallel computing) from Central Queensland University, Rockhampton, QLD, Australia, in 2016. From 2012 to 2013, he was a research assistant with the State Key Laboratory of Traction Power, Southwest Jiaotong University. From 2016 to 2018, he was a senior research officer with the School of Engineering and Technology, Central Queensland University, Rockhampton, QLD, Australia, where he is currently a principal research fellow. His research interests include mechanical system dynamics, parallel computing, multiobjective optimization, dynamics control, and their applications in road and railway vehicles. He was the recipient of an Australian Research Council Discovery Early Career Award. He is on the Editorial Boards of the *Journal of Traffic and Transport Engineering* and *Railway Engineering Science* and the guest editor of several special issues.

Nicolò Zampieri is an associate professor at the Department of Mechanical and Aerospace Engineering of Politecnico di Torino. He was awarded his Ph.D. in 2014 with a dissertation on the development of monitoring systems for railway applications. His research interests include railway vehicle dynamics and monitoring, modeling of wheel-rail/roller contact, wear, and RCF. His current research activity concerns the design of test benches for railway applications and the development of specific applications for railway vehicle monitoring. In recent years, he started exploring digital twin solutions to combine longitudinal train dynamics and multibody simulations. He is a co-author of more than 60 scientific publications.

Lulai Zhu is a postdoctoral research fellow in the Department of Mechanical Engineering at Politecnico di Milano (Italy). He mainly works on interdisciplinary research topics across software and performance engineering like system discovery.

Shengyang Zhu received his Ph.D. in vehicle operation engineering from SWJTU in 2015. He is currently a professor with SWJTU. He has presided over more than 20 scientific research projects and has published more than 120 academic papers. He is the recipient of the National Natural Science Fund for Excellent Young Scholars and has been selected as Elsevier's Most Cited Chinese Researchers and one of the World's Top 2% Scientists. His achievements have been awarded the First-class Natural Science Awards by the Ministry of Education of China (ranked 2) and the National Railway Science and Technology Innovation Award for Young Scholars. His research interests include train and track dynamic interaction and safety control, virtually coupled train system dynamics and safety control. He is also the editorial board member of the *International Journal of Rail Transportation* and *Journal of Southwest Jiaotong University*, a part-time doctoral supervisor of Central Queensland University, and a vice-chairman of a discipline of the World Transport Convention.

Introduction

This handbook consists of 20 chapters, including an introductory chapter, 18 chapters on specialized topics, and a concluding chapter.

Chapter 1 introduces digital twin and artificial intelligence technologies as overarching concepts applicable across the multidisciplinary initiatives of Industry 4.0 and Industry 5.0. This chapter investigates the evolution of relevant definitions and terminology from their historical roots to the present day. While these technologies have broad applicability across various disciplines, the chapter focuses on their transition to tasks specific to the rail domain.

In *Chapter 2*, the original idea of digital twins and their relation to product lifecycle management processes is revisited and applied to categorize a review of selected digital twin applications in the rail domain. It becomes apparent that digital twins designed for different objectives and lifecycle stages require distinct architecture concepts. These concepts are discussed from various architectural perspectives, focusing on key elements of digital twins regarding components, software, and data. The chapter concludes with a discussion of appropriate architecture patterns for digital twins in rail applications, based on the required interaction with their environment and/or other digital twins.

Chapter 3 reviews several technologies which have been recently introduced to enhance the availability of real-time data streams from industrial processes and services, coupled with significant advancements in computing capabilities. The increased visibility of system conditions in real time has stimulated the development of methods to optimize system performance in the short term. Central to these advancements is the tight integration of digital models with physical production systems, making the study and development of digital twin architectures critical. This chapter presents a state-of-the-art review of digital twin evolution focusing on data collection and the elaboration of data into digital models. Additionally, this chapter addresses the research gaps and implementation challenges that must be overcome to achieve widespread adoption of digital twins in the rail industry.

Chapter 4 introduces the key principles, technologies, and distinctive features used in digital twin modeling and simulation for both traditional and complex systems. It explores the implementation of digital twin simulations in a computer environment using theory-based thinking. It reviews co-simulation techniques and methodologies necessary for modeling complex, multidisciplinary systems, as commonly required by digital twin specifications. The chapter's objective is to provide a theoretical foundation for understanding the implementation of a methodological digital twin framework, considering traditional simulation, multi-domain co-simulation, software and hardware integration and implementation, cloud computing, and platform deployment principles.

Chapter 5 provides the theoretical concept and methodology for consideration in building data-driven digital twins in an entire rail asset lifecycle. The presented systematic knowledge is based on the analysis of developments and findings published recently. An extensive literature review on data-driven digital twins conducted in this chapter allows conclusions on digital twin technologies, design requirements, and classification. All findings are summarized in a series with further consideration of development approaches. Digital twin design methodologies are analyzed in accordance with the rail asset lifecycle to address the asset management need and enable features of data-driven digital twins. Evaluation and validation techniques and approaches based on a systems engineering concept have also been presented and discussed. At the end of this chapter, there is a list of references related to data-driven digital twin technologies to ensure that the presentation is sufficiently extended to cover the wider literature for readers.

Chapter 6 presents the concept of a physics-based digital twin that is often confused with the application of physics-based models. It is important to enlist the characteristic feature of a digital twin (DT) to distinguish it from physics-based models. This chapter discusses the design requirements and classification of a physics-based DT system that can be used for various applications within a railway system. Based on a literature search, various aspects of development considerations for DTs have been included. The variation in tasks required at the design, development, and deployment stages of data-driven and physics-based DTs have been added. An eight-step DT development framework has been proposed that includes the definition of the requirement, identifying technological tools, development of the model, analysis of results, decision-making algorithms, verification, validation, and evaluation of various steps of DT. A design methodology for a DT system has been proposed and an application of a physics-based DT system has been presented. Quantitative and qualitative evaluation tools of DTs have been presented. Finally, barriers to implementing a physics-based DT system have been discussed.

Chapter 7 considers key design requirements in human-driven systems substantive to human digital twins and human interaction design methodologies. The first part includes key considerations for digital twins in human performance modeling and an awareness of the broader socio-technical context. It also considers the need for real-time responsivity and how humans interact with multiple data sources. This leads to considerations for intuitive and accessible interface design, recognizing the importance of user experience, trust, and reliability, which invariably resonates with implications for safety, security, ethics, and privacy. The second part of this chapter presents current concepts and approaches for integrating human behavior in digital twins and describes a series of human-driven approaches relevant to emerging technologies which have methodological value for interacting with digital twins. The last part of the chapter covers broader considerations. Applications for digital twins in the rail environment are examined and explored throughout.

Chapter 8 describes the application of digital twin techniques in the design phase of railway vehicles. The design phase can gain important benefits from the application of digital twin techniques that allow the creation of reliable numerical models to predict the behavior of the vehicle during service. This allows the design to be correctly addressed to achieve the requested performance and safety objectives. For this reason, the development of advanced numerical modeling techniques should be introduced very early both to evaluate the dynamic behavior of the vehicle and to evaluate the structural behavior, initially as simple numerical models but later extended to include different manufacturing and verification stages of the vehicle's production. The recent introduction of active components and onboard monitoring systems has allowed the use of more advanced types of digital twins capable of communicating with onboard sensors and determining the behavior of the actuators effectively. The chapter traces the evolution of digital twins used in vehicle design and analyzes the most commonly used types.

Chapter 9 aims to outline the potential of digital twins to advance railway infrastructure asset management and maintenance. The chapter first discusses the major opportunities and challenges that are expected as a result of the adoption of technologies and techniques to build and operate digital twins in the context of advanced railway systems. The current state of the art is then classified by three application areas, namely, condition assessment and predictive maintenance, operations and maintenance management, and railway infrastructure lifecycle management. Different levels of maturity for the digital twins are evident for each area based on the literature findings. Building on the current status, the chapter also foresees the evolution of digital twins in railway infrastructures of the future; a foresight is provided as a set of conjectures that lead to assume the creation of a network of digital twins of the railway infrastructure as a viable solution; the progressive and scalable deployment of the network of digital twins; and the smooth integration of humans in the loop of the digital twin-based decisional processes. This finally results in augmented decision-making applied in the complex and heterogeneous context of railway infrastructure assets.

Chapter 10 aims to the development of detailed digital twin operational train models. The practice of train simulation for evaluating trains and providing train operator training is well developed. The resulting data from these simulations has been used in many ways ranging from simple trip time analysis to the development of complex control strategies for the management of coupler forces in very large trains. The tradition of simulation has typically been focused on the understanding of issues from an overall fleet perspective, and parameters are typically an 'average' of the fleet. Digital twins offer a new way of using modeling and simulation to achieve comprehensive and specific modeling of individual trains. Once achieved, such a model allows management of aspects of the overall train operation: its speed control, its energy use, and a variety of track and operating limits can be optimized to achieve multiple objectives. The detailed digital twin model can also be updated regularly with condition monitoring data so that the model continues to mirror the real train system and allows up-to-date assessments and optimizations to be completed.

Chapter 11 explores the integration of advanced digital twin technologies into railway systems to enhance operational safety and reliability. Digital twins serve as comprehensive virtual replicas of physical trains and their operating environments, enabling monitoring, analysis, and optimization of train performance. By simulating various operational scenarios and stress-testing the virtual coupling system, digital twins identify vulnerabilities and optimize response strategies. Details presented include methods for the modeling of individual components such as train dynamics, track dynamics, overhead power systems, and control systems. Challenges in developing high-fidelity digital twins are addressed by leveraging data-driven surrogate models and parallel computing. Surrogate models approximate complex systems using machine learning techniques, enabling quick predictions and analyses. Parallel computing divides computational tasks across multiple cores, significantly speeding up processing.

Chapter 12 explores the role of digital twins (DTs) in the rail network infrastructure sector, examining their potential applications, components, and integration with existing engineering tools. The chapter also contrasts building information modeling (BIM) and digital twin, clarifying their distinct roles and how they complement each other in managing complex infrastructure projects. The chapter concludes by discussing the challenges and opportunities in implementing DTs in the rail industry, considering factors such as data management, interoperability, and scalability.

Chapter 13 explores the integration of artificial intelligence (AI) within digital twins (DTs) for rail operations. DTs operate as virtual replicas of physical rail assets and infrastructure, continuously updated through real-time data to mirror the actual system's state. When integrated with AI techniques like machine learning, natural language processing, and generative AI, DTs could support various applications in rail operations, including predictive maintenance, operational optimization, strategic decision-making, and real-time monitoring. Thus, the role of AI is investigated in different application areas, such as train scheduling and dispatching, anomaly detection and fault diagnosis in railway signaling systems, energy efficiency, passenger information, and ticketing. The key challenges of AI and DTs in rail operations are also addressed, including the need for standardized protocols and regulations, enhanced operational safety, and improved passenger satisfaction. The integration of AI into rail operations not only enhances efficiency and safety but also contributes to creating more sustainable and resilient transportation systems. As AI technologies continue to evolve, their applications in railways are expected to expand, offering new opportunities for innovation and improvement.

Chapter 14 explores how artificial intelligence (AI) can automate human-oriented tasks and enhance human capabilities in rail vehicle management, maintenance, and network operations, promising increased efficiency and reduced errors. It is the firm belief of the authors of this chapter, given the global lack of skilled rail expertise and the increased complexity and efficiency requirements of the operational environment, that AI will become a complementary companion to humans in that work environment with the potential for making railways more efficient. The aim of this chapter is therefore to provide a backgrounding regarding some of the beneficial aspects of AI and, based on concrete examples, give recommendations on how to avoid typical pitfalls when using and implementing AI in a rail organization.

Chapter 15 provides an overview of the use of artificial intelligence (AI) methods for the monitoring and maintenance of railway infrastructure. Topics addressed are the monitoring of track geometry and rail defects, the monitoring of railway bridges, the detection of faults in the overhead equipment, and the use of AI for the planning of infrastructure maintenance. Emphasis is placed on the use of AI in the context of new methodologies for track monitoring, particularly from in-service trains, and the use of AI in combination with non-conventional sensors.

Chapter 16 discusses the use of artificial intelligence (AI) in transportation systems that represent a promising research direction for improving the performance of railway hubs and upgrading the criteria and methods for their design and optimal management. The advent of big data sources and machine learning techniques has expanded the capabilities of traditional simulation tools, paving the way for real-time estimated models (online) and automated control of real-world systems (closed-loop optimization). This chapter delves into the concept of digital twins of multimodal hubs, combining integrated micro- and macro-simulation models with real-time data collection and policy implementation. The use of digital twins for railway hub planning and management is examined, and their functional architecture and the challenges related to real-world implementation are discussed.

Chapter 17 explores the ethical considerations in developing and deploying artificial intelligence (AI) in the rail industry, highlighting principles such as fairness, transparency, accountability, privacy, and safety. It references ethical AI frameworks from NASA and the European Commission that offer guidelines to prevent bias, build trust, and ensure responsible AI outcomes. Human oversight and comprehensive training are vital for fair and sustainable AI deployment. In this chapter, organizations are advised to adopt ethical guidelines for AI, incorporating interdisciplinary perspectives and stakeholder input to address privacy, safety, cybersecurity, and accountability. Fairness and transparency in AI systems are crucial to avoid biases and discrimination, with regular audits and clear explanations of AI decisions necessary to build trust. Privacy and data governance are emphasized, with robust data protection measures ensuring control over passenger and employee data and the technical resilience of AI systems to prevent service disruptions. AI development must consider social responsibility, address diverse population needs without bias, and promote environmental sustainability by reducing carbon footprints. Public engagement and education are critical in fostering trust and incorporating diverse perspectives in AI ethics and governance. The chapter summarizes that addressing challenges in AI implementation requires continuous adaptation, interdisciplinary collaboration, and strong ethical frameworks.

Chapter 18 focuses on the integration of artificial intelligence (AI) into digital twins (DTs), which is a promising approach to enhance railway systems. DTs generate new knowledge through interactive data exchange between physics-based and data-based models and physical assets, enabling optimization across planning, operations, maintenance, and management. AI, and particularly machine learning, plays a crucial role in processing complex and vast streams of data to extract actionable insights. However, developing and deploying AI-powered DTs poses significant challenges. This chapter explores machine learning operations as a key framework for integrating AI into railway DTs. Furthermore, the organizational and technical challenges in implementing AI and DT technologies are also discussed, highlighting the need for experimentation, knowledge integration, data enrichment, and change management. Successful adoption requires interdisciplinary collaboration, investment in data infrastructure, and a comprehensive strategy to address potential challenges.

Chapter 19 delves into the rapidly evolving concept of digital twins and their transformation from static systems of systems to dynamic ones. It analyzes relevant technologies and trends driving this shift, defines new directions, and explores possible applications. The chapter identifies emerging functionalities of digital twins, highlighting existing conceptual solutions and potential challenges. It then discusses the further evolution and development of fundamental theoretical aspects and frameworks within a digitalized environment. Finally, the chapter presents a roadmap for rail digital twin development, considering potential directions and comparing them to state-of-the-art advancements in other industries.

Chapter 20 provides concluding remarks based on the experience obtained from preparing and publishing this handbook.

1 Introduction to Digital Twins and Artificial Intelligence

Maksym Spiryagin, Stefano Bruni, and Colin Cole

1.1 HISTORICAL REVIEW

Prior to the German government's Industry 4.0 initiative in 2011, manufacturing had undergone three industrial revolutions. The initiative, also known as the Fourth Industrial Revolution, is fundamentally focusing on integrating information and communication technologies in industrial production by means of the application of cyber-physical systems and the joining together of the newly developed digital and existing advanced technologies and multidisciplinary subjects. This convergence eliminates the boundaries between physical and digital elements in manufacturing and transportation. In essence, Industry 4.0 emphasizes the implementation of real-time data for advanced analysis and processing. This, in turn, enhances traditional control systems, leading to significant improvements in productivity and performance. In addition, Industry 4.0 introduces new terminology within cross-disciplinary fields. These terms are crucial for the digital transformation of design and lifecycle evaluation of physical products through the introduction of the concept of a "digital twin" (DT), which is directly relevant to this handbook. The earliest evidence of the DT concept dates back to 2002 when Michael Grieves of the University of Michigan introduced it to industry participants while discussing the idea of a Product Lifecycle Management Center [1]. However, a formal definition of DT wasn't established at that time; it was merely a descriptive term [1]. It is generally accepted that the first definition of a DT came from NASA researchers and was published in [2]:

> A digital twin is an integrated multi-physics, multi-scale, probabilistic simulation of a vehicle or system that uses the best available physical models, sensor updates, fleet history, etc., to mirror the life of its flying twin. The digital twin is ultra-realistic and may consider one or more important and interdependent vehicle systems, including propulsion/energy storage, avionics, life support, vehicle structure, thermal management/TPS, etc. In addition to the backbone of high-fidelity physical models, the digital twin integrates sensor data from the vehicle's on-board integrated vehicle health management (IVHM) system, maintenance history, and all available historical/fleet data obtained using data mining and text mining. The systems on board the digital twin are also capable of mitigating damage or degradation by recommending changes in mission profile to increase both the life span and the probability of mission success.

While Industry 4.0 is still at the developing stage, the European Commission introduced in 2020–2022 its own novel concept of Industry 5.0 that provides a vision of an industry that "aims beyond efficiency and productivity as the sole goals, and reinforces the role and the contribution of industry to society" [3]. The concept does not represent any new industrial revolution yet, but it significantly extends Industry 4.0 concept by "specifically putting research and innovation at the service of the transition to a sustainable, human-centric and resilient European industry" [3]. A comparative analysis of technology pillars versus enablers for Industry 4.0 and Industry 5.0, as published in [4–7] and summarized with questions in Table 1.1, highlights the significance of artificial intelligence (AI) alongside DT technologies across both industries. The term "artificial intelligence" was introduced by US computer scientist John McCarthy in 1956; however, the conceptual idea of machines possessing intelligence had been around and discussed for many centuries before this. Considering the key areas of Industry 4.0, AI was introduced into manufacturing and production, which increased

DOI: 10.1201/9781003492146-1

TABLE 1.1
Questions on Technological Pillars and Enablers for Industry 4.0 vs. Industry 5.0

Industry 4.0		Industry 5.0	
Pillars	**Question**	**Enablers**	**Question**
Big data	How can we effectively manage and analyze vast amounts of data?	Human-centric solutions and human–machine interaction technologies	What are the best combinations of human strengths with machine capabilities to achieve optimal levels of interconnection and collaboration?
Augmented reality	What are the most effective ways to showcase products using augmented reality, eliminating the need for physical prototypes?	Bio-inspired technologies and smart materials	What are the potential application areas for recyclable materials with embedded sensors and enhanced functionalities?
Simulation	What level of precision is required for simulations to be most beneficial?	Real-time-based digital twins	For DT technologies to replicate entire systems, what level of precision is necessary?
Internet of Things	What is the optimal number of devices to connect within a physical system for efficient data transmission and processing?	Artificial intelligence	To generate actionable insights, what level of autonomy is needed for AI to detect causal relationships within complex and dynamic systems?
Cloud computing	What is the ideal cloud network configuration for storing, managing, and processing data?	Technologies for energy efficiency and trustworthy autonomy	What are the energy limitations for these technologies, and how can we ensure their reliable and trustworthy operation?
Cyber security	How can we ensure comprehensive data and equipment protection?		
Systems integration	How can we achieve seamless integration between different computer systems to enable data exchange and collaborative operation as a unified physical system?		
Additive manufacturing	In which applications can 3D design and printing be most effectively utilized?		
Autonomous systems	Should autonomous systems operate solely on pre-programmed tasks, or should we leverage industrial AI to enable them to think, act, and react autonomously?		

productivity and decreased costs. However, considering the existing developments in cutting-edge technologies such as 5G, edge computing, DTs, quantum computing, and human augmentation, AI's implementation in Industry 5.0 should allow further steps to a new age of human–machine cooperation.

Based on historical trends and questions, continued co-development of DT and AI technologies has the potential to drive Industry 5.0. This transformation aligns perfectly with two key areas of this handbook: using DTs and simulation to model entire systems, and embedding AI to generate actionable intelligence within processes, not as an external tool. In addition, there is a historical concern regarding the lack of specific guidelines, standards, and unique requirements for using these two key areas and for responding to the relevant questions in Table 1.1 within the rail domain, its complex systems, and applications.

1.2 DEFINITION AND APPLICATION OF DTs

While the concept of DT has been extensively defined in various recent publications, these definitions often lack a systematic structure and clear application context within specific domains. To address this, it requires starting with analyzing established definitions found in existing standards or normative documents. This analysis leads to exploring how these definitions are transformed and adapted when applied to the railway sector, drawing insights from real-world observations within the rail domain, which is a subject of this handbook.

1.2.1 What Is a DT?

The previously provided NASA definition in Section 1.1 is specifically tailored to the aerospace engineering field, and it may not be directly applicable in that form to other industries. The recent ISO/IEC 30173:2023 "Digital twin – concept and terminology" tries to resolve the specific industry-related issue by providing a more general definition of a DT, formulated as follows:

- *Digital twin*: digital representation of a target entity with data connections that enable convergence between the physical and digital states at an appropriate rate of synchronization [8].

To clarify this definition, we need to use the following definitions from the standard [8]:

- *Digital representation*: digital entity representing either a set of properties or behavior or both of one or more observable elements.
- *Target entity*: entity providing a functional purpose in reality which is the subject of digital representation. The target entity, which provides some functional purpose in reality, can be either physical or digital under consideration.
- *Entity*: thing (physical or non-physical) having a distinct existence.
- *Digital entity*: computational entity computing data elements and procedural elements.
- *Physical entity*: entity in the physical world that can be subject of sensing and actuating.

It is necessary to mention that ISO/IEC 30173:2023 also distinguishes the definition of "DT system" that is defined as [8]:

- *Digital twin system*: system providing functionalities for the digital twin composed of inter-operating target entities, data connections, models, data and interfaces involved in the data connection process.

However, there are many other definitions that have been described over time relating to the idea or concept of DTs. It is written in the publication [1] that "the idea of digital twin is to be able to design, test, manufacture, and use the virtual version of the system". Meanwhile, others assume that the concept should be built considering that "the digital twin is a collection of all digital artifacts that accumulate during product development linked with all data that is generated during product use" [9]. To gain a clearer understanding, the review of numerous definitions from publications and patents, focusing on the purpose, application area, and key characteristics of DTs, was outlined in [10]. Within the context of Industry 4.0, a focus on virtual replicas led to the definition in [11]:

"A digital twin is a one-to-one virtual replica of a 'technical asset'". Alternatively, emphasizing dynamic representation for fault diagnostics, the definition in [12] states: "Digital twin represents physical entities with their functions, behaviors, and rules dynamically".

The definitions of modeling and simulation within the concept of digital entities, as defined by ISO/IEC 30173:2023, are a particular interest of this handbook. For instance, reference [13] defines a DT as "a multi-domain, ultra-high-fidelity digital model integrating various disciplines such as mechanical, electrical, hydraulic, and control systems". In contrast, reference [14] proposes that a "digital twin can be viewed as a paradigm where selected online measurements are dynamically integrated into the simulation environment, with the running simulation model guiding the real world adaptively in return".

In the context of rail domain applications, [15] provides a general definition that builds upon the definitions found in [16–18]. The authors indicate that the structure of a DT definition should include four important components:

- a model of the object built based on an integrated multiphysics, multiscale, probabilistic simulation approach,
- an evolving set of data relating to the object,
- a means of dynamically updating or adjusting the model in accordance with the data,
- mirror and predict activities/performance over the life of its corresponding physical twin.

However, the specific criteria for a DT depend heavily on the field of application. In the case of autonomous vehicles (AV), as published in [16], a DT must meet the technical requirements summarized in the following four criteria:

- an accurate and updatable model of a real-world driving environment, including other vehicles and pedestrians and some description of weather and other atmospheric and environmental effects,
- models that simulate the response of the sensors deployed on the vehicle based on a careful choice of data from tests of these sensors,
- a model of the vehicle response to driving commands (e.g., steering changes, braking, etc.) that takes road surface conditions into account,
- and the AV control algorithms.

However, when defining DTs for connected and automated vehicles, it's crucial to consider their characterization as outlined in [19]. The article indicates that the DTs can be characterized using the following dimensions: "the digital model, the physical system, the communication between them, the big data that is created from the process, and the services that can be deployed to add value".

The analysis of the existing definitions justifies the conclusion that there is a lack of consensus on the definition of a DT among practitioners and researchers. Existing terminology varies considerably based on the technical requirements of each specific type of DT. This diversity, however, may lead to new and innovative applications in DT technologies. This observation is supported by [16], where the authors also referred to the simplistic definition provided in [20]: "A Digital Twin is an executable virtual model of a physical thing or system", where the physical thing can be anything from a manufactured object, like a car or rail vehicle, to its operational update in the field. It further emphasizes the ongoing debate surrounding the definition of a DT, with each interpretation having its own established practices [21].

As summarized in [7], the DT definitions have specific relevant features that can be associated with a DT application. It leads us to understand that DTs require a process of making the digital representation specific to each real-world physical system [22].

1.2.2 How DTs Are Used in the Rail Domain

As defined in [7], there are multiple application scenarios that in general cases can be classified into three general categories:

1. DT for the initial system design, modeling, and assessment
2. DT of the existing physical system for fault detection and prognosis
3. DT for the real-time interaction with the existing physical system for fault detection, prognosis, or component degrading notifications

Those categories can be further extended considering the characterization and analysis of the application areas specifically within the rail domain based on the performed literature review of 60 DT railway-related publications for the period of 2018–2022 as defined in [23]:

- Maintenance and inspection, including predictive maintenance – 33.3% of reviewed publications
- Safety (operational safety and other types of safety) – 10% of reviewed publications
- Planning (navigation services, production planning, and other forms of planning activities) – 11.7% of reviewed publications
- Management and policies to enhance the resilience of railways and facilitate better asset management – 6.7% of reviewed publications
- Passenger service (predicting the flow of passengers) – 6.7% of reviewed publications
- Optimization, including operational optimization, design optimization, and other optimization approaches – 18.3% of reviewed publications
- Monitoring (structural health monitoring and similar topics) – 8.3% of reviewed publications
- Other (DT generation, interface standardization, etc. that discusses railways in general) – 5% of reviewed publications

An alternative analysis, performed in [22], outlines the main areas of DT technologies in the rail domain with the usage of rail vehicle dynamics studies:

- Design studies that look at the design aspects, which are generally related to the design of rolling stock and the design of track infrastructure and bridges
- Safety analysis studies that have a specific objective of safety analysis, which should be to apply the knowledge of the system safety process in a DT to provide design proposals for improving safety in the railway system
- Maintenance studies in which the DTs are commonly related to three maintenance problems: wheel-rail interface, track components, and rolling stock components
- Monitoring studies that focus on the application of DT technologies for the condition assessment of railway vehicles, tracks, and bridges – one of the critical tasks in railway operational practice
- Optimization studies that focus on the usage of DTs are progressing in two areas – energy analysis of train operational scenarios and train driving strategies

This confirms the significant relevance of DTs in the railway industry. The introduction and development of DT concepts and architectures for specific applications will continue to progress in the future. In the rail domain, advanced DT techniques have enabled the analysis of entire network operations. For new rail networks with existing digital survey data, all the necessary features are present for implementing DT applications. However, taking DTs to the next level and providing more advanced systems require gathering even more specific application data. Data at this level of detail presents significant challenges due to the high resolution and large volume required. Therefore, it is expected that many railway applications will utilize offline DTs, while real-time monitoring and detailed physical data from modern surveys will facilitate the creation of DTs for manufacturing or entire network operations.

1.2.3 The Benefits and Challenges of Using DTs in the Rail Domain

The benefits of using DTs in the rail domain can be formulated considering the following general advantages and benefits published in ISO/IEC 30173:2023 [8]:

- *Improved decision-making*: Test and simulate various scenarios (including system enhancements) to identify optimal solutions and troubleshoot issues before real-world implementation.
- *Optimized operations*: DTs enable real-time monitoring and online or real-time simulation of the rail network, leading to improved scheduling and resource allocation.
- *Reduced operational costs*: Minimize on-site interventions and lower overall maintenance expenses.
- *Enhanced efficiency*: Optimize productivity gains and unlock value across the supply chain.
- *Increased transparency and trust*: Gain long-term and real-time insights into system performance, fostering trust and accountability.
- *Reduced project risk and program delivery*: Simulate project plans to mitigate risks associated with delivery.
- *Innovation, improved design, and development*: Drive innovation by facilitating experimentation in a safe, virtual environment that leads to more efficient and cost-effective development processes.
- *Sustainability considerations*: Integrate social and environmental factors into planning and operations.

According to [24], despite the numerous benefits, the full potential of DT technologies in the rail sector has yet to be fully realized due to a lack of exploration of their underlying technologies. The reference [25] also identified additional DT challenges in general engineering fields, which are typically dependent on the specific business case and context, and they can be summarized as challenges in:

- defining scope and model transparency;
- moving from models to actions (uncertainty propagation, fast inference, model error quantification, identifiability, causality, optimization and control, surrogates and reduced-order models, and multi-fidelity information);
- end user needs to understand what information end users want from a DT and how often that information should be updated;
- security and privacy with a focus on how to secure data and models to prevent unauthorized access and maintain prediction validity and how to address liability for operational failures;
- digital model fidelity and validation;
- data collection and computation with a focus on minimization of computation time and cost, while ensuring reliable and affordable collection of asset state and usage data;
- deployment and analytics that focus on migrating the analytics into production or operational scenarios (for example, how the DT is capable of identifying analytic degradation and uncertainty with time);
- scalability and team composition (i.e., a scalable data platform, a balanced team with data scientists, software engineers, domain experts, and business owners);
- non-technical barriers such as cultural, financial, and managerial hurdles that can hinder the full potential of DTs.

According to [26], the biggest challenges faced by organizations in the transportation industry domain arise during the implementation of DT technology. The study suggests that the main obstacles to adoption and deployment include a lack of buy-in from owners or clients, insufficient senior management support, and the perceived high cost of setting up and maintaining a DT.

This analysis dictates that not only should the DT systems consider the topics presented in this handbook, but they should also extend beyond its scope by incorporating multidisciplinary approaches used in complex system engineering.

1.3 INTEGRATION OF AI INTO A DT

The definition of AI in use today is quite broad, often focusing on replicating human intelligence to a limited extent. Within Industry 4.0, AI applications have been deployed across various stages of design, manufacturing, and operations, including design and virtual testing, manufacturing quality control, predictive maintenance, production optimization and planning, process automation, and safety performance monitoring. In these applications, AI doesn't aim for a perfect replica of human intelligence. Instead, it leads to solutions that augment human capabilities, not replace them. With Industry 5.0 initiative, the situation with AI applications is different. Industry 5.0 takes a more advanced approach, emphasizing human-centric automation. This approach focuses on using computational power to handle tasks that are complex for humans, while still leveraging human expertise in the decision-making process. One key concept introduced in Industry 5.0 is the DT Intelligent System (DTIS), also known as the Intelligent DT (IDT). This technology creates a digital replica of a physical system, infused with AI capabilities for enhanced analysis and decision-making. While reference [27] suggests AI for IDTs doesn't require human-level intelligence, the true value of IDTs lies in augmenting human intelligence. It means that IDTs have a significant potential to process massive datasets and present relevant information for the complex system to consider in their application tasks. As defined in [27], the core concept of DTs from their inception has been to augment, not replace, human intelligence. This approach is good practice to avoid increasing the complexity of systems, which necessitates organizing and managing their information through DTs because the advancements in AI will further contribute to this complexity. These complex systems will exhibit true emergent behavior as AIs modify aspects based on goal-seeking and learning algorithms. Consequently, DTs will need to become more capable, adopting intelligent approaches with the growth in the development of artificial technologies. This section discusses the general evolution concepts and tasks of AI integration in IDTs and its role in assisting with the development and management of complex systems and touches on some general questions about their applicability in the rail domain.

1.3.1 A Review of AI Techniques and Their Integration in Model-Based Approaches

Since both AI and DT systems are reliant on data, it looks quite reasonable for them to be naturally integrated. This convergence, termed DTIS or IDT as mentioned earlier, utilizes AI for data-driven predictions across various domains, where the AI component in the IDT configuration commonly "makes predictions based on data stemming from the DT" [28]. Why is such integration necessary? The two fundamental questions formulated in [29], "Is it possible to get such accurate models as frequently desired? And is it possible to compute and optimize them quickly enough?", provide some specific directions to answer the integration question – if the level of design is highly complex for a DT, realizing the DT concept in a precise way may require novel approaches like AI techniques. This can be achieved by leveraging lower-fidelity models while mitigating the mismatch between the DT and the physical system through continuous feedback loops, where these feedback loops can facilitate adaptation or self-adjustment of the DT's parameters.

The analysis of AI applications in transportation performed in [30] simplistically divides all AI techniques/methods into two categories:

1. symbolic AI, which focuses on the development of knowledge-based systems; and
2. computational intelligence, which includes such methods as neural networks, fuzzy systems, and evolutionary computing.

However, in the report [31], the AI techniques in the rail domain are defined as:

- Evolutionary computing (EC) is a subfield of computer science inspired by biological evolution. In EC, a population of candidate solutions evolves through various operations across generations, aiming to find the optimal solution to a given problem. This approach leverages the power of natural problem-solving mechanisms like evolution to discover solutions [32,33]. Evolutionary algorithms (EAs) are the core of EC, encompassing techniques like genetic algorithms, evolution strategies, multi-objective EAs, genetic programming, and swarm intelligence algorithms. EAs typically operate iteratively, progressively removing low-quality solutions. They are well-suited for optimization problems in dynamic environments where the algorithms should be "continuously and efficiently adapting to the solution to a changing environment" [34]. The success of an EA is usually measured by performance criteria. Since each individual in an EA population has a different fitness value, typically evaluated simultaneously for the entire population, the DT framework implementation should consider a parallel computing architecture to make fitness function evaluation less time-consuming [35].
- Logic programming (LP), also known as logic-based approaches in AI applications (or computational logic), is a method used to represent and manipulate complex and abstract concepts. These include knowledge representation, reasoning, model checking, and verification. Knowledge representation plays a leading role in implementing expert systems and intelligent systems. These systems aim to mimic human-like reasoning capabilities. On the contrary, model checking and verification are designed to significantly increase the confidence level in AI-based systems [36]. LP can also handle uncertainty, inconsistency, and incompleteness of information. It achieves this by employing techniques such as probabilistic logic, fuzzy logic, or non-monotonic logic.
- Machine learning (ML) is used by scientists to provide the ability to learn from examples and improve with experience, as humans do, without being explicitly programmed [30]. When provided with sufficient data, ML methods "discover and exploit structural relations among the data, and this structural inference underlies both the strengths and weaknesses of machine learning algorithms" [37]. Systems supported by ML methods can improve the performance in specific tasks, based on previous experience, provided that the data used to train the machine to perform the desired tasks is sufficiently accurate. ML methods are "often regarded with a certain degree of suspicion' as 'they are frequently presented as 'black boxes' that take data and, without any guidance, somehow learn part of the true structure of the world" [37].

As defined further in the analysis performed in [38,39], the common AI-related techniques used in the railway systems can be classified by their general AI application fields: expert systems (*rule-based expert systems*, *fuzzy expert systems*, *neuro-fuzzy expert systems*), data mining (DM) (*big data analytic*), pattern recognition, adversarial search (*deep reinforcement learning, agent-based modeling*), EC (*evolution strategies*, *genetic programming*, *genetic algorithms*, *multi-objective EAs*, *swarm intelligence*), ML (*deep learning*, *deep neural networks*, *convolutional neural networks, supervised learning, unsupervised learning*), operations research and scheduling (*linear programming, integer programming, nonlinear programming, dynamic programming, constraint programming, network flows and graph theory, scheduling*), LP (*abductive LP, meta LP, constraints LP, concurrent LP*), natural language processing and speech recognition (*knowledge extraction, word segmentation, lemmatization, stemming, sentiment analysis*), computer vision and image processing (*object recognition, optical character recognition, movement tracking, image acquisition, feature extraction, image understanding*), autonomous systems and robotics (*robot behavior control, active sensory processing and control, sensor data integration*). In a common application sense, all these AI fields can be applicable in the intellectual DT systems operating or designated for the rail domain.

The success of integrating AI techniques with DTs depends on two factors: the capabilities of the DT systems themselves and the difficulty of achieving high accuracy in digital models to capture complex dynamics. Given the significant uncertainties and inconsistencies inherent in current DT techniques and platforms, AI can be a powerful tool to address these performance limitations across various rail applications. The choice of AI techniques in such cases is the responsibility of a DT developer or integrator and it would be dependent on the complexity of intellectual DT systems.

1.3.2 Data Analysis with the Application of AI Techniques

Data analysis (DA) is commonly classified as DM within DT technologies, and it aims to process the existing data and discover useful information. DM often uses AI techniques to analyze data in an automatic or semiautomatic manner, producing knowledge that can be represented, for example, as a set of rules easy to understand by humans [40]. However, DM is only a subset of DA. The definition of DA is significantly wider, and it focuses on the entire data lifecycle and includes extraction, cleaning, transformation, modeling, and visualization of data with a DT to lead to deriving conclusions and support decision-making.

DA in DT systems is associated with a complex data management process that includes several key stages: data transmission, storage, processing, fusion, and visualization. The AI techniques like fuzzy set theory, rough set theory, neural networks, wavelet theory, and support vector machines are employed for data fusion at both feature and decision levels [41].

In industry, DA is closely tied to the concept of big data. Big data refers to massive and complex datasets that are difficult to process using traditional methods [42]. Given the volume of data generated and processed by DT systems and the need for high-speed processing, the application of AI techniques becomes highly logical.

In some DT design cases, the DT framework offers another approach. Data from both the physical and virtual twins can be fed into AI models to achieve specific industrial goals. The results from this approach can then be used to further improve and update both the physical and virtual twins. Choosing the right AI technique for DA can be challenging. Different AI techniques and methods have varying levels of accuracy and efficiency, depending on the specific application and dataset [42]. In some cases, there is a chance of utilizing a combination of AI techniques to improve the overall performance of intelligent systems.

1.3.3 Current Applications, Limitations, and Challenges

Current AI algorithms hold promise across all three application areas outlined in [43]: infrastructure, operations, and vehicles. In infrastructure asset management, AI methodologies can pave the way for condition-based or predictive maintenance, ultimately reducing costs. AI also plays a role in railway operations, with current research and development focusing on scheduling and dispatch. This includes strategic and tactical planning, analysis of completed journeys, and dynamic rescheduling in response to delays or changes, all aimed at enhancing operational efficiency. AI is also a key technology for automating driving functions within the vehicle category.

Based on the existing literature reviews related to the usage of AI methods in railway systems, it is possible to classify in a more general way the existing areas of current applications [31,39]: maintenance and inspection, safety and security, autonomous driving and control, traffic planning and management, revenue management, transport policy, and passenger mobility. It is important to note that this is just a list of AI applications that should not be limited because, as AI and DT technologies are continuously developing and evolving, more innovative applications might emerge in the railway industry after the publication of this handbook.

The limitations and challenges for the general implementation of AI in railway systems can be summarized in a similar manner as published in [44]:

- The difficulty of acquiring a wide variety of class information for certain components or events results in insufficient and imbalanced data for model training. While data from the physical system is abundant, data on defects is scarce. This leads to a situation where the training data for AI models is severely inconsistent and insufficient.
- Selecting appropriate AI methodologies requires understanding the inherent dynamics and physical characteristics of railway components through various monitoring systems. This knowledge is crucial to identifying the strengths and limitations of different AI approaches considering the design limitations of each monitoring system.
- Railway systems are inherently dynamic and stochastic with distributed parameters. Critical parameters can vary significantly across locations and time due to differences in operational cycles, environmental conditions, and other factors. Consequently, there's no single optimal AI technique; the choice depends on the specific physical system and the digital framework employed.
- The railway industry's conservative nature presents a barrier to AI deployment. A thorough understanding of system hazards and associated risks is crucial for integrating new systems. However, not all this information can be readily captured by existing monitoring techniques or AI methods.

From the analysis presented in [45] and other practical thoughts, the main common challenges for integrating AI and DTs can be summarized as follows:

- Developing accurate and comprehensive models for DTs requires vast amounts of data to be processed by the AI component. Gathering and managing this data can be complex.
- Safety is paramount, especially for applications like autonomous trains. Integrating AI introduces new regulatory hurdles that need to be addressed.
- Real-world testing of autonomous systems is expensive and risky. DTIS can help create simulated scenarios, but these might not capture all possibilities due to the limitations of the AI components.
- There is a concern that the virtual environment might not perfectly replicate the real world, potentially leading to inaccurate test results and serious risk hazard issues in operations.
- There is a lack of standards in this field, not only in the rail domain but also in other industry clusters.

To effectively assess the real-world applicability of new digital solutions built on DT and AI technologies in the rail domain, a thorough understanding of their backgrounds and integration aspects is crucial. The next chapters in this handbook should guide readers in this direction, while readers should also consider other important basic concepts and the theoretical framework of railway systems [46], as well as factors like implementation costs, safety concerns, and regulatory requirements, especially for safety-critical applications [47].

REFERENCES

1. Grieves M, Vickers J. Digital twin: Mitigating unpredictable, undesirable emergent behavior in complex systems. In *Transdisciplinary Perspectives on Complex Systems*, FJ Kahlen, S Flumerfelt, and A Alves (Eds.), Springer-Verlag, Cham, 85–113, 2017.
2. Shafto M, et al. Modeling, simulation, information technology & processing roadmap. National Aeronautics and Space Administration, Washington, DC, 2012.
3. European Commission. Industry 5.0. Available at https://research-and-innovation.ec.europa.eu/research-area/industrial-research-and-innovation/industry-50_en.
4. Silvestri L, Forcina A, Introna V, et al. Maintenance transformation through Industry 4.0 technologies: A systematic literature review. *Computers in Industry* 2020, Vol. 123, 103335.
5. Müller J. Enabling technologies for Industry 5.0. Results of a workshop with Europe's technology leaders. Independent Expert report. European Commission, Brussels, September 2020, ISBN 978-92-76-22048-0.
6. Vyhmeister E, Castane GG. When industry meets trustworthy AI: A systematic review of AI for Industry 5.0. arXiv preprint arXiv:2403.03061. 2024. Available at https://arxiv.org/pdf/2403.03061.
7. Viola J, Chen Y. Smart control engineering enabled by digital twin. In *Digital-Twin-Enabled Smart Control Engineering: A Framework and Case Studies*, Springer, Berlin/Heidelberg, Germany, pp. 73–102, 2023. https://doi.org/10.1007/978-3-031-22140-8_4
8. International Standard BS ISO/IEC 30173:2024. *Digital Twin – Concepts and Terminology*, Edition 1.0, BSI Standards Limited, 2023, ISBN 978 0 539 15696 6.
9. Haag S, Anderl R. Digital twin – Proof of concept. *Manufacturing Letters* 2018, Vol. 15, 64–66.
10. Liu M, Fang S, Dong H, et al. Review of digital twin about concepts, technologies, and industrial applications. *Journal of Manufacturing Systems* 2021, Vol. 58 (Part B), 346–361.
11. Schluse M, Priggemeyer M, Atorf L, et al. Experimentable digital twins—Streamlining simulation-based systems engineering for Industry 4.0. *IEEE Transactions on Industrial Informatics* 2018, Vol. 14 (4), 1722–1731.
12. Xu Y, Sun Y, Liu X, et al. A digital-twin-assisted fault diagnosis using deep transfer learning. *IEEE Access* 2019, Vol. 7, 19990–19999.
13. Luo W, Hu T, Zhang C, et al. Digital twin for CNC machine tool: Modeling and using strategy. *Journal of Ambient Intelligence and Humanized Computing* 2019, Vol. 10, 1129–1140.
14. Wang P, Yang M, Peng Y, et al. Sensor control in anti-submarine warfare—A digital twin and random finite sets based approach. *Entropy* 2019, Vol. 21 (8), 767.
15. Spiryagin M, Wu Q, Polach O, et al. Problems, assumptions and solutions in locomotive design, traction and operational studies. *Railway Engineering Science* 2022. https://doi.org/10.1007/s40534-021-00263-w.
16. Wright L, Davidson S. How to tell the difference between a model and a digital twin. *Advanced Modelling and Simulation in Engineering Sciences* 2020, Vol. 7 (13), 1–13.
17. West T, Blackburn M. Is digital thread/digital twin affordable? A systemic assessment of the cost of DoD's latest Manhattan project. *Procedia Computer Science* 2017, Vol. 114, 47–56.
18. Defense Acquisition University. Digital twin. DAU glossary of defense acquisition acronyms and terms. Available at https://www.dau.edu/glossary. Accessed on May 25, 2022.
19. Schwarz C, Wang Z. The role of digital twins in connected and automated vehicles. *IEEE Intelligent Transportation Systems Magazine* 2022. https://doi.org/10.1109/MITS.2021.3129524.
20. Hughes A. *Forging the Digital Twin in Discrete Manufacturing: A Vision for Unity in the Virtual and Real Worlds*. LNS Research, 2018. https://members.lnsresearch.com/research-library/research-articles/ebook-forging-the-digital-twin-in-discrete-manufacturing-a-vision-for-unity-in-the-virtual-and-real-worlds
21. Tao F, Qi Q. Make more digital twins. *Nature* 2019, Vol. 573, 490–491.
22. Spiryagin M, Edelmann J, Klinger F, et al. Vehicle system dynamics in digital twin studies in rail and road domains. *Vehicle System Dynamics* 2023, Vol. 61 (7), 1737–1786.
23. Dirnfeld R. Digital twins in railways: State of the art, opportunities, and guidelines. VT2022. Bachelor Degree Project. Linnaeus University, Faculty of Technology, Department of Computer Science and Media Technology, 2022. Available at https://www.diva-portal.org/smash/get/diva2:1674046/FULLTEXT01.pdf.
24. Ghaboura S, Ferdousi R, Laamarti F, et al. Digital twin for railway: A comprehensive survey. *IEEE Access* 2023, Vol. 11, 120237–120257.

25. National Academies of Sciences, Engineering, and Medicine, Division on Engineering and Physical Sciences. Opportunities and challenges for digital twins in engineering. In *Proceedings of a Workshop—In Brief*, L Casola (Ed.), National Academies Press (US), Washington, DC, 18 September 2023.
26. Cockerell S. Is the transportation industry ready for digital twins? Available at https://it.bentley.com/wp-content/uploads/industry-trends-digital-twins-in-rail-stevecockerell.pdf. Accessed on May 25, 2024.
27. Grieves M. Intelligent digital twins and the development and management of complex systems [version 1; peer review: 4 approved]. *Digital Twin* 2022, Vol. 2 (8). https://doi.org/10.12688/digitaltwin.17574.1.
28. Kreuzer T, Papapetrou P, Zdravkovic J. Artificial intelligence in digital twins—A systematic literature review. *Data & Knowledge Engineering* 2024, Vol. 151, 102304.
29. Cronrath C. On reinforcement learning and digital twins for intelligent automation. PhD thesis, Department of Electrical Engineering, Chalmers University of Technology, Gothenburg, Sweden, 2023, ISBN 978-91-7905-838-8.
30. Sadek AW. Artificial intelligence applications in transportation. Transport Research Circular, Number E-C113, January 2007, Transportation Research Board of the National Academies, Washington, DC, 1–6.
31. Flammini F, Tang T, Liu R, et al. RAILS Deliverable D 1.1: Definition of a reference taxonomy of AI in railways. Project: RAILS-Roadmaps for AI integration in the rail sector (Horizon 2020-Shift2Rail JU), Report, 2020(D1):1, 2020.
32. Rudolph G. Convergence of evolutionary algorithms in general search spaces. In *Proceedings of IEEE International Conference on Evolutionary Computation*, Nagoya, Japan, 1996, 50–54.
33. Eiben AE, Smith JE. *Introduction to Evolutionary Computing*, 2nd edition. Springer, Berlin, Heidelberg, 2015, ISBN 978-3-662-44874-8.
34. Branke J. *Evolutionary Optimization in Dynamic Environments.* Springer US, New York, 2002, ISBN 978-1-4615-0911-0. https://link.springer.com/book/10.1007/978-1-4615-0911-0
35. Eklund M, Sierla SA, Niemistö H, et al. Using a digital twin as the objective function for evolutionary algorithm applications in large scale industrial processes. *IEEE Access* 2023, Vol. 11, 24185–24202.
36. Calegari R, Ciatto G, Denti E, et al. Logic-based technologies for intelligent systems: State of the art and perspectives. *Information* 2020, Vol. 11 (3), 167.
37. Danks D. Learning. In *The Cambridge Handbook of Artificial Intelligence*, K Franklish and WM Ramsey (Eds.), Cambridge University Press, Cambridge, UK, 151–167, 2014.
38. Bešinović N, Tang R, Lin Z, et al. RAILS Deliverable D1.2 Summary of existing relevant projects and state-of-the-art of AI application in railways (Horizon 2020-Shift2Rail JU). Report. 2021(D1):2, 2021.
39. Tang R, De Donato L, Bešinović N, et al. A literature review of Artificial Intelligence applications in railway systems. *Transportation Research Part C: Emerging Technologies* 2022, Vol. 140, 103679.
40. Witten IH, Frank E, Hall MA, et al. *Data Mining. Practical Machine Learning Tools and Techniques.* Morgan Kaufmann Publishers, 2017, ISBN 978-0-12-804291-5. https://www.sciencedirect.com/book/9780123748560/data-mining-practical-machine-learning-tools-and-techniques
41. Qi Q, Tao F, Hu T, et al. Enabling technologies and tools for digital twin. *Journal of Manufacturing Systems* 2021, Vol. 58 (part B), 3–21.
42. Rathore MM, Shah SA, Shukla D, et al. The role of AI, machine learning, and big data in digital twinning: A systematic literature review, challenges, and opportunities. *IEEE Access* 2021, Vol. 9, 32030–32052.
43. Sinnemann J, Franzen J, Meirich C, et al. Artificial intelligence for railroads: Potential and challenges for application. ETR-Eisenbahntechnische Rundschau, 2023, 22–24. Available at https://elib.dlr.de/197606/1/2023_Sinnemann_Franzen_INT.pdf.
44. Phusakulkajorn W, Nunez A, Wang H, et al. Artificial intelligence in railway infrastructure: Current research, challenges, and future opportunities. *Intelligent Transportation Infrastructure* 2023, Vol. 2, 1–23.
45. Dirnfeld R, De Donato L, Somma A, et al. Integrating AI and DTs: challenges and opportunities in railway maintenance application and beyond. *Simulation* 2024. https://doi.org/10.1177/00375497241229756.
46. Jägare V. A challenge-framework for innovations in railways. PhD thesis, Division of Operation and Maintenance Engineering, Department of Civil, Environmental and Natural Resources Engineering, Luleå University of Technology, Luleå , Sweden, 2022.
47. Flammini F, De Donato L, Lin Z, et al.. RAILS Deliverable D 3.2 WP3 Report on AI approaches and models (Horizon 2020-Shift2Rail JU). Report 2022(D3):2, 2022.

2 Architecture Concepts of Digital Twins

Florian Klinger and Johannes Edelmann

2.1 INTRODUCTION

More than two decades after the idea of a digital twin was first presented by Michael Grieves in 2002, a consensus has been found across various sectors in academia and industry regarding what defines a digital twin (cf. Chapter 1). However, concepts for the design and realization of digital twins are less systematically explored [1]. This chapter aims to discuss architecture concepts and related aspects relevant to the design and implementation of digital twins in rail applications. First, however, we refer back to the origins of the idea of a digital twin as a concept for product lifecycle management.

In [2], Grieves and Vickers distinguish two types of digital twins: the *digital twin prototype* and the *digital twin instance*; both are implemented in a *digital twin environment*. The respective characteristics are:

- The *digital twin prototype* is a virtual representation of a product that has not yet been created in the real world; it contains the informational sets required to design and manufacture its physical counterpart, the *physical twin*.
- The *digital twin instance* is a virtual representation of a product that already exists in the real world; it contains the informational sets required to comprehensively describe and model the actual and individual properties of the physical twin.
- The *digital twin environment* can be understood in two ways: in [2] it is defined as "an integrated, multi-domain physics application space for operating on digital twins"; on the contrary, in [3], it is interpreted as "the virtual representation of the environment within which the physical product exists." In the former case, the digital twin environment represents the IT systems in which the digital twin is implemented; in the latter case, it contains the informational sets needed to describe the interaction between the physical twin and its environment and the effects the environment has on the behavior of the physical object.
- Although not directly stated in [2], the term *digital twin aggregate* appears regularly in literature (see, e.g., [3–5]), denoting the combination of multiple digital twin instances. Here, we prefer to use the more general term *system of systems* for a combination of (virtual) systems "that interact to provide a unique capability that none of the constituent systems can accomplish on its own" [6]. Appropriate interfaces are required to allow for the interoperability of the individual systems.

It is noteworthy that the definition of the digital twin prototype neither refers to a physical twin nor a data connection between the virtual world and the physical world, which are often claimed prerequisites for digital twins in the literature [7–11]. Nevertheless, we will use the above terminology throughout this chapter when the precise distinction is appropriate. Otherwise, we simply use the term *digital twin* to refer to the entirety of "the physical component, the logical component(s) and the relation between the physical and logical entities" [12].

The remainder of this chapter is organized as follows: in Section 2.2, the architectural concepts of digital twins in rail application scenarios are presented, using the above-introduced definitions

DOI: 10.1201/9781003492146-2

to categorize the respective approaches. Section 2.3 refers to architectural concepts from an application-independent perspective; therein, the essential components of digital twins and their relation to the lifecycle stages are discussed, along with software-building solutions and concepts for data space sharing and data flow. In the final Section 2.4, architecture patterns are revisited from the literature, again first from a general perspective and then focusing specifically on rail applications of digital twins.

2.2 REVIEW OF DIGITAL TWIN ARCHITECTURE CONCEPTS BASED ON THE RAIL APPLICATION SCENARIOS

While a general review of digital twins in the rail domain is provided in Chapter 1, here we focus on scientific literature addressing the underlying architecture of the digital twin.

Spiryagin et al. [13] discuss the implementation of the digital twin concept in locomotive design. The development of modern locomotives is no longer based on detailed specifications for the vehicle provided by the railway operators only but has evolved to a list of performance expectations for the vehicle on a particular railway network, constrained by applicable legislation and track access standards. In addition, requirements set by the infrastructure owners regarding the wear expected to be introduced to the infrastructure need to be considered. The authors argue that "to solve this task at the design stage [...] the assessment of all these design, behavior and maintenance aspects for various locomotive operational scenarios should be performed through computer simulations" [13]. The simulations require a multidisciplinary approach since models of different engineering disciplines are encompassed, ranging from mechanical multibody system models of locomotives (as an individual vehicle or as a locomotive consist of multiple vehicles), mechatronic traction models, wheel–rail contact interface characteristics considering tribological aspects, to train operational scenarios requiring a virtual train driver. The methodology may be categorized as a digital twin prototype, with the model of the track as a separate model representing the digital twin environment.

To run these models simultaneously and to handle the computational demands, co-simulation and parallel computing schemes are required, along with high-performance computing (HPC) clusters. Figure 2.1 shows a potential simulation architecture design introduced in [14]. The simulation process commonly uses one independent processor core on the HPC to simulate each vehicle in the train as it travels over the railway route. Individual cores then communicate with the longitudinal train dynamics simulator via a co-simulation interface. Significant computational resources are required (since, for example, a typical heavy haul train may consist of 3 locomotives and up to 240 wagons [13]). Considerable computing times may be expected; however, real-time capabilities are not required in the development stage and for the use of digital twin prototypes.

However, for digital twin instances in the utilization and support stage, real-time and near real-time computations are required. One option to reduce the computational effort is to use local models, which consider just the key parts of the system required to fulfill the objectives of the digital twin at the necessary level of detail and replace the other components with a simpler modeling approach or even just by appropriate boundary conditions [15]. Local models can be divided into *sub-models* and *surrogate models*. The former typically refers to a physics-based representation of the parts of interest, which may map the behavior of the real system with a high level of accuracy. The latter are usually data-driven models, which are constructed based on results either from experiments or from physics-based models and map the respective input–output behavior without providing information on the inner physical processes.

Bernal et al. [16] and Bosso et al. [17] investigate the use of surrogate models to calculate railway vehicle dynamics safety indexes from longitudinal train dynamics simulations. In both papers, the output of complex multibody simulations is used to derive regression models, which allow for an estimate of main safety indexes describing the running behavior of railway vehicles. The idea of integrating the surrogate model in a digital twin framework according to [16] is illustrated in Figure 2.2. Based on

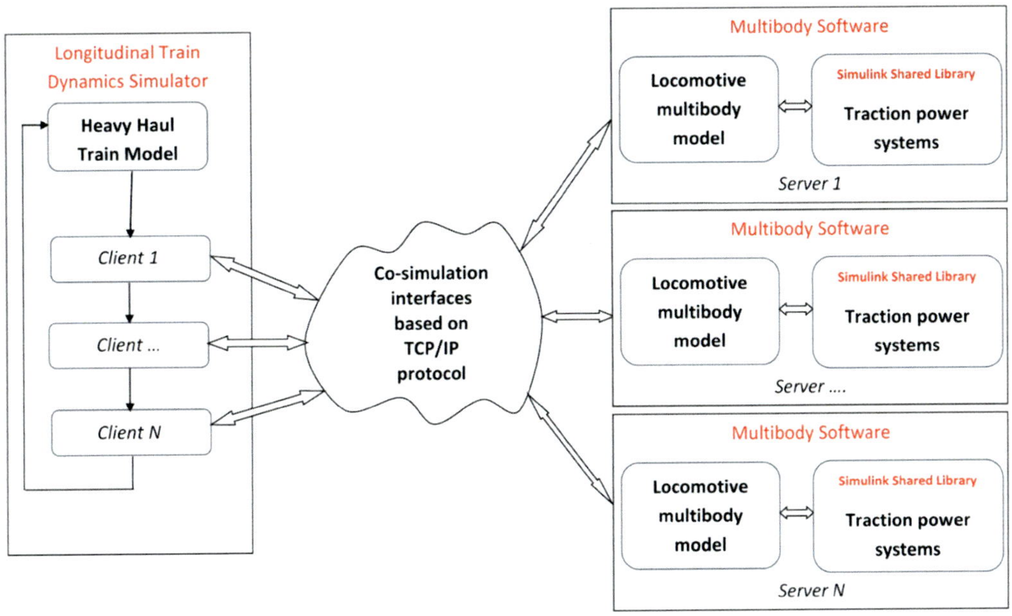

FIGURE 2.1 Architecture of a multi-locomotive co-simulation approach for locomotive/track damage studies. (Courtesy of Spiryagin M, Persson I, Wu Q, et al. A co-simulation approach for heavy haul long distance locomotive-track simulation studies. *Vehicle System Dynamics* 2019; 57(9):1363–1380.)

measured coupler forces, train speed, curve radius, and wagon loading conditions, its actual derailment risk is predicted by the surrogate model and fed to the longitudinal train control system. The authors refer to the surrogate model as an *augmented digital twin*, since it is "augmenting the results of an on-board sensor into more complex vehicle dynamic behavior parameters in real-time, with the computing power available on-board the vehicle" [16]. This surrogate model may be categorized as a digital twin instance, and the locomotive and the instrumented coupler may resemble the physical twin.

Artificial intelligence (AI) and its integration in digital twins in the railway sector are reviewed in [18,19], with an emphasis on predictive maintenance, which the authors consider to be the most promising service benefiting from the combination of digital twins and AI. Based on their literature review, the authors provide generic guidelines for the implementation of digital twins, together with a proposal for a generic high-level architecture for AI-assisted digital twins for the predictive

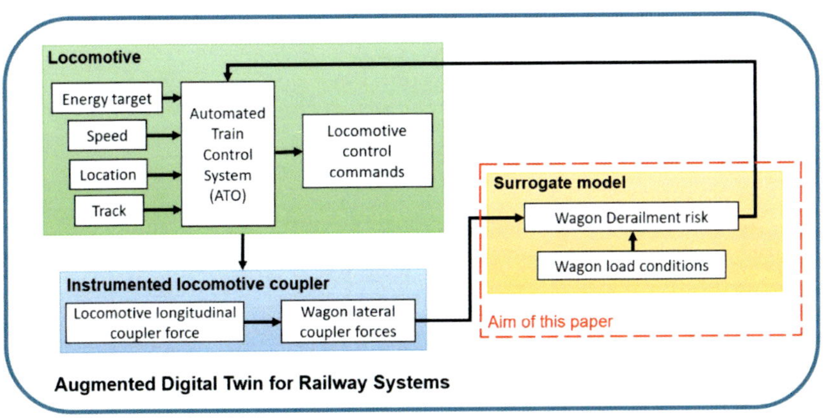

FIGURE 2.2 Augmented digital twin for railway systems. (Adapted from Bernal E, Wu Q, Spiryagin M, et al. Augmented digital twin for railway systems. *Vehicle System Dynamics* 2024; 62(1):67–83.)

maintenance of railway assets. Both, the implementation strategy and the architecture pattern, will be discussed in Section 2.4 of this chapter.

Digital twins of railway infrastructure elements are discussed in literature from different perspectives (see, e.g., [20]). Here it is distinguished between two approaches: (1) where a digital twin of an infrastructure element is used to mirror the physical element itself and (2) where a digital twin of an infrastructure element interacts with other physical or virtual objects within a broader railway scenario. The first approach is related to building information modeling (BIM) and is discussed in Chapter 12. An example of the second is found in [21]. The paper describes the optimization of an automated operation of rail level crossings by means of a digital twin. The architecture of the system is depicted in Figure 2.3. Two physical layers are used, one comprising the track, including the level crossing, and the second the train. The virtual layer includes the (mathematical) model of the level crossing, along with a database and a monitoring module, providing communication with a wayside controller. The system is primarily used to optimize the operating strategies for the level crossing but also to design the location of the detection devices, by taking the minimum braking distances of the train into account.

This approach may be categorized as a system of systems approach, which may be applied both as a digital twin prototype in the design stage of a railway system or as a digital twin instance in the operational service and support stage.

A "conceptual model-based digital twin platform that aims to integrate measurement data, numerical models, and machine learning models from different railway subsystems into a holistic large-scale railway infrastructure platform" is proposed in [22]. Facilitated by the digitization of railway infrastructure and based on digital twin approaches, services such as monitoring, diagnostics, and prognostics to optimize the performance of the railway system should be provided throughout the entire lifespan of the infrastructure. To separate the complex interrelations between the main elements of the railway system into smaller clusters, a layered architecture is proposed as shown in Figure 2.4. Briefly summarized, executable models (physics-based and data-driven) are stored together with measurement data in the asset layer. The integration layer manages the cooperation between those models and data, and simulations are allocated to suitable hardware and executed by the infrastructure layer. The coordination of the simulation flow and the processing of the simulation data according to the demanded services are performed within the function layer, while the visualization layer assists to explore and understand the simulation results. The interaction layer provides different roles for different operators of the platform, managing accessibility and related privileges.

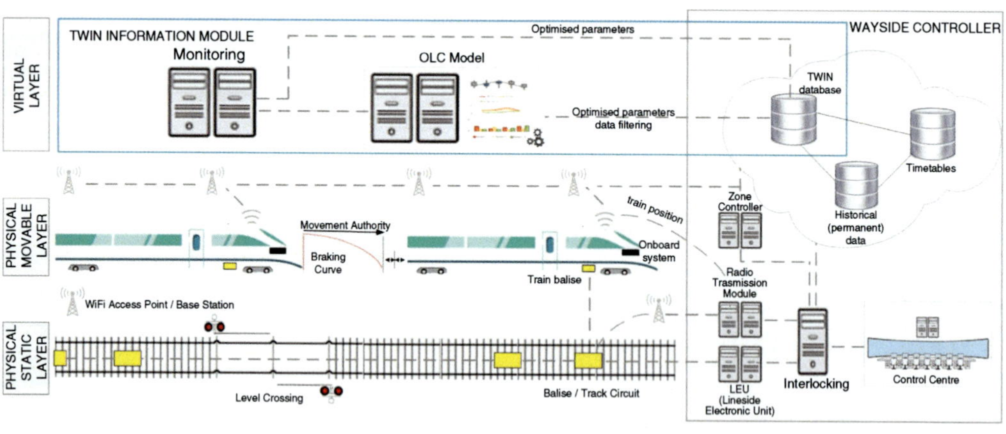

FIGURE 2.3 Architecture of the digital twin system of a level crossing. (From Grieves M, Vickers J. Digital twin: Mitigating unpredictable, undesirable emergent behavior in complex systems. In: Kahlen J, Flumerfelt S, Alves A (eds) *Transdisciplinary Perspectives on Complex Systems*. Springer, Cham, 85–113, 2017, CC BY 4.0.)

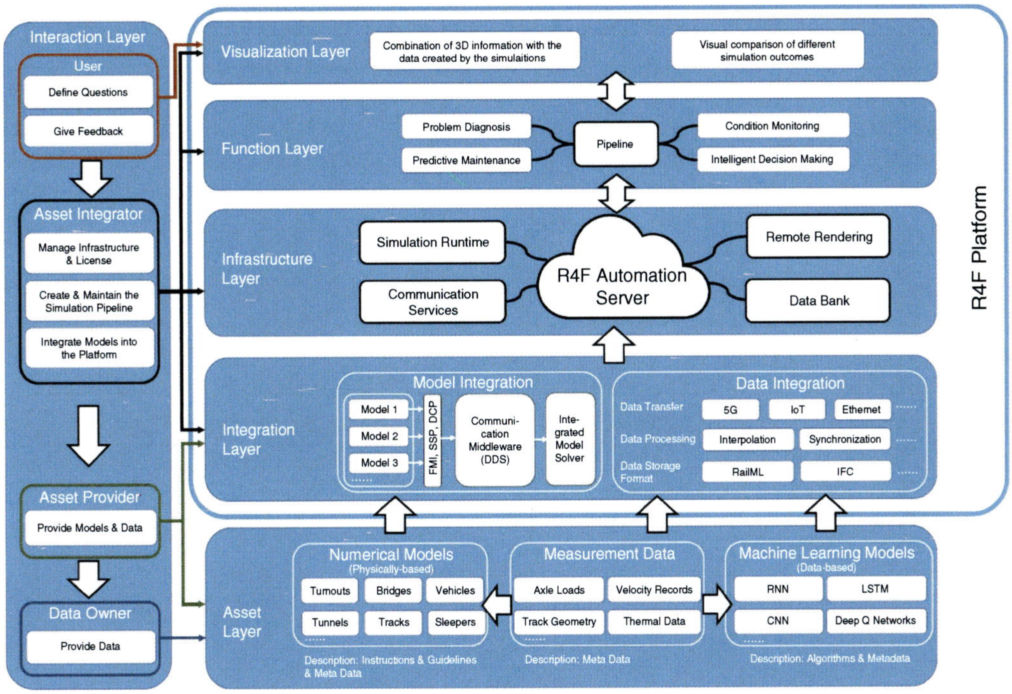

FIGURE 2.4 The six layers of the Rail for Future Platform as presented in [22]. (From Zhou S, Dumss S, Nowak, et al. A conceptual model-based digital twin platform for holistic large-scale railway infrastructure systems. *Procedia CIRP* 2022; 109:362–367, with permission.)

2.3 DIGITAL TWIN ARCHITECTURE CONCEPTS

The design and implementation of a digital twin is a complex and multidisciplinary project. As for any technical system, the development process of a digital twin typically starts with a definition of its objectives in terms of expected services and field of application. The collection of these objectives is then transformed into a list of (functional and non-functional)[1] requirements. This process is commonly referred to as *requirements engineering*, and its outcomes form the foundation for the design of the architecture of the digital twin. In [23], the term *system architecture* is defined as

> a representation of a system in which there is a mapping of functionality onto hardware and software components, a mapping of the software architecture onto the hardware architecture, and a concern for the human interaction with these components. That is, system architecture is concerned with the totality of hardware, software, and humans.

Although originally from software engineering, this definition seems appropriate for digital twins in the rail domain, as it explicitly includes the human, either as an operator or as an integral component (e.g., as a virtual train driver) within the digital twin system, besides hardware and software.

In [6], the process of defining a system architecture is described as follows:

> The system architecture definition process focuses on defining an architecture that addresses the stakeholder concerns and is applied iteratively and concurrently with the business or mission analysis, stakeholder needs and requirements definition, and system requirements definition processes to determine the best solution to address stakeholder concerns. [...] Architecture focuses on suitability, viability, and desirability, whereas design focuses on compatibility with technologies and other design elements and on feasibility of construction and integration. An effective architecture is as design-agnostic as possible to allow for maximum flexibility in the design trade space.

The outputs of an architecture definition process are *architectural views* and *architectural models* that describe how the system is organized as a set of communicating components, including their interfaces [6,24,25], as discussed in the subsequent sections.

2.3.1 Digital Twin Components

According to Tao's five-dimension concept model of digital twins [8], the main components of a digital twin comprise (1) the *physical entity*, existing in the physical space; (2) the *virtual entity*, consisting of a set of models; (3) *services*, for both the physical and the virtual entity; (4) *data*; and (5) the *connection* that ties the different parts of the digital twin together. Josifovska [26] provides a similar partition including a detailed specification of the components. Physical entities, for instance, are further distinguished as *physical objects*, *physical nodes*, and *humans*. Physical objects refer to entities with no communication abilities; they simply are "observable" objects within the physical environment. Physical nodes, in contrast, are able to observe and communicate with other entities. The human is considered a separate entity, with the modeling of their behavior strongly depending on the specific application context. Similar considerations are made for the other components, which are then completed by respective interrelations and combined into a reference framework for the development of digital twins within cyber-physical systems, as shown in Figure 2.5.

From an informatics point of view, Schroeder et al. [27] list these digital twin components:

- *Storage*: To maintain a historical log of operational data, events, model updates, etc.; implemented as a local file record, local database, or via connection to a remote database;
- *Events*: To indicate that a given condition has been met;
- *Methods*: To provide functionalities that are accessed by applications or users, with the purpose to control the device, execute simulations, run diagnostics, etc.;
- *Application-programming interface (API)*: To connect external applications with potentially different programming languages;
- *Access control*: To provide security measures and access restrictions;
- *Human–machine interface (HMI)*: to enable interaction with the human user.

An alternative approach to analyzing the components of a digital twin is based on the distinction between the digital twin prototype and the digital twin instance, as introduced at the beginning of this chapter. According to [12,28,29], the purpose of a digital twin application – and thus its components and functional scope – largely depends on the lifecycle stage of the physical twin.

In Figure 2.6, the consecutive lifecycle stages are depicted, as defined in the systems engineering handbook [30] and ISO/IEC/IEEE standard [6], respectively; the digital twin and its components evolve along these stages as follows:

- In the *concept stage*, the digital twin is created in the virtual world as a digital twin prototype. Initially, it may consist of just a list of requirements (e.g., in the form of semantic data) expected from the final product (Chapter 8); additionally, concepts for the digital twin environment need to be composed.

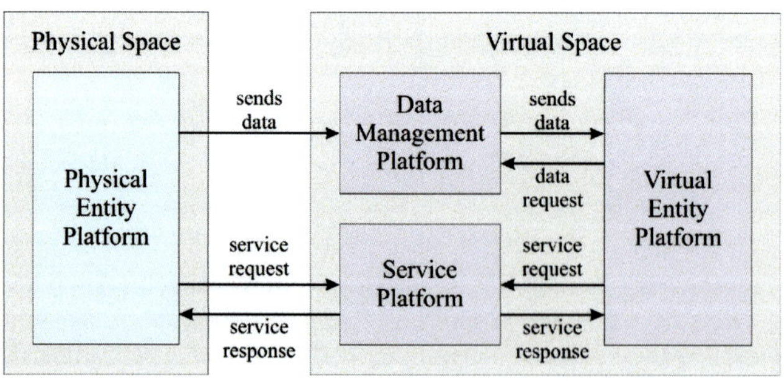

FIGURE 2.5 Interrelations and components of a digital twin framework. (Based on Josifovska K, Yigitbas E, Engels G. Reference framework for digital twins within cyberphysical systems. *2019 IEEE/ACM 5th International Workshop on Software Engineering for vSmart Cyber-Physical Systems (SEsCPS)*, Montreal, QC, Canada, May 28, 2019.)

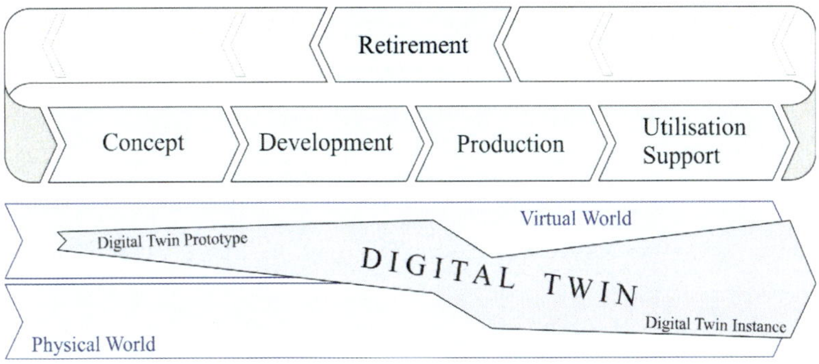

FIGURE 2.6 Evolution of a digital twin along the lifecycle stages. (Idea based on Boschert S, Rosen R. Digital twin—the simulation aspect. In: Hehenberger P, Bradley D (eds) *Mechatronic Futures*. Springer, Cham, 2016, pp. 59–73.)

- In the *development stage*, engineering data and executable simulation models extend the informational set of the digital twin. Parameter studies and optimizations are used to tune the behavior of the digital twin prototype. Multiple digital twin prototypes may be maintained simultaneously to compare different design variants. They also support verification and validation processes; eventually, the first realizations of components of the product are manufactured and tested, e.g., via hardware-in-the-loop simulation. In this way, the digital twin not only grows in volume but also starts to extend into the physical world.
- In the *production stage*, the physical twin is created based on the informational set of the digital twin prototype. During this transition stage, the information flow reverses, and the digital twin is fed with information from the physical product to mirror its actual configuration (also referred to as as-built data). The digital twin prototype becomes the digital twin instance; in this step, it may be reasonable to transfer not all but just the relevant data, models, etc., such that the volume of the digital twin can be temporarily reduced.
- In the *utilization* and *support stage*, the digital twin instance is operated in the digital twin environment and provides the services it is intended for. These services may focus on different timescales. A digital twin may support the (real-time) optimization of the operation of the physical twin. In this case, persistent data exchange is required, and the digital twin instance immediately affects the physical world through its actions. Another digital twin may support predictive maintenance, where data of a long timescale are processed. This digital twin instance interacts with the physical world rather rarely and punctually. However, through the exchange of information such as load cycles, wear, repairs, etc., a unique pairing of digital twin and physical twin is established.
- In the *retirement stage*, the product reaches the end of its useful life and is taken out of operation. Within the digital twin instance, however, data and models from the previous lifecycle stages are stored, which can be used for a predecessor system: "While the physical system may need to be retired, the information about it can be retained at little cost" [2]. For the latter, a systematic processing of these data is required.

2.3.2 Digital Twin Software-Building Solutions

The architecture of a system is often modeled using basic block diagrams (see, e.g., Figure 2.2). While such a representation is helpful in highlighting the key components of the system, more elaborate and systematic ways of modeling and documenting an architecture are required when additional information needs to be involved. Then, *architecture views* are "perhaps the most important concept" [23] associated with (software) architecture documentation. Since a single diagram cannot capture all the relevant details of a system architecture, different views of the system can be applied to focus on different aspects, such as the functional and behavioral aspects, the logical and technical structure, and the decomposition into components [31].

A representative assembly of views is described in [24] consisting of:

1. *A logical view*, which shows the key abstractions in the system as objects or object classes; this view relates requirements to entities of the architecture;
2. *A process view*, which shows how, at runtime, the system is composed of interacting processes; this view allows assessment of non-functional requirements such as performance and availability;
3. *A development view*, which shows how the software (or system) is decomposed for development; this view is especially useful from a management perspective;
4. *A physical view*, which shows the system hardware and how software components are distributed across the processors in the system; this view is, in particular, of interest to system engineers;
5. *A conceptual view*, which shows an abstract view of the system that can be the basis for decomposing high-level requirements into more detailed specifications; this view may be the starting point for a more detailed specification of the architecture.

Referring to the architecture concepts discussed in Section 2.2, one may categorize Figure 2.1 as the process view, Figure 2.2 as the conceptual view, Figure 2.3 as the physical view, and Figure 2.4 as the development view. Other views discussed in [6,23] involve functionalities, performance, environment, validation, security, components and connectors, and allocation. It is finally up to the user to decide which views fit best to the actual project and help to ensure that the architecture aligns with the overall goals of the system and integrates well with (existing) components. The latter aspect is of particular interest regarding the software implementation of a digital twin environment [12].

For some implementation steps of digital twins, pre-designed software solutions are available. Those cover a broad spectrum from rather general software packages to assist Industry 4.0 applications (cf. Chapter 3) to unique IT solutions designated for specific application scenarios. Commercial examples for the former category are *General Electric Predix OPM*, *Microsoft Azure Digital Twins*, *PTC ThingWorx*, *Dassault Système 3DEXPERIENCE*, *Siemens Insights Hub*, and others [32]. Open-source frameworks are provided, e.g., by *Eclipse Ditto* [33] and *FIWARE* [34]. On the other side of the spectrum, Pool [35] provides a list of 40 software solutions currently existing in the field of rail freight.

From an *application perspective*, the following software solutions dedicated to digital twin implementations in the rail domain are available:

- *Siemens Mobility Railigent X* offers a cloud-based platform that combines data from various rail assets via standardized APIs; it provides data-driven solutions for optimized operations, predictive maintenance, and lifecycle management [36].
- *Alstom HealthHub* offers a web-based platform that captures data from trains, signaling, and infrastructure; it provides real-time monitoring and advanced data analytics for predictive maintenance [37].
- *Hitachi HMAX* offers a digital asset management platform that assists in integrating data from trains, signaling, and infrastructure; it provides performance monitoring capabilities based on key performance indicators and condition-based maintenance [38].
- *IBM Maximo for Transportation* offers a transportation asset management platform for consists, locomotives, and rail vehicles; it enables predictive maintenance and optimization of asset performance and reliability and helps to ensure compliance with regulatory requirements [39].

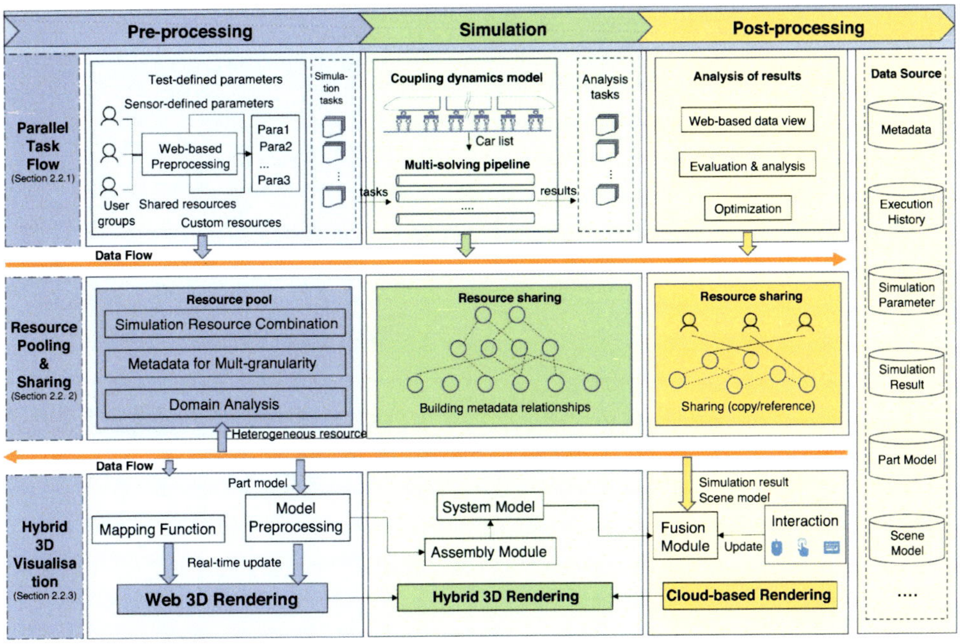

FIGURE 2.7 Architecture of the cloud-based framework for the development of vehicle dynamics simulation software for digital twin trains. (From Tang Z, Ling L, Zhang T, et al. Towards digital twin trains: Implementing a cloud-based framework for railway vehicle dynamics simulation. *International Journal of Rail Transportation* 2024; 1–24. https://doi.org/10.1080/23248378.2024.2355578, with permission.)

- *Bentley iTwin* offers a range of digital twin solutions with a focus on infrastructure; these solutions combine BIM technology, models, and real-time data to monitor and manage rail systems from planning to maintenance [40].

From a railway vehicle dynamics *simulation perspective*, Tang et al. [41] identify several shortcomings of traditional simulation software packages for use as a digital twin environment. These refer to resources required for real-time computing, on the one hand, and to the lack of scalability to handle digital twins of larger railway systems, on the other hand. The authors conclude that "it is necessary to explore new frameworks in terms of software architecture and the organization and management of computing resources for railway vehicle dynamics simulation software" [41]. To overcome the above shortcomings, a cloud-based framework is proposed, which separates the sequence of pre-processing, simulation, and post-processing into single units, as shown in Figure 2.7. This separation allows the individual units to be developed, maintained, and scaled independently of each other. Based thereon, a train-track simulation software was developed and compared to traditional railway vehicle dynamics software in terms of practical applicability and to results of field tests in terms of accuracy of simulation data. A continuative and comprehensive discussion of digital twin modeling and simulation aspects is provided in Chapter 4.

2.3.3 Digital Twin Data Space Sharing and Flow

Digital twin data originate from the digital twin, the physical twin, or the environment. Communication in terms of data flow may appear between two or all three components. In a digital twin system of systems, more complex communication scenarios need to be accounted, for example, by means of a separate connectivity layer in the architecture, which allows respective "inter-twin" and "intra-twin" communication [42], as shown in Figure 2.8. Alam and El Saddik [43] propose a digital twin architecture framework to accommodate cloud-based communication.

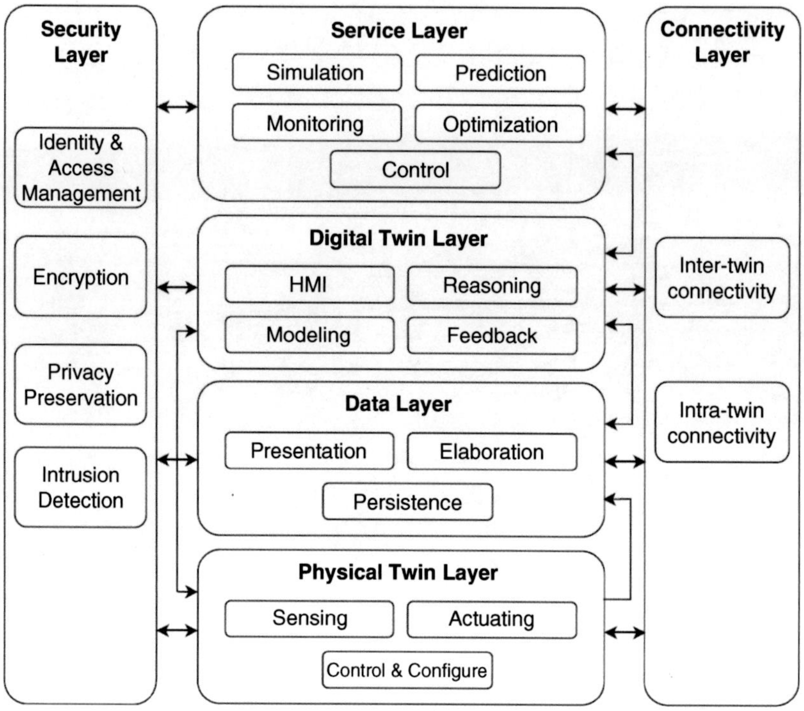

FIGURE 2.8 Layered architecture for a digital twin system of systems in healthcare. (From De Benedictis A, Mazzocca N, Somma A, et al. Digital twins in healthcare: An architectural proposal and its application in a social distancing case study. *IEEE Journal of Biomedical and Health Informatics* 2023; 27(10):5143–5154, CC BY 4.0.)

Interaction among the physical and virtual components can occur either through direct "ad-hoc" communication in the physical layer or through "peer-to-peer" communication in the cyber layer of the architecture. Additional layers are proposed to organize communication groups, data exchange, storage and visualization, and services. The framework addresses many of the conceptual challenges of a cloud-based digital twin.

Data may be transferred in real time or offline. From a practical perspective, a data flow is considered to appear in real time when the sending interval and latency of the messages of the data flow are negligible compared to the time between changes of states of the physical and the digital twin, respectively. In contrast, in the offline case, the data flow may happen periodically or just occasion-related. The main architectures of real-time analysis include parallel processing clusters using traditional relational databases, on the one hand, and memory-based computing platforms, on the other hand [44]; see Chapter 4 for a more detailed discussion of real-time computing. To reduce communication latency, data may be compressed and/or filtered to reduce the amount of data to be transmitted; in the case of cloud-based analyses, one may use edge or fog computing techniques to move the data processing closer to the physical hardware [9].

Not every data flow is intended to serve as communication between components; data may also be transferred to a storage component (either local hardware or cloud-based storage solutions). Minerva [12] recommends keeping the meaningful data together with an indication of location and time such that the respective state of the system can be reproduced and analyzed on demand. Ontologies are well-established approaches to structure and formalize such indications and additional semantic information in a machine-understandable vocabulary [45].

The management of data and data spaces from an architectural point of view is discussed in [46], which the authors argue is a topic that has been paid little attention to in digital twin applications so far, such that actual "architectural models lack details on how data are exchanged between digital twin components and especially between different digital twins." As a potential remedy, the integration of the *data space* concept as presented in [47,48] is discussed; therein, a reference architecture for data spaces is provided, which enables addition of virtual data spaces for managing and aggregating heterogeneous data as building blocks to a digital twin environment and describes the underlying principles of data sharing, data usage policies, data asset description, and a standardized data exchange API [48].

A thorough discussion of current communication and data exchange protocols frequently used in current digital twin applications is found in [4,9]. A discussion of respective standards and overlapping aspects of digital twins, Internet of Things, and Industry 4.0 technologies is performed by [49–51], aiming to stimulate a consolidation of the standards to facilitate interoperability, which according to the authors "is presumably the most important aspect of Industry 4.0" [50].

2.4 DIGITAL TWIN ARCHITECTURE PATTERNS

Several literature studies have been performed to extract common strategies and guidelines for the design and implementation of digital twins. Segovia and Garcia-Alfaro [9] analyze the necessary steps to methodologically build a digital twin by considering multiple phases from requirement engineering and architecture planning to integration and verification of the final model. Fett et al. [52] explore systematic approaches for the development of digital twins by also considering related concepts such as cyber-physical systems and product-service systems. Regarding the architecture of digital twins, the study revealed that most contributions are limited to the formal description of the architecture (e.g., as a conceptual view), while tools, methods, or guidelines for an architecture definition process receive less attention. Ferko et al. [53] review the crucial aspects of software architectures for digital twins including architectural solutions, quality attributes, and architectural patterns. It was found that the majority of the software solutions for digital twins are based either on a layered architectural pattern or on a combination of layered and service-oriented patterns.[2]

Considering digital twins and in the context of systems engineering, Tekinerdogan and Verdouw [1] derive a catalog of architecture design patterns that can be reused for building digital twins; the patterns are customized for the different stages of the product lifecycle (cf. Section 2.3.1) and build on each other as follows:

- *For the concept and design stages*:
  - The *digital model* pattern is tailored for the case that several digital twin prototypes appear simultaneously; the digital twin is considered to be operated by humans; the pattern is shown in Figure 2.9.

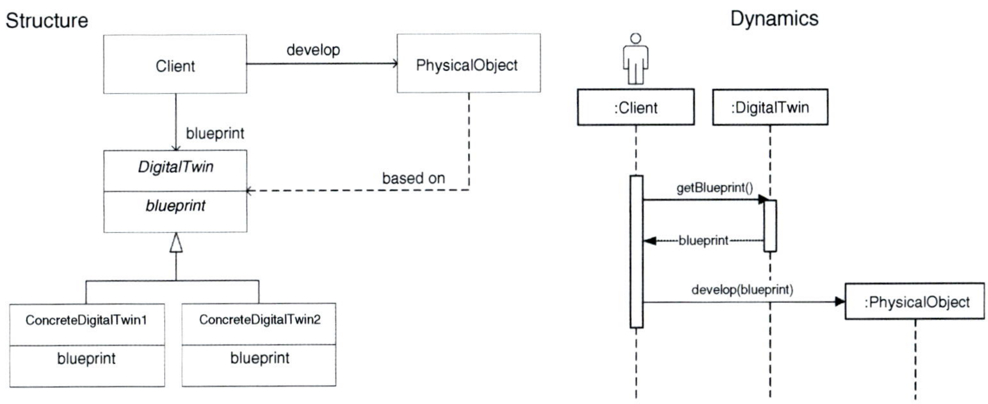

FIGURE 2.9 Digital model pattern. (From Tekinerdogan B, Verdouw C. Systems architecture design pattern catalog for developing digital twins. *Sensors* 2020; 20(18):5103, CC BY 4.0.)

- *For the production stage*:
  - The *digital generator* pattern is tailored for the manufacturing process of the product based on the data provided by the digital twin prototype; the manufacturing process is considered to be executed in an automated fashion.
  - The *digital shadow* pattern[3] is tailored for the creation of the digital twin instance based on the specification of the manufactured product; the process is considered to be performed either manually or automatically.
- *For the utilization and support stage*:
  - The *digital matching* pattern is designed for the process of adaption of the parameters of the digital twin instance to the actual properties of the physical twin (e.g., predictive maintenance); the recognition and adaption process is considered to be performed automatically.
  - The *digital restoration* pattern is tailored to enable a restoration of an earlier state of the physical twin.
  - The *digital monitor* pattern is designed to enable monitoring of the physical twin, considering a many-to-many relationship; this allows for scenarios where a single digital twin instance monitors multiple physical objects and vice versa, where one physical object is monitored by multiple (distinct) digital twins.
  - The *digital control* pattern extends the pattern above by an additional control and actuation layer.
  - The digital autonomy pattern finally extends the digital control pattern by self-adapting capabilities.
- *For the utilization, support, and retirement stage*:
  - The *digital proxy* pattern is designed for the case that, for communication with the environment, the physical twin is (fully) replaced by the digital twin instance; this may provide protection from unauthorized access in the utilization and support stage but may also be used for inquiries regarding operational data in the retirement stage.

2.4.1 Digital Twin Architecture Patterns for Rail Applications

The above patterns can be used for digital twins across a wide range of industries and application scenarios. In this last section, the railway sector is focused again. De Donato et al. [18] derive guidelines to implement a digital twin with a focus on data-driven predictive maintenance. The guidelines are formulated as a flowchart, providing a sequence of tasks and steps to go through in an appropriate sequence. These tasks are further divided into design guidelines (involving requirement specification, process planning, architectural design, digital representation, digital twinning, and service planning) and subsequent deployment guidelines (data flow, conversion, training process, prediction process, decision process, and evaluation). An architectural pattern is proposed as well (see Figure 2.10), which provides a logical view (cf. Section 2.3.2) on the main components, functions, and data flow required for "AI-assisted digital twins for predictive maintenance of railway assets" [18].

Jeschke and Grassmann [54] address the development of a generic implementation strategy of digital twins in rail logistics systems. The strategy was initially developed for German rail transport, and the objective of the digital twin system was to optimize operations and enable intelligent management of rail transport assets. The strategy is intended to serve as a reference for future implementations of digital twins in similar scenarios. The strategy focuses on the creation of digital twin systems for already existing assets. Since existing guidelines do not provide the necessary level of detail to use them as a reference for actual implementation, the authors develop a comprehensive strategy, wherein – besides the technical aspects – requirements originating from the expectations of business stakeholders and considering the human–machine interaction are regarded as well.

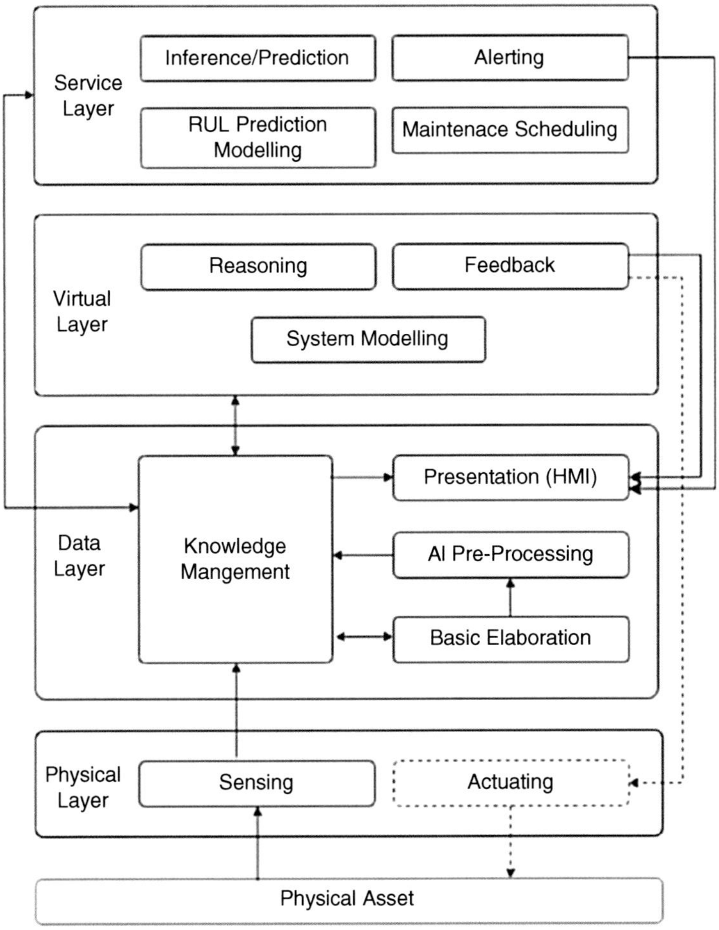

FIGURE 2.10 High-level architecture pattern for AI-assisted digital twins for predictive maintenance for railway assets. (From De Donato L, Dirnfeld R, Somma A, et al. Towards AI-assisted digital twins for smart railways: Preliminary guideline and reference architecture. *Journal of Reliable Intelligent Environments* 2023; 9(3):303–317, CC BY 4.0.)

Based on the development processes of system engineering, the authors define a sequence of 15 individual steps (divided into design, development, operation, and decommissioning phases) as an implementation strategy for digital twins. Key features thereof are (1) the digital twin environment is integrated already at an early stage; (2) the digital twin system is intended to adapt and evolve over time; and (3) data relevant for the purpose of the digital twin is defined at an early stage such that the storage and processing of unnecessarily high data volumes are avoided.

Standardization, compatibility of data, definition of interfaces, data sovereignty, and the legal use of data across multiple system operators and companies are considered future challenges. As an outline and to increase (cost) effectiveness, the authors propose that for

> tendering procedures of cost-intensive assets, a kind of digital twin prototype exists between manufacturer and operator, which makes the design and development more attractive by creating a win–win situation for both parties. [...] The aim of this procedure for all manufacturers would be to submit their offer including the further developed digital twin prototype, which would reflect the iterative further development of the asset on the one hand and the basis for a digital twin instance in the operational phase of the asset on the other. [54]

NOTES

1 Functional requirements describe the services the system should provide (monitoring, optimisation, etc.); non-functional requirements relate to system properties such as performance, reliability, latency, memory use, security, etc.; typically, the latter may affect the architecture of the system rather than its individual components [24,25].

2 A detailed discussion of architecture patterns specifically for software systems can be found in [23,24].

3 The term *digital shadow* is introduced in [7] to distinguish a system with an (automatic) one-way data flow from the physical to the digital object from a digital twin, i.e., a system with an automatic data flow in both directions.

REFERENCES

1. Tekinerdogan B, Verdouw C. Systems architecture design pattern catalog for developing digital twins. *Sensors* 2020; 20(18):5103.
2. Grieves M, Vickers J. Digital twin: Mitigating unpredictable, undesirable emergent behavior in complex systems. In: Kahlen J, Flumerfelt S, Alves A (eds) *Transdisciplinary Perspectives on Complex Systems.* Springer, Cham, 85–113, 2017.
3. Jones D, Snider C, Nassehi A, et al. Characterising the digital twin: A systematic literature review. *CIRP Journal of Manufacturing Science and Technology* 2020; 29:36–52.
4. Juarez MG, Botti JV, Giret AS. Digital twins: Review and challenges. *Journal of Computing and Information Science in Engineering* 2021; 21(3):030802-1–030802-23.
5. Singh M, Fuenmayor E, Hinchy EP, et al. Digital twin: Origin to future. *Applied System Innovation* 2021; 4(2):36.
6. ISO/IEC/IEEE. 15288*:2023(E): Systems and Software Engineering – System Life Cycle Processes*, ISO, Vernier, Switzerland, 2023.
7. Kritzinger W, Karner M, Traar G, et al. Digital twin in manufacturing: A categorical literature review and classification. *IFAC-PapersOnLine* 2018; 51(11):1016–1022.
8. Tao F, Zhang M, Nee AYC. Chapter 3 – Five-dimension digital twin modeling and its key technologies. In: Tao F, Zhang M, Nee AYC (eds) *Digital Win Driven Smart Manufacturing.* Academic Press, London, UK, 2019, pp. 63–81.
9. Segovia M, Garcia-Alfaro J. Design, modelling and implementation of digital twins. *Sensors* 2022; 22(14):5396.
10. Schwarz C, Wang Z. The role of digital twins in connected and automated vehicles. *IEEE Intelligent Transportation Systems Magazine* 2022; 14:41–51.
11. Spiryagin M, Edelmann J, Klinger F, et al. Vehicle system dynamics in digital twin studies in rail and road domains. *Vehicle System Dynamics* 2023; 61:1737–1786.
12. Minerva R, Lee GM, Crespi N. Digital twin in the IoT context: A survey on technical features, scenarios, and architectural models. *Proceedings of the IEEE* 2020; 108(10):1785–1824.
13. Spiryagin M, Wu Q, Polach O, et al. Problems, assumptions and solutions in locomotive design, traction and operational studies. *Railway Engineering Science* 2022; 30(3):265–288.
14. Spiryagin M, Persson I, Wu Q, et al. A co-simulation approach for heavy haul long distance locomotive-track simulation studies. *Vehicle System Dynamics* 2019; 57(9):1363–1380.
15. Wright L, Davidson S. How to tell the difference between a model and a digital twin. *Advanced Modeling and Simulation in Engineering Sciences* 2020; 7(13):1–13.
16. Bernal E, Wu Q, Spiryagin M, et al. Augmented digital twin for railway systems. *Vehicle System Dynamics* 2024; 62(1):67–83.
17. Bosso N, Magelli M, Trinchero R, et al. Application of machine learning techniques to build digital twins for long train dynamics simulations. *Vehicle System Dynamics* 2024; 62(1):21–40.
18. De Donato L, Dirnfeld R, Somma A, et al. Towards AI-assisted digital twins for smart railways: Preliminary guideline and reference architecture. *Journal of Reliable Intelligent Environments* 2023; 9(3):303–317.
19. Dirnfeld R, De Donato L, Somma A, et al. Integrating AI and DTs: Challenges and opportunities in railway maintenance application and beyond. *Simulation* 2024; 100:903–917.
20. Doubell GC, Kruger K, Basson AH, et al. The potential for digital twin applications in railway infrastructure management. In: Pinto JOP, Kimpara MLM, Reis RR, et al. (eds) *15th WCEAM Proceedings*; 15–18 August, Campo Grande, Springer, Cham, 2022, pp. 241–249.

21. Djordjević B, Krmac E, Lin CY, et al. An optimisation-based digital twin for automated operation of rail level crossings. *Expert Systems with Applications* 2024; 239:122422.
22. Zhou S, Dumss S, Nowak, et al. A conceptual model-based digital twin platform for holistic large-scale railway infrastructure systems. *Procedia CIRP* 2022; 109:362–367.
23. Bass L. Software architecture in practice. 4th ed. Addison-Wesley, Boston, MA, 2022. https://www.oreilly.com/library/view/software-architecture-in/9780136885979/
24. Sommerville I. *Software Engineering*, 10th ed. Pearson, Boston, München, 2016.
25. Ozkaya I. Architectural concerns of digital twins. *IEEE Software* 2022; 39(2):3–6.
26. Josifovska K, Yigitbas E, Engels G. Reference framework for digital twins within cyberphysical systems. *2019 IEEE/ACM 5th International Workshop on Software Engineering for vSmart Cyber-Physical Systems (SEsCPS)*, Montreal, QC, Canada, May 28, 2019.
27. Schroeder GN, Steinmetz C, Rodrigues RN, et al. A methodology for digital twin modeling and deployment for Industry 4.0. *Proceedings of the IEEE* 2021; 109(4):556–567.
28. Boschert S, Rosen R. Digital twin—the simulation aspect. In: Hehenberger P, Bradley D (eds) *Mechatronic Futures*. Springer, Cham, 2016, pp. 59–73.
29. Wagg D, Worden K, Barthorpe R, et al. Digital twins: State-of-the-art future directions for modelling and simulation in engineering dynamics applications. *ASCE-ASME Journal of Risk and Uncertainty in Engineering Systems, Part B: Mechanical Engineering* 2020; 6:030901.
30. Walden DD. *Systems Engineering Handbook: A Guide for System Life Cycle Processes and Activities*. 5th ed. John Wiley & Sons, Hoboken, NJ, 2023.
31. Bauer T, Antonino PO, Kuhn T. Towards architecting digital twin-pervaded systems. *2019 IEEE/ACM 7th International Workshop on Software Engineering for Systems-of-Systems (SESoS) and 13th Workshop on Distributed Software Development, Software Ecosystems and Systems-of-Systems (WDES)*, Montreal, QC, Canada, May 28, 2019.
32. Available from: https://dashdevs.com/blog/product-owner-talks-20-digital-twins-service-companies/ (accessed in October2024).
33. Available from: https://www.eclipse.org/ditto/ (accessed in October2024).
34. Available from: https://www.fiware.org/ (accessed in October 2024).
35. Pool A. Digital twins in rail freight - The foundations of a future innovation. Master Thesis, University of Twente, The Netherlands, 2021, 137p.
36. Available from: https://www.mobility.siemens.com/global/en/portfolio/digital-solutions-software/digital-services/railigent-x.html (accessed in October 2024).
37. Available from: https://www.alstom.com/press-releases-news/2023/2/healthhub-intelligent-way-improve-maintenance (accessed in October 2024).
38. Hitachi Rail STS S.p .A. White paper: Introducing HMAX, 2024. Available from: https://go.hitachirail.com/l/993992/2024-09-20/q8fdl/993992/17270902736FzctFOR/HMAX___White_Paper.pdf (accessed in October 2024).
39. Available from: https://www.ibm.com/products/maximo/travel-transportation (accessed in October 2024).
40. Available from: https://www.bentley.com/software/infrastructure-digital-twins (accessed in October 2024).
41. Tang Z, Ling L, Zhang T, et al. Towards digital twin trains: Implementing a cloud-based framework for railway vehicle dynamics simulation. *International Journal of Rail Transportation* 2024; 1–24. https://doi.org/10.1080/23248378.2024.2355578
42. De Benedictis A, Mazzocca N, Somma A, et al. Digital twins in healthcare: An architectural proposal and its application in a social distancing case study. *IEEE Journal of Biomedical and Health Informatics* 2023; 27(10):5143–5154.
43. Alam KM, El Saddik A. C2PS: A digital twin architecture reference model for the cloud-based cyber-physical systems. *IEEE Access* 2017; 5:2062.
44. Chen M, Mao S, Liu Y. Big data: A survey. *Mobile Networks and Applications* 2014; 9(2):171–209.
45. Barricelli BR, Casiraghi E, Fogli D. A survey on digital twin: Definitions, characteristics, applications, and design implications. *IEEE Access* 2019; 7:167653–167671.
46. Somma A, Benedictis AD, Zappatore M, et al. Digital twin space: The integration of digital twins and data spaces. *IEEE International Conference on Big Data*, Sorrento, Italy, 15–18 December 2023, pp. 4017–4025.
47. Usländer T, Baumann M, Boschert S, et al. Symbiotic evolution of digital twin systems and dataspaces. *Automation* 2022; 3(3):378–399.

48. Volz F, Sutschet G, Stojanovic L, et al. On the role of digital twins in data spaces. *Sensors* 2023; 23(17):7601.
49. Profanter S, Tekat A, Dorofeev K, et al. OPC UA versus ROS, DDS, and MQTT: Performance evaluation of Industry 4.0 protocols. *2019 IEEE International Conference on Industrial Technology (ICIT)*, Melbourne, Australia, 13–15 February 2019, pp. 955–962.
50. Jacoby M, Usländer T. Digital twin and internet of things—Current standards landscape. *Applied Sciences* 2020; 10(18):6519.
51. Schmidt C, Volz F, Stojanovic L, et al. Increasing interoperability between digital twin standards and specifications: Transformation of DTDL to AAS. *Sensors* 2023; 23(18):7742.
52. Fett M, Wilking F, Goetz S, et al. A literature review on the development and creation of digital twins, cyber-physical systems, and product-service systems. *Sensors* 2023; 23(24):9786.
53. Ferko E, Bucaioni A, Behnam M. Architecting digital twins. *IEEE Access* 2022; 10:50335–50350.
54. Jeschke S, Grassmann R. Development of a generic implementation strategy of digital twins in logistics systems under consideration of the German rail transport. *Applied Sciences* 2021; 11(21):10289.

3 Data Collection and Management for Digital Twins

Giovanni Lugaresi, Lulai Zhu, and Andrea Matta

3.1 INTRODUCTION

The growing application of digitization is enhancing the productivity and efficiency of manufacturing systems [1,2]. Through the incorporation of smart machines and embedded sensors in equipment, different elements can be interconnected and real-time data can be collected, achieving the integration of the physical and the cyber world [3]. The digital twin (DT) is recognized as a particularly promising technology to enable the convergence of the physical and virtual realms via the real-time replication of a physical system behavior.

The first introduction of the DT concept comes from science fiction as found in the book titled *Mirror Worlds* by David Gelertner (1991). It has then been elaborated by Michael Grieves in 2002 in the field of product lifecycle management. The concept involves establishing a digital counterpart to a physical entity and linking these two components across all stages of the lifecycle. This connection is established through the exchange of data and information, facilitating a comprehensive twinning of the physical product with its digital representation [4]. The first definition of a DT is attributed to a presentation from NASA, which considers a DT to be an integrated multi-physics, multi-scale, probabilistic simulation of a vehicle or system that utilizes the best available physical models, sensor updates, fleet history, etc., to mirror the life of its flying twin. The DT is ultra-realistic and may consider one or more important and interdependent vehicle systems. [5]

Presently, the DT concept is garnering increased interest across various sectors, ranging from manufacturing to smart cities and healthcare. Notably, the DT concept has been featured in Gartner's Top 10 strategic technology trends from 2017 to 2019. Also, in 2022, the global market size for DTs reached 8.6 billion USD, and it is anticipated to surge to 137.7 billion USD by 2030, demonstrating a compounded average growth rate of 42.6% [6]. A recent survey indicates that 86% of companies consider DTs to be a vital solution within their strategic approach [7]. Furthermore, international standardization efforts, such as the ISO 23247:2021, are being directed toward DTs [8]. The architecture of a DT relies on a diverse array of enabling technologies, which is why the concept has seen significant growth only with recent advancements, experiencing exponential growth [9]. According to the review conducted by Tao et al. [9], the most common industrial application areas include (1) the design of an industrial system, in which a DT has the capacity to enhance both product and production systems design, offering virtual commissioning services; (2) the operation of the system, since the DT can contribute to improvements in the production process, production optimization, maintenance support, and more accurate predictions of health conditions. The applications of DTs in industry are vast, from monitoring and analyzing production processes, identifying areas for improvement, and optimizing production schedules, to monitoring the performance of machines and equipment, detecting potential problems, and alerting operators when maintenance is needed. DTs can also be used to track inventory levels, predict demand, and optimize supply chains [10,11]. In the manufacturing field, the DT serves as a virtual representation of a manufacturing system, monitoring its performance, forecasting, and optimizing its behavior [12]. Additionally, it can offer feedback on improvements based on the state of the system. Most of the mandatory characteristics of a DT that have been noted in the literature are synchronization, bidirectional communication,

DOI: 10.1201/9781003492146-3

control, anomaly detection and diagnostic, and optimization [3,13,14]. An article by van der Valk et al. [15] described a DT as

> A virtual construct that represents a physical counterpart, integrates several data inputs with the aim of data handling, data storing, and data processing, and provides an automatic, bi-directional data linkage between the virtual world and the physical one. Synchronization is crucial to the DT to display any changes in the state of the physical object. Additionally, a DT must comply with data governance rules and must provide interoperability with other systems.

As highlighted in the latter definition, DTs heavily rely on the availability of data and data processing capabilities to provide value to a real system. This chapter aims to provide an overview of the main data collection techniques and how data is used and managed by significant contributions in DTs. The remainder of this chapter is organized as follows: Section 3.2 introduces the most common data types and their role within a DT framework; Section 3.3 illustrates how data is used to generate the most common digital models used in DTs; Section 3.4 lists significant applications, while Section 3.5 highlights some promising directions that have been identified by the authors as a result of this work; concluding remarks are to be found in Section 3.6.

3.2 DATA COLLECTION IN DTs

The DT is a dynamic concept, representing an evolution of simulation that integrates modeling and optimization capabilities. Simulation has been employed for several years in applications such as system validation, testing, and design, as well as decision-making. These applications typically occur in the early stages of the lifecycle, but simulation can extend into subsequent stages for offline decision-making [1]. The primary distinction between simulation and DT lies in the type and volume of data. Most DT architectures are characterized by the ability to gather and utilize real-time data from the field [16], which is the focus of Section 3.2.1.

3.2.1 Data Streaming from Sensors

Typically, a DT acquires data from its real-world entity and manipulates it to provide a set of services, which have the potential of improving both its digital models and its real-world counterparts. To improve or effectively reflect the behavior of its real-world entity, a DT should have the capability to automate data acquisition [17], and this should be a fundamental element of a DT architecture [18]. Figure 3.1 shows an example of a DT architecture for a generic manufacturing system with its main data flows. Given that the DT acquires data from multiple sources, including offline data sources (e.g., databases, stored data, external data, and maintenance logs), there is an emerging case for semi-manual data acquisition [15].

Crucial to the functioning of a DT is the real-time synchronization between the physical system and its virtual counterpart, facilitated through sensors positioned on the equipment for data collection. For instance, in a manufacturing system real-time data collection is made possible by a sensory infrastructure in the shop floor and information systems such as manufacturing execution system (MES). A distinguishing feature of a DT lies in its ability to autonomously impact the physical system, achieved through a bidirectional flow of data. Optimization and control algorithms play a key role in determining the optimal solution, which is then fed back to the physical component. Consequently, the virtual system gains the capability to control the physical counterpart and support real-time decision-making.

Another noteworthy characteristic of the DT is its capacity to dynamically update the digital model, enabling it to adapt to any changes occurring in the physical system. Compared to the use of simulation in isolation, the implementation of a DT leads to more efficient decision-making processes.

The communication or integration medium of a DT is typically implemented using communication protocols with the capability of providing a bidirectional communication between the PT and

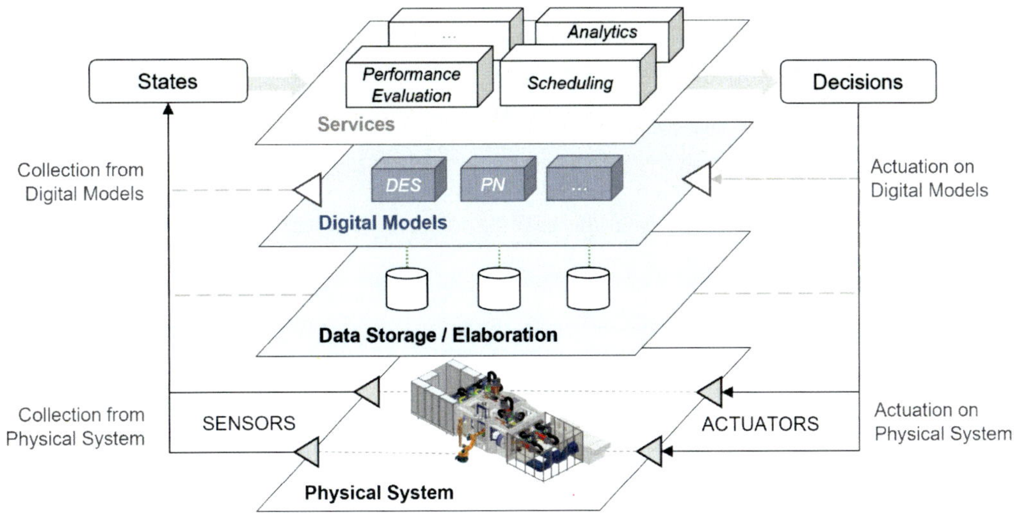

FIGURE 3.1 Example of a digital twin architecture with its main data flows.

DT. The requirements of the use case, including the type of data to be transferred, determine the type of devices used and the communication strategy and protocols that apply. Such protocols vary extensively. For instance, machine-to-machine (M2M) communication is typically achieved via hypertext transfer protocol (HTTP), message queue telemetry transport, and open platform communication unified architecture (OPC UA), among others [19].

3.2.2 Data Streaming from Simulation Models

Simulation models generate vast amounts of data, and a DT framework can foresee the continuous collection of digitally generated data, allowing for dynamic adjustments, predictive analysis, and responsive decision-making.

In early works, the data collection from the simulation models is identified as essential to gather digitally produced results in time for making a decision on a physical system. In [20], for instance, if the simulation cannot provide a system policy result before a due date, the station controller would apply a pre-defined policy to avoid slowing down production. The data collected from a digital model is also used to align and synchronize alternative models of a DT, e.g., representing different production policies to be evaluated [21]. Further, data from the digital models can be used to train learning approaches (e.g., RL and Q-learning) that identify actions and optimal behaviors based on the exploration of states not yet observed in the real system [22].

The data streaming from a digital model naturally requires a robust infrastructure for data ingestion, processing, and storage to ensure the integrity and accuracy of the DT's performance, especially in complex, multi-layered simulations. The data acquisition module is often cited as a crucial component of a DT [18]. Simulation models are not the only components generating data in a DT. Indeed, several contributions in the literature identify value in the integration with the existing data infrastructure of an enterprise (e.g., MES, ERP, RTLS systems). Information fusion techniques based on data from enterprise systems can also be used in automated model adaptation techniques, allowing for a model to be continuously updated during operations [23].

The data requirements of the DT influence the implementation of the respective modules in the DT architecture, for instance, choosing between a web service application [24,25] or a mobile application [26]. The data exchange can be either volatile (e.g., sensors transmitting the part's location) or static (e.g., a list of production resources) [27]. In general, database modules represent the internal repository or any form of data storage mechanism that houses the data and models of a DT [5,28].

3.2.3 System Knowledge Graphs for Data Structures

In 2012, Google introduced to their main search engine the *knowledge graphs*, a formalism that has become increasingly popular in representing knowledge. A knowledge graph encompasses facts into entities and relations annotated with semantic information. Entities can be real-world objects or abstract concepts, while relations define the connections between these entities. Being easily accessible and interpretable by both humans and machines, knowledge graphs have been widely applied to language representation learning, question answering, and recommender systems [29].

Event knowledge graphs are a specific type of temporal knowledge graph for representing dynamic and procedural knowledge, centering on event-related information [30]. Unlike static knowledge graphs where facts are assumed to be invariant over time, event knowledge graphs aim to capture temporal, causal, and conditional relations between events and the roles of entities within those events. They are often used to model and link events in a way that helps with a more comprehensive understanding of real-world scenarios and processes. In the following paragraphs are listed significant works that made use of data to design and enrich digital models based on knowledge graphs.

Fahland [31] uses event knowledge graphs to address the limitations of traditional process mining, which relies on unique case identifiers. The proposed methodology includes constructing knowledge graphs from event tables, inferring relations between events and entities, and querying the graphs to analyze the behavior. It is shown through a retailer case study that event knowledge graphs can capture well the behavior over multiple interacting entities.

Klijn et al. [32] introduce a novel set of aggregation operations on event knowledge graphs to support process mining. These operations enable the aggregation of similar task instances, facilitating the analysis of task executions and actor behaviors within processes. The resulting data summaries are useful in answering questions of how work is organized and divided among actors in distinct stages of the process.

Swevels et al. [33] propose an approach that can add missing case identifiers to incomplete event data based on physical constraints and contextual information. They extend event knowledge graphs with concepts for context and rule-based inference and define five rules to infer the missing identifiers for physical objects within a process. This approach makes it possible to efficiently find missing identifiers even when no normative control-flow model is available or applicable. In [34], the authors present a method to automate the translation of legacy data into object-centric event data through event knowledge graphs. The method relies on a semantic header and a property graph schema to define the mappings of a given dataset to the object-centric concepts and the domain-specific ontology. An extract-load-transform procedure is employed to convert raw data into event knowledge graphs, which are then stored and queried in a graph database.

Klijn et al. [35] investigate the application of graph-based event data representations to financial auditing. They aim to tackle the limitations of traditional process mining methods, which rely on a single case notion and may misrepresent relationships between business and accounting objects, with event knowledge graphs. The case study demonstrates how a more accurate and flexible visualization of processes and related accounting entries can help auditors better understand and assess the internal controls of a company.

3.3 DATA-DRIVEN MODELING FOR DISCRETE-EVENT DTs

As data represents the partial knowledge of the state of a system at the time of its collection, data collection gathers the understanding of its behavior through the evolution of its states in time. Data-driven model generation allows for an unbiased view of the system behavior and for quick adaptation in case of system changes, effectively enabling the twinning. This section collects the most common data-driven model creation techniques and their applications in DTs.

3.3.1 Process Mining

Process mining is a collection of approaches that bridge the gap between traditional data analysis and business process management. Initially, process mining has been developed to gather insights from event logs to visualize, analyze, and improve business processes and enable organizations to understand how their processes function in practice as opposed to their designed behavior [36]. Process mining techniques revealed to be successful in identifying inefficiencies, bottlenecks, and deviations in workflows, providing a data-driven foundation for optimizing operations and driving continuous improvement across various application domains.

In process mining, data is the cornerstone that drives the analysis of business processes. The primary data source used in process mining is event logs, digital records that capture the sequence of events occurring in the execution of a business process, making it possible to reconstruct and analyze the flow of activities. The main data types used are (1) a case ID, namely the unique identifier for each process instance or case, such as an order number or customer ID; (2) an activity ID, which identifies the specific action or step in the process; and (3) timestamps, which record the exact date and time when each event has occurred. Recent research has demonstrated that the data types used for traditional process mining can also be used for generating models of manufacturing systems [24,37].

Most process mining techniques generate Petri Nets as output models, but also other model formalisms can be used or derived from the results of a process mining algorithm (e.g., directly follows graphs). Also, hybrid approaches have recently been proposed to exploit the best characteristics of each discovery algorithm in modeling specific parts of a system [38].

3.3.2 Data Analysis with the Application of Artificial Intelligence

Identifying discrete-event systems (DESs) from external observations of their behavior is a challenging problem that has attracted significant attention over the past decade. Interest in DES identification often arises from the need for reverse engineering, fault diagnosis, and property verification of a partially unknown system. The goal of DES identification is to derive a mathematical model, such as a Petri Net or a finite state automaton [39,40] that captures the behavior of a DES system from the input and/or output sequences observed during its operation. If the resulting model is a Petri Net, both the structure of the model (including places, transitions, and arcs) and its initial marking must be provided for the sake of completeness.

Research on Petri Net learning is conducted in two directions that differ in their requirements for the input data. The first direction, Petri Net synthesis, involves determining if a given transition system corresponds to a Petri Net whose reachability or coverability graph is isomorphic to it [41]. Existing synthesis methods expect the whole knowledge of the state space or the language of the system being studied [42–45]. However, obtaining the entire state space or the full language of a real-world system is often impractical. Petri Net identification thus emerges as an alternative direction, which focuses on finding a suitable Petri Net model for a DES system from observed data (i.e., a subset of the state space and/or part of the language). Techniques mentioned in [46–49] are notable in this direction.

Hiraishi [42] proposes an algorithm for constructing a Petri Net model from a finite set of firing sequences. The proposed algorithm operates in two phases. In the first phase, a language is identified in the form of a finite state automaton from the given firing sequences. In the second phase, the structure of the Petri Net is guessed using dependency relations extracted from the language. The algorithm is specifically tailored for safe Petri Nets and has a running time bounded by a polynomial function of the inputs.

Cortadella et al. [43] introduce a method to derive Petri Nets based on the theory of regions for elementary transition systems. They extend the class of synthesizable transition systems to excitation-closed transition systems by relaxing the requirement of isomorphism to bi-simulation and use irredundant region sets to minimize the number of places in the resulting Petri Net. The method shows good effectiveness in creating safe, place-irredundant Petri Nets from various specification models.

Darondeau [44] presents methods for synthesizing Petri Nets from finite transition systems and languages using regions. These methods take advantage of linear algebra to solve synthesis problems efficiently. The resulting nets are guaranteed to meet separation axioms and can be distributed across different sites without conflicts. It is also possible to apply the methods to the decentralized controller synthesis problem by reformulating it in the settings of Petri Nets.

Cabasino et al. [45] provide a linear algebraic approach for discovering the structure and initial marking of a Petri Net system based on a finite set of strings generated by it. They first tackle the problem of synthesizing free-labeled Petri Nets, where all transition labels are distinct, by solving an integer programming problem. This approach is then extended to include additional structural information, such as conservative components or stationary sequences, and to labeled Petri Nets where transitions may share labels, ensuring the resulting system is deterministic.

Dotoli et al. [46] propose a method that can identify DES systems using Petri Nets in real time. Two scenarios are considered. When both the sets of places and transitions are known, an integer linear programming problem is formulated to determine the Petri Net model. When only the upper bound on the number of places is given, the Petri Net model is derived by dynamically updating the transition and place sets and recursively solving the integer linear programming problem based on the observed events and output vectors. This method allows for identification of DES behavior without complete system knowledge.

El Mehdi et al. [47] present an approach for discovering the structure and parameters of a stochastic Petri Net or a deterministic stochastic Petri Net. The approach utilizes the measurable sequences of events and states observed from the DES system to identify the Petri Net model in two steps. It first infers the optimal structure of the Petri Net through integer linear programming and then estimates the timing parameters of the Petri Net by analyzing the underlying Markov chain.

Basile et al. [48] introduce an approach for discovering the structure and parameters of a time Petri Net. They formulate logical conditions from timed sequences observed from the DES system and transform these conditions into a mixed-integer linear programming problem to solve for the structure of the Petri Net model, the initial markings of places, and the timing intervals of transitions. This approach addresses practical constraints and assumptions relevant to fault detection and model repair of manufacturing systems.

Denno et al. [49] propose a methodology for dynamically identifying and updating Petri Net models of manufacturing systems based on operational logs. The approach exploits genetic programming to evolve the models, combining probabilistic neural networks and causal analysis to manage exceptional events and evaluate model accuracy. This methodology aims to improve production scheduling and control by continuously adapting the Petri Net model to changes in the manufacturing environment.

3.4 SIGNIFICANT APPLICATIONS

It is time-consuming and error-prone for human experts to build DTs manually for complex physical systems. As a result, a considerable amount of research effort has been recently put into the investigation of data-driven approaches that can help to automatically generate discrete-event DTs. Table 3.1 reports a list of example approaches and their applications to real-world systems. In these examples, the input data types are either system specification models, such as piping and instrumentation diagrams, or event logs extracted from the databases of physical systems, while the output model types are diverse and include simulation models, extended directed graphs, Petri Nets, and dynamic process maps. Model-to-model transformation and process mining are the two core techniques in use.

TABLE 3.1
Application Examples of Data-Driven Approaches for Discrete-Event DT Generation

Reference	Input Data	Output Model	Technique	Physical System
[50]	Piping and instrumentation diagrams	Simulation models	Model-to-model transformation	Lab-scale water processing plant
[37]	Event logs	Extended directed graphs	Process mining	Assembly line for electric motor stators
[53]	Event logs	Petri nets	Process mining	Artificial order handling system
[38]	Event logs	Petri nets	Process mining	Lab-scale assembly line for quadcopter drone parts
[55]	Automation markup language models	Simulation models	Model-to-model transformation	Wood prefabrication system
[56]	Event logs	Dynamic process maps	Process mining	Fully automated manufacturing facility
[51]	Piping and instrumentation diagrams	Simulation models	Model-to-model transformation	Pilot fiber processing plant
[52]	Event logs	Extended directed graphs	Process mining, state reconstruction	Flow shop

Sierla et al. [50] propose a semi-automatic methodology that creates simulation-based DTs for brownfield industrial plants. They represent information extracted from piping and instrumentation diagrams in a graph format and then convert it into a simulation model. The methodology is initially demonstrated on a lab-scale water processing plant and later applied by Azangoo et al. [51] to a pilot fiber processing plant, showing good capability of increasing the engineering productivity of experts.

Lugaresi and Matta [37] introduce a method for the automated generation of graph-based DTs for manufacturing systems using event logs. The method not only provides a solution to align digital models with physical systems in a time-efficient manner but also enables the tuning of details captured by the generated models to avoid unnecessary complexity. The effectiveness of the method is validated on an assembly line for electric motor stators. This work is further extended by Zhu et al. [52] with state reconstruction.

Park and Van Der Aalst [53] investigate how to create a DT of an organization for the improvement of business processes. They present a framework that continuously monitors processes to check constraints and automatically triggers actions to resolve issues based on object-centric Petri Nets. A case study is conducted on an artificial order handling system, of which the process efficiency is enhanced by the framework.

Friederich et al. [54] present a generic framework that employs a data-driven approach to automatically build and continuously improve Petri Net-based DTs through process mining. The resulting DTs model the states of manufacturing systems and enable the analysis of their reliability. A case study of the framework is conducted on a lab-scale assembly line for quadcopter drone parts, exemplifying its applicability to real systems.

Kaiser et al. [55] propose a methodology for automatically generating simulation-based DTs of reconfigurable timber construction systems. The automation markup language is adopted to describe the components and behavior of the system under study and then translated into a simulation model for the validation of system configurations during the planning phase. When applied to a wood prefabrication system, the proposed methodology exhibits good ability to streamline validation, speed up iterations, and reduce errors.

Kumbhar et al. [56] develop a DT framework to detect, diagnose, and improve throughput bottlenecks in complex manufacturing systems. They extract event logs from real-time enterprise data and rely on process mining to generate dynamic process map DTs, which can help with the detection of sole and shifting bottlenecks and the validation of various what-if scenarios to improve throughput. The framework is evaluated in a fully automated manufacturing facility, leading to a minimum throughput increase of 10%.

3.5 FURTHER RESEARCH DIRECTIONS

DTs are an emerging technology in industry, and they enable manufacturers to gain insight into their operations and make better decisions about how to improve efficiency and reduce costs. The use of DTs in smart manufacturing systems presents both challenges and opportunities. For instance, DTs require the integration of multiple data sources and the development of sophisticated algorithms to make sense of the data. This can be a difficult and time-consuming task. Also, the design and deployment of DTs in industry and services is still in its pilot stages. There is a need for further research to develop the technology and make it more widely applicable. A selection of relevant research directions is listed as follows:

- *Data integration*: DTs require the integration of multiple data sources to provide meaningful insights. Research is needed to develop algorithms and systems to effectively integrate data from diverse sources.
- *Data-driven modeling and simulation*: DTs require sophisticated models and simulations to accurately represent physical systems. Research is needed to develop data-driven modeling techniques that accurately produce valid models to be used by the DT.
- *Balancing data and model granularity*: The extensive collection of data can gather a precise understanding of the system state. However, if all data sources are reproduced in the digital models, the level of detail often risks being excessive for the application purpose. Future research should design proper methods to adapt the DT model granularity for the intended application.
- *User interfaces*: DTs require user-friendly interfaces that enable users to easily access and interact with the data. Research is needed to develop user-friendly interfaces that make it easy for users to access and interact with digital information.
- *Integration of artificial intelligence*: The analysis of data and some modeling phases can be improved via the integration of artificial intelligence. However, research is still needed to develop algorithms and systems that can effectively analyze data and identify areas for improvement.
- *Security and privacy*: The use of DTs requires the secure storage and transmission of sensitive data. Research is needed to develop secure systems that can protect data from unauthorized access.

3.6 CONCLUSIONS

This chapter provided an overview of the role of data in DTs, focusing on data-driven modeling methods. This work is affected by several limitations that can be considered for future research. For instance, the relationship between the data requirements of a DT and an existing information system should be investigated. Future work should validate the suitability and viability of the proposed architectures by tests on a diverse spectrum of use cases. Several challenges hinder a DT from providing useful services to a system with the processed data: for instance, the ability to integrate and share data is aggravated by the heterogeneity, sources, volume, mode, and frequency of exchange of data, which need to be adequate to enable a DT to synchronize with its physical counterpart. Also, the development of a DT remains dependent on the requirements of its intended applications. In each application, several choices are complex, for instance, the choice of communication standards or the level of integration with the existing information systems and data management systems.

REFERENCES

1. R. Rosen, G. Von Wichert, G. Lo, and K. D. Bettenhausen, "About the importance of autonomy and digital twins for the future of manufacturing," *IFAC-Pap.*, vol. 48, no. 3, pp. 567–572, 2015.
2. F. Tao and Q. Qi, "New IT driven service-oriented smart manufacturing: framework and characteristics," *IEEE Trans. Syst. Man Cybern. Syst.*, vol. 49, no. 1, pp. 81–91, 2017.
3. F. Tao, Q. Qi, A. Liu, and A. Kusiak, "Data-driven smart manufacturing," *J. Manuf. Syst.*, vol. 48, pp. 157–169, 2018.
4. M. Grieves and J. Vickers, "Digital twin: Mitigating unpredictable, undesirable emergent behavior in complex systems," in *Transdisciplinary Perspectives on Complex Systems: New Findings and Approaches*, Springer, Cham, pp. 85–113, 2017. https://link.springer.com/chapter/10.1007/978-3-319-38756-7_4#citeas
5. M. Shafto, M. Conroy, E. Glaessgen, and R. Doyle, "Modeling, simulation, information technology & processing roadmap," *Natl. Aeronaut. Space Adm.*, vol. 32, pp. 1–38, 2012.
6. Fortune Business Insights, "Digital Twin Market Report," 2023. https://www.fortunebusinessinsights.com/digital-twin-market-106246
7. S. Dertien and C. Macmahon, "State of Digital Twin 2022," PTC, 2022. https://www.ptc.com/en/resources/iiot/manufacturing/white-paper/state-of-digital-twin
8. International Organization for Standardization, *Automation Systems and Integration—Digital Twin Framework for Manufacturing—Part 1: Overview and General Principles*, International Organization for Standardization, Geneva, Switzerland, 2021.
9. F. Tao, H. Zhang, A. Liu, and A. Y. C. Nee, "Digital twin in industry: State-of-the-art," *IEEE Trans. Ind. Inform.*, vol. 15, no. 4, pp. 2405–2415, 2019, doi:10.1109/TII.2018.2873186.
10. D. Ivanov, "Intelligent digital twin (iDT) for supply chain stress-testing, resilience, and viability," *Int. J. Prod. Econ.*, vol. 263, p. 108938, 2023.
11. G. Lugaresi, Z. Jemai, and E. Sahin, "Digital twins for supply chains: Main functions, existing applications, and research opportunities," in *2023 Winter Simulation Conference (WSC)*, IEEE, San Antonio, Texas, US, 2023, pp. 2076–2087.
12. E. Negri, L. Fumagalli, and M. Macchi, "A review of the roles of digital twin in CPS-based production systems," *Procedia Manuf.*, vol. 11, pp. 939–948, 2017.
13. W. Kritzinger, M. Karner, G. Traar, J. Henjes, and W. Sihn, "Digital twin in manufacturing: A categorical literature review and classification," *IFAC-Pap.*, vol. 51, no. 11, pp. 1016–1022, 2018.
14. J. Moyne, Y. Qamsane, E. C. Balta, I. Kovalenko, J. Faris, K. Barton, and D. M. Tilbury, "A requirements driven digital twin framework: Specification and opportunities," *IEEE Access*, vol. 8, pp. 107781–107801, 2020.
15. H. van der Valk, H. Haße, F. Möller, and B. Otto, "Archetypes of digital twins," *Bus. Inf. Syst. Eng.*, vol. 64, pp. 1–17, 2021.
16. M. Liu, S. Fang, H. Dong, and C. Xu, "Review of digital twin about concepts, technologies, and industrial applications," *J. Manuf. Syst.*, vol. 58, pp. 346–361, 2021.
17. S. Dittmann, P. Zhang, A. Glodde, and F. Dietrich, "Towards a scalable implementation of digital twins-A generic method to acquire shopfloor data," *Procedia CIRP*, vol. 96, pp. 157–162, 2021.
18. T. H. J. H.-J. Uhlemann, C. Schock, C. Lehmann, S. Freiberger, and R. Steinhilper, "The digital twin: Demonstrating the potential of real time data acquisition in production systems," *Procedia Manuf.*, vol. 9, pp. 113–120, 2017, doi:10.1016/j.promfg.2017.04.043.
19. A. Talaminos-Barroso, M. A. Estudillo-Valderrama, L. M. Roa, J. Reina-Tosina, and F. Ortega-Ruiz, "A Machine-to-Machine protocol benchmark for eHealth applications – Use case: Respiratory rehabilitation," *Comput. Methods Programs Biomed.*, vol. 129, pp. 1–11, 2016.
20. O. Cardin and P. Castagna, "Using online simulation in Holonic manufacturing systems," *Eng. Appl. Artif. Intell.*, vol. 22, no. 7, pp. 1025–1033, 2009, doi:10.1016/j.engappai.2009.01.006.
21. G. Lugaresi, V. V. Alba, and A. Matta, "Lab-scale models of manufacturing systems for testing real-time simulation and production control technologies," *J. Manuf. Syst.*, vol. 58, pp. 93–108, 2021.
22. B. Osinski, A. Jakubowski, P. Zięcina, P. Miłoś, C. Galias, S. Homoceanu, and H. Michalewski, "Simulation-based reinforcement learning for real-world autonomous driving," in *2020 IEEE International Conference on Robotics and Automation (ICRA)*, IEEE, Paris, France, 2020, pp. 6411–6418. https://ieeexplore.ieee.org/document/9196768?denied=
23. T. Ruppert and J. Abonyi, "Integration of real-time locating systems into digital twins," *J. Ind. Inf. Integr.*, vol. 20, p. 100174, 2020.

24. L. Rhodes-Leader, B. S. S. Onggo, D. J. Worthington, and B. L. Nelson, "Airline disruption recovery using symbiotic simulation and multi-fidelity modelling," In *Proceedings of the Operational Research Society Simulation Workshop*, 2018, pp. 146–155.
25. N. Mustafee, J. Powell, and A. Harper, "Right hospital-right time: A business analytics framework for analysing urgent care/A&E wait time data," in *Proceedings of the 2018 Simulation Workshop*, The Operational Research Society, Stratford-upon-Avon, UK, 2018, pp. 105–115.
26. B. S. Onggo, N. Mustafee, A. Smart, A. A. Juan, and O. Molloy, "Symbiotic simulation system: hybrid systems model meets big data analytics," in *Presented at the 2018 Winter Simulation Conference (WSC)*, Institute of Electrical and Electronics Engineers Inc., Gothenburg, Sweden, January 2018, pp. 1358–1369, doi:10.1109/WSC.2018.8632407.
27. V. Damjanovic-Behrendt and W. Behrendt, "An open source approach to the design and implementation of Digital Twins for Smart Manufacturing," *Int. J. Comput. Integr. Manuf.*, vol. 32, pp. 366–384, 2019.
28. K. M. Alam and A. El Saddik, "C2PS: A digital twin architecture reference model for the cloud-based cyber-physical systems," *IEEE Access*, vol. 5, pp. 2050–2062, 2017.
29. S. Ji, S. Pan, E. Cambria, P. Marttinen, and S. Y. Philip, "A survey on knowledge graphs: Representation, acquisition, and applications," IEEE *Trans. Neural Netw. Learn. Syst.*, vol. 33, no. 2, pp. 494–514, 2021.
30. S. Guan, X. Cheng, L. Bai, F. Zhang, Z. Li, Y. Zeng, X. Jin, and J. Guo, "What is event knowledge graph: A survey," *IEEE Trans. Knowl. Data Eng.*, vol. 35, no. 7, pp. 7569–7589, 2022.
31. D. Fahland, "Process mining over multiple behavioral dimensions with event knowledge graphs," in Wil M. P. van der Aalst, Josep Carmona (eds), *Process Mining Handbook*, Springer, 2022, pp. 274–319. https://link.springer.com/book/10.1007/978-3-031-08848-3
32. E. L. Klijn, F. Mannhardt, and D. Fahland, "Aggregating event knowledge graphs for task analysis," in *International Conference on Process Mining*, Springer, 2022, pp. 493–505.
33. A. Swevels, R. Dijkman, and D. Fahland, "Inferring missing entity identifiers from context using event knowledge graphs," in *International Conference on Business Process Management*, Springer, Bolzano, Italy, 2023, pp. 180–197.
34. A. Swevels, D. Fahland, and M. Montali, "Implementing object-centric event data models in event knowledge graphs," in *International Conference on Process Mining*, Springer, Utrecht, The Netherlands, 2023, pp. 431–443.
35. E. L. Klijn, D. Preuss, L. Imeri, F. Baumann, F. Mannhardt, and D. Fahland, "Event knowledge graphs for auditing: A case study," in *International Conference on Process Mining*, Springer, Rome, Italy, 2023, pp. 84–97.
36. W. M. P. Van Der Aalst, *Process Mining: Data Science in Action*. Springer, Rome, Italy, 2016, p. 467, doi:10.1007/978-3-662–49851-4.
37. G. Lugaresi and A. Matta, "Automated manufacturing system discovery and digital twin generation," *J. Manuf. Syst.*, vol. 59, pp. 51–66, 2021, doi:10.1016/j.jmsy.2021.01.005.
38. J. Friederich, G. Lugaresi, S. Lazarova-Molnar, and A. Matta, "Process mining for dynamic modeling of smart manufacturing systems: Data requirements," *Procedia CIRP*, vol. 107, pp. 546–551, 2022.
39. A. P. Estrada-Vargas, E. Lopez-Mellado, and J.-J. Lesage, "A comparative analysis of recent identification approaches for discrete-event systems," *Math. Probl. Eng.*, vol. 2010, no. 1, p. 453254, 2010.
40. M. P. Cabasino, P. Darondeau, M. P. Fanti, and C. Seatzu, "Model identification and synthesis of discrete-event systems," in The Institute of Electrical and Electronics Engineers, Inc (ed.), *Contemperory Issues in Systems Science and Engineering*, pp. 343–366, 2015. https://onlinelibrary.wiley.com/doi/book/10.1002/9781119036821
41. E. Badouel, L. Bernardinello, and P. Darondeau, *Petri Net Synthesis*. Springer, 2015.
42. K. Hiraishi, "Construction of a class of safe petri nets by presenting firing sequences," in Mengchu Zhou, Han-Xiong Li, Margot Weijnen (eds), *International Conference on Application and Theory of Petri Nets*, Springer, Sheffield, UK, 1992, pp. 244–262.
43. J. Cortadella, M. Kishinevsky, L. Lavagno, and A. Yakovlev, "Deriving petri nets from finite transition systems," *IEEE Trans. Comput.*, vol. 47, no. 8, pp. 859–882, 1998.
44. P. Darondeau, "Region based synthesis of P/T-nets and its potential applications," in *International Conference on Application and Theory of Petri Nets*, Springer, Aarhus, Denmark, 2000, pp. 16–23.
45. M. P. Cabasino, A. Giua, and C. Seatzu, "Identification of petri nets from knowledge of their language," *Discrete Event Dyn. Syst.*, vol. 17, pp. 447–474, 2007.
46. M. Dotoli, M. P. Fanti, and A. M. Mangini, "Real time identification of discrete event systems using petri nets," *Automatica*, vol. 44, no. 5, pp. 1209–1219, 2008.

47. S. O. El Mehdi, R. Bekrar, N. Messai, E. Leclercq, D. Lefebvre, and B. Riera, "Design and identification of stochastic and deterministic stochastic petri nets," *IEEE Trans. Syst. Man Cybern.-Part Syst. Hum.*, vol. 42, no. 4, pp. 931–946, 2011.
48. F. Basile, P. Chiacchio, and J. Coppola, "Identification of time petri net models," *IEEE Trans. Syst. Man Cybern. Syst.*, vol. 47, no. 9, pp. 2586–2600, 2016.
49. P. Denno, C. Dickerson, and J. A. J. A. Harding, "Dynamic production system identification for smart manufacturing systems," *J. Manuf. Syst.*, vol. 48, pp. 192–203, 2018, doi:10.1016/J.JMSY.2018.04.006.
50. S. Sierla, L. Sorsamäki, M. Azangoo, A. Villberg, E. Hytönen, and V. Vyatkin, "Towards semi-automatic generation of a steady state digital twin of a brownfield process plant," *Appl. Sci.*, vol. 10, no. 19, p. 6959, 2020.
51. M. Azangoo, L. Sorsamäki, S. A. Sierla, T. Mätäsniemi, M. Rantala, and K. Rainio, "A methodology for generating a digital twin for process industry: A case study of a fiber processing pilot plant," *IEEE Access*, vol. 10, pp. 58787–58810, 2022.
52. L. Zhu, G. Lugaresi, and A. Matta, "Automated generation of digital models for production lines through state reconstruction," in *IEEE 19th International Conference on Automation Science and Engineering*, IEEE, Auckland, New Zealand, 2023, pp. 1–8.
53. G. Park and W. M. P. Van Der Aalst, "Realizing a digital twin of an organization using action-oriented process mining," in *3rd International Conference on Process Mining*, IEEE, Eindhoven, The Netherlands, 2021, pp. 104–111.
54. J. Friederich, D. P. Francis, S. Lazarova-Molnar, and N. Mohamed, "A framework for data-driven digital twins of smart manufacturing systems," *Comput. Ind.*, vol. 136, p. 103586, 2022.
55. B. Kaiser, A. Reichle, and A. Verl, "Model-based automatic generation of digital twin models for the simulation of reconfigurable manufacturing systems for timber construction," *Procedia CIRP*, vol. 107, pp. 387–392, 2022.
56. M. Kumbhar, A. H. C. Ng, and S. Bandaru, "A digital twin based framework for detection, diagnosis, and improvement of throughput bottlenecks," *J. Manuf. Syst.*, vol. 66, pp. 92–106, 2023.

4 Digital Twin Modeling and Simulation

Maksym Spiryagin, Yan Quan Sun, Thomas Mond, Chris Bosomworth, and Leron Postelnik

4.1 INTRODUCTION

A digital twin (DT) is a technology to virtually represent a physical system or process, which can collect real data and information, validate, simulate, and predict the present and future behaviors of its counterpart in the real world [1]. As a developing technology, it has been widely applied and implemented in many areas such as agriculture, industry, construction, cities, medicine, aerospace, and transportation [2]. The concepts, methods, and technologies used in DT can vary significantly between application areas. This is primarily driven by the unique characteristics of data and information in each domain. Therefore, the DT classifications can be divided into two categories according to their maturity and degree of evolution, depending on the perspective of consideration [3]. Based on the level of data integration between the physical object and its digital representation, DTs can be categorized into three main types [4]: digital models, which are built based on the modeling component of a physical object and a digital counterpart, and they lack any automated exchange of data or information with their physical counterparts; digital shadows/mirrors, which presents a one-way automated data flow from the physical object to its digital counterpart, and they act as real-time (RT) observers, constantly monitoring the state of the physical system; DTs, which are the most advanced among the types, enabling two-way data flow between the physical and digital objects that allows the DT to act as a "controlling instance," influencing and optimizing the physical counterpart. However, the basic framework for a DT technology is consistent across most applications. It consists of three elements: the physical system, the virtual system, and the connection between them.

The implementation of DTs in the reality of the physical environment faces a lot of obstacles and challenges, such as RT communication and data exchanges, system complexities and uncertainties, and data collection and computation accuracy and integration. The challenges of implementing the DTs are not only the demand on running computing codes and reiterating various algorithms for the purpose of optimization but also the communication and data exchanges among the physical system and its DT. Due to the high variabilities and uncertainties for some systems and with no availability of all the required measured data, the applications of the DTs require the usage of more effective platforms and tools built on the modeling and simulation approaches that are greatly in demand while continuously evolving. Specific to the integration of DTs with simulation technologies, some challenges include the integration of complex simulation environments and algorithms and cloud/edge parallel computing in DT frameworks with the incorporation of big data analytics in the digital optimization process linked to operational tasks and practices that focus on obtaining, maintaining the accuracy of, and integrating diverse data streams for effective DT operation.

4.2 HISTORY TRANSITION IN SIMULATION TECHNOLOGIES

The computer simulation concept was originally proposed in 1979 by the Society for Computer Simulation Technical Committee on Model Credibility [5]. It contains three major, cross-connected components: reality, conceptual model, and computerized model. In more current terms, "reality"

DOI: 10.1201/9781003492146-4

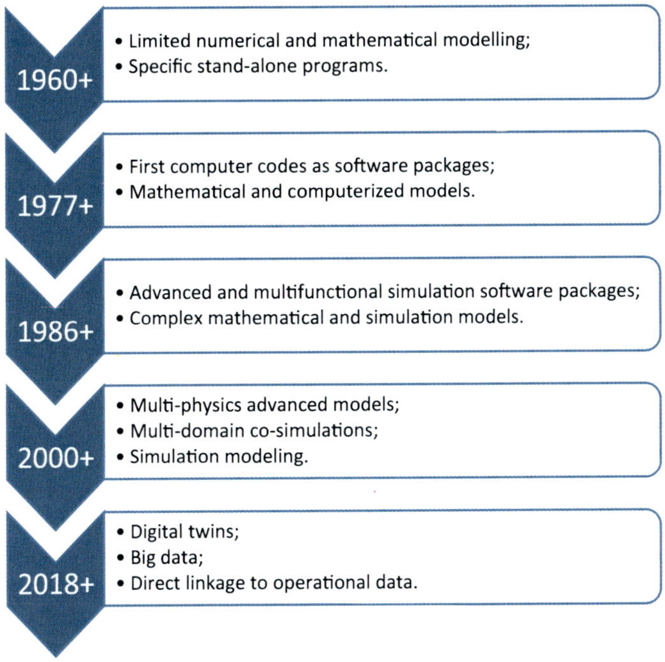

FIGURE 4.1 History of simulation-based technologies. (Adopted from M. Spiryagin, J. Edelmann, F. Klinger, C. Cole. Vehicle system dynamics in digital twin studies in rail and road domains. *Vehicle System Dynamics* 2023, 61(7), 1737–1786, with permission.)

refers to a physical system. The term "conceptual model" can be defined as a system specification (also called technical specification) that outlines the model. The term "computerized model" refers to an operational computer program that implements a conceptual model [5], essentially acting as a replica of a physical system [6]. To restate these definitions for a simulation process used in vehicle system dynamics, we can adopt the following definition: "A model is conceived as any physical, mathematical, or logical representation of a system, entity, phenomenon, or process" [7]. Following this definition, a computerized model can be considered

> as a set of instructions, i.e., a model needs some agent capable of actually activating and obeying the instructions and generating behavior. We call such an agent a simulator. Thus, a simulator is any computation system (such as a single processor, a processor network, the human mind, or more abstractly an algorithm) capable of executing a model to generate its behavior. [7]

Therefore, simulation itself represents a process that can work with a model to deliver output measures of a physical system's performance. As shown in Figure 4.1 from the historical timeline published in [8], all simulation technologies have a strong connection with the transition from simplistic developments to more complex ones.

4.3 UNDERSTANDING THE DIFFERENCE BETWEEN A MODEL AND A DT

A physics-based model (or a virtual model) is a two- or three-dimensional (2D or 3D) mathematical representation of a physical object or system in the virtual environment, generated by various computing software packages based on knowledge about physics and on experimental observations and then used for simulation and analysis. For a long time, simulations have been widely used to imitate the operations or processes within a physical system, to analyze the performance of the system, and to test system improvements by inputting new ideas. Therefore, a virtual model is built through digitization by using computer-aided design and analysis software packages, which can be created in 2D or 3D to represent a physical system or system components, validated through the model validation process, and then run by introducing variables into the digital environment or interface for simulation. Typical examples of such simulations in engineering are finite element analysis and stress analysis. The models can be built using various professional software packages, e.g., ABQUAS and ANSYS. These numerical tests can be performed by high-performance computing (HPC) systems to analyze the effect of external forces or pressure on mechanical components to enhance their design and features or other aspects. However, a validated virtual model can only provide a snapshot of the behavior of an object or system at a specific moment and cannot automatically incorporate or exchange the data and information from or to its counterpart physical system and can only conduct the manual exchange and modification. Therefore, it cannot demonstrate the RT state of a physical system. Meanwhile, it is typically impossible for a virtual model to simultaneously replicate all behaviors of a complicated physical system happening at all scales. Therefore, multiple virtual models are required to capture different physical behaviors at different scales for such a system. There can be significant differences between virtual and digital models. The transition between them depends heavily on the input parameters defined within the technical specifications of a DT framework, particularly for communication interfaces.

Advanced technologies like digital shadows or mirrors extend beyond stand-alone virtual or digital models. These digital shadows are built upon a foundation of RT data continuously collected from the physical systems they represent. These technologies create dynamic digital replicas of physical objects or systems that can interact with people and their environments in real time.

To extend their applicability beyond these stated interaction points, it is necessary to shift to the upper level of DT technologies. As defined in [9,10], DTs are virtual representations of physical objects or systems, created by integrating digital models and data. However, these advanced DTs go beyond stand-alone static models. They can dynamically adapt based on RT data from the physical twin, establishing dynamic models that can interact with other DTs and models and even enable autonomous decision-making for the physical system with limited human intervention. This suggests that a DT can closely mirror the behavior of its corresponding physical system, functioning with a high degree of autonomy if the physical system is fitted with sensors that measure data on the system performance. The data is then relayed to a processing system and applied to the digital model. The DT can run simulations to evaluate the performance and create potential improvements that can then be applied back to the physical system in real time. Meanwhile, the digital model is updated according to the data. The data used by DTs is usually collected from Internet of Things (IoT) platforms and other embedded devices, allowing for the capture of high-level information that can then be integrated into the virtual model. A DT, on the contrary, has fully integrated data flow, allowing it to properly and consistently reflect the actual state of the physical system. For a DT, not only can the data and information be interchanged between a physical system and its virtual one in a bidirectional way, but also the DT can be an exemplification of the physical model or system and hence be used for managing its life cycle [4]. There is a simultaneous response and change of its counterpart when a change occurs in either the physical or virtual model. In addition, there could be more than one DT for a physical system. A DT can be composed of several sub-DTs or a collection of DTs. As we know, a DT could be that of a physical system or the system components. Usually, it could be more feasible and convenient to divide the system or system components into subsystems or sub-components, generating several DTs and making connections among them. Furthermore, in some actual scenarios where creating the physical prototype is expensive or simulating their actual conditions in a lab is not possible, several DTs for the same physical system can be created, which could be highly beneficial.

While both conventional simulations and DTs leverage virtual models to represent physical systems, key distinctions exist between them. The most important difference lies in the ability of a DT to create a virtual environment for simulating various scenarios. This environment is continuously updated with RT data, facilitated by a one- or two-way information flow between the DT and the data collection devices. This enhanced data integration significantly bolsters the accuracy of predictive analytical models, leading to a deeper understanding of system performance and facilitating more effective operational monitoring.

4.4 INTEGRATION OF SIMULATION MODELING IN DTs

One initial step in developing DTs is creating virtual advanced models for simulations. However, simulations themselves have limitations. By the existing common sense definition, they are simply computerized models of physical systems used for studying their behavior. For a more comprehensive approach, we need to move to simulation modeling, as defined in [11]: "Simulation modeling and analysis is the process of creating and experimenting with a computerized mathematical model of a physical system." In other words, simulation modeling involves designing a virtual model of a physical system to not only describe its behavior but also analyze it. As shown in Figure 4.2a, the main components to analyze a physical system include a system that is a subject for analysis, modeling (the process of creating a model by collecting interrelated elements); a model (the digital representation of a system which is composed of a set of assumptions and mathematical or logical relationships (e.g., differential equations of a dynamic model for an engineering application)); and simulation – the operation of a model over time or the result of solving the equations by using a computer. A key step in simulation modeling for establishing a DT framework is the model transition process. As shown in Figure 4.2b, compared to a simulation box, which typically focuses on a single process, a DT offers a key distinction. It can integrate multiple simulations within itself. This allows for a more comprehensive analysis of the system's behavior under various operating scenarios and

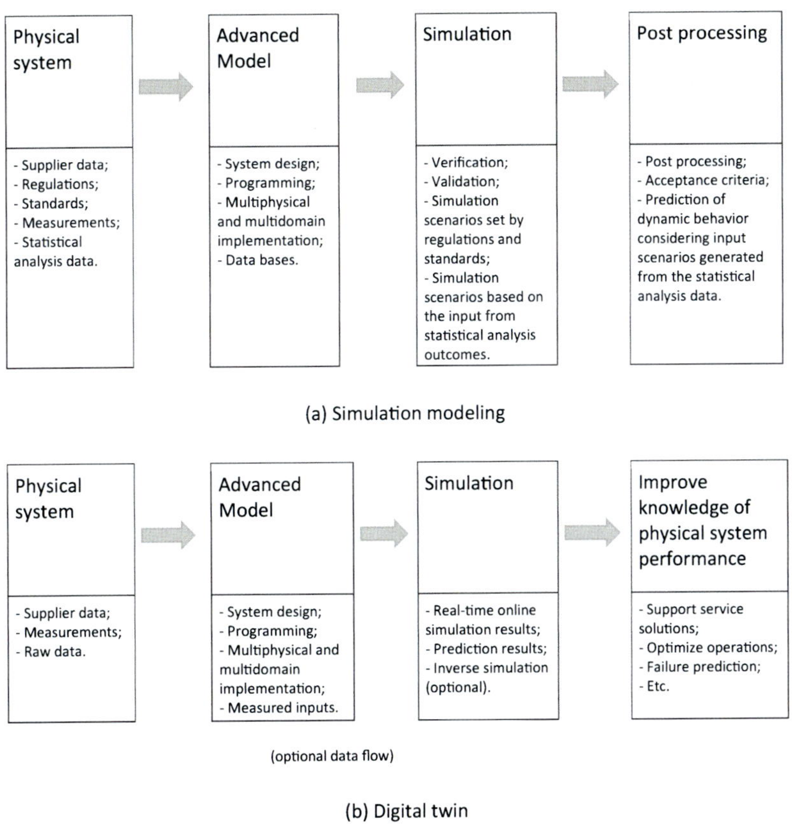

FIGURE 4.2 Conceptual overview of simulation modeling and digital twin techniques. (a) Simulation modeling; (b) Digital twin. (From M. Spiryagin, J. Edelmann, F. Klinger, C. Cole. Vehicle system dynamics in digital twin studies in rail and road domains. *Vehicle System Dynamics 2023,* 61(7), 1737–1786, with permission.)

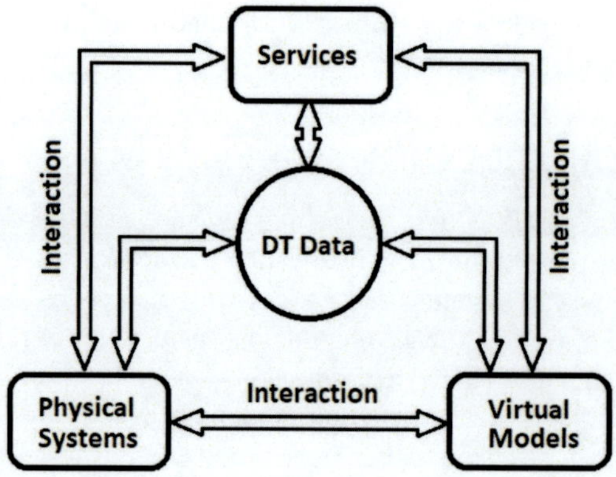

FIGURE 4.3 Concept of a five-dimension digital twin model.

processes. The DT process considers the formation of the mechanism for knowledge improvement of physical system performance and one or two-way connection interface requirements with a transformation of a virtual (computerized) model into a digital model suitable for DT simulations and operational studies that establish a base for a relevant DT framework. A DT framework leverages a digital twin, a virtual replica of a physical system, to simulate its behavior. This simulation can predict how the system will perform after implementing structural changes. Additionally, it can model interactions between the digital twin and other DTs on the platform. In essence, a DT creates a digital model to simulate a physical system's behavior, monitor its performance, recognize environmental conditions, detect anomalies, and forecast future trends. One concept of a five-dimensional DT framework is discussed in [2]. As shown in Figure 4.3, this framework can be logically represented with five dimensions that describe a DT's architecture, not the geometry of the physical system.

In Figure 4.3, the first dimension represents physical systems that implement activities based on physical laws and operate in uncertain environments. The second dimension consists of virtual models, which are accurate replicas of physical systems. These models are typically built using various software packages and computing codes to reproduce the geometries, properties, behaviors, and rules of the physical system. However, unlike traditional computerized models, DT virtual models incorporate rule models with logical abilities such as reasoning, evaluation, judgment, and decision-making. The third dimension encompasses DT data, a key driver of the DT, obtained from physical systems. The fourth dimension focuses on services, another essential component of a DT because DTs commonly require users to implement a range of services related to simulating, verifying, optimizing, monitoring, and diagnosing physical systems. However, building a DT also requires external services such as data provision, knowledge management, algorithm development, and various software development and model-building platforms. Therefore, this explains why the fifth dimension is interaction, which dynamically links a virtual model with its real-world counterpart to enable simulation, operation, and analysis. These interactions between physical systems, virtual models, services, and DT data facilitate information and data exchange, enabling collaboration.

4.5 CONCEPTS OF SIMULATION TECHNOLOGIES

Traditional simulation technology typically relies on a single software product for analyzing systems and informing design, operation, or resource policy decisions. In contrast, advanced simulation technologies leverage a combination of different simulations built on theory-based principles. The functional view of DTs depicted in Figure 4.4 highlights the critical role of simulation technologies within this architectural framework. The simulation principles are then adjusted for specific processing application tasks to enhance individual simulation process performance. The commonalities underlying all simulation technologies will be discussed in the following sections.

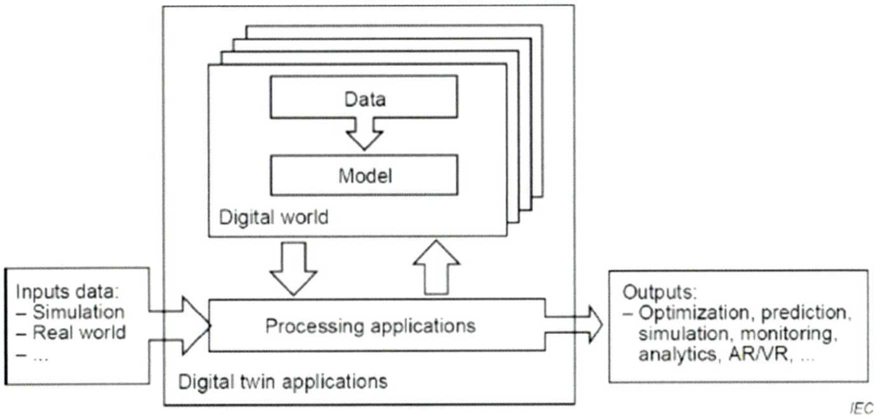

Key

AR augmented reality

VR virtual reality

FIGURE 4.4 Functional view of digital twin. (ISO/IEC 30173 ed. 1.0. Copyright © 2023 IEC Geneva, Switzerland. www.iec.ch, with permission.)

4.5.1 Simulation Models

Basically, the simulation models in DTs can be categorized into the physics-based, the data-driven, and the big data-based hybrid models. Physics-based modeling is based on the observation of the behavior and performance of a physical system, the assumptions necessary due to uncertainties or restraints, and then their expression in mathematical equations.

Data-driven modeling is mainly focused on computational intelligence (CI) and machine learning (ML) methods, which can be used to build models to complement or replace physics-based models. The CI methods, mainly including neural networks, fuzzy systems, and evolutionary computing, are used to establish the relationships between inputs and outputs of a physical system. The ML methods are used to determine the relationship between a physical system's inputs and outputs by using a training data set that is representative of all the behaviors found in the system. Once such a model is created or trained, it can be tested using an independent data set to determine how well it can generalize its outputs to unseen data. Data-driven modeling is becoming more and more popular due to the increasing amount of data processing capability, the RT prediction requirement, efficient training methods, and the long-time usage in computing solutions of physics-based models.

Usually, the physics-based models are less biased than the data-driven methods because they use natural laws, which can be easily interpreted and generalized. However, the main disadvantages of physics-based modeling are possible numerical instability, possible high demand of computational time, and the likelihood of uncertainties in the models. Sometimes, understanding a physical process is hard and may not be possible in complex physical systems. In contrast, the data-based models are more flexible and can consider new data and are more suitable for complex systems where understanding the physics is not possible, or even in systems where physics models cannot be applied.

Big data-based hybrid modeling combines physics-based modeling and data-driven modeling with big data approaches and allows the inclusion of more physics by increasing model complexity. Modern instruments and the Internet have a great capacity to generate, transfer, and store enormous amounts of data, and the tasks involved in processing, data mining, analyzing, and visualizing the data need to be well realized. The big data approach can provide better estimates of the related quantities. As mentioned previously, both the physics-based and the data-driven methods have advantages and disadvantages. The hybrid models can obtain the strong points from both methods. For example, the integration of physics-based models with ML to design a DT can allow the use of an interpretable model (physics-based) to build a fast DT (ML) that will be connected to the physical system to support RT prediction. The approach uses a physics-based model for data processing and response predictions and a data-driven ML model for the time evolution of the physical system parameters.

4.5.2 Principles of Complex Simulations

Those building DTs have often overlooked some basic principles of the modeling. The ambition to develop a DT often leads to the DT being too complex with a lot of unnecessary redundant information [12]. Actually, a DT is impossible to be exactly the same as its physical counterpart being modeled. Generally, there are two kinds of physical systems. One is the discrete system in which the variables change instantaneously at separate points in time, and the second is the continuous one where the state variables change continuously with respect to time. Some questions will arise, especially for a complex system. How are the number and definition of events for a system determined for modeling? How are the state variables specified that are needed to keep the simulation running in the correct event sequence and to obtain the desired output measures? There would likely be no complete answers to these questions. Perhaps different researchers might create different ways of representing a model in terms of events and variables, and they all may be correct. But there are some principles and techniques to facilitate complex simulations and to avoid logical errors. A DT, with simulation as its main purpose, can and should draw many valuable theories and methods from the field of modeling and simulation.

In a DT, multiple models can typically be used, capturing different physical processes at different length scales, and with a range of fidelities, for the same system. The essence of the DT concept is that these models can be augmented with available data, and beyond that with each other. The primary purpose of this data and model augmentation is to increase confidence in the simulated results. Co-simulations will be often implemented for a DT. The DT workflow will coordinate various software packages or bespoke pieces of code to perform required subtasks, and the multiple subtasks can then be run in parallel, or cross-coupled to create new super-tasks, some of which may not have been previously achievable [9].

As the fundamental task of a DT is to conduct the predictions for its counterpart physical system's performance, the verification and validation of simulated results must be necessary. There are two categories for simulation verification and validation, one is the software quality assurance to make sure to correctly employ any commercial software and to check the interactions of code, and the other is the algorithm verification to ensure the correctness of the implementation of mathematical representations. The validation should be satisfied regarding the quantitative accuracy of the simulation results against measured testing data, the prediction confidence, and the accuracy for the intended use.

Finally, the simulation output visualization is necessary, which requires a DT to have the capability to organize and present a huge amount of information and data through well-established visualization and animation techniques. The interactive 3D visualization for a DT is needed to show the promised capabilities for RT monitoring and prediction, or other specific tasks set by end users.

4.5.3 Construction of the DT Considering the Simulation Aspects

The construction of the simulation component of a DT can be rate limited in the rail context for a number of reasons. Outside of the urban environment encompassing passenger and metro networks, freight and heavy haul trains can travel large distances through unpopulated areas with limited communication capability. The physical size and weight of trains often means they are mechanically stiff systems and electrically have high voltages and frequencies that can impact monitoring systems and limit measurement and diagnostic systems. These limitations can affect data rates for simulators, the complexity of the models produced, and the physical location where the simulation is executed. This section will explore some of the general challenges in constructing the simulation components for a DT and where possible give some rail-specific examples. This section assumes a physics-based model is being simulated.

4.5.3.1 Data Availability

The concept of data availability for a DT and more specifically the simulation aspect is a broad brush covering a number of areas including communication synchronization, proprietary platform interoperability, and domain-specific constraints. Communication synchronization is whereby the DT and the physical asset bidirectionally transmit data to remain in sync with each other. The criticality of data transferred will determine the reliability of the communication medium required and data latency. The latency of the data can range from milliseconds to days, weeks, or even months depending on the type of DT. For example, low-powered wireless sensor nodes may only communicate sporadically. Other influences may include internal or external telecommunication providers.

Proprietary platform interoperability and domain-specific constraints can often influence the data rates and quality of data available as input to the simulation. For example, a proprietary system may not make data available in real time, place usage constraints on it, or not be available at the data rate required for the *preferred* simulation. These data availability aspects will dictate the usage method of the data, the execution location, and constraints on simulation modeling.

4.5.3.2 Usage Methods

Where telecommunication availability is constrained, or there is a hard RT data constraint, this may dictate the execution location of the simulation for the DT. Often when people think of DTs, it's in the context of the physical asset having physical separation from the DT and communicating through the cloud. However, for the aforementioned, this must be executed on the physical asset in RT abiding by the hard RT constraints of the input measurement system. An example of this type of system is automation or early warning failure devices. Other instantiations with lower data criticality may conform to the traditional DT approach of transferring the data in real time to a shore-based simulation system or recording the data on-board the physical asset for downloading at a shore base station such as a loading or unloading terminal. If required, the DT will transmit the result of the simulation back to the physical asset.

4.5.3.3 System Engineering

The system engineering of the simulation covers the modifiability and reusability of the simulation system. During the early phase of development, a domain-specific expert helps define the physical asset, its subsystems, and simulation derivable modules. Usually, these subsystems are broken down into parameter-defined systems to provide for future modifiability. This enables the software engineering team to design and implement the simulation for the DT with an adaptable and hopefully interoperable interface, sub-system decoupling, and execution pathways to allow for easily modifying input data types and asset components and outputs during run time. The final stage is the verification and validation (V&V) of the system to ensure it is generating the correct results for the given input set. More detailed requirements on this fundamental topic can be found in [13].

4.5.3.4 Continual Improvement and Integration

The final aspect of simulation development for a DT is the ability to continually improve and converge. Convergence is the process of identifying differences between the physical asset and simulation and executing algorithms to adjust simulation parameters to increase the coefficient of determination between the two systems. Continual improvement is a process undertaken by a data analyst or ML algorithm to identify new features in the data or adapt existing models.

4.6 SIMULATION PLATFORMS

The simulation component of DTs necessitates sophisticated platforms capable of integrating diverse computational techniques and software and hardware tools to efficiently mimic complex multiphysics systems. This section explores the critical components and technological frameworks that underpin DT simulation platforms, focusing on co-simulation environments, RT computing, HPC, and cloud-based simulation techniques. Understanding these components is essential for developing robust, efficient, and scalable DT simulators that can drive innovation and optimization across various engineering and scientific domains.

4.6.1 Co-Simulation Environment and Platform

In the real world, all physical systems within a system of systems ideally communicate with each other. However, in the simulation world, finding a single software product that can handle multidisciplinary areas and encompass all relevant domains is quite challenging. Therefore, it's necessary to utilize different software products tailored to specific disciplines, while enabling them to communicate effectively with each other. The communication process between software packages is commonly defined as a co-simulation process, and it is governed by specific rules and limitations. In essence, it allows for the simulation of complex systems by treating them as collections of interacting subsystems. Arnold et al. [14,15] define a co-simulation process as one in which "the subsystems are handled or integrated by different integration methods or solvers and each of these methods or solvers can be or is tailored to the corresponding subsystems." Another definition can be found in [16], which states that "co-simulation is used to solve a coupled system by simulating each part with its own couplable simulation tool." A further comprehensive definition is presented in [17] as "co-simulation exploits the modular structure of coupled problems in all stages of the simulation process beginning with the separate model setup and pre-processing for the individual subsystems in different simulation tools."

A co-simulation environment is primarily defined by the chosen data exchange mechanism/process that facilitates communication between two or more software packages. As outlined in [18], these data exchange processes typically rely on three main communication techniques:

1. Integrated memory shared communication between software products
2. Network data exchange
3. Exporting code from one package to another

All these techniques have their own strengths and weaknesses, and the choice of communication technique hinges on initial requirements and existing hardware or software limitations. Below are some scenarios where the integrated shared memory communication technique can be chosen:

- It is ideal when working with in-house developed software products. It requires programming knowledge and access to source code that can provide some advantages in the unique development of this technique for efficient communication considering specific system requirements. However, for commercial software products, this is usually not the case.
- In the case when required commercial software packages come from the same software development company, shared memory communication can be a viable option if it is implemented or can be implemented by an end-user request.
- When a commercial software product allows users to implement a user-specific subroutine that can be compiled and integrated into the software code. However, this technique often comes with restrictions on programming capabilities and strict implementation requirements.

Network data exchange is a valuable approach when commercial or in-house software packages have the possibility to communicate with each other through a well-established network platform. For DT applications, various network protocols are used including TCP/IP (transmission control protocol/Internet protocol) protocol, UDP (user datagram protocol), and RT Ethernet. The choice of protocol depends on the system's technical requirements, such as data verification algorithms, low-latency needs, or tolerance for data loss, which are defined in software packages. For a general co-simulation technique, the client-server architecture is commonly used. It is implemented through the concept of computer engineering architecture, where a server is a resource provider and a client is a service requester. In more complex system environments with multiple client agents for various subsystems, a common approach is to use a single main software package as a server to provide computational services for all interconnected clients. In such an architecture, the data exchange mechanism between each client and the server can significantly slow down the co-simulation process due to data delivery delays and other computational or associated issues.

In these cases, it's preferable to use a code export technique to minimize communication speed and interface processing overhead. This technique allows for both importing from and exporting to digital models within a shared library file. All necessary communication data processing actions can be performed within the shared library. For example, one variation of this approach can be built using the functional mock-up interface (FMI) [16,17]. This interface enables software packages to export their models to both a model description script written in eXtensible markup language and C code or binary code containing the model's equations and predefined parameters. It is important to note that the FMI specification covers not only FMI for model exchange (import and export) but also FMI for co-simulation. The key distinction between these approaches lies in solver usage. In FMI for co-simulation, each software uses its own solver, whereas the importing software's solver is used in the other approach. This latter requirement creates a dependency between software products that might be an issue in the development, validation, and verification of complex DT platforms.

Choosing the right co-simulation environment requires careful consideration of all software products and operating system compatibility. There is no one-size-fits-all solution. This highlights the complexity and multidisciplinary nature of building a co-simulation DT platform. In essence, this platform is a software–hardware environment that enables the integration of simulations from diverse programs, operating systems, and hardware systems, ultimately creating more intricate and realistic simulation models required for DT tasks.

4.6.2 RT Computing

RT computing platform is needed in cases when the DT reacts simultaneously with the physical counterpart. This enables the generation of unmeasurable data alongside measured sensor data, allowing for parallel processing and immediate analysis. Furthermore, it facilitates RT decision-making and adjustments in the physical system through simulation model updates. It is understandable that complex simulation models are well-suited for intricate systems where numerous factors interact. However, in the case of RT computing, not all objectives can be achieved, and it requires setting some reasonable limitations.

For example, as defined in [19], the railway RT DT mainly comprises a physical twin, a DT platform, and the twin projections generated from their interaction. The platform requires the deployment of the data-sharing space that enables the interaction between the physical and virtual entities of the DT, and it includes modeling and analysis spaces. Where the virtual twin is considered a digital replica of a railway system that integrates multiphysics models and RT feedback from on-board and wayside sensors. The RT modeling space can be implemented in a similar manner to hardware-in-the-loop (HiL) or software-in-the-loop (SiL) simulation platforms that enable the communication and synchronization between the different models comprising the virtual twin. When correctly implemented, the RT DT based on [19,20] should satisfy the following general requirements:

- *Bidirectional synchronization*: Enables the physical system and the digital models to remain connected and exchange information.
- *Convergence*: The platform capacity to identify and correct differences between the behavior of the physical system and its DT.
- *V&V:* Both refer to the capability of the DT platform functionalities to be tested on historical data and, through the virtual twin, to guarantee and verify the safety of the system before deployment of new solutions.
- *RT behavior*: The capacity of the platform to enable the physical system and the DT to mirror each other's behavior in RT following the RT operation system, multidisciplinary, and simulation model requirements defined in [21].
- *Automation protocols:* Inclusion of general-purpose procedures for automated systems.
- *Platform interoperability:* The capacity of the DT platform to interact in RT with other hardware, digital, and software products such as data visualization tools, dashboards, data from wayside and on-board systems, and other value-adding services.
- *System Interoperability:* The DT platform allows communication and interaction between several DTs forming the entire railway system.
- *Connection and data security:* Data protection and prevention of unauthorized access to the DT platform and from external cyber-physical attacks to the railway subsystems.
- *Reusability:* The DT platform should allow for new rail vehicles, vehicle components, or sensors to be easily integrated and updated.
- *Continuous integration and continuous deployment:* This requirement refers to the DT platform's capacity to update the models with new validated information without stopping the system.
- *Provisioning:* The DT platform should be hosted in several places to guarantee the system's robustness and redundancy performance.

The computer hardware requirements to run RT models can be similar to what is in use in the HiL or SiL simulation platforms. Under the definition of RT computing, a physical system can be represented as a simulated component through one or more RT models running in the RT operational system (RTOS) environment. The simulation component running on RTOS should run faster than real time [18]. In other words, for DT models written using RT code principles, their simulation execution total time should be faster than real time or in the expected time period designated in a software component, as shown in Figure 4.5.

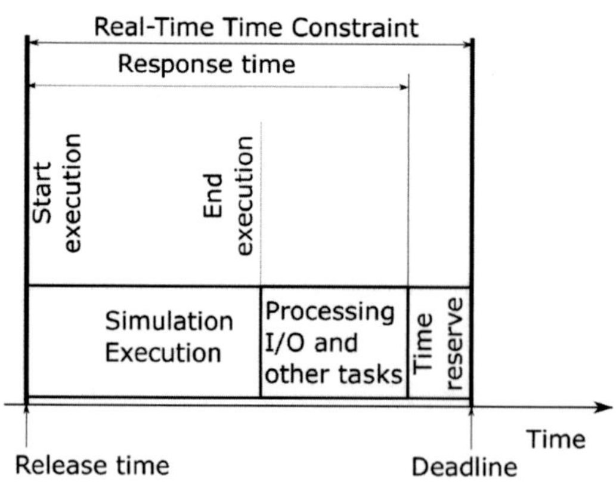

FIGURE 4.5 Real-time execution requirements for execution of a simulation component. (From M. Spiryagin, S. Bruni, C. Bosomworth, P. Wolfs, C. Cole. *Rail Vehicle Mechatronics* (1st ed.). CRC Press, Boca Raton, FL, 2022, with permission.)

In the case of RT computing code development and testing for railway applications, there are two standards: EN 50128 [22] and IEC 62279 [23]. The content of both publications is identical and requires the use of coding standards such as MISRA and CERT [24,25].

4.6.3 High-Performance Computing

HPC involves the use of numerous powerful computer components in combination with advanced software engineering techniques to make effective use of the hardware solutions built on the basis of high-performance workstation(s) or high-performance cluster architectures. Such methodologies, applied to DT simulation, have significantly enhanced the capabilities of numerical solvers and support the processing of vast amounts of data collected on the physical system. These technological advances are in fact essential for the simulation of complex multiphysics, multiscale, big-data-driven DT models. Indeed, such simulations require a high amount of computing power, which can only be provided by high-specification industrial workstations or servers. Industrial workstations, unlike personal computers, are engineered specifically to handle specialized and resource-intensive tasks such as complex calculations and ML performed on large datasets. They incorporate enterprise-grade hardware, including processors, graphics cards, memory, and storage systems that are not only more performant compared to personal computers but also designed to higher reliability standards. Thus, the architecture of an industrial workstation, with high-end examples such as the HP Z8 Fury Workstation currently available in configurations with up to 50+ processor cores, 1+ TB of error-correcting code memory, and multiple graphics processing units (GPUs), can ensure stability and reliability during continuous intensive simulation. The use of HPC computer clusters, composed of multiple workstations or servers (nodes), is also considered for models of increased scale and complexity.

To harness the full capabilities of the computer hardware, it is essential to focus on HPC aspects (e.g., parallelism, concurrency, and single instruction multiple data (SIMD) extensions) within simulation frameworks. This includes developing specific HPC algorithms, exploiting heterogeneous systems with GPU accelerators, ensuring load balancing, optimizing data locality, and utilizing chip-level vectorization [26]. These techniques are indeed crucial for improving the computational performance and efficiency of simulation platforms, thereby preventing resource wastage and enhancing the reliability of DT simulations. Effective parallel computing can be achieved at multiple levels (intra-node, i.e., within nodes using multi-core CPUs and GPUs, and inter-node, i.e., system-wide using networked nodes) and requires sophisticated software engineering strategies summarized below [26]:

- *Inter-node parallelism* involves partitioning the computational mesh into subdomains, each handled by different processes running on various computer cores. These subdomains, which can be either disjointed or overlapping, require synchronization at their interfaces to maintain solution coherence. Since nodes do not share common memory, data exchange at interfaces relies on distributed memory techniques.
- *Intra-node parallelism* leverages shared memory within a computing node, where both loop and task parallelism are supported in shared memory. Loop parallelism divides main loops into chunks, using explicit annotations, for concurrent execution. Task parallelism, on the contrary, allows for more general concurrent execution of tasks on different cores.
- *Chip-level parallelism,* i.e., within a single processor, involves techniques such as vectorization, which uses the SIMD model to perform operations in parallel. This can then be optimized by compilers to increase arithmetic intensity (defined as the number of floating-point operations per second (FLOPs) per total bytes transferred). Memory usage can also be optimized by focusing on spatial and temporal data locality ensuring that, as loops progress, data access occurs on memory locations close to each other.

- *Thread-level parallelism,* utilizing the single instruction multiple threads model, can also be harnessed with GPU chips. In this approach, code is designed to run in parallel as thread blocks. This method efficiently accelerates operations with high arithmetic intensity by emphasizing throughput through parallel execution.

A growing trend in high-performance computers is the heterogeneity of node architectures, with GPUs becoming common due to their efficiency and higher FLOPs per watt ratios compared to CPUs [26]. The post-Moore era is expected to see increased use of domain-specific devices, such as tensor-processing unit deep learning chips [27], which will require applications like DT simulations to adapt to leverage these specialized processors for enhanced performance and efficiency.

4.6.4 Cloud Platform Simulations and Computing Techniques

Cloud-based simulations and computing is a derivative complementary technology from the software virtualization concepts and it represents the next step from server or HPC technologies because of the capability to provide combined additional services like streaming, data storage and processing analysis, and IoT-related functionalities. The cloud platform represents an integrated software-hardware solution that can be flexibly configured for specific cloud implementation DT tasks. This should benefit in deployment of cloud simulations, cybersecurity-related aspects, and data exchange of big data generated by the application processing during the simulation on the cloud servers. Basically, the cloud platform concept for DT simulation and computational tasks can be represented by a functional diagram as shown in Figure 4.6. The DT models in this case are sitting inside of the lower layer of the cloud layout through platform as a service, software as a service, and direct cloud services (Web, IoT, and computational and data storage). All these services should be chosen for implementation based on the DT technical specification requirements and rules for connecting embedded systems to the cloud platform. As summarized in [28], there are many commercial service providers for the cloud platform technologies, and cloud platform environment implementation can bring the following advantages for end users in the DT development and deployment tasks [18]:

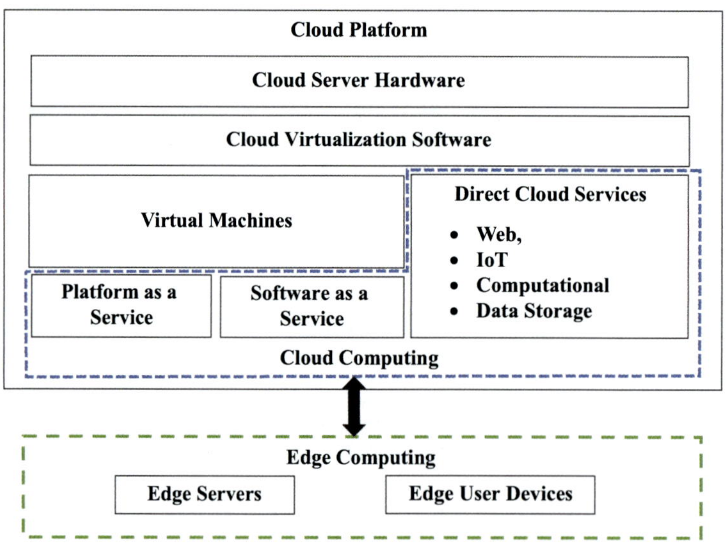

FIGURE 4.6 Typical layout of cloud platform concept for digital twin (DT) tasks.

- Significant savings in hardware infrastructure that traditionally required huge investments in the form of capital expense.
- Reduction in hardware and software maintenance cost, software license ownership cost, and IT-related services.
- Simplification of software upgradability and scalability.
- Digitization of software licensing.

The case of computing techniques [28,29] can be narrowed to two types: cloud computing and edge computing. The latter can be considered as some kind of an extension for cloud computing, which allows devices in remote locations to process data at the "edge" of the network, either by the end-user device or by local servers as shown in Figure 4.5, and it is used for specific tasks that focus on RT or near-RT control tasks with low latency and strict data bandwidth requirements. Meanwhile, cloud computing is more dedicated to computational target tasks with more heavy-loaded processing algorithms. This leads us to an understanding that cloud computing is more efficient for asset performance and prediction tasks, while edge computing is needed for RT analytics and control of a physical system.

REFERENCES

1. D.M. Botín-Sanabria, A.-S. Mihaita, R.E. Peimbert-García, M.A. Ramírez-Moreno, R.A. Ramírez-Mendoza, J.d.J. Lozoya-Santos. Digital twin technology challenges and applications: A comprehensive review. *Remote Sensing* 2022, 14, 1335.
2. Q. Qi, F. Tao, T. Hu, H. Anwer, A. Liu, Y. Wei, L. Wang, A. Nee. Enabling technologies and tools for digital twin. *Journal of Manufacturing Systems* 2021, 53, 3–21.
3. J.F. Uhlenkamp, J.B. Hauge, E. Broda, M. Lütjen, M. Freitag, K.D. Thoben. Digital twins: A maturity model for their classification and evaluation. *IEEE Access* 2022, 10, 69605–69635.
4. W. Kritzinger, M. Karner, G. Traar, J. Henjes, W. Sihn. Digital Twin in manufacturing: A categorical literature review and classification. *IFAC-Paper Online*, 2018, 51(11), 1016–1022.
5. Computer Simulation Technical Committee on Model Credibility. Terminology for model credibility, Reports of SCS Technical Committees. *Simulation*, 1979, 32(3), 103–104.
6. W.D. Cameron. Terminology for model credibility, Technical notes and comments. *Simulation*, 1980, 35(6), 206–207.
7. B.P. Zeigler, A. Muzy, E. Kofman. Chapter 1 – Introduction to systems modeling concepts, In: *Theory of Modeling and Simulation: Discrete Event and Iterative System Computational Foundations* (3rd ed.). Zeigler BP, Muzy A, Kofman E (Eds.), Academic Press, London, UK, 2019, pp. 3–25.
8. M. Spiryagin, J. Edelmann, F. Klinger, C. Cole. Vehicle system dynamics in digital twin studies in rail and road domains. *Vehicle System Dynamics* 2023, 61(7), 1737–1786.
9. D.J. Wagg, K. Worden, R.J. Barthorpe, P. Gardner. Digital twins: State-of-the-art and future directions for modeling and simulation in engineering dynamics applications. *ASCE-ASME Journal of Risk and Uncertainty in Engineering Systems, Part B: Mechanical Engineering*, 2020, 6(3), 030901.
10. L. Wright, S. Davidson. How to tell the difference between a model and a digital twin. *Advanced Modeling and Simulation in Engineering Sciences*, 2020, 7, 1–13.
11. C.A. Chung. *Simulation Modeling Handbook: A Practical Approach*. CRC Press, Boca Raton, FL, 2004.
12. L. Zhang, L. Zhou, B.K Horn. Building a right digital twin with model engineering. *Journal of Manufacturing Systems*, 2021, 59, 151–164.
13. ISO/IEC/IEEE 29119. *Software and Systems Engineering – Software Testing – Part 1: General Concepts* (2nd ed.), Taylor & Francis, Colchester, UK, 2022.
14. M. Arnold, A. Carrarini, A. Heckman, G. Hippmann. Simulation techniques for multidisciplinary problems in vehicle system dynamics, In: *Computational Mechanics in Vehicle System Dynamics, Supplement to Vehicle System Dynamics*. M. Valášek (Ed.), Vol. 40, Taylor & Francis, London, UK, 2004, pp. 17–36.
15. M. Arnold, B. Burgermeister, C. Führer, G. Hippmann, G. Rill. Numerical methods in vehicle system dynamics: State of the art and current developments. *Vehicle System Dynamics*, 2011, 49(7), 1159–1207.
16. MODELISAR. Functional mock-up interface for co-simulation, Document version 1.0, 21 October 2010.

17. MODELISAR. Functional mock-up interface for model exchange and co-simulation, Document version 2.0 Beta 4, 10 August 2012.
18. M. Spiryagin, S. Bruni, C. Bosomworth, P. Wolfs, C. Cole. *Rail Vehicle Mechatronics* (1st ed.). CRC Press, Boca Raton, FL, 2022.
19. E. Bernal, M. Spiryagin, J. Santa, A. Toro, Q. Wu, C. Cole. Real-time digital twin for railway systems. In: *Proceedings of the 28th Symposium of theInternational Association of Vehicle System Dynamics, IAVSD 2023*, Ottawa, Canada, 21–25 August 2023.
20. M. Spiryagin, S. Bruni, B. Fu, E. Bernal, S. Ahmad, Q. Wu, C. Cole, I. Persson. Principles of the framework development of a physics-based rail vehicle digital twin in innovative decision support and prediction complex systems. In: *Proceedings of the Third International Conference on Rail Transportation (ICRT2024)*, Shanghai, China, 7–9 August 2024. Paper No. J0012.
21. J. Belanger, P. Venne, J.N. Paquin. The what, where and why of real-time simulation, OPAL-RT Technologies, October 2010, pp. 37–49. Available from: https://blob.opal-rt.com/medias/L00161_0436.pdf. Accessed on July 16, 2024.
22. CENELEC – EN 50128. Railway applications – Communication, signalling and processing systems – Software for railway control and protection systems, 2020.
23. IEC/TC 9. IEC 62279: Railway applications – Communications, signalling and processing systems – Software for railway control and protection systems, 2015.
24. SEI CERT C Coding Standard. Available at: https://wiki.sei.cmu.edu/confluence/display/c/SEI+CERT+C+Coding+Standard. Accessed on July 16, 2024.
25. SEI CERT C++ Coding Standard. Available at: https://wiki.sei.cmu.edu/confluence/pages/viewpage.action?pageId=88046682. Accessed on July 16, 2024.
26. D. Mira, E.J. Pérez-Sánchez, R. Borrell, G. Houzeaux. HPC-enabling technologies for high-fidelity combustion simulations. *Proceedings of the Combustion Institute*, 2023, 39 (4), 5091–5125.
27. C.E. Leiserson, N.C. Thompson, J.S. Emer, B.C. Kuszmaul, B.W. Lampson, D. Sanchez, T.B. Schardl. There's plenty of room at the Top: What will drive computer performance after Moore's law? *Science*, 2020, 368 (6495), eaam9744, doi:10.1126/science.aam9744.
28. N. Khaled, B. Pattel, A. Siddiqui. *Digital Twin Development and Deployment on the Cloud* (1st ed.). Academic Press, Elsevier, London, UK, 2020. ISBN: 978-0-12-821631-6.
29. M. Mahbub, R.M. Shubair. Contemporary advances in multi-access edge computing: A survey of fundamentals, architecture, technologies, deployment cases, security, challenges, and directions. *Journal of Network and Computer Applications*, 2023, 219, 103726.

5 Development of Data-Driven Digital Twins

Jacek Mocki, George Stojkovski, and Jake Pisano

In the rapidly evolving landscape of digital transformation, the concept of digital twins (DTs) has emerged as a disruptive technology driving innovation across various industries. A DT is a virtual replica of a physical entity (PE), system, or process, enriched with real-time data and advanced analytics to simulate, predict, and optimize performance. While the foundational idea of DTs has been around for years, the advent of data-driven approaches has significantly enhanced their capabilities and applications.

The core of a data-driven DT lies in its ability to harness vast amounts of data from sensors, Internet of Things (IoT) devices, and other sources to create an accurate and dynamic model of the physical counterpart. This model not only mirrors the current state of the PE but also provides predictive insights and actionable intelligence for decision-making. The integration of machine learning (ML), artificial intelligence, and advanced data analytics allows these DTs to evolve continually, improving their accuracy and effectiveness over time.

In this chapter, we will discuss the key considerations for developing data-driven DTs, exploring the technological enablers that make them possible, the standards that govern their implementation, and the processes for validation and verification. By understanding these elements, stakeholders can leverage DTs to drive efficiency, innovation, and competitive advantage in their respective fields.

5.1 DESIGN REQUIREMENTS AND CLASSIFICATION

The DT is a concept recognized in the literature for more than two decades [1]. The concept defines the construct being a digital representation of a physical system, its appearance, and behavior.

5.1.1 Overview

The complete functionality of a physical system will be implemented in its digital equivalent. In fact, the concept assumes a linkage of real information coming from the physical system to the DT to ensure the DT receives the same information. In data-driven DTs, the data from a physical asset (entity) is captured using a variety of tools and sensors, and then the data is processed to implement the PE's functionality and clone its real-life performance into a virtual entity. The data-driven DT can be then used for analysis and simulation/modeling of improvements/rectifications of the PE.

The concept was first presented at the University of Michigan in 2002 mirroring a product lifecycle management (PLM) center for the production of cars and car parts. Translating this concept to the railway field, it can be presented as shown in Figure 5.1. This model was referred to as a "mirrored spaces model". In 2011, the term "digital twin" was introduced. On the left-hand side of the figure the symbols of Real Space items represent physical train manufacturing while the symbols presented in the Virtual Space represent a DT of the physical plant. There can be other Virtual Spaces linked with each other (VS_1, VS_2, VS_n), where n denotes the set of natural numbers. The "Data" line represents the information flow from the physical system to be input into the DT. The "Information process" dotted line indicates that there can be some feedback from the DT that

DOI: 10.1201/9781003492146-5

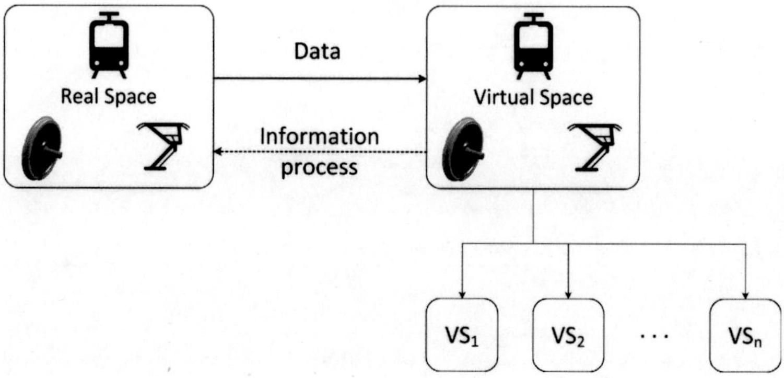

FIGURE 5.1 Conceptual ideal for PLM translated to rail domain.

may influence the Real Space, e.g., process improvement initiatives based on experimentation in the Virtual Space.

The concept of the DT was applied in the development of a next-generation Fighter aircraft and by the National Aeronautics and Space Administration (NASA) in the development of NASA vehicles. Formalizing, a DT is a set of virtual information constructs that fully describes a potential or actual physical manufactured product from the micro atomic level to the macro geometrical level. The following types of DT can be distinguished [1]:

- *Digital twin prototype (DTP)*: This type describes the digital prototype. The artifact's examples include but are not limited to a set of requirements, bills of materials, a description of the 3D model, etc.
- *Digital twin instance (DTI)*: An individual DT that remains linked to the physical product through the entire product lifecycle.
- *Digital twin aggregate (DTA)*: This is a construct of all DTIs that may not contain an independent data structure.

DTs exist in a digital twin environment (DTE) that is an integrated, multi-domain physical application space to enhance additional functionality of the DTs. The additional functionality can be grouped as follows:

- *Predictive*: Where the DT would predict the future functioning and performance of the physical asset. For example, a future asset update is being designed to enhance its performance. The design can be first modeled into the DT and tested in the DTE. Data captured for the future asset update can support the certification process before any physical update is undertaken.
- *Interrogative*: DT instances could be interrogated for their current and past histories. Irrespective of where their physical counterpart resided in the world, individual instances could be interrogated for their current system state: fuel amount, throttle settings, geographical location, structure stress, or any other characteristic that was instrumented. Multiple instances of products would provide data that would be correlated for predicting future states.

5.1.2 Lifecycle

The literature review provides some evidence that digital twinning is still in its infancy. This may indicate the DT applications are rather individual use cases than universal solutions that fit all [2].

It is vital from the theory development perspective that a classification is introduced. Then the theories can be grouped, developed, and applied consistently, increasing the chance for future standardization and comparison to achieve maximum accuracy in the cloning behavior of real assets.

5.1.3 Model Classification

In the realm of DTs, three primary classification models emerge from the literature: analytical, data-driven, and hybrid [3]. Each model type, as shown in Figure 5.2, presents distinct characteristics and applications, catering to diverse needs and objectives. The models are classified as follows:

- *Analytical models*: Analytical models form the cornerstone of DT development, relying on fundamental principles of physics and engineering. These models are built upon predefined equations that describe the behavior and interactions within the system. By combining known physical laws and parameters, analytical models offer a structured approach to understanding system dynamics. They provide a theoretical framework for predicting system responses under varying conditions.

  This type of model is an excellent tool for emulating systems exactly without any reliance on data and provides high levels of accuracy and reliability. However, over time as physical assets degrade and change, these models no longer accurately reflect an asset.
- *Data-driven models*: In contrast to analytical models, data-driven models harness the power of artificial intelligence (AI) and ML techniques. Instead of relying solely on predefined equations, these models extract patterns and relationships from empirical data. By processing large datasets, data-driven models identify complex correlations and nonlinearities that may not be captured by traditional analytical approaches. Through iterative learning and refinement, these models adapt and evolve to improve their predictive capabilities.

  When using a data-driven approach it is necessary to have vast amounts of data to be trained which can be computationally expensive. It is also important to be cautious of the data used to train the model as poorly selected data can develop inaccurate and misleading models. Consequently, care should be taken in the training of any model to ensure that the model does not overfit resulting in poor performance when introduced to new data.
- *Hybrid models*: Hybrid models bridge the gap between analytical and data-driven approaches, combining the strengths of both models. Initially, these models leverage foundational physics-based equations to establish a baseline understanding of the system. Subsequently, data-driven techniques are employed to enhance the model's fidelity and accuracy. By integrating empirical data, hybrid models capture nuances and deviations from idealized assumptions present in analytical models.

An integrated approach of both the analytical and data-driven methods appears to mitigate against the issues suffered by them. For analytical models the issue of the model diverging from the actual

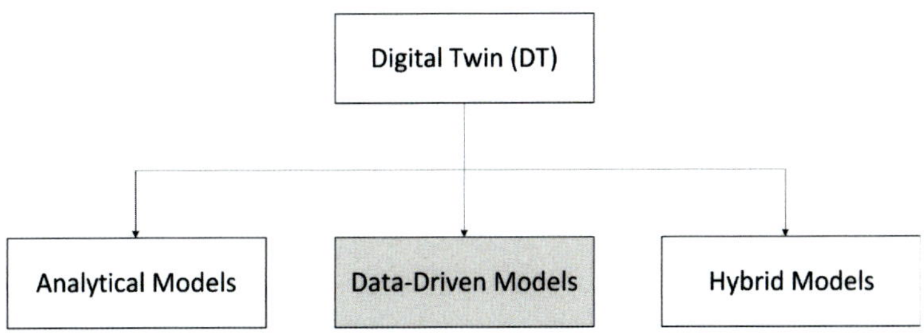

FIGURE 5.2 Digital twin model classification.

behavior is significantly reduced as a result of being updated and trained with live real-time data. This also solves the problem of large data requirements and vast computational resources needed for a data-driven model as the physics equations can create the bulk of the model reliably and accurately with no data and minimal computational resources.

5.2 DEVELOPMENT CONSIDERATIONS

Rail has a strong connection to the Architecture, Engineering, and Construction (AEC) industry, particularly during the design, construction, and operating and maintenance (O&M) of many rail assets including stations, depots, tunnels, bridges, and offices. DT implementation in AEC is further explored in [4], where the authors reviewed 1,000 publications and found that 83% of the publications focused on the application of DT technology during the O&M phase of the asset lifecycle, with the next most significant being the construction stage which represents only 11%. The study, relevant to data-driven DTs, identified that there were very few publications between 2016 and 2018, and then a steep increase in DT publications from 2018, peaking at circa 390 publications in 2022. Although the data for 2023 has shown 200 articles published, the data was incomplete and only represented the first few months of 2023. Therefore, it is recommended that any developer of data-driven DTs start the process by seeking an applicable theory from the most recent publication to ensure full guidance is accessed.

5.2.1 Rail Overview

DTs assist in the challenge of modeling, monitoring, prototyping, and managing the structural health of buildings, bridges, underground structures, and other rail systems during the O&M stage. They model, monitor, and assess the health status of structures, using IoT, BIM, and 3D scanning technologies. Building Information Modeling or Building Information Management (BIM) is a highly collaborative process that enables architects, engineers, real estate developers, contractors, manufacturers, and other construction professionals to plan, design, and construct a structure or building within a single 3D model. BIM can also extend to the operation and management of buildings using data that building or structure owners have access to (hence the Building Information Management component). This data enables decision making based on information derived from the model, even after the building has been constructed.

A DT model can be created for a railway bridge to monitor its structural integrity, detect any signs of deterioration or damage, and provide early warnings for maintenance and repair. The model can incorporate data from sensors installed on the bridge, such as strain gauges or accelerometers, to continuously monitor the bridge's behavior and identify any anomalies. This allows for proactive maintenance planning and helps ensure the safety and reliability of the railway infrastructure.

The use of DT models combined with BIM also improves facility monitoring and management during the O&M stage. For instance, a DT model can be created for a railway station building to monitor the performance of Heating, Ventilation, and Air Conditioning (HVAC) systems, lighting, and other utilities. The model can integrate data from sensors installed in different areas of the station to provide real-time information on energy consumption, occupancy levels, and equipment performance. This enables efficient facility management, optimization of energy usage, and timely maintenance interventions. The most effective benefits are delivered by implementing a data-driven DT model. These models are, as the name suggests, heavily reliant on data flowing between the PE and the DT. For instance:

- In the case of rollingstock, bi-directional communication between the rollingstock and the data-driven DT, as well as the communication among other rollingstock is maintained over a connected vehicle network.

- DT models enable visibility of the live rail network to the railway operators, allowing them to understand the health of the operation and to take corrective measures when operational disruptions occur.

5.2.2 Model Elements

A data-driven DT is indeed a virtual representation with a strong mathematical foundation dependent on many elements which are considered parts of a PE. The DT elements collectively enable an integrated DT model (M_{DT}) that provides a comprehensive and dynamic representation of a PE. In [4], it is formulated as

$$M_{DT} = (\text{PE}, \text{VM}, \text{Ss}, \text{DD}, \text{CN}) \quad (5.1)$$

The following attributes as graphically represented in Figure 5.3 must be considered when developing DT models:

Ⓐ	The physical entity (PE)	The actual physical object or system that the DT represents. In the context of the AEC industry, the PE could be a building, a bridge, a tunnel, a city, or any other engineering object.
Ⓑ	The virtual model (VM)	The virtual representation of the PE in the DT. It includes the 3D model, simulation, and visualization of the PE. The virtual model allows for the monitoring, analysis, and manipulation of the PE in a digital environment.
©	The DT data (DD)	The data collected from the PE and used to populate the DT. The data can be obtained from various sources such as sensors, IoT devices, and other data collection methods. The DD includes information about the current state, performance, and behavior of the PE.
Ⓓ	The Services (SS)	The functionalities and capabilities provided by the DT. The SS can include real-time monitoring, predictive analytics, decision support, maintenance planning, and other value-added features. The SS enable users to interact with and make informed decisions based on the DT.
Ⓔ	The Connections (CN)	The CN and interactions between the different components of the DT. It includes the integration of data from various sources, the communication between the physical and virtual entities, and the interoperability between different systems and platforms. The CN ensure that the DT is up-to-date, accurate, and reliable.

5.2.3 Segmentation

To enable mitigation against the barriers identified in Section 5.2.2 and their associated risks, the PE, the virtual entity (VM), the DD, the SS, and the CN can be further segmented for more accurate consideration as presented in Table 5.1.

As demonstrated by the segments, DD is a substantial contributor to the DT model. Therefore, data and data management are pivotal in implementing accurate, issue-free, and value-driven DTs.

5.2.4 Data Management

Research by Anshan Zhang et al. [4] identifies five key phases in the data management lifecycle. Figure 5.4 identifies those phases with some primary areas of technological consideration, forming the foundational framework for a comprehensive data management approach.

The framework is further explained below in a "bottom-to-the-top" approach, starting from Perception as a natural way to capture data.

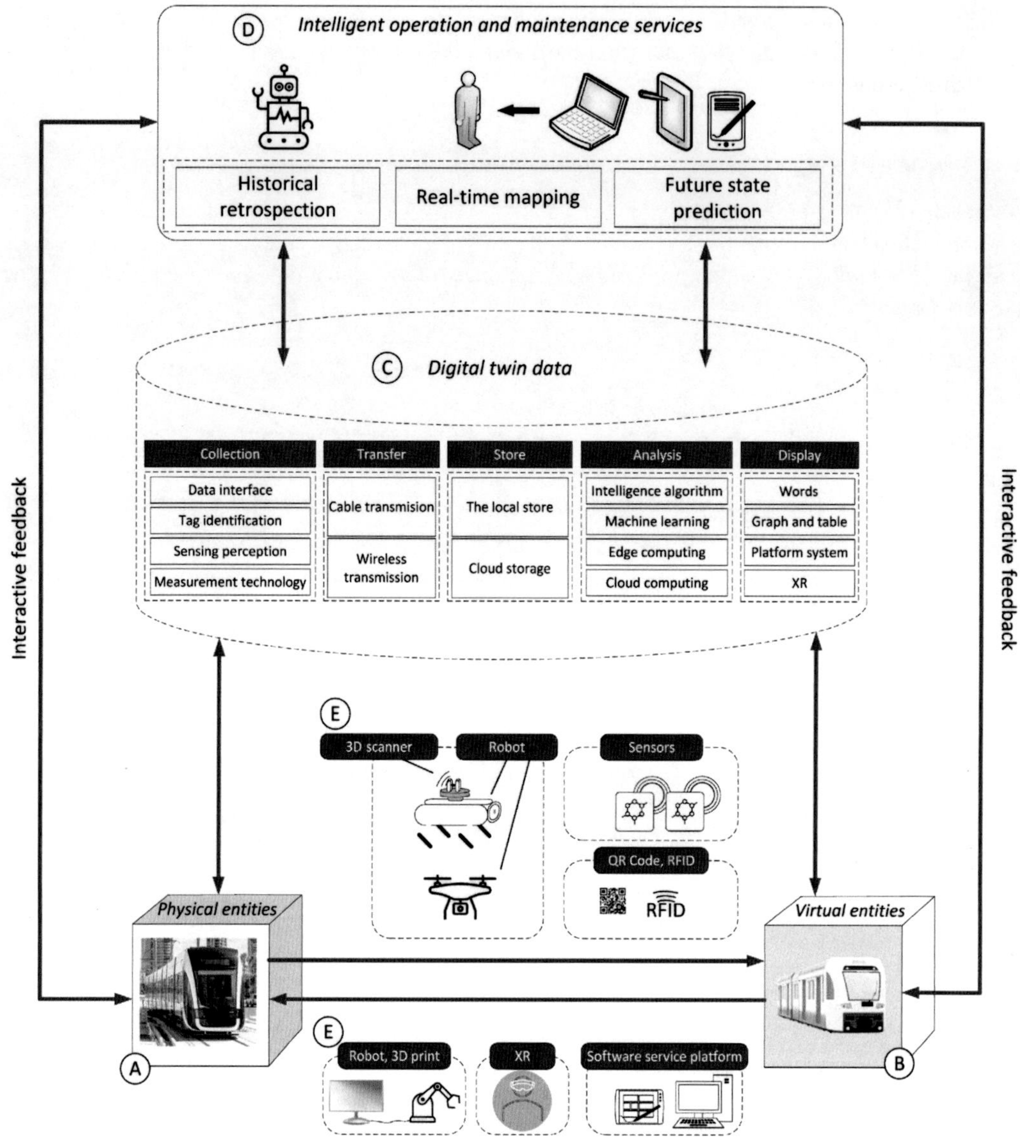

FIGURE 5.3 Example of the graphical representation of digital twin elements.

TABLE 5.1
Segmentation of the Digital Twin Elements

Segments	Description
Geometry and visualization	The digital twin should include a geometric representation of the physical entity, such as a 3D model or CAD drawings. This allows for visualizing and interacting with the digital twin in a realistic manner.
Sensor data integration	The digital twin should integrate real-time and historical data from sensors and IoT devices deployed on the physical entity. This data provides insights into the current state, performance, and behavior of the entity.

(Continued)

TABLE 5.1 (*Continued*)
Segmentation of the Digital Twin Elements

Segments	Description
Physical properties and attributes	The digital twin should capture the physical properties and attributes of the entity, such as dimensions, materials, components, and specifications. This information is essential for accurate representation and analysis.
Behavior and dynamics	The digital twin should simulate and represent the behavior and dynamics of the physical entity. It should capture how the entity responds to different inputs, environmental conditions, and operational parameters.
Connectivity and interactions	The digital twin should model the connectivity and interactions between different components, systems, and processes within the entity. This includes the flow of energy, materials, and information.
Operational data and context	The digital twin should incorporate operational data and contextual information related to the entity. This includes operational constraints, historical performance, maintenance records, and operational procedures.
Control and actuation	The digital twin should enable control and actuation of the physical entity. It should allow for sending commands, adjusting parameters, and monitoring the effects of control actions on the entity.
Analytics and simulation capabilities	The digital twin should include analytics and simulation capabilities to analyze data, predict behavior, and simulate scenarios. This enables performance optimization, predictive maintenance, and decision support.
Lifecycle information	The digital twin should capture information and data throughout the lifecycle of the entity, from design and construction to operation and maintenance. This includes design models, as-built information, maintenance records, and performance history.
User interface and visualization tools	The digital twin should provide a user-friendly interface and visualization tools to interact with and explore the digital twin. This allows users to monitor, analyze, and make informed decisions based on the digital twin's insights.
Integration with external systems	The digital twin should be capable of integrating with external systems, such as enterprise resource planning systems, asset management systems, and data analytics platforms. This enables seamless data exchange and collaboration.
Security and privacy measures	The digital twin should incorporate security and privacy measures to protect sensitive data and ensure the integrity and confidentiality of the digital twin and its associated information.

5.2.4.1 Perception

Perception is defined as a data acquisition (also known in the literature as Collection – for example, see Figure 5.3). It plays a crucial role in translating real-world phenomena into digital representations, with traditional methods in rail network data gathering being laborious, time-consuming, and costly. However, advancements in data-gathering technologies, such as track-recording cars equipped with sensors, offer rapid and efficient alternatives. Despite their efficacy, such methods only provide a snapshot of the network at a single point in time, making them suitable for assets that change or degrade slowly over time. For assets with rapidly evolving behavior, such as points, electrical transformers, and communication lines, the implementation of IoT facilitates the deployment of fixed sensors for continuous real-time data feedback. This constant stream of data enables regular updates to the model, ensuring an accurate representation of the PE amid network changes. Therefore, while track-recording cars remain suitable for gathering data on assets like bridges and tunnels, the use of fixed sensors proves more appropriate for dynamic assets requiring frequent monitoring and model updates.

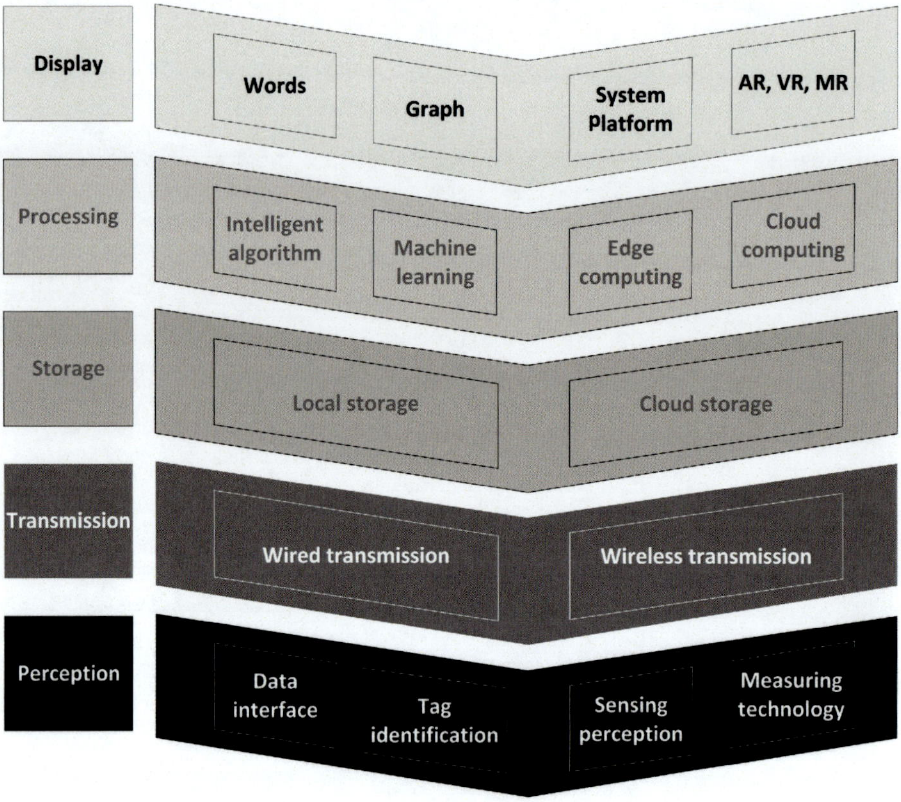

FIGURE 5.4 Data management phases.

5.2.4.2 Transmission

The transmission of data (also known as Transfer – for example, see Figure 5.3) from the network to digital storage is a vital step, as it directly impacts the accuracy of the model, particularly if they intend to model in real time. Minimizing data transmission time and ensuring data integrity are crucial factors, especially for highly dynamic systems prone to frequent changes. Designing infrastructure to facilitate efficient data movement is essential, requiring careful consideration to accommodate the requirements of the DT. Failure to address these concerns can result in significant discrepancies between the model and real-world system behavior.

5.2.4.3 Storage

Once data is extracted from the PE, the next step involves deciding on the storage (also known as Store – for example, see Figure 5.3) method, typically either cloud-based or local storage. Cloud storage offers accessibility from different geographical locations and can be accessed from anywhere with an internet connection, making it ideal for organizations with dispersed teams. Scalability is a key advantage, allowing users to tailor storage to their exact needs, albeit with potential financial implications. Additionally, cloud storage eliminates the need for hardware maintenance and offers automated data backup for enhanced protection. However, it requires a stable internet connection and raises security and privacy concerns, with less control over the geographic location of data storage. On the contrary, local storage ensures faster data access and provides higher levels of security and privacy. Yet, it necessitates responsibility for hardware maintenance, introducing the risk of data loss and increased energy consumption from hardware operation. Careful consideration of these factors is essential in determining the most suitable storage solution for DD management.

5.2.4.4 Processing

Data processing (also known as Analysis – for example, see Figure 5.3) plays an instrumental role in extracting insights and value drivers from network data. In a data-driven approach, parameter classification is commonly achieved by setting thresholds to categorize the current state. Each category corresponds to specific actions, such as maintenance tasks or safety interventions. ML and AI techniques automate data processing, providing new insights and enabling predictions. Access to historical data enhances predictive model reliability. This iterative process empowers operators to make informed decisions and optimize system performance.

5.2.4.5 Display

In managing the vast quantities of data generated by DTs, it's crucial to determine which information is important to users and how best to visualize it. This step serves as the interface between humans and machines, keeping the human-in-the-loop element intact. Extended Reality (XR) is a broad term referring to Augmented Reality (AR), Mixed Reality (MR) and Virtual Reality (VR). A group of emerging visual based technologies that are playing a significant role in effective visualisation. Other technologies being utilised include geometric-based display platforms, and graphing. Typically, the final solution integrates multiple technologies to cater to diverse use cases, optimizing the user experience and facilitating informed decision-making. Equally important facilitators that drive the successful development and implementation of DTs are enabling technologies.

5.2.5 Enabling Technologies

Research by Anshan Zhang et al. [4] documents the roles some technologies play in different aspects of DT. These technologies have been categorized and displayed in Table 5.2.

TABLE 5.2
The Roles Technologies Play in Different Aspects of DT

		Extended Reality	Robot	Smart Device/ Terminal	Building Information Model/ Geographic Information System	Finite Element Analysis	Artificial Intelligence
DT physical entity	Perception		☑	☑	☑		
	Control		☑				
DT virtual entity	Geometry modeling				☑	☑	
	Physical modeling					☑	
	Behavior modeling					☑	
	Role modeling						☑
DT data management	Data collection			☑	☑		
	Data transmission						
	Data storage						
	Data processing						☑
	Data display	☑					
Services		☑	☑	☑			
Connection				☑	☑		

A research study performed by Mateo Paterno et al. [5] has identified some enabling factors for DTs.

5.2.5.1 Artificial Intelligence

AI is the bread and butter of a data-driven model. It is more than likely that any data-driven model will rely heavily on ML, which provides the ability to develop algorithms and models without explicitly being programmed. The power of ML lies in its ability to self-learn and process large amounts of data. There are three generic categories of ML:

- *Supervised learning*: In supervised learning, the model is trained on a labeled dataset, meaning each training example is paired with an output label. The goal is to learn a mapping from inputs to outputs to make predictions on new, unseen data.
- *Unsupervised learning*: Unsupervised learning involves training a model on data without labeled responses. The model tries to identify patterns, relationships, or structures within the data, such as clustering similar items together or reducing dimensionality.
- *Reinforcement learning*: Reinforcement learning focuses on training an agent to make a sequence of decisions by interacting with an environment. The agent learns to achieve a goal by receiving rewards or penalties based on the actions it takes and aims to maximize cumulative rewards over time.

In a data-driven model for DTs, supervised learning and reinforcement learning are often the most relevant categories. Unsupervised learning may also play a role, particularly in identifying patterns and anomalies within the data that could inform the behavior and adjustments of the DT, but it is less central compared to supervised and reinforcement learning [6].

5.2.5.2 Hardware

New developments in computational hardware such as graphics processing units and tensor processing units, alongside cloud and edge computing, are undeniably the primary driving forces behind technological advancements.

5.2.5.3 Cloud, Edge, and Fog Computing

To mitigate against the increasing demand for storage and processing power, cloud, edge, and fog computing is beginning to play a crucial role in how DTs store and process vast quantities of data [5].

- Cloud computing refers to the storage of data on remote servers. This generally will be an enterprise service that relieves the owner of operating and maintaining their own hardware and software management. Data processing can be applied in the cloud where there is access to greater computing power for computationally complex simulations.
- Edge computing is the partial processing and or storage of data locally at the edge/nodes of a network. This enables the computational load to be distributed, helping to reduce strain on other parts of the network.
- Fog computing combines both cloud and edge computing. The user is unaware of where the data is stored and processed. In this instance, network data and processing are allocated automatically between local and remote servers to optimize resource utility.

5.2.5.4 Communications

For any DT to function reliably, timely information from various components must reach its intended target. Bandwidth and protocols have greatly improved recently. 5G facilitates high speed, low latency, and the ability to support a vast number of devices and is poised to overcome these challenges but requires attention to address its limitations in range and signal interference. An alternative when applicable that has been building in popularity is long-range wide area network technology, which has had success in supporting IoT networks while keeping the network structures and management simple. There are many solutions to dealing with communication between various devices that may not natively interact with each other. The general standards used are MTConnect, MQTT, and Constrained Application Protocol (CoAP). For machine-to-machine communication, when it comes to industrial applications the most popular protocol being used is the open platform communication – unified architecture (OPC – UA) [5].

5.2.5.5 Virtual Reality/Augmented Reality (AR)/Extended Reality (XR)

In the context of DTs, the distinction between human and machine is beginning to blur, and extended reality (XR) offers a solution to maintaining human involvement. XR, encompassing both AR and virtual reality, enables humans to integrate directly with large and complex systems [6]. Three primary areas where XR can be notably influential include data visualization, natural language processing, and gesture control. These technologies empower users to interact with DTs more intuitively and effectively, bridging the gap between humans and machines in the DT ecosystem.

5.2.5.6 IoT/IIoT

With the convergence of high-speed internet, enhanced communication capabilities, affordable sensors, and a plethora of enterprise software solutions, the IoT has become an indispensable component of DTs. Today, it is feasible to establish sensor networks and smart devices/terminals for monitoring and, to some extent, controlling various assets in the real world. In industries like rail, which boast vast networks, the deployment of IoT typically involves leveraging enterprise software platforms such as Amazon Web Services, Microsoft Azure, and other service providers. These platforms offer tailored solutions to effectively manage IoT data and facilitate seamless integration into DT frameworks.

5.2.5.7 Development/Simulation Technologies

Development technologies are the category of tools and techniques used in DT development. The most popular technique used is simulation. The terms development and simulation are often used interchangeably. A discussion of importance is becoming increasingly highlighted around the advantages and disadvantages of developing DTs using enterprise software (FEM, BIM, GIS, etc.) as opposed to utilizing open-source software. When developing a DT, the cost and risk of enterprise lock-in should be considered [5].

5.3 DESIGN METHODOLOGY

Adopting a design methodology in the development of a data-driven DT provides a standardized and structured approach that significantly increases the chances of a successful implementation which delivers the desired results.

5.3.1 Design Approach

Figure 5.5 presents a structured approach to data-driven DT design/acceptance that lists key steps that must be followed to minimize the risks of not considering all data-driven DT elements, introducing human errors, increasing data-driven DT development time, and/or implementing a model that functions differently to the PE. The steps of the structured approach (see Figure 5.5) are further expanded in Table 5.3 to enable a greater understanding of the DT methodology. Further enhancement can be achieved with an application of available standards and theories.

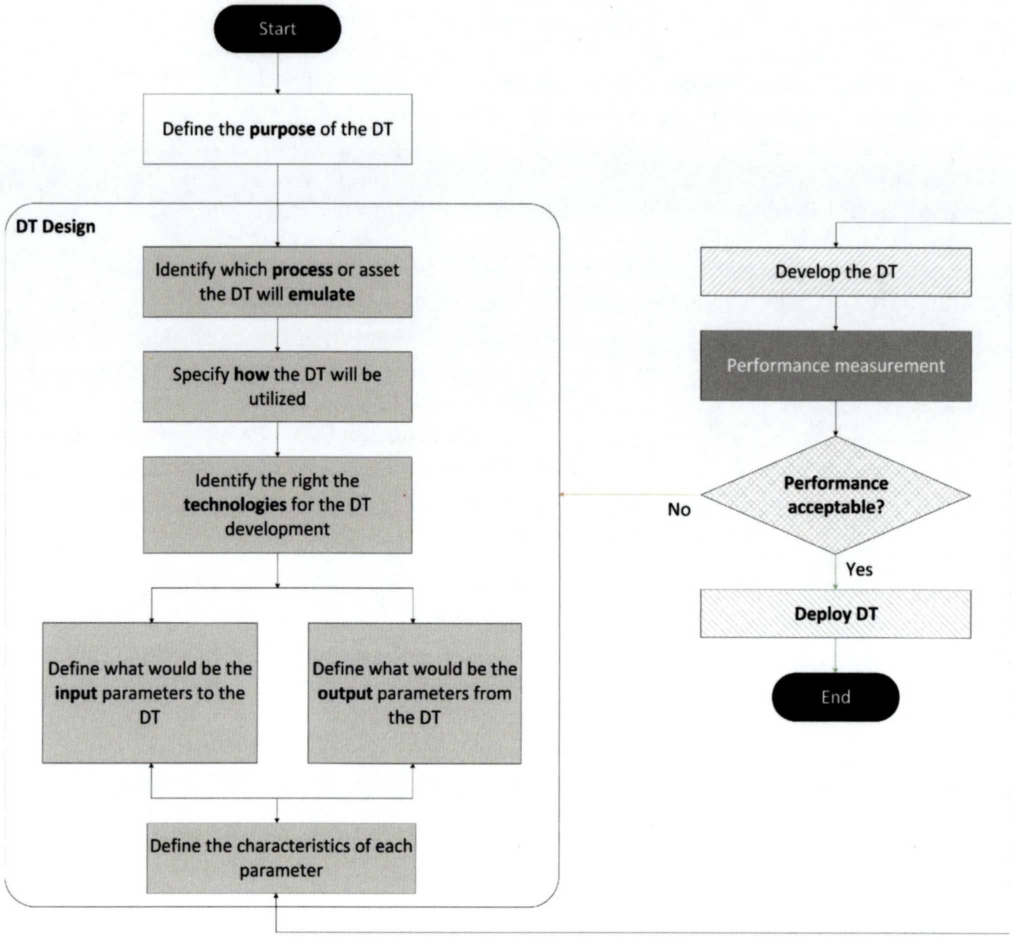

FIGURE 5.5 A structured approach to DT design and acceptance.

5.3.2 Applicable and Developing Standards

Considering that the DT concept is relatively recent, such as the literature demonstrates that there is a lack of standardized frameworks and guidelines for implementing data-driven DTs. However, there are many established standards that are relevant, and bespoke standards are currently in development. Table 5.4 provides a general classification for data-driven DT-related standards that should be considered when developing a data-driven DT. As a result, the subsequent four tables (Tables 5.5–5.8) contain the established standards shown in alignment with the classifications.

5.3.3 Theories Related to Data-Driven DTs

The most recent literature review concludes that DT theories are growing substantially. However, with a lack of DT standardization, it is expected the number of theories will continue growing because they are based on specific application cases. As a result, when searching for the appropriate relevant theory, it is crucial to be specific on a DT model application as well as the model classification. Below there are examples of applications/theories presented and described for each of the model classifications.

TABLE 5.3
Description of DT Methodology

Item	DT Methodology Step	Alignment to the Integrated DT Model (M_{DT})	Description	Rail Example
1.	Purpose of the DT		The overall goal or objective of the DT, such as improving product and process quality or product design.	Improving operational efficiency and maintenance effectiveness in rail systems.
2.	Process or asset that the DT emulates	The physical entity (PE)	The specific process or asset that the DT will model, such as milling machine, cutting process, scheduling tool, or defect detection.	Point machines, rolling stock, overhead structures, and civil structures such as bridges and lift and escalators.
3.	Type of use	The virtual entity (VM) The DT data (DD) The services (SS) The connections (CN)	Whether the DT will be used continuously (data-driven), when needed, or in a mixed mode (hybrid). It also specifies whether the DT will be dynamic (data-driven) or static (analytical).	Continuous monitoring and analysis of real-time data from rolling stock sensors.
4.	Technologies used for the DT	The DD The SS	The technologies that will be utilized for the development of the DT, such as machine learning (ML), artificial intelligence (AI), or design of experiments.	BIM models, Internet of Things (IoT) sensors, data analytics, machine learning algorithms, and predictive maintenance tools.
5.	Input parameters	The CN	The input parameters of the DT, including the parameter name, units (if applicable), and type of data (single value, historical data, real-time data, etc.).	Parameters could include point machine power consumption, speed, and temperature. Trackside sensors measure the train wheel-bearing temperature.
6.	Output parameters	The CN	The output parameters of the DT, including the parameter name, units (if applicable), type of data, and interval rate of new values (if applicable).	Output parameters could include train performance metrics (e.g., energy consumption, on-time performance), track condition assessments, maintenance recommendations, and predictive failure alerts.
7.	Parameters characteristics	The VM The CN	Determine the applicable inputs and outputs to be used in the DT to enable implementation of the DT functionality as required. The parameters should be set according to each use case in order not to increase the complexity of the DT without reason.	Alarms defined in the PE may not be needed in the virtual entity, e.g., measuring the temperature.
8.	DT development	The VM The DD The SS The CN	In this step the digital twin is physically created.	No example required.

(Continued)

TABLE 5.3 (*Continued*)
Description of DT Methodology

Item	DT Methodology Step	Alignment to the Integrated DT Model (M_{DT})	Description	Rail Example
9.	Performance measures	The PE The VM	Establish DT performance criteria that will enable the DT functionality acceptance in the validation and verification process (the DT needs to be able to demonstrate that it performs/clones the physical entity's functions/behavior, and it meets its output parameters). It might be that the PE's results are slightly different from the virtual entity's ones. This situation can still be acceptable; however, tolerance parameters need to be defined as part of this step.	Previously mentioned output parameters: i.e., train performance metrics (e.g., energy consumption, on-time performance) will be used as performance criteria. Variation tolerances will be established as a percentage of the output parameters.
10.	Model validation and verification	The PE The VM The DD The SS The CN	In this step, the behavior of VM is directly correlated with its PE equivalent (that may include the connections and services wherever applicable). Acceptance is confirmed after the VM meets the performance criteria (within the tolerance, if applicable).	No example required.
11.	Digital twin – improvement loop (NO)	The PE The VM The DD The SS The CN	If the acceptance is denied, the process of DT development needs to be recommenced and most likely repeated to ensure the VM meets the performance criteria. The process begins from Step 2 as defined in this table.	No example required
12.	Digital Twin – improvement loop (YES)	The Virtual Entity (VM)	The VM, with the other applicable elements, meets the performance criteria with the tolerances (if applicable). It is confirmed and documented as acceptable and is ready for the DT deployment.	No example required.
13.	DT deployment	The VM	The deployment of DT often requires a definition of the application and a process of deployment. Those should be defined in an implementation plan (implementation plan is excluded from the DT development process).	No example required.

TABLE 5.4
Standards Related to the Classification of Data-Driven DT

	Descriptor	Definition
Overarching classification	Conceptual	Standards to describe ideal, conceptual, and abstract views, including application considerations
	Functional	Standards to list functional elements involved
	Holistic	Standards to introduce overall and architectural views
	Landscape	Standards to show an ecosystem of associated parties, including general interoperability matters
Specification	Zero Spec	Standards to have no implementation specifics
	Interface	Standards to specify data exchanging procedures
	Data	Standards to specify data syntaxes, formats, and schemas
	System	Standards to specify systematic configuration, integration, and relation views to build a resulting system
Implementation	Logical	Standards to show functional entities to be implemented
	Functions	Standards to specify functional constituents, unit-level execution steps, etc., to be implemented
	Structure	Standards to specify functional procedures, configurations, relations, etc., to be implemented
	O&M	Standards to describe operating models and processes, management schemes, quality assurance, performance evaluation, cost analysis, maintenance, etc., to be implemented

TABLE 5.5
Standards Related to the Concept of DT

Specification	Implementation	Standard	Description
Zero – Spec	Logical	ISO/IEC 30173 [7]	Digital twin – concepts and terminology.
Interface		No standard identified	
Data		No standard identified	
System		ISO 10303-1 [8]	Industrial automation systems and integration – product data representation and exchange.
	Structure	ISO/IEC 29182-6 [9]	Describes and provides A compilation of sensor network applications for which International Standardized Profiles are needed Guidelines for the structured description of sensor network applications Examples for structured sensor network applications.
		ISO/IEC TR 30172 [10]	This document provides a collection of representative use cases of digital twin applications in a variety of domains, for example, smart manufacturing and smart cities. This document is applicable to all types of organizations (for example, commercial enterprises, government agencies, and not-for-profit organizations).
		ISO 10303-1 [11]	Industrial automation systems and integration – product data representation and exchange
		IEC 62832-1 [12]	Industrial-process measurement, control, and automation – Digital factory framework – Part 1: General principles

TABLE 5.6
Standards Related to the Functionality of DT

Specification	Implementation	Standard	Description
Zero – Spec	Logical	ISO/TR 24464 [13]	This document analyzes visualization elements that are key components of the interface between the physical asset and the avatar (digital replica of the physical asset).
Interface	Functions	IEEE 1451.0 [14]	Defines the functions that are to be performed by a transducer interface module (TIM) and the common characteristics for all devices that implement the TIM.
			Specifies the formats for transducer electronic data sheets (TEDS).
			Defines a set of commands to facilitate the setup and control of the TIM as well as reading and writing the data used by the system.
			Application programming interfaces (APIs) are defined to facilitate communications with the TIM and with applications.
		IEEE 1451.2 [15]	Standard for a Smart Transducer Interface for Sensors and Actuators – Transducer to Microprocessor Communication Protocols and TEDS Formats.
		IEEE 1451.5 [16]	Standard for a Smart Transducer Interface for Sensors and Actuator – Wireless Communication Protocols and TEDS Formats.
		IEEE 1451.7 [17]	Standard for Smart Transducer Interface for Sensors and Actuators–Transducers to Radio Frequency Identification Systems Communication Protocols and TEDS Formats.
		IEEE P2888.1 [18]	Standard for Specification of Sensor Interface for Cyber and Physical Worlds.
		IEEE P2888.2 [19]	Standard for Actuator Interface for Cyber and Physical Worlds.
		MTConnect-3 [20]	Establishes the rules and terminology that describes the information returned by an MTConnect Agent from a piece of equipment.
		MTConnect-5 [20]	Defines a structured data model used to organize information required to coordinate inter-operations between pieces of equipment.
		OPC-UA Part 14	Open Platform Communications – Unified Architecture that promotes interoperability of loosely coupled PubSub Applications.
		OPC-UA Part 100	Defines an OCP UA information model to represent the models from Ethernet powerlink.
		OMA DDS	Defines a 3GPP Packet Switched (e.g., WCDMA) bearer specific parameters used together with the standardized connectivity management object to have a complete standardized Network Access Point definition for 3GPP Packet Switched connectivity settings in the OMA DM management tree.
		AMQP	An application layer protocol that serves as an open standard for messaging middleware.

(*Continued*)

TABLE 5.6 (*Continued*)
Standards Related to the Functionality of DT

Specification	Implementation	Standard	Description
		ISO/IEC 20922 [21]	A Client Server publish/subscribe messaging transport protocol.
		RFC 6120 [22]	An Extensible Messaging and Presence Protocol that defines XMPP's core protocol methods: setup and teardown of XML streams, channel encryption, authentication, error handling, and communication primitives for messaging, network availability ("presence"), and request-response interactions.
		RFC 6121 [23]	This document defines extensions to core features of the Extensible Messaging and Presence Protocol (XMPP) that provide basic instant messaging and presence functionality in conformance with the requirements in RFC 2779.
		RFC 7252 [24]	The Constrained Application Protocol (CoAP) is a specialized web transfer protocol for use with constrained nodes and constrained (e.g., low-power, lossy) networks.
	Structure	IEEE 1451.3 [25]	EEE Standard for a Smart Transducer Interface for Sensors and Actuators – Digital Communication and TEDS Formats for Distributed Multidrop Systems. A digital interface for connecting multiple physically separated transducers to a single processor over a single pair of wires.
		IEEE 1451.4 [26]	A new standard for adding plug-and-play capabilities to analog transducers.
		IEEE P2888.3 [27]	A Standard for Holographic Visualization for Interfacing Cyber and Physical Worlds.
		IEC 63278-1 [28]	Defines the structure of a standardized digital representation of an asset, called Asset Administration Shell (AAS).
		ISO/IEC 29182-5 [29]	Provides the definitions and requirements of sensor network interfaces of the entities in the Sensor Network Reference Architecture.
Data	Functions	IEEE 1451.1 [30]	This standard defines an object model with a network-neutral interface for connecting processors to communication networks, sensors, and actuators.
		MTConnect-2 [31]	Establishes the rules and terminology to be used by designers to describe the function and operation of a piece of equipment and to define the data that is provided by an *agent* from the equipment.
		MTConnect-4 [32]	Provides semantic models for entities that are used in the manufacturing process but are not considered to be a piece of equipment. These entities are defined as *Assets*. These assets may be removed from a piece of equipment without detriment to the function of the equipment and can be associated with other pieces of equipment during their lifecycle.
		OPC-UA Part 5	Defines the information model of the OPC Unified Architecture. The information model describes standardized nodes of a server's address space.

(*Continued*)

TABLE 5.6 *(Continued)*
Standards Related to the Functionality of DT

Specification	Implementation	Standard	Description
	Structure	IEC 63278-2 [33]	AAS for Industrial Applications – Part 2: Information meta model
		ISO 10303-242 [34]	Specifies the application protocol for managed model-based 3d engineering.
		ISO/IEC 26580 [35]	This document is a specialization of the more general reference model for software and systems product line engineering and management described in ISO/IEC 26550. The specialization defined herein addresses a class of methods and tools referred to as feature-based software and systems product line engineering.
		ISO 15836-1 [36]	Establishes 15 core metadata elements for cross-domain resource description. These terms are part of a larger set of metadata vocabularies maintained by the Dublin Core Metadata Initiative.
		ISO 15836-2 [37]	This document establishes a vocabulary for cross-domain resource description, known as the Dublin Core metadata terms (hereafter DCMI Metadata Terms).
System	Logical Functions	OPC-UA Part 1	Presents the concepts and overview of the OPC Unified Architecture (OPC-UA).
		OPC_UA Part 3	UA Part 3: Address Space Model.
		ML Schema	The ML Schema is a simple shared schema that provides a set of classes, properties, and restrictions that can be used to represent and interchange information on data mining and machine learning algorithms, datasets, and experiments.
	Structure	ISO/IEC 29182-4 [38]	Presents models for the entities that enable sensor network applications and services according to the Sensor Network Reference Architecture.
		IEC 63278-1 [28]	Defines the structure of a standardized digital representation of an asset, called Asset Administration Shell (AAS).
		IEC 63278-3 [39]	Asset Administration Shell for Industrial Applications – Part 3: Security provisions for Asset Administration Shells
		IEC 62832-2 [40]	Industrial-process measurement, control, and automation – digital factory framework – Part 2: Model elements.
		ISO/IEC 19514 [41]	The purpose of **ISO/IEC 19514**:2017 is to specify the Systems Modeling Language (SysML), a general-purpose modeling language for systems engineering.
		ISO/PAS 19450 [42]	This document specifies Object-Process Methodology (OPM) with detail sufficient for enabling practitioners to utilize the concepts, semantics, and syntax of OPM as a modeling paradigm and language for producing conceptual models at various extents of detail and for enabling tool vendors to provide application modeling products to aid those practitioners.
	O&M	IEC 62541-9:2015 [43]	Alarms and Conditions.

TABLE 5.7
Standards Related to DT as a Holistic Approach

Specification	Implementation	Standard	Description
Zero – Spec	Logical	ISO 23247-1 [44]	This document provides an overview and general principles of a digital twin framework for manufacturing including: Terms and definitions Requirements of the digital twin framework for manufacturing
		IEEE P2888.4 [45]	This standard defines the evaluation methods and frameworks for virtual training systems.
Interface	Logical	ISO 23247-4 [46]	This document identifies technical requirements for information exchange between entities within the reference architecture.
	Structure	ISO 10303-209 [47]	Withdrawn.
Data	Logical	ISO 23247-3 [48]	Digital twin framework for manufacturing – Automation systems and integration.
	Functions	ISO 10303-45 [49]	Industrial automation systems and integration – Product data representation and exchange – specifies the integrated resource constructs for material and other engineering properties.
		ISO 10303-46 [50]	Industrial automation systems and integration – Product data representation and exchange – specifies the integrated resource constructs for Visual presentation.
		ISO 10303-235 [51]	This document specifies the use of the integrated resources necessary for the scope and information requirements for the representation of engineering property data that are used for product design, product validation, and other purposes.
	Structure	ISO 16400-2 [52]	Industrial automation systems and integration – This document specifies a formal structure and building rules for an equipment behavior catalog (EBC) template.
		ISO 16400-3 [53]	Automation systems and integration – This document provides requirements and recommendations on how to construct an equipment instance model using an EBC.
System	Logical	ISO 23247-2 [54]	This document provides a reference architecture for the digital twin in manufacturing including. Reference model from domain and entity point of view. Functional view specifying functional entities supported by the entity-based reference model.
		ISO/IEC 30141 [55]	This document provides a standardized IoT Reference Architecture using a common vocabulary, reusable designs, and industry best practices.
		ISO/IEC 30188 [56]	Digital twin – reference architecture – still under development

(Continued)

TABLE 5.7 *(Continued)*
Standards Related to DT as a Holistic Approach

Specification	Implementation	Standard	Description
		ANSI/MTC1.4-2018 [57]	The MTConnect standard is a domain-specific semantic vocabulary for manufacturing equipment. Developers and system integrators use this information model to provide structured, contextualized data with no proprietary format.
		ISO 10303-233 [58]	Specifies an application protocol for the representation of systems engineering data.
	Structure	ISO/IEC 29182-1 [59]	Provides a general overview of the characteristics of a sensor network and the organization of the entities.
		ISO 16400-1 [60]	This document specifies the concept and structure of the ISO 16400 series. This series specifies a methodology for preparing a template and items of an equipment behavior catalog (EBC) as a basis for structuring a virtual production system.
	O&M	ISO/IEC/IEEE 12207 [61]	Provides processes that can be employed for defining, controlling, and improving software lifecycle processes within an organization or a project.
		ISO/IEC/IEEE 15288 [62]	Establishes a common framework of process descriptions for describing the lifecycle of systems created by humans
		ISO/IEC/IEEE 24748-1 [63]	This document provides guidelines for the lifecycle management of systems and software, complementing the processes described in ISO/IEC/IEEE 15288.
		ISO/IEC 30186 [64]	Digital twin – Maturity model and guidance for a maturity assessment – This document is in draft.

TABLE 5.8
Standards Related to the DT Landscape

Specification	Implementation	Standard	Description
Zero – Spec	Logical	ISO/IEC 30173 [65]	This document establishes terminology for digital twin (DT) and describes concepts in the field of digital twin, including the terms and definitions of digital twin, concepts of digital twin (e.g., digital twin system context, lifecycle process for digital twin, types of digital twin), functional view of digital twin, and digital twin stakeholders.
		JTC1-SC41-7 [66]	Develop maritime, underwater, and inland waterways things, digital twin, and related technologies standards for the maritime sector.
Interface		No standard identified	
Data	Structure	ISO 10303-239 [67]	Industrial automation systems and integration – Product data representation and exchange.
		ISO 10303-439 [68]	Industrial automation systems and integration – Product data representation and exchange – Part 439: Application module: AP239 product lifecycle support.
System	Structure	ISO/IEC 29182-7 [69]	Provides a general overview and guidelines for achieving interoperability between sensor network services and related entities.
		ISO 19450 [70]	This document specifies object-process methodology (OPM) with detail sufficient for enabling practitioners to utilize the concepts, semantics, and syntax of OPM as a modeling paradigm and language for producing conceptual models at various extents of detail and for enabling tool vendors to provide application modeling products to aid those practitioners.
		IEC 62832-3 [71]	Industrial-process measurement, control, and automation – digital factory framework – Part 3: Application of Digital Factory for lifecycle management of production systems
		IEC 62890 [72]	Establishes basic principles for lifecycle management of systems and components used for industrial-process measurement, control, and automation. These principles are applicable to various industrial sectors. This standard provides definitions and reference models related to the lifecycle of a product type and the lifetime of a product instance. It defines a consistent set of generic reference models and terms.

5.3.3.1 Predictive Maintenance for Switch Machine Based on DTs

Yang et al. [73] have proposed a predictive maintenance DT model for monitoring switch machines. The architecture is designed around the same 5D DT approach outlined in Section 5.2.2 where the PE is the switch machine, virtual entity is the DT model and visualization, DD is the DT static data collected from the switch machine, connections are the connection of the IoT devices that are collecting the data, and the services are the prediction of current state and visualization for maintenance operators to use.

The DT proposed intends to measure the displacement and therefore indirectly calculate the degree of contact of the switch and stock rail. As the switch machine gap value is an indicator of the operating state, analyzing and predicting this value provides insightful data to maintenance personnel to respond to any anomalies or predictions that exceed standards.

The virtual entity consists of two main components, the behavior model and the rule model. The behavior model is a three-dimensional representation of the switch machine according to the specific parameters of the machine. The model is then imported to Unity3D where it is finalized based on mechanical constraints. In Unity3D the model is updated with operating data and dynamic parameters that drive and update the simulation. The rule model is constructed by a data-driven approach. This approach allows for the capture of real-time data and training of historical data. This ability allows the system to learn the healthy and unhealthy states of the switch machine that can predict the possibility of asset degradation. The authors used data sampled from September to November 2020. The sample period was 2 hours and yielded 648 time series data points. Seven days' worth of data were used to test the system and all other data was used in training. The rule model leverages two single prediction models combined with a third weighting model to produce a final combined prediction outcome. The first prediction model utilizes a long short-term memory (LTSM) model. LTSM algorithms are a specific type of recurrent neural network that analyses time series data and can extract long-term dependency trends. The second model used is an autoregressive integrated moving average (ARIMA) model, a popular model used in linear statistics. The values in an ARIMA model assume linearity, therefore the predictive outcomes will be described as a linear relationship between the current and historical values. These two models can be run concurrently as they are not dependent on each other. Before the two models can be combined the weighting needs to be defined in which the entropy weight method was utilized. The idea of entropy information was introduced by Shannon in 1984 [74] and states that it is the objective measure of the degree of disorder of information. An increase in the amount of information in an indicator corresponds to a decrease in entropy and therefore an increase in the weight value. It is described that the information entropy method has strong operability and objectivity. The combination model then integrates the two single prediction models and applies the previous weighting calculation to produce a predictive model that is more accurate than either single prediction model.

The LSTM, ARIMA, and combination model were all tested on a switch machine with the ARIMA model yielding the least accurate predictions followed by an improvement in the LTSM with the combination model providing the highest degree of accuracy. Table 5.9 outlines the quantitative results by observing three indicators: root mean square error (RMSE), mean absolute error (MAE), and mean absolute percentage error (MAPE).

TABLE 5.9
Performance of the Prediction Models

Method	RMSE (mm)	MAE (mm)	MAPE
LSTM model	0.1998	0.1370	3.9855%
ARIMA model	0.2067	0.1418	4.1420%
Combination model	0.1955	0.1318	3.8435%

5.3.3.2 An Optimization-Based DT for Automated Operation of Rail-Level Crossings

Boban et al. [75] proposed a model that was put forth that aims to improve the efficiency of operating level crossings (LC) through the use of an optimized level crossing (OLC) model with an emphasis on reducing wait time for vehicles. The concept was developed to meet the following requirements:

- Real-time monitoring and updating
- Fault and failure detection
- Correct operation of LC based on historical data from road traffic volume and characteristics of rail traffic and trains

The architecture for this DT is divided into two main sections: one being the virtual layer and the other the physical layer. The virtual layer comprises the monitoring equipment, databases for storing DD and historical data, communication network, and the OLC model. This allows for the collection of real-time data from both the static and dynamic layers described below. The collected data and information are then used as input to the OLC model that optimizes the necessary parameters to drive the system. This layer acts as the engine that drives the DT.

The physical layer is categorized into two components: a dynamic and static layer. The dynamic layer refers to the movable assets, i.e., the train and the static layer is the train tracks, balise, and other immovable assets.

In developing the OLC model the team had developed a set of equations that would dictate the operation of the LC. A linear programming model was used to find the minimum operation time of the LC. The objective function of the optimization problem is defined by the time required to open and close the barriers, the warning time for road users, and the time it takes for the train to reach the LC from the balise.

$$\min T = c_{ij}t_{i,j}^{c} + c_{ij}t_{i,j}^{o} + c_{ij}t_{j}^{\text{warning}} + c_{ij}d_{i}^{b} + c_{ij}d_{i,j}^{B} \tag{5.2}$$

Equation (5.3) shows that the LC operation time is equal to the warning, closing, and opening time of barriers:

$$a_{ij}t_{i,j}^{c} + a_{ij}t_{i,j}^{o} + a_{ij}t_{j}^{\text{warning}} = b_{ij}t_{j}^{\text{oper}} \tag{5.3}$$

The full list of parameters used to model the LC can be found in Table 5.10. The DT was tested on two reference models, one based on a Swedish network and the other on a Taiwanese network; this is the source of operation and infrastructure data that was used to train and test the OLC model. The model successfully determined the minimum required values to minimize operation time while adhering to relevant safety standards. The outcomes found that, based on mutual information transmission between layers and DT modeling, the operation of LCs can be improved. It should be noted that this specific application is reliant on the speed and accuracy of the data transmitted.

5.3.3.3 Augmented DT for Railway Systems

The DT design theory published in [76] states that determining real-time calculations of train dynamics using a traditional DT may not be a feasible option due to the cost of instrumentation that would need to be employed and processing power availability to operate the DT. An augmented DT for railway systems (ADTR), as shown in Figure 5.6, has been proposed by the authors that negates these issues.

A hybrid model has been developed that utilizes training data from 2,100 simulations in GENSYS on a high-performance cluster computer to be used in a surrogate model. Figure 5.7 highlights the four inputs to the surrogate model and the output. Onboard measurements are used as inputs to the

TABLE 5.10
Description of Variables Used in Equations (5.2) and (5.3)

Variable	Description
a_{ij}, b_{ij}, c_{ij}	Parameters of variables
$t_{i,j}^{c}$	Time of closing LC_j
$t_{i,j}^{o}$	Time of opening LC_j
$D_{i,j}^{B}$	Total time for train i (from arming loop, trigger loop) to reach LC_j
d_i^b	Time of train to run from first (arming loop) to second balise (trigger loop)
$t_{i,j}^{\text{block}}$	Blocking time of LC_j for train i
t_j^{oper}	Time of jth LC operating
$t_{i,j}^{\text{road}}$	Waiting time road users of LC_j for train i
t_j^{warning}	Warning time for road users at LC_j
t_j^{minimum}	Minimum warning time at LC_j
$t_j^{\text{additional}}$	Additional warning time at LC_j
T_i^c	Train's clearance time
T_i^d	Traversing time of the train to complete the journey from its current location to the level crossing

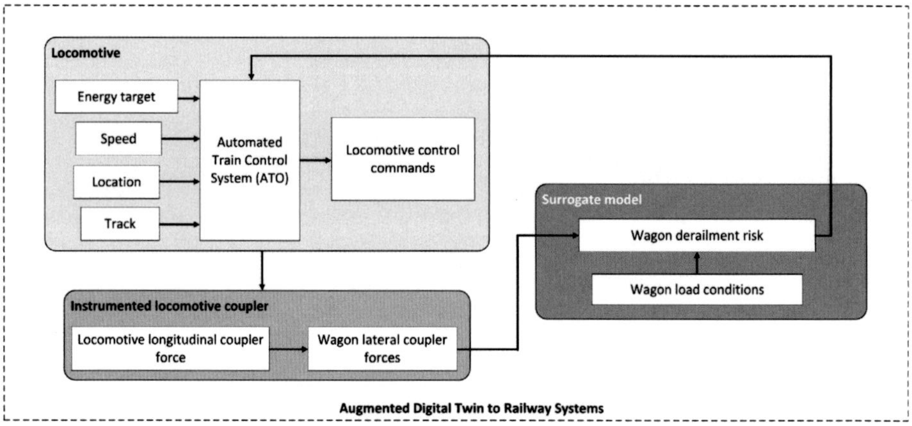

FIGURE 5.6 Augmented digital twin architecture. (From E. Bernal, Q. Wu, M. Spiryagin, and C. Cole, "Augmented digital twin for railway systems," *Vehicle System Dynamics*, vol. 62, no. 1, pp. 67–83, 2023, doi: 10.1080/00423114.2023.2194543, with permission.)

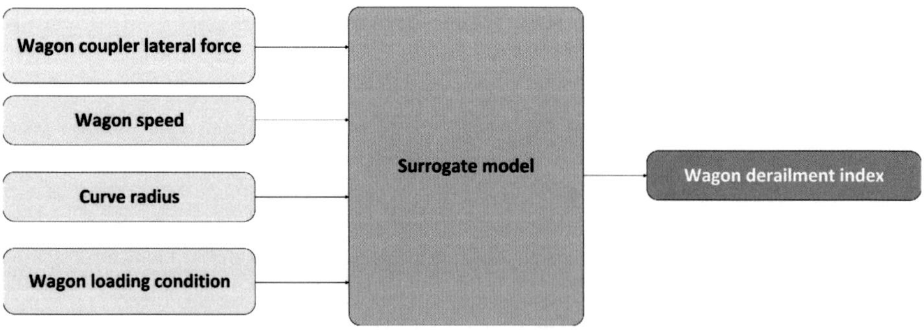

FIGURE 5.7 Surrogate model inputs and output. (From E. Bernal, Q. Wu, M. Spiryagin, and C. Cole, "Augmented digital twin for railway systems," *Vehicle System Dynamics*, vol. 62, no. 1, pp. 67–83, 2023, doi: 10.1080/00423114.2023.2194543, with permission.)

surrogate model to calculate the derailment index (DI) as an output. The DI output is then fed into the automated train controller (ATO), which then uses this input, among others, to calculate train control commands to reduce DI, as shown in Figure 5.6.

In developing the ADTR, two main functions were considered. To develop the surrogate model 2,100 vehicle dynamic simulations were performed. The data taken from the simulations was post-processed, and DI was determined for specific operating conditions. This step improves the accuracy of the final model and vastly reduces the computing power needed on board. This preparation of data to be used in training also circumvents the need to collect data in real time, which can be economically expensive reducing the need for extra instrumentation on board.

The surrogate model receives four inputs: wagon coupler lateral force, wagon speed, curve radius, and wagon loading condition. In this study four ML techniques were tested: linear regression (LR), polynomial regression (PR), ensemble forest (EF), and decision tree. When the DI had values greater than 0.8, the LR and PR methods greatly overestimated the true DI and had unreliable mixed results when DI values were below 0.8. The EF and decision tree techniques yield the most consistent results, with EF having the lowest residuals. For the case study, an EF technique was trained with 80% of the 2,100 multi-body simulations conducted in GENSYS, with the remaining 20% used for testing.

A case study was conducted using the Centre for Railway Engineering longitudinal train simulator. This tested for various conditions such as loading conditions, rail curve radius, and elevation, to name a few. The ADTR was able to reasonably predict the DI with higher values in unloaded conditions compared to lower values in loaded conditions with varying results according to different lateral force, vehicle speed, and curve radius values.

As a result, the ATO receives the DI, which can be used to optimize train control, allowing for higher speeds, mass, and increased efficiency in operation. Although this chapter only accounted for DI, it appears that the methodology employed would be successful in determining other critical rail indicators.

5.4 MODEL AUDITING

In the context of the DT development and auditing, the auditing process can be simplified by clearly differentiating between pre-evaluation (validation) and post-evaluation (verification) activities. This approach divides the auditing process into distinct phases, enabling more focused management and setting clearer objectives for each stage of assessing the DT models and/or systems.

5.4.1 Validation

Validation determines whether the developer's conceptual descriptions and design specifications accurately reflect the intentions, objectives, requirements, and use cases of the client (including stakeholders) in performing DT modeling and designing activities. Model validation refers to the confirmation that the model of the PE conforms to the expected system performance requirements.

It is one key step that is largely under-addressed in the literature, although more work is being currently published in this space. Validating DTs is crucial as they utilize constantly changing data and occasionally abstract data originating from real-world objects. Translating and transmitting such information can pose challenges, emphasizing validation to ensure the reliability of all parameters and data. An important and common theme that has been identified in the most recent literature is the importance of ensuring high fidelity in a system. The concept of fidelity relates to the quantity of parameters exchanged between the virtual and physical environments, alongside their precision and degree of abstraction. Fidelity in the virtual model indicates a highly precise replication of the physical object. This should be taken into consideration when implementing any system validation.

Validation is a quality assurance process that occurs before or during the development phase, ensuring designs and plans align with user requirements and intended use cases.

Validation serves as a pre-evaluation activity, selectively applied based on the development project's complexity and stakeholder requirements. For example, simple projects may bypass this validation-based quality assurance process, whereas sophisticated projects may necessitate validation. This validation aims to assess whether the proposed and designed DT aligns with stakeholders' expectations and requirements before its actual implementation process starts, effectively answering the question, "Are we planning to build the right thing?" By doing so, validation ensures that a high-complexity development project can be conducted as expected, required, and planned.

5.4.2 Verification

Verification aims to confirm to what extent the implemented DT models and systems accurately represent the target PE or real-world systems. The primary objective is to ensure that the model has been constructed correctly according to the intended use cases, established requirements, design specifications, and any relevant standards. This process is crucial for validating that the DT aligns with its intended purpose and functions as expected.

Following the implementation phase, verification acts as a post-evaluation activity to determine if the implemented DT meets the established requirements and adheres to the original design specifications. This process addresses the critical question: "Have we constructed the system as planned?". Verification involves a thorough examination of the DT to ensure it faithfully replicates the PE it represents.

Verifying the implemented DT's adherence to its initial design specifications is vital for affirming its operational performance and functionality. This step includes rigorous testing and analysis to identify any discrepancies between the model and the real-world system. It ensures that all components, interactions, and behaviors of the DT are correctly implemented and performed as intended.

Moreover, verification serves as a quality assurance measure, providing stakeholders with confidence that the DT will deliver accurate and reliable insights. By confirming that the DT has been built correctly, organizations can leverage it to optimize operations, predict outcomes, and enhance decision-making processes.

The verification process involves several critical steps to ensure the accuracy and reliability of the implemented models. Initially, the auditor must develop a verification plan that includes strategic analysis and risk assessment to identify potential issues and determine the scope of evidence-gathering activities. This plan, once approved, guides the auditor through a series of activities such as document review, site visits, and direct assessments of the DT's functionalities. Evidence is gathered through various techniques like observation, analytical testing, and confirmation to verify that the DT adheres to its design specifications and accurately represents its physical counterpart. The process concludes with compiling the audit findings into a comprehensive report which includes feedback for the development team and recommendations for improvement. Follow-up actions are also planned to address any new findings or changes post-verification, ensuring ongoing accuracy and compliance.

REFERENCES

1. M. Grieves, "Origins of the Digital Twin Concept," Working Paper, August 2016. Available: https://www.researchgate.net/publication/307509727_Origins_of_the_Digital_Twin_Concept.
2. H. van der Valk, H. Haße, F. Möller, and B. Otto, "Archetypes of digital twins," *Business & Information Systems Engineering*, vol. 64, no. 3, pp. 375–391, 2022.
3. A. Wunderlich, K. Booth, and E. Santi, "Hybrid analytical and data-driven modeling techniques for digital twin applications," in *Proc. 2021 IEEE Electric Ship Technologies Symposium (ESTS)*, Washington, DC, August 2021, pp. 1–7, doi:10.1109/ESTS49166.2021.9512364.

4. A. Zhang, J. Yang, and F. Wang, "Application and enabling digital twin technologies in the operation and maintenance stage of the AEC industry: A literature review," *Journal of Building Engineering*, vol. 80, Art. no. 107859, 2023, doi:10.1016/j.jobe.2023.107859.
5. M. Perno, L. Hvam, and A. Haug, "Implementation of digital twins in the process industry: A systematic literature review of enablers and barriers," *Computers in Industry*, vol. 134, Art. no. 103558, 2022, doi:10.1016/j.compind.2021.103558.
6. A. Rasheed, O. San, and T. Kvamsdal, "Digital twin: Values, challenges and enablers from a modeling perspective," *IEEE Access*, vol. 8, pp. 21980–22012, 2020, doi:10.1109/ACCESS.2020.2970143.
7. ISO/IEC 30173:2022, *Information Technology – Security Techniques – A Framework for Identity Management*, International Organization for Standardization, Geneva, Switzerland, 2022.
8. ISO 10303-1:1994, *Industrial Automation Systems and Integration—Product Data Representation and Exchange—Part 1: Overview and Fundamental Principles*, International Organization for Standardization, Geneva, Switzerland, 1994.
9. ISO/IEC 29182-6:2014, *Information Technology – Sensor Networks: Sensor Network Reference Architecture (SNRA) – Part 6: Applications*, International Organization for Standardization, Geneva, Switzerland, 2014.
10. ISO/IEC TR 30172:2021, *Information Technology – Artificial Intelligence (AI) – Overview of Computational Approaches for AI Systems*, International Organization for Standardization, Geneva, Switzerland, 2021.
11. SO 10303-1:1994, *Industrial Automation Systems and Integration—Product Data Representation and Exchange—Part 1: Overview and Fundamental Principles*, International Organization for Standardization, Geneva, Switzerland, 1994.
12. IEC 62832-1:2020, *Industrial-Process Measurement, Control and Automation – Digital Factory Framework – Part 1: General Principles*, International Electrotechnical Commission, Geneva, Switzerland, 2020.
13. ISO/TR 24464:2023, *Automation Systems and Integration – Industrial Data – Visualization Elements of Digital Twins*, International Organization for Standardization, Geneva, Switzerland, 2023.
14. IEEE 1451.0-2007, *Standard for a Smart Transducer Interface for Sensors and Actuators – Common Functions, Communication Protocols, and Transducer Electronic Data Sheet (TEDS) Formats*, Institute of Electrical and Electronics Engineers, New York, 2007.
15. IEEE 1451.2-2017, *Standard for a Smart Transducer Interface for Sensors and Actuators – Transducer to Microprocessor Communication Protocols and Transducer Electronic Data Sheet (TEDS) Formats*, Institute of Electrical and Electronics Engineers, New York, 2017.
16. IEEE 1451.5-2019, *Standard for a Smart Transducer Interface for Sensors and Actuators – Wireless Communication Protocols and Transducer Electronic Data Sheets (TEDS) Formats*, Institute of Electrical and Electronics Engineers, New York, 2019.
17. IEEE 1451.7-2010, *Standard for a Smart Transducer Interface for Sensors and Actuators – Transducers to Radio Frequency Identification (RFID) Systems Communication Protocols and Transducer Electronic Data Sheet (TEDS) Formats*, Institute of Electrical and Electronics Engineers, New York, 2010.
18. IEEE P2888.1-2023, *Standard for Virtual Reality and Augmented Reality: Computational Framework and Methods*, Institute of Electrical and Electronics Engineers, New York, 2023.
19. IEEE P2888.2-2023, *Standard for Virtual Reality and Augmented Reality: Enabling Concepts and Technologies*, Institute of Electrical and Electronics Engineers, New York, 2023.
20. MTConnect Institute, "MTConnect Standard Part 1.0 – Overview and Fundamentals," MTConnect Institute, Version 1.8, 2021. Available: https://docs.mtconnect.org/MTC_Part_1-0-Overview_and_Fundamentals_1-6-0.pdf.
21. ISO/IEC 20922:2016, *Information Technology – Message Queuing Telemetry Transport (MQTT) v3.1.1*, International Organization for Standardization, Geneva, Switzerland, 2016.
22. P. Saint-Andre, "Extensible Messaging and Presence Protocol (XMPP): Core," IETF RFC 6120, March 2011. Available: https://datatracker.ietf.org/doc/html/rfc6120.
23. P. Saint-Andre, "Extensible Messaging and Presence Protocol (XMPP): Instant Messaging and Presence," IETF RFC 6121, March 2011. Available: https://datatracker.ietf.org/doc/html/rfc6121.
24. Z. Shelby, K. Hartke, and C. Bormann, "The Constrained Application Protocol (CoAP)," IETF RFC 7252, June 2014. Available: https://datatracker.ietf.org/doc/html/rfc7252.
25. IEEE 1451.3-2003, *Standard for a Smart Transducer Interface for Sensors and Actuators – Transducer to Network Communications Protocols and Transducer Electronic Data Sheet (TEDS) Formats*, Institute of Electrical and Electronics Engineers, New York, 2003.

26. IEEE 1451.4-2004, *Standard for a Smart Transducer Interface for Sensors and Actuators – Mixed-Mode Communication Protocols and Transducer Electronic Data Sheet (TEDS) Formats*, Institute of Electrical and Electronics Engineers, New York, 2004.
27. IEEE P2888.3-2023, *Standard for Virtual Reality and Augmented Reality: Communication Protocols and Media Formats*, Institute of Electrical and Electronics Engineers, New York, 2023.
28. IEC 63278-1:2020, *Asset Administration Shell for Industrial Applications – Part 1: Administration Shell Structure*, International Electrotechnical Commission, Geneva, Switzerland, 2020.
29. ISO/IEC 29182-5:2013, *Information Technology – Sensor Networks: Sensor Network Reference Architecture (SNRA) – Part 5: Interface Definitions*, International Organization for Standardization, Geneva, Switzerland, 2013.
30. EEE 1451.1-1999, *Standard for a Smart Transducer Interface for Sensors and Actuators – Network Capable Application Processor (NCAP) Information Model*, Institute of Electrical and Electronics Engineers, New York, 1999.
31. MTConnect Institute, "MTConnect Standard Part 2.0 – Overview and Fundamentals," MTConnect Institute, Version 2.0, 2022. Available: https://docs.mtconnect.org/MBSD_MTConnect_Part_2_2_0_0.pdf.
32. MTConnect Institute, "MTConnect Standard Part 4.0 – Overview and Fundamentals," MTConnect Institute, Version 4.0, 2023. Available: https://docs.mtconnect.org/MTConnect_Part_4-0_Assets_Information_Model_1-8-0.pdf.
33. IEC 63278-2:2022, *Asset Administration Shell for Industrial Applications – Part 2: Application of Asset Administration Shell Elements*, International Electrotechnical Commission, Geneva, Switzerland, 2022.
34. ISO 10303-242:2020, *Industrial Automation Systems and Integration – Product Data Representation and Exchange – Part 242: Application Protocol: Managed Model-Based 3D Engineering*, International Organization for Standardization, Geneva, Switzerland, 2020.
35. ISO/IEC 26580:2020, *Software and Systems Engineering – Methods and Tools for the Feature-Based Approach to Software and Systems Product Line Engineering*, International Organization for Standardization, Geneva, Switzerland, 2020.
36. ISO 15836-1:2017, *Information and Documentation – The Dublin Core Metadata Element Set – Part 1: Core Elements*, International Organization for Standardization, Geneva, Switzerland, 2017.
37. ISO 15836-2:2019, *Information and Documentation – The Dublin Core Metadata Element Set – Part 2: DCMI Properties and Classes*, International Organization for Standardization, Geneva, Switzerland, 2019.
38. ISO/IEC 29182-4:2013, *Information Technology – Sensor Networks: Sensor Network Reference Architecture (SNRA) – Part 4: Entity Models*, International Organization for Standardization, Geneva, Switzerland, 2013.
39. IEC 63278-3:2022, *Asset Administration Shell for Industrial Applications – Part 3: Industrial Requirements*, International Electrotechnical Commission, Geneva, Switzerland, 2022.
40. IEC 62832-2:2020, *Industrial-Process Measurement, Control and Automation – Digital Factory Framework – Part 2: Model Elements*, International Electrotechnical Commission, Geneva, Switzerland, 2020.
41. ISO/IEC 19514:2017, *Information Technology – Object Management Group – Systems Modeling Language (OMG SysML)*, International Organization for Standardization, Geneva, Switzerland, 2017.
42. ISO/PAS 19450:2015, *Automation Systems and Integration – Object-Process Methodology*, International Organization for Standardization, Geneva, Switzerland, 2015.
43. IEC 62541-9:2015, *OPC Unified Architecture – Part 9: Alarms and Conditions*, International Electrotechnical Commission, Geneva, Switzerland, 2015.
44. ISO 23247-1:2021, *Automation Systems and Integration – Digital Twin Framework for Manufacturing – Part 1: Overview and General Principles*, International Organization for Standardization, Geneva, Switzerland, 2021.
45. IEEE P2888.4-2023, *Standard for Virtual Reality and Augmented Reality: Communication Protocols and Media Formats*, Institute of Electrical and Electronics Engineers, New York, 2023.
46. ISO 23247-4:2022, *Automation Systems and Integration – Digital Twin Framework for Manufacturing – Part 4: Information Exchange*, International Organization for Standardization, Geneva, Switzerland, 2022.
47. ISO 10303-209:2014, *Industrial Automation Systems and Integration – Product Data Representation and Exchange – Part 209: Application Protocol: Multidisciplinary Analysis and Design*, International Organization for Standardization, Geneva, Switzerland, 2014.

48. ISO 23247-3:2022, *Automation Systems and Integration – Digital Twin Framework for Manufacturing – Part 3: Digital Representation*, International Organization for Standardization, Geneva, Switzerland, 2022.
49. ISO 10303-45:2021, *Industrial Automation Systems and Integration – Product Data Representation and Exchange – Part 45: Integrated Generic Resource: Materials*, International Organization for Standardization, Geneva, Switzerland, 2021.
50. ISO 10303-46:2022, *Industrial Automation Systems and Integration – Product Data Representation and Exchange – Part 46: Integrated Generic Resource: Visual Presentation*, International Organization for Standardization, Geneva, Switzerland, 2022.
51. ISO 10303-235:2010, *Industrial Automation Systems and Integration – Product Data Representation and Exchange – Part 235: Application Protocol: Engineering Properties for Product Design and Verification*, International Organization for Standardization, Geneva, Switzerland, 2010.
52. ISO 16400-2:2019, *Automation Systems and Integration – Framework for Capability of Equipment – Part 2: Methodology for Profile Matching of Capability of Equipment*, International Organization for Standardization, Geneva, Switzerland, 2019.
53. ISO 16400-3:2022, *Automation Systems and Integration – Framework for Capability of Equipment – Part 3: Verification Methods for Capability of Equipment*, International Organization for Standardization, Geneva, Switzerland, 2022.
54. ISO 23247-2:2022, *Automation Systems and Integration – Digital Twin Framework for Manufacturing – Part 2: Reference Architecture*, International Organization for Standardization, Geneva, Switzerland, 2022.
55. ISO/IEC 30141:2018, *Internet of Things (IoT) – Reference Architecture*, International Organization for Standardization, Geneva, Switzerland, 2018.
56. ISO/IEC 30188:2023, *Information Technology – Artificial Intelligence – Framework for the Governance of AI*, International Organization for Standardization, Geneva, Switzerland, 2023.
57. ANSI, *ANSI/MTC1.4-2018: American National Standard for Metadata and Testing for Conformance to Metadata*, American National Standards Institute, New York, NY, 2018.
58. *ISO 10303-233: Industrial Automation Systems and Integration – Product Data Representation and Exchange – Part 233: Systems Engineering Data Representation*, International Organization for Standardization, Geneva, Switzerland, 2020.
59. *ISO/IEC 29182-1: Information Technology – Sensor Networks: Sensor Network Reference Architecture (SNRA) – Part 1: General Requirements and Reference Model*, International Organization for Standardization/International Electrotechnical Commission, Geneva, Switzerland, 2013.
60. *ISO 16400-1: Automation Systems and Integration – Equipment Behaviour Catalogues for Manufacturing Systems – Part 1: Overview and Fundamental Principles*, International Organization for Standardization, Geneva, Switzerland, 2018.
61. *ISO/IEC/IEEE 12207: Systems and Software Engineering – Software Life Cycle Processes*, International Organization for Standardization/International Electrotechnical Commission/Institute of Electrical and Electronics Engineers, Geneva, Switzerland, 2017.
62. *ISO/IEC/IEEE 15288: Systems and Software Engineering – System Life Cycle Processes*, International Organization for Standardization/International Electrotechnical Commission/Institute of Electrical and Electronics Engineers, Geneva, Switzerland, 2015.
63. *ISO/IEC/IEEE 24748-1: Systems and Software Engineering — Life Cycle Management – Part 1: Guidelines for Life Cycle Management*, International Organization for Standardization/International Electrotechnical Commission/Institute of Electrical and Electronics Engineers, Geneva, Switzerland, 2018.
64. *ISO/IEC 30186: Information Technology – Object Oriented BioAPI (OOBioAPI)*, International Organization for Standardization/International Electrotechnical Commission, Geneva, Switzerland, 2020.
65. ISO/IEC, *ISO/IEC 30173: Information Technology – Automatic Identification and Data Capture Techniques – Face Recognition Application Programming Interface*, International Organization for Standardization/International Electrotechnical Commission, Geneva, Switzerland, 2023.
66. *JTC1-SC41-7: Internet of Things (IoT) – Reference Architecture*, International Organization for Standardization/International Electrotechnical Commission, Geneva, Switzerland, 2022.
67. *ISO 10303-239: Industrial Automation Systems and Integration – Product Data Representation and Exchange – Part 239: Application Protocol: Product Life Cycle Support*, International Organization for Standardization, Geneva, Switzerland, 2020.

68. *ISO 10303-439: Industrial Automation Systems and Integration – Product Data Representation and Exchange – Part 439: Application Module: Design Data Management*, International Organization for Standardization, Geneva, Switzerland, 2021.
69. *ISO/IEC 29182-7: Information Technology – Sensor Networks: Sensor Network Reference Architecture (SNRA) – Part 7: Interoperability Guidelines*, International Organization for Standardization/International Electrotechnical Commission, Geneva, Switzerland, 2015.
70. *ISO 19450: Automation Systems and Integration – Object-Process Methodology*, International Organization for Standardization, Geneva, Switzerland, 2020.
71. *IEC 62832-3: Industrial-Process Measurement, Control and Automation – Digital Factory Framework – Part 3: Application of Digital Factory for Life Cycle Management of Production Systems*, International Electrotechnical Commission, Geneva, Switzerland, 2020
72. *IEC 62890: Industrial-Process Measurement, Control and Automation – Life-Cycle Management for Systems and Components*, International Electrotechnical Commission, Geneva, Switzerland, 2016.
73. J. Yang, Y. Sun, Y. Cao, and X. Hu, "Predictive maintenance for switch machine based on digital twins," *Information*, vol. 12, no. 11, Art. no. 485, 2021, doi:10.3390/info12110485.
74. C. E. Shannon, "A mathematical theory of communication," *Bell System Technical Journal*, vol. 27, pp. 379–423, 1948.
75. B. Djordjević, E. Krmac, C.-Y. Lin, O. Fröidh, and B. Kordnejad, "An optimization-based digital twin for automated operation of rail level crossings," *Expert Systems with Applications*, vol. 239, Art. no. 122422, 2024, doi:10.1016/j.eswa.2023.122422.
76. E. Bernal, Q. Wu, M. Spiryagin, and C. Cole, "Augmented digital twin for railway systems," *Vehicle System Dynamics*, vol. 62, no. 1, pp. 67–83, 2023, doi:10.1080/00423114.2023.2194543.

6 Development of Physics-Based Digital Twin

Sanjar Ahmad and Farzaneh Tahmoorian

6.1 INTRODUCTION

In railway applications, the physics-based digital twins (DTs) can be used on both vehicle and infrastructure systems. The vehicle system includes mechanical and electrical elements, which can be represented by mass, spring, and damping systems. On the contrary, a good representative of the infrastructure system is the track system that includes sleepers, rail, ballast, and foundations, which resemble a construction environment. In this chapter, the conceptual principles of development and building such physics-based DTs are discussed.

6.2 DESIGN REQUIREMENTS AND CLASSIFICATION

Grieves [1] identified that a DT model consists of three main parts: (1) physical products in real space, (2) virtual products in virtual space, and (3) connections of data and information that connect virtual and real products [1]. To represent the behavioral characteristics, it is necessary to understand the dynamics of the physical entity (PE) and measured data of various components of the PE to validate the DT. In the cases of structural dynamical systems, Ganguli et al. [2] extended the nominal equation of a single degree of freedom spring, damper, and mass model to demonstrate the characteristics of a DT (Equation 6.1). While the nominal dynamic equation represents the mass, damping, stiffness, and forcing parameters to be a function of time, in a DT system, these parameters can be represented at a delayed or slow time 'ts' that represents a degraded state of structural components [2].

$$m(t_s)\frac{\partial^2 u(t,\, t_s)}{\partial t^2}+c(t_s)\frac{\partial u(t,t_s)}{\partial t}+k(t_s)u(t,t_s)=f(t,ts) \tag{6.1}$$

where m = mass, c = damping, k = stiffness, f = force, t = system time, t_s = slow time or service time.

A similar analogy of delayed time response can be used in the case of prediction of wear maintenance in wheel–rail contact, which is subject to post-processing of a large number of high-fidelity models, which cannot communicate with the PE in real time due to requirements of high computation time.

To ensure accuracy and reliability in physical models, it is essential to set clear design requirements for creating a DT that can effectively replicate the actual behavior of components of a vehicle, machine, structure, or infrastructure. Many research studies have discussed the main design requirements for developing physics-based DTs in the construction industry, including:

- Accuracy and reliability of data to ensure accurate representation of the materials (e.g., strength, stiffness, elasticity) or the structure and infrastructure conditions [3–6]. Additionally, ensuring accuracy through using up-to-date data obtained from multiple sources, regular updates, and validation to ensure maintaining reliability and eliminating inconsistencies in the data have been recognized as a vital requirement in physics-based DTs [7].

DOI: 10.1201/9781003492146-6

- Scalability to model the entire system as well as its components. Scalability is particularly important in those systems or networks which are made up of various components or, in other words, systems of systems to provide cohesion between the systems (components) within a network to ensure achieving different levels of details from localized analyses to the entire system evaluations [8].
- Standardization and interoperability are essential for the implementation of a DT. This provides compatibility with different systems, domains, and common tools and software used in different industries. The compatibility facilitates the easy exchange and sharing of data across different platforms such as building information modeling, geographic information systems, and CAD [9].
- Data integration is a vital requirement for the successful implementation of a DT. The data integration ensures a synchronized representation of the system for effective and continuous monitoring. The real-time data collected from sensors, cameras, IoT, and other monitoring systems allows in-time response to evolving conditions and enhances the operational efficiency and decision-making processes [10,11].

DTs can be classified based on the integration of the PE [12], the fidelity of the target entity [13], or a state of lifecycle [1,14,15]. Focusing on the level of integration of the PE, Pronost et al. [12] classified DTs into four groups – pre-DT, digital model, digital shadow, and DT. The pre-DT represents a concept development part where there is no physical counterpart. The other three terms, namely, digital model, digital shadow, and digital twin, are based on communication levels between the physical and digital components using the definitions proposed in [16]:

- *Digital model* – in the digital model data from the physical part is used as a manual input to the digital model.
- *Digital shadow* – in the digital shadow system data is transferred automatically in one direction from the physical part to the digital part only.
- *Digital twin* – in the DT arrangement data is transferred automatically in both directions between the physical and digital parts.

The International Standard (ISO) on the concept and terminology of DTs [13] provides a classification of DTs based on the target entity:

- *Component DT*: A component DT can be developed for an influential element of a system that can affect the full system. This type of DT can be a complex element of a component such as a motor or pump in a piece of equipment.
- *Asset DT*: Asset DTs are defined as the collection of multiple component twins.
- *System DT*: A system DT includes interconnected or interdependent target and digital entities.
- *Process DT*: When the target entity of a system DT is a process rather than an individual component, it is termed a process DT.

The relationships between the four types of DTs presented in the ISO standard have been depicted graphically in Figure 6.1. The system DT can include component, asset, and process DTs. The system DT includes relationships between target entities which are not visible in component DTs. The system DT can also include the relationships or interdependencies of different assets and processes. Some of the applications have multiple systems that communicate among themselves such as multiple systems in a plant. In such cases, a further aggregation of the system DT would form a multi-system DT. While the system DT appears to cover aspects of all situations in an industrial plant, it does not imply whether it records all instances during the lifecycle of a product or the plant.

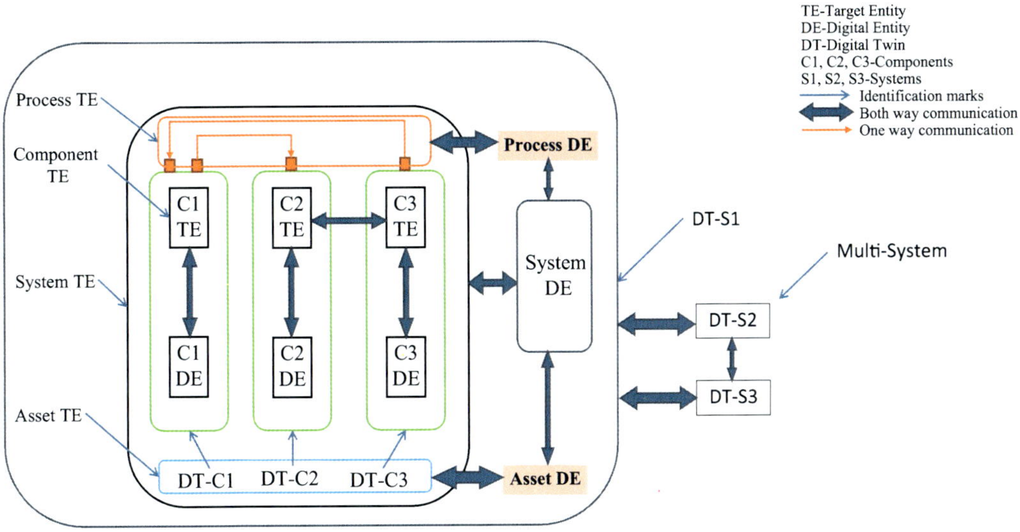

FIGURE 6.1 Digital twin classifications based on component, asset, process, system, and multi-system.

In the cases of a product, DTs can be developed for mainly four lifecycle phases, namely, the design phase, manufacturing phase, service phase, and retire phase [1,14]. Simulation models need to be developed for different levels of detail, disciplines, and lifecycle phases [15].

The presence of physics-based models in the DT system requires that the model be tested adequately before implementing the DT. Two terms are widely used in terms of model testing – verification and validation. In the context of simulation, verification refers to checking a model's accuracy in representing underlying mathematical equations and solutions [17]. Spiryagin et al. [18] added that verification should include checking the full workflow in a DT.

In the simulation context, validation refers to assessing the degree to which a model represents the corresponding PE. In a DT system, the validation term needs to be extended to include validation of all sub-models at the system level [18].

In the domain of traffic and transportation engineering, DTs are extensively used in the planning, design, simulation, validation, construction, monitoring, and operation stages [19–23]. However, the actual application is mostly at the prototype stage and also for the creation of DTs for individual assets [24]. There is currently no established standard for the development and use of DTs in transportation areas. Additionally, the requirement for a large amount of data and their collection, storage, processing, and analysis has made the use of DT challenging to some extent [25,26].

In transportation, physics-based DTs play a crucial role across those stages. They can be used to simulate and analyze different traffic scenarios, forecasting travel demands, creating 3D models of transportation infrastructure such as roads, bridges, intersections, and tunnels. Additionally, they can be used for simulating traffic flow under different conditions such as road closures and accidents. They can also be used for impact assessments to evaluate the effect of new infrastructures on traffic patterns and travel times. Moreover, validation of proposed designs, predicting the maintenance requirements, and real-time traffic management are some other applications of DTs in traffic and transportation engineering [27–31].

6.3 PHYSICS-BASED DT vs. PHYSICS-BASED MODEL

Physics-based models have long been used to simulate various situations of a system or component. To nominate a physics-based model as a DT, the following conditions need to be met:

1. The model needs to be updated using measured data from the PE over its lifecycle
2. Availability of data from the PE over time
3. Presence of a corresponding PE all the time. In the case of the design stage, the PE would be a prototype that can provide information to the digital entity (DE) for regular updating. However, it is inconvenient to use a design stage DT for the remaining lifecycle as the prototype is unable to replicate the behavior over the lifecycle. Hence, for a bigger system, the requirement of a PE can be replaced by the availability of supplier data that can be digitally coupled with the model.

The model for DT purposes differs from models for other purposes such as safety verification or performance evaluation where high accuracy is required and is run less frequently. Wright et al. [32] provided the below characteristics of the physics-based model to be used in a DT:

1. The model is capable of replicating physics based on updated measured data.
2. The outputs generated by the model are accurate enough for applications of interest.
3. The model can produce outputs based on the need for a DT within a required timeframe.

6.4 DEVELOPMENT CONSIDERATIONS

The development of a DT needs to take into account several aspects, including modeling, the degree of twinning, the design level of components, application fields, and the level of communication. Aivaliotis et al. [33] proposed that an advanced physics-based model should be developed in three phases:

- *Phase 1*: Modeling of the dynamic behavior of the physical system
- *Phase 2*: Modeling of virtual sensors
- *Phase 3*: Definition of modeling parameters

In the first phase, machine components that represent the crucial functionality of the physical system need to be identified to be digitally modeled. The digital model of each component of a system needs to represent the main physical phenomena such as static deformations, vibrations, etc. At this phase, the relationship among all components is also defined using rules and restrictions. The modeling information should include the level of information available for each component, namely, black box (without any knowledge of internal operation), grey box (assumed theoretical calculation), and white box (fully described parameter sets).

In the second phase, virtual sensors are modeled to estimate the data that are not available directly from the physical system. Virtual sensors provide calculated results based on the mathematical equations that reflect the system's behavior.

In the third phase, the DT components that can be reliably twinned between the PE and DE are selected based on the influence of a component in a system and its modeling information levels (white, grey, black). At this stage, a parameter set is identified which can be tuned between the PE and DE. To implement the change based on the tuning, the PE would be fitted with external sensors and a controller which will collect rule-based data from the digital sensors and implement tuning to the PE as required. The communication between the physical and digital systems requires a twining stage where component-wise rules and analysis methods are implemented to decide on any update of parameters on either side. In a complex machine, the amount of synchronization can be significant and interdependent with other components.

The method of twining and implementation can be different based on the requirement such as optimizing the operation and performance of the PE based on data from the digital model [34].

A five-dimensional architecture – a physical entity, virtual model, connection, DT data, and services – has been proposed in [35]. The PE includes all equipment in a system. The virtual model includes the geometry model, physics model, behavioral model, and rule model. The service model includes services for PE and DE such as function, input, output, quality, and state of services. The DT data model includes data from the PE, DE, services, and knowledge bases, as well as the outcome of the fusion of all data within a DT. The connection model includes interconnection among the PE, DE, data, and services.

In industrial applications, four phases of application of the DT at design, manufacturing, service, and retire phases have been identified in [14]. In the design phase, the application of the DT includes iterative optimization, evaluation of data integrity, virtual evaluation, and verification. The manufacturing phase of the DT includes real-time monitoring, production control, asset management, production planning, workplace performance predictions, human–robot collaboration and interaction, process evaluation, and optimization. The service phase application of the DT includes predictive maintenance, fault detection and diagnosis, monitoring of states, virtual tests, and performance prediction. The retire phase is intended to store the whole lifecycle information of a PE in a virtual space to help find solutions for a similar product in the future from the historical database.

An eight-step DT development framework is proposed in [34] as follows: (1) define requirements; (2) develop information flow from data collection to model interfaces; (3) build the virtual representation based on the requirements of stakeholders; (4) develop and select the data exchange system between the DE and PE; (5) carry out analysis based on defined requirements; (6) develop actions based on decision-making rules; (7) verify that the DE represents the PE and that the communication between the PE and DE meets specified requirements and that the DT is capable of delivering the required outcomes for decision-making; an (8) validate that the DT serves its purpose based on a required level of acceptance by the stakeholders.

The availability of a DE for all phases of the lifecycle of a product allows iterative testing of a product for various situations in the lifecycle, which allows optimization of the product at the early design stage [36]. In the cases of asset degradation, it is often required that the DE can adapt to progressive changes in the PE and can represent updated results based on those changes in the PE. An example of a model updating strategy based on parameter sensitivity analysis is available in [37].

Stark et al. [38] listed eight categories or dimensions of a DT based on context and environment (integration breadth, connectivity modes, update frequency, product lifecycle), and behavior and capability richness (cyber-physical system intelligence, simulation capabilities, digital model richness, human interaction). Based on particular needs, DTs have different influence levels on tools or technologies which can dictate requirements for specific DTs.

Donato et al. [39] identified 12 different tasks and 33 steps on a framework for the design and development of DTs focusing on data-driven predictive maintenance. In the area of predictive maintenance, data from many physics-based models can be used to provide predictive assessment advice. In such a case the physics-based model approach can be incorporated with the data-driven DTs. Tables 6.1 and 6.2 include the task list for physics-based DT systems in comparison with the data-driven DT systems at the design and development stages included in [39]. The main difference between the data-driven and physics-based DT is the presence of a high-fidelity multibody simulation (MBS) model in the physics-based DT. While the physics-based models are deterministic, the data-driven models rely on big data sets to accurately model the interrelations among most of the operation parameters.

In a simulation study, Ahmad et al. [40] used three methods, namely advanced simulation, digital shadow, and DT approaches to calculate rail surface damage. The locomotive model used in [40] includes a multibody model of the locomotive and a traction model. The communication between the MBS model and the traction model was performed based on a shared library technique [41]. The advanced simulation approach replicates the conventional physics-based model only where all data were collected and used as input to a high-fidelity multibody model. In the case of the absence of measured data from a PE, synthetic data obtained from longitudinal train simulation (LTS) software was used as input to the MBS model.

TABLE 6.1
Tasks Required at the Design Stage of a DT

		Scope of Tasks	
Tasks No	**Description**	**Data-Driven Digital Twin [39]**	**Physics-Based Digital Twin**
1	Requirement specification	Identify industry, purpose of DT, process or asset, PT development or realization, type of DT use, specify needed technologies, define input, and output parameters	Specify the need for the DT
2	DT process planning	Identify functionalities, define required PE data, and design communication infrastructure	Same as data-driven DT
3	DT architectural design	Combination of layered and service-oriented patterns of connectivity	Same as data-driven DT
4	Digital representation	What to represent: behavior or structure	Mathematical model of the physics-based property of the system
5	Digital twinning	Generation, tuning, and validation of PT models	Tune and validate the DT model for the required purpose
6	Service planning	Supervised or unsupervised learning algorithms	Choose or design the required algorithm for specific analysis for the chosen DT

TABLE 6.2
Tasks Required at Development and Deployment Stages of a DT

		Scope of Tasks	
Tasks No	**Description**	**Data-Driven Digital Twin [39]**	**Physics-Based Digital Twin**
7	Data management	Data collection and storage	Data collection and storage
8	Conversion of raw data	Process, clean, fuse, and reduce data from multiple sources	Conversion of raw data to reduce inconsistency and uncertainty
9	Training process/ model selection	Predictive model training and deployment	Manual or automated selection of models for the chosen service
10	Prediction process	Based on AI-generated outcome	Based on developed algorithms and historical data from the PE and DE
11	Decision process	Implement decision-making tool	Generate and communicate information obtained from DE and associated decision-making algorithms
12	Evaluation	Evaluation by examples	Select an evaluation method suitable for the chosen DT

In the other two methods of digital shadow and digital twin, the LTS and MBS were synchronized using a transmission control protocol/internet protocol interface [42]. A co-simulation method was implemented to synchronize the speed, and the positions of vehicles and realized traction effort in LTS and MBS. The co-simulation allows data transfer in both ways between the LTS and MBS (Figure 6.2). It can be noted that often the co-simulation system requires small communication intervals for numerical stability, which decreases the simulation speed [43].

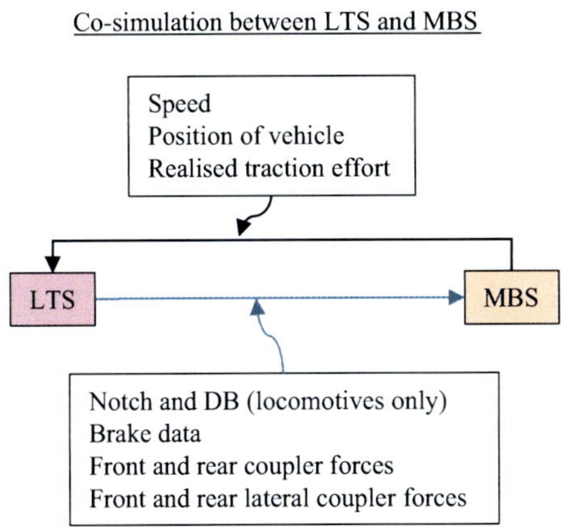

FIGURE 6.2 Co-simulation between longitudinal train simulation and multibody simulation.

Noting that the communication between two software products can be computationally expensive, the one-way communication from LTS to MBS was implemented in the digital shadow method [40] where only the coupler force components, position, and speed of the vehicles from the LTS were transferred to the MBS. Of the three methods, the DT model approach implemented the data communication between both the LTS and MBS and is expected to provide more accurate results compared to the other two methods. However, communication between two different software environments was found computationally expensive. The digital shadow and digital twin approach consumed 5% and 20% more time compared to the advanced simulation method. In all three cases, the computation time was much higher (more than 13.8 hours for a 1.7 km track section) than the real-time physical operation. It therefore follows that the availability of results data from the DE would take more time than the actual operation time. Spiryagin et al. [44] proposed a fast-computing model for real-time application by modifying the friction modeling approach in the wheel–rail contact that achieved a computation time of 0.04 seconds for 0.1 seconds of operation time. The computation time achieved using a faster friction model was found to provide a negligible variation of 3 kN in tractive effort compared to a more precise friction model using the modified Fastsim approach. The computation time for the modified Fastsim approach was found up to 0.22 seconds for 0.1 seconds of simulation time which is about 5.5 times slower than the real-time modeling approach. The computation time is however dependent on the specific purpose of using the DTs. For non-critical systems such as maintenance predictions, the output can be post-processed and analyzed to suit the need and delivered to the PE for any change. On the contrary, for real-time critical systems such as control systems, it is necessary to reduce computation time.

Spiryagin et al. [18] identified five main areas where vehicle dynamic studies can be utilized in DT studies, namely design, safety analysis, maintenance, monitoring, and optimization. The development consideration of the DT would be dependent on the intended areas of application.

The design aspect of the application would require a high-fidelity assessment of the dynamic behavior of the vehicle and track interaction. The simulation aspect of the design phase is well known, and the V-model system approach is a widely acceptable norm to improve design and adaptive validation approaches. Spiryagin et al. [45] proposed an additional stage to convert the simulation approach in design to the DT approach in the locomotive design process (Figure 6.3).

The main difference between the simulation approach and the DT approach lies in the spread of the application. While the simulation approach tries to solve a specific model for a specific set of information, the DT can combine a few environments that influence the object. In the case of locomotive design, in addition to the mechanical and mechatronics components of the locomotive, the track, train, and operation conditions need to be incorporated into the design aspect. The integration of different types of models and multiple components requires advanced technological

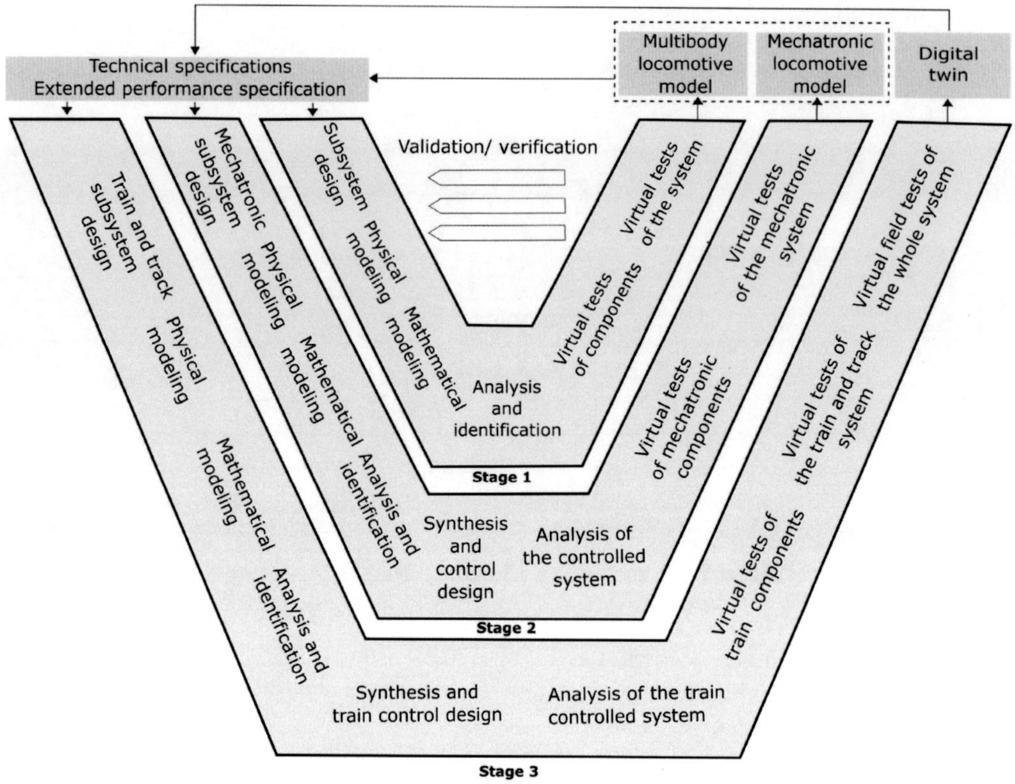

FIGURE 6.3 V-model system integration in a locomotive product design using digital twin approach [45]. (Adapted from Spiryagin, M., Q. Wu, O. Polach, et al., Problems, assumptions and solutions in locomotive design, traction and operational studies. *Railway Engineering Science*, 2022. **30**: pp. 265–288, permissions granted under the creative commons license: https://creativecommons.org/licenses/by/4.0/.)

tools such as co-simulation within different software packages such as longitudinal train dynamics, multibody models of mathematical models of vehicles, detail track models, mechatronic modeling, etc., [41,42,46,47] and parallel computing in a high-performance computing (HPC) system [48]. The V-model system approach in a DT also requires that all components be tested individually as well as a full system. The results from the digital test need to be communicated with the input data at the design stage to fulfill the requirement of both-way communication of the DT.

The concept of applying safety analysis such as factors affecting the safety of train operation can be incorporated at the design stage as discussed in [49]. The safety aspect requires identification of the system, objectives of the system, identification of hazards, control of hazards, and monitoring of the system. In terms of the safety of train operation, the purpose of the DT would be to predict accidents and act automatically to reduce the chance of accidents or control hazards.

The development of a DT should also consider the field of application. Spiryagin et al. [18] identified three maintenance problems in the railway field that can be studied using DT tools, namely the wheel–rail interface, track components, and rollingstock components. The wheel–rail interface study involves advanced simulation techniques to apply theories of wheel and rail contact mechanics and contact damage models that can predict rail damage with respect to tonnage passes and operation conditions [50–53]. Such tools can advise rail grinding intervals [51,54] and modification in operation conditions to reduce maintenance needs [50] or evaluate wheelset selection based on material properties and shakedown map analysis [52].

In the case of monitoring applications, it is necessary to collect live data from the physical system. An example of the application of DT in maintenance for railway switches and crossings is added in [55]. Bustos et al. [56] demonstrated the application of a DT to monitor a bogie system and provide an advance warning on upgrades if necessary by comparing the natural and characteristic frequencies of the system obtained in a digital model and physical measurement of vibration from the accelerometers installed on bogies. The main components of the monitoring DT proposed in [56] included a 3D model of the bogie, 'numerical (FEM) models of key parts of the bogie, a multi-body model and computation method of theoretical characteristic frequencies'.

Two of the most appealing fields in the application of DTs in optimization studies are energy analysis of train operational scenarios and train-driving strategies [18,41,57]. Meng et al. [57] used a five-dimensional DT architecture (PE, DE, Connection, Service model, and database) to formulate a DT structure to provide advice on train operation based on comfort level evaluation.

Based on the discussion in this section, an eight-step framework for physics-based models in DT applications is proposed in Figure 6.4.

The first step includes defining the purpose of the DT, types of data, constraints on data availability, specific needs for stakeholders and areas of application, and qualitative or quantitative expectations or acceptance criteria of DT outcomes. The purpose of the DT can be detection, judgment, and prediction [35]. Depending on the purpose of the DT, the relevant development steps will be different. In the next step, it is necessary to identify the all technological tools needed for the DT system which includes the selection of data collection processes, preprocessing of data to make it suitable for digital application in the specific software or toolbox, data handling and storage of the PE and DE, communication of data between the PE, DE and cloud servers, and all associated network issues. At step 3, a physics-based model can be developed based on the need specified at step 1. In step 4, the result from the DE is required to be post-processed to allow the presentation of data in a suitable format for stakeholders. The decision-maker tool in the system allows the selection of actions based on the results of the physics-based model, data from the PE, and the requirements of the DT. At step 5 of the DT development framework, it is proposed that the decision-maker tool would integrate the results of the model, input data, and relevant inconsistency or uncertainty in data and methods. Step 6 involves verification of technology selection, modeling methods, post-processed results, and decision-making algorithm. Once verification is achieved to a

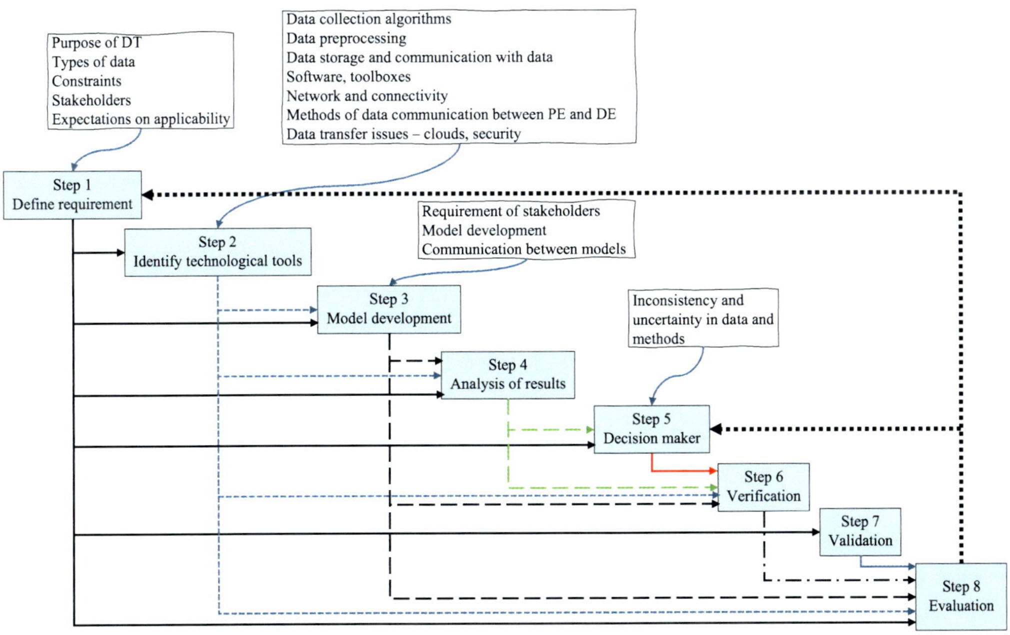

FIGURE 6.4 Steps of DT development framework.

specified standard, at step 7 an overall validation of the DT needs to be performed to check if the outcome matches the defined requirement in step 1. At the final step 8, a suitable evaluation technique needs to be chosen to perform an assessment of the DT system based on the applicability and maturity of the method and tools. The evaluated outcome can be used to redefine the requirements and decision-making characteristics if required.

6.5 DIGITAL TWIN DESIGN METHODOLOGY

A methodology for the design of a DT has been proposed in this chapter (see Figure 6.5). In any physics-based model, the measurement of data and preprocessing of data to make it suitable for the model is the first step.

The data collected from sensors or surveys need to be processed along with relevant information on trustworthiness, provenance, and traceability. It is noted that, in a complex dynamic application like a railway system, all data of the PE such as the wheel–rail friction along the whole rail track is unavailable. In such cases, assumptions are made to develop the model based on physics-based laws or a historical data-driven physics-based approach [18,58]. It is expected that an assessment of the assumptions is included in the analysis and decision-making stages. A list of uncertainties listed in [17] is added below:

- Errors in measurement process due to sensor characteristics and data processing methods
- Errors in modeling due to the selection of mathematical models, numerical integrators
- Scattering and variability in data due to manufacturing and operational variations

Since simulation is a deterministic approach, the handling of uncertainty by simulation would need to be assessed. Uncertainty can be handled both by quantification such as probability theories, sensitivity analysis, and general toolbox methods [59–61] and by qualitative analysis [17]. In addition to measuring uncertainty, it is beneficial to estimate the correlation between various uncertainties [32].

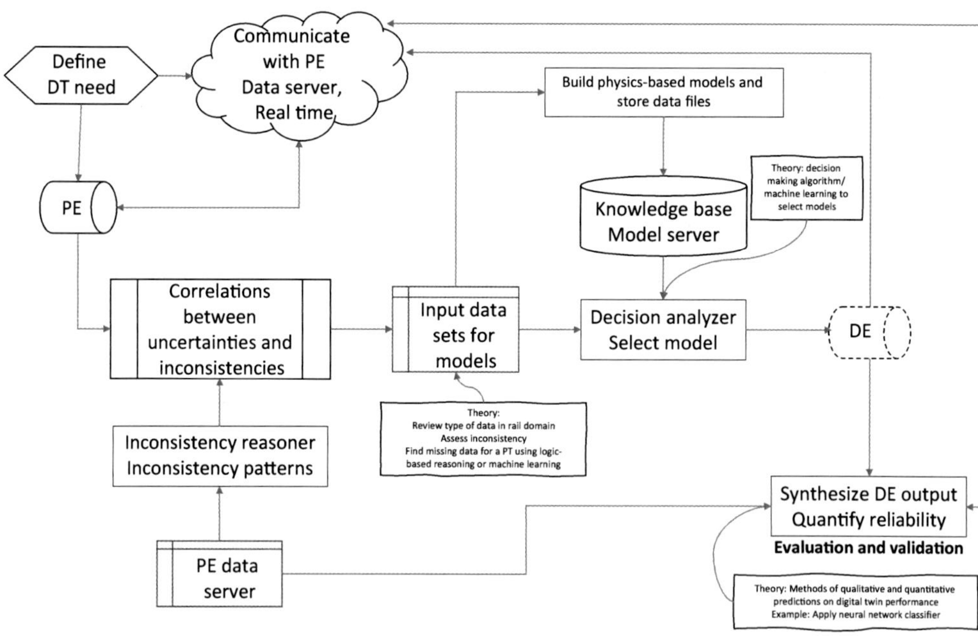

FIGURE 6.5 Digital twin design architecture for a physics-based model.

Depending on the need, a qualitative ranking of handling the uncertainty is defined in [17]. At level one, a safety factor is used to handle the uncertainty in data and test methods. At level two, the variability is characterized and modeled within a parameter. Level two provides a reliability factor, which is an assessment of variability. Level three is assigned to situations when many parameters that have a significant impact on outcomes are tested and the variability of the overall response of the system is measured. Level four constitutes the highest ranking where all significant uncertainties are identified and rectified. At level four, the uncertainties between physical tests and virtual tests are assessed and an adaptation of acceptance criteria is proposed if there is a significant difference between uncertainties obtained by physical and virtual tests.

For a physics-based model, it is necessary to identify correlations among inputs and provide a robust input data processing methodology for the enhanced performance, reliability, and accuracy of DT systems. Machine learning and other data-based algorithms can be used to integrate advanced data science in the DT system to provide optimized data storage and use data-based classification systems and deep learning techniques to quantify the performance of a DT.

The model development includes verification and validation of the model components and behavior with the respective PE. In a developed DT system for a railway, it is expected that the model would be developed for multiple input conditions. Once the information on the requirement of the DT and relevant input data is available, a decision analyzer to select the appropriate model for the chosen DT objective is proposed. It is proposed that decision-making algorithms can be logic-based or machine learning-based. Once the model is selected and the DE is developed for the chosen DT purpose, it is necessary to carry out qualitative and quantitative evaluations of DT performance.

The results from the DE along with the evaluation of the respective DT need to be stored in a data server, where tools on decision-maker algorithms are implemented to communicate with the PE and implement changes based on decision-maker algorithms.

A design of a DT system for a railway operation and maintenance application is added in Figure 6.5. Railway operation and maintenance systems would require data on physical entities in a system such as track infrastructure, train configuration, structural and mechatronic data of locomotives, and suspension parameters of vehicles in a train. The operation data includes the train-driving strategy or actual train-driving log data. Maintenance data such as historical records of rail grinding, tonnage data, and corresponding wheel–rail profiles can be used as inputs to the DT system. In the first step, it is necessary to collect and store data on the PE and process them as required. As measurement of the PE is performed using various sensors, it is necessary to pre-process the data to make it suitable for the chosen digital technology. In the case of vehicle dynamic simulations, the chosen technologies are LTS, MBS, co-simulation interface, Simulink, etc.

Due to utilizing high-fidelity models, the MBS of all vehicles in a train cannot be run on a single CPU. The availability of HPC systems now allows parallel simulations of all vehicles in a train, which reduces total simulation time significantly [48]. However, the communication between two software packages can be significant, as discussed in Section 6.2. Hence, the users need to decide which type of performance and computation time is reasonable. In Figure 6.6, a few levels of DT have been assigned based on the data transfer approach between software and model complexity. Of the DT levels, four levels have been demonstrated in published papers [40,44] and the remaining ones are expected to evolve based on requirements.

All results from the DE would then be post-processed to make them suitable for application in the prediction and decision-making algorithms. The predictive assessment and decision-making algorithms take the post-processed data to perform relevant analysis and provide feedback to the stakeholders; examples are available in [50,54]. Following the ISO/IEC 30173:2023 Standard [13], the stakeholders are separated into two groups: DT system stakeholders (developers, resource providers, integrators, users, and operators) and ecosystem partners (infrastructure provider, service provider, standards development organization, government, and community). The DT system stakeholders decide mostly on the validation of the DT and, based on the advice from the predictive assessment, decide on adjustment to the DE by changing some parameters such as wheel–rail profile

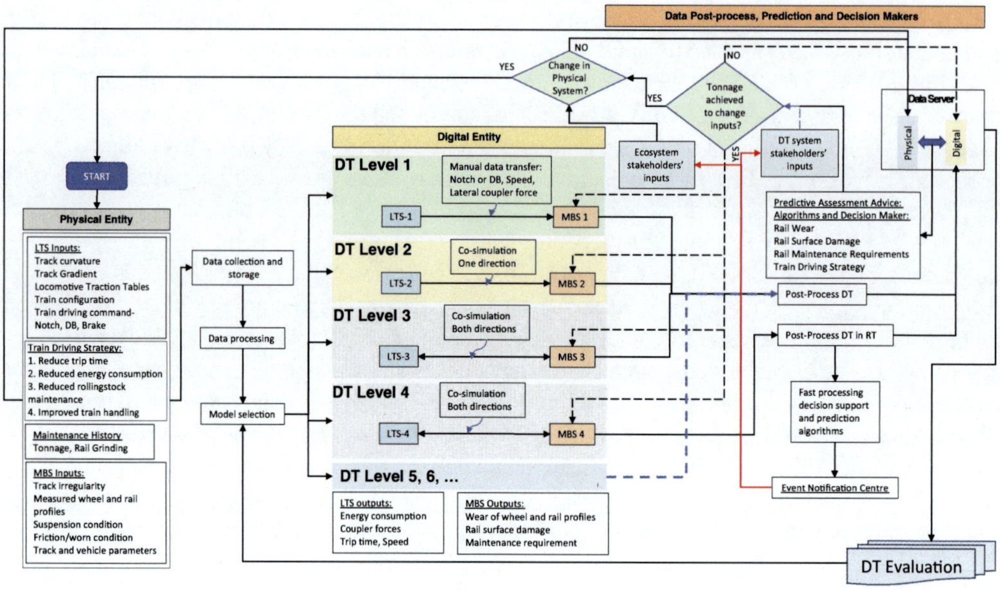

FIGURE 6.6 Design of DT system for a railway operation application. (Modified from Ahmad, S., M. Spiryagin, Q. Wu, et al., Development of a Digital Twin for prediction of rail surface damage in heavy haul railway operations. *Vehicle System Dynamics*, 2024. **62**(1): pp. 41–66, permissions granted under the Creative Commons license: https://creativecommons.org/licenses/by/4.0/.)

and train-driving strategy to make it suitable for the most recent mode of operation. Change in the physical system based on prediction and decision-making algorithms would require the ecosystem stakeholder's input.

Summaries at the end of this section show how physics-based DTs can be updated over time.

6.6 EVALUATION AND VALIDATION APPROACHES

A DE would represent a PE up to a degree that the stakeholders need to be aware of for implementing a change if required. The term 'Evaluation' refers to an assessment of the DT that allows the stakeholders to make informed decisions.

Liu et al. [53] enlisted 21 references that include various strategies for the evaluation of a DT, which cover six features:

1. Number of aspects or dimensions of the DT
2. Number of rubrics or more detailed sub-criteria
3. Status of qualitative checks
4. Status of quantitative checks
5. Weightage on the importance of different aspects and criteria
6. Overall (all aspects together) or partial (separate evaluation for each aspect) aspects of DT.

Liu et al. [53] identified ten aspects of a DT for evaluation purposes based on a literature search method:

- *Function*: Assess capabilities and services provided by DTs
- *Reliability*: Measures the accuracy and dependability of DTs, related to maturity level
- *Temporal scope*: Temporary state/s within the lifecycle of a DT
- *Modeling scope*: Measures the applicability of modeling at various levels

- *Coverage scope*: Measures the comprehensiveness of DT while following its physical counterpart
- *Fidelity*: Measures the accuracy of the presentation of the DE with respect to its PE
- *Scalability*: Measures the ability of a model to adapt to the new addition in the model
- *Autonomy*: Measures the ability to operate independently in a system
- *Update frequency*: Assess the need for a DT to update the data based on the use case
- *Data*: Assess the workability with the variability in the data structure

DT evaluation depends on a range of factors including technological sophistication, scale, and precision of the model [53]. Four aspects of assessment are used in [53] to evaluate and provide a quantified approach to calculate the maturity of the DT, namely, Capability (Cap), Cooperability (Cor), Comprehensiveness (Com), and Lifecycle (Lc).

The 'Cap' dimension indicates the DT's capability in performing data collection, real-time visualization, prediction, application, broader impact, interoperability and accuracy in simulation, prediction, and decision-making process. Four capability levels (Cap) were identified:

a. *Cap1*: when only real-time data processing and control is applied.
b. *Cap2*: DT can provide analysis based on historical states.
c. *Cap3*: DT can predict future states based on technologies like ML.
d. *Cap4*: DT can provide reasoning behind the decision it made.

The 'Cor' dimension provides insight into the DT's ability to cooperate with other DT networks reflecting the modeling scope and scalability of the DT.

a. *Cor1*: DT represents a single product.
b. *Cor2*: DT represents one to multiple objects of a certain level of similarity in the same environment.
c. *Cor3*: DT represents one to multiple objects of a certain level of similarity in different environments.

The 'Com' dimension refers to how thoroughly the DE represents its PE. It can be divided into three levels:

a. *Com1*: the DT model represents one single isolated aspect of an object.
b. *Com2*: the DT models represent multiple aspects of an object.
c. *Com3*: DT can represent a system of systems.

The 'Lc' dimension assesses the DT's ability over different phases of lifecycle. Three lifecycle levels were selected for evaluation:

a. *Lc1*: DT is applicable for one single phase throughout different phases in the lifecycle.
b. *Lc2*: DT can incorporate more than one phase (such as design and operation) in the lifecycle.
c. *Lc3*: DT operates over the entire lifecycle of the product.

A scoring system on the maturity of different evaluation aspects based on normalized maturity score and adjustment of the score to incorporate the stakeholders' view on the significance of a particular aspect has been proposed in [53]. Based on the scoring system, the maturity score of a DT's use can be evaluated by using Equations (6.2)–(6.4).

$$m_i = \frac{n_i}{N_i} \tag{6.2}$$

$$W_i = \frac{w_i}{\sum w_i} \tag{6.3}$$

$$L_{DT} = \sum m_i W_i \tag{6.4}$$

where m_i is the normalized maturity score of the DT; n_i is the maturity level in dimension I; N_i is the total number of levels in dimension I; W_i is the normalized weight of dimension I; w_i is the weight score of dimension i chosen by stakeholders; i is the chosen dimension for the evaluation of the DT; and L_{DT} is the overall maturity score of the DT use case.

A qualitative ranking-based approach is added to a project by Garner [17] on introducing simulations in the certification process of rolling stock.

The principles of DTs in civil engineering were established by the *Gemini Principles* and were formulated by the National Infrastructure Commission and the UK Government's Digital Framework Task Group, ensuring that the DTs are effective, reliable, and beneficial for the public and decision-makers [62]. Arup [63] has presented four metrics of autonomy (ability of the system to act without human input), intelligence (ability of the DT to replicate human cognitive processes and perform tasks), learning (ability to learn from data without being programmed), and fidelity (the level of details of a system) for evaluating the current state of DTs based on five levels as presented in Table 6.3.

TABLE 6.3
Levels of DT in the Civil and Construction Industry

	Metrics				
Level	**Autonomy**	**Intelligence**	**Learning**	**Fidelity**	**Example**
Level 1 – virtual-real connection but with limited functionality	No – user controlled	No intelligence	No learning component	Low accuracy	Basic model or map
Level 2 – capable of some feedback and control	Limited – user assisted	Cannot use previously gained experience	Programmed	Low to medium accuracy	House temperature sensors
Level 3 – allows for predictive maintenance and analytics	Partial – DT controls in certain ways	Capable of learning from historical data	Trained using a supervised learning approach for example labels	Medium accuracy	Predicting the life expectancy of railway
Level 4 – able to learn about the environment and has autonomous decision-making for a particular task	High – DT is able to perform the tasks and monitor the conditions with no/limited human assistance	Understand the needs of other intelligent systems	Trained using an unsupervised learning approach	Can provide precise measurement	Real-time route recommendation
Level 5 – capable of reacting to unpredicted scenarios	Full –DT can operate safely with no human intervention	Human-like intelligence	Uses reinforcement learning from previous experiences	High accuracy and can be used in critical operational decisions	Model of a neighborhood in a smart city

Source: Adapted from An, A., S. Blakeney, R. Bowden, J. Burgess, et al., Digital twin TOWARDS A MEANINGFUL FRAMEWORK. 2019, ARUP.

Specific examples of Level 4 and Level 5 in civil engineering are not yet widespread; however, some projects, including smart infrastructure projects, smart bridges and tunnels, and advanced building management systems, are pushing their boundaries toward these levels [19–23,63–65].

Validation refers to checking the applicability of the DT for the intended purpose. For evaluation purposes it is often necessary to measure parameters. Ariansyah et al. [34] identified five steps of the validation process:

1. Define hypotheses or expected improvements when adopting a DT.
2. Identify metrics to measure performance differences between DT and non-DT systems.
3. Use the exact condition for comparison purposes.
4. Selection of appropriate statistical method to analyze results of empirical data.
5. Statistical evaluation of the relative improvements in a DT compared to a non-DT system.

A term called digital twin certified (DTC) has been introduced in [66] to suggest the application of a DT throughout the lifecycle of a product to demonstrate that both cost and risk are minimized while improving the quality of a product. It is also promised that, in a DTC, the future experience of a customer can be predicted by testing products passively with data and actively with simulation to destruction.

6.7 INTEGRATING PHYSICS-BASED DT IN THE HYBRID DIGITAL TWIN SOLUTIONS

A hybrid DT system works on a closed-loop system between physics-based and data-driven models.

Physics-based DTs face challenges of accurately presenting the complex physics in digital models corresponding to a physical system which resulted from the approximation of complex behavior of a physical system. Data-driven models require a huge amount of data, which is difficult to curate or unavailable for some systems. To overcome the shortcomings of data handling, a smart-data paradigm is gaining interest that uses model reduction and sampling methods. In a hybrid model, physics-based models evaluate the need for data, while data-driven models identify the ignorance in the model and help improve the physics model.

The approximation in the physics-based DT contains both biased and unbiased deviations between data of a physical and a digital system [67]. The biased deviation refers to the lack of knowledge on combinations of a multi-physics system that has interconnections within a system. One way to deal with biased uncertainties is to correct the model using measured data. The unbiased deviation comes from modeling or measurement noise that can be managed by using appropriate filtering approaches. Furthermore, additional control actions can be applied on a physics-based DT to improve model solutions to the physical system. Chinesta et al. [67] proposed a governing Equation (6.5) to represent a hybrid twin:

$$\dot{X}(t;\ \mu) = A(X,\ t; \mu) + B(X,t) + C(t) + R(t) \tag{6.5}$$

where $\dot{X}(t;\ \mu)$ is the rate of change of the system with time t subject to parameter μ; $A(X,\ t;\mu)$ is the rate of change contribution along time (t) subject to parameter μ from the model; $B(X,t)$ is the output of a data-based model that dictates the difference between prediction and measurement; $C(t)$ is the external actions that can be applied in the system dynamics to improve the model predictions; $R(t)$ is the unbiased noise that comes from measurement or models.

6.8 AGGREGATION OF PHYSICS-BASED DTS AND THE EXISTING BARRIERS

In a big system, a DT can be developed for a component or specific operation, which are interconnected with other components and operations in a system. As an example, the rail wear prediction DT primarily aims to relate the rail wear information with the gross axle load tonnage passing over the rail. However, various combinations of vehicle type, track profile, wheel and rail profiles, variations in vehicle dynamic conditions, and operation conditions can affect the wear on the rail. A physics-based DT is developed for a vehicle and track condition. To better predict rail wear, it is necessary to aggregate multiple DTs representative of the chosen track, vehicle, and operation conditions. It is also possible to separate the most time-consuming component in a model and produce a lookup library based on extensive simulation cases on the components. It therefore follows that multiple DTs can be developed in a system which can be aggregated by using data links.

One of the challenges in implementing the physics-based DT is its computation time. In addition to that, the physics-based models of an element or system take time to reach a reliable state considering variances, uncertainties, and errors in a model. Hence it is not uncommon to use a generic model of a product for wider application. The generic model needs to be calibrated with measured data from a PE whenever available.

Despite taking all actions at the time to make the physics-based model and DT accurate, the DE can never be a complete representation of the PE. In a holistic approach, it is expected that the DT would accurately represent the current situation of PE. However, to allow that to happen it is necessary to take account of changes in information at various stages of design, development, manufacturing, commissioning, and operation of the asset. It is also noted that the availability of data and behavior information can be exhaustive as one can develop more detailed models and sensors to extract an exhaustive combination of DT instances. It follows that an exact definition of a physics-based DT may not be specified; rather it can be taken as a paradigm. The developer needs to define the restrictions and applicability of a physics-based DT.

Two key barriers to implementing a DT system have been identified: the cost of development of a DT for a specific purpose and the perception of stakeholders on the purpose/s of the DT [68]. As a DT can be specific to a requirement, stakeholders must have the vision, expertise, and knowledge of the systems engineering approach to provide necessary inputs to a DT development.

REFERENCES

1. Grieves, M., Digital Twin: Manufacturing Excellence through Virtual Factory Replication, in Digital Twin White Paper. 2014. p. 1–7.
2. Ganguli, R., S. Adhikari, S. Chakraborty, and M. Ganguli, *Digital Twin A Dynamic System and Computing Perspective*. 2023: CRC Press.
3. Smarslok, B., A. Culler and S. Mahadevan. Error quantification and confidence assessment of aerothermal model predictions for hypersonic aircraft. in *53rd AIAA/ASME/ASCE/AHS/ASC Structures, Structural Dynamics and Materials Conference 20th AIAA/ASME/AHS Adaptive Structures Conference 14th AIAA*. 2012. Honolulu, Hawai.
4. Redelinghuys, A.J.H., K. Kruger and A. Basson. A Six-Layer Architecture for Digital Twins with Aggregation. in *Service Oriented, Holonic and Multi-agent Manufacturing Systems for Industry of the Future*. SOHOMA 2019. Cham, Springer International Publishing. 2020.
5. Lutters, E. and R. Damgrave, The development of pilot production environments based on digital twins and virtual dashboards. *Procedia CIRP*, 2019. **84**: pp. 94–99.
6. Human, C., A. Basson, and K. Kruger, A design framework for a system of digital twins and services. *Computers in Industry*, 2023. **144**: p. 103796.
7. Meža, S., A.M. Pranjić, R. Vezočnik, et al., Digital twins and road construction using secondary raw materials. *Journal of Advanced Transportation*, **2021**. **2021**(1): p. 8833058.
8. O'donovan, P., K. Leahy, K. Bruton, et al., Big data in manufacturing: A systematic mapping study. *Journal of Big Data*, 2015. **2**: pp. 1–22.

9. Minerva, R., G.M. Lee, and N. Crespi, Digital twin in the IoT context: A survey on technical features, scenarios, and architectural models. *Proceedings of the IEEE*, 2020. **108**(10): pp. 1785–1824.
10. Kuhn, T., F. Schnicke and P.O. Antonino. Service-based architectures in production systems: Challenges, solutions & experiences. in *2020 ITU Kaleidoscope: Industry-Driven Digital Transformation (ITU K)*. 2020. Ha Noi, Vietnam: IEEE.
11. Jones, D., C. Snider, A. Nassehi, et al., Characterising the Digital Twin: A systematic literature review. *CIRP Journal of Manufacturing Science and Technology*, 2020. **29**: pp. 36–52.
12. Pronost, G., F. Mayer, B. Marche, et al., Towards a framework for the classification of digital twins and their applications, in *2021 IEEE International Conference on Engineering, Technology and Innovation (ICE/ITMC)*, Cardiff, United Kingdom, 2021.
13. International Standard. Digital twin-Concepts and terminology. BS 1S0/IEC 30173:2023. ISBN 978-2-8322-7680-8. 2023, International Standard Organisation, International Electrotechnical Commission.
14. Liu, M., S. Fang, H. Dong, et al., Review of digital twin about concepts, technologies, and industrial applications. *Journal of Manufacturing Systems*, 2021. **58**: pp. 346–361.
15. Boschert, S. and R. Rosen, Digital twin—The simulation aspect, in *Mechatronic Futures: Challenges and Solutions for Mechatronic Systems and Their Designers*, P. Hehenberger and D. Bradley, Editors, 2016. Springer International Publishing, Cham, pp. 59–74.
16. Kritzinger, W., M. Karner, G. Traar, et al., Digital Twin in manufacturing: A categorical literature review and classification. *IFAC-PapersOnLine*, 2018. **51**(11): pp. 1016–1022.
17. Garner, P., D. Jakobi, M. Juris, J. Murua, and M.-L. Nguyen-Tajan, D5.2 Virtual Certification: Recommendation for Introducing Simulations in the Certification Process of Rolling Stock. Horizon 2020 - the Framework Programme for Research and Innovation (2014-2020)/ PLASA-2, Deutsche Bahn AG, 10.3030/826151, 2021 European Commission.
18. Spiryagin, M., J. Edelmann, F. Klinger, et al., Vehicle system dynamics in digital twin studies in rail and road domains. *Vehicle System Dynamics*, 2023. **61**: pp. 1737–1786.
19. Batty, M., Digital twins. *Environment and Planning B:* Urban Analytics and City Science, 2018. **45**(5): pp. 817–820.
20. Ma, H., C. Qian, L. Li, et al., Progressive virtual risk-based vehicle trajectory optimization in mixed traffic flow. *Transportation Research Part C: Emerging Technologies*, 2024. **165**: p. 104701.
21. Li, Z., S. Li, A. Abdelraouf, et al., Digital twin-based cooperative driving at roundabouts for connected and automated vehicles, in *2024 Forum for Innovative Sustainable Transportation Systems (FISTS)*, Riverside, CA, 2024. IEEE.
22. Ran, B., Y. Zheng, K. Luo, et al., Architecture design of a vehicle–road-cloud collaborative automated driving system. *Urban Lifeline*, 2023. **1**(1): p. 9.
23. Mardan, A. and S. Zhu, Using Crowdsourced Data to Inform Transportation Planning: A Case Study on Safety of Cyclists and Infrastructure Needs. CEIE, George Mason University, DTRT13-G-UTC48, 2017.
24. Evans, S., Beyond buzzwords: The true meaning and value of "digital twins", in *Digital Twin Hub*, 2020. Online, Australia.
25. Gao, Y., S. Qian, Z. Li, et al., Digital twin and its application in transportation infrastructure, in *2021 IEEE 1st International Conference on Digital Twins and Parallel Intelligence (DTPI)*, Beijing, China, 2021. IEEE.
26. Jafari, M., A. Kavousi-Fard, T. Chen, et al., A review on digital twin technology in smart grid, transportation system and smart city: Challenges and future. *IEEE Access*, 2023. **11**: pp. 17471–17484.
27. Kušić, K., R. Schumann, and E. Ivanjko, A digital twin in transportation: Real-time synergy of traffic data streams and simulation for virtualizing motorway dynamics. *Advanced Engineering Informatics*, 2023. **55**: p. 101858.
28. Bao, L., Q. Wang and Y. Jiang. Review of digital twin for intelligent transportation system. in *2021 International Conference on Information Control, Electrical Engineering and Rail Transit (ICEERT)*. 2021. Lanzhou, China: IEEE.
29. El Marai, O., T. Taleb, and J. Song, Roads infrastructure digital twin: A step toward smarter cities realization. *IEEE Network*, 2020. **35**(2): pp. 136–143.
30. Hu, C., W. Fan, E. Zeng, et al., Digital twin-assisted real-time traffic data prediction method for 5G-enabled internet of vehicles. *IEEE Transactions on Industrial Informatics*, 2021. **18**(4): pp. 2811–2819.
31. Mahmoodian, M., F. Shahrivar, S. Setunga, et al., Development of digital twin for intelligent maintenance of civil infrastructure. *Sustainability*, 2022. **14**(14): p. 8664.

32. Wright, L. and S. Davidson, How to tell the difference between a model and a digital twin. *Advanced Modeling and Simulation in Engineering Sciences*, 2020. **7**(1): p. 13.
33. Aivaliotis, P., K. Georgoulias, Z. Arkouli, et al., Methodology for enabling Digital Twin using advanced physics-based modelling in predictive maintenance. *Procedia CIRP*, 2019. **81**: pp. 417–422.
34. Ariansyah, D., I. Fernàndez del Amo, J.A. Erkoyuncu, et al., in *TESConf 2020 - 9th International Conference on Through-life Engineering Services*. 2020: Cranfield UK.
35. Tao, F., M. Zhang, Y. Liu, et al., Digital twin driven prognostics and health management for complex equipment. *CIRP Annals*, 2018. **67**(1): pp. 169–172.
36. Tao, F., F. Sui, A. Liu, et al., Digital twin-driven product design framework. *International Journal of Production Research*, 2019. **57**(12): pp. 3935–3953.
37. Wang, J., L. Ye, R. Gao, et al., Digital Twin for rotating machinery fault diagnosis in smart manufacturing. *International Journal of Production Research*, 2019. **57**(12): pp. 3920–3934.
38. Stark, R., C. Fresemann, and K. Lindow, Development and operation of Digital Twins for technical systems and services. *CIRP Annals*, 2019. **68**(1): pp. 129–132.
39. De Donato, L., R. Dirnfeld, A. Somma, et al., Towards AI-assisted digital twins for smart railways: Preliminary guideline and reference architecture. *Journal of Reliable Intelligent Environments*, 2023. **9**(3): pp. 303–317.
40. Ahmad, S., M. Spiryagin, Q. Wu, et al., Development of a Digital Twin for prediction of rail surface damage in heavy haul railway operations. *Vehicle System Dynamics*, 2024. **62**(1): pp. 41–66.
41. Spiryagin, M., I. Persson, Q. Wu, et al., A co-simulation approach for heavy haul long distance locomotive-track simulation studies. *Vehicle System Dynamics*, 2019. **57**(9): pp. 1363–1380.
42. Wu, Q., M. Spiryagin, Y. Sun, et al., Parallel co-simulation of locomotive wheel wear and rolling contact fatigue in a heavy haul train operational environment. *Proceedings of the Institution of Mechanical Engineers, Part F: Journal of Rail and Rapid Transit*, 2021. **235**(2): pp. 166–178.
43. Vaculín, O., W.R. Krüger, and M. Valášek, Overview of coupling of multibody and control engineering tools. *Vehicle System Dynamics*, 2004. **41**(5): pp. 415–429.
44. Spiryagin, M., S. Bruni, B. Fu, E. Bernal, S. Ahmad, Q. Wu, C. Cole, and I. Persson, Principles of the Framework Development of a Physics-Based Rail Vehicle Digital Twin in Innovative Decision Support and Prediction Complex Systems, in The Third International Conference of Rail Transportation (ICRT). 2024: Shanghai.
45. Spiryagin, M., Q. Wu, O. Polach, et al., Problems, assumptions and solutions in locomotive design, traction and operational studies. *Railway Engineering Science*, 2022. **30**: pp. 265–288.
46. Cole, C., M. Spiryagin, Q. Wu, et al., Modelling, simulation and applications of longitudinal train dynamics. *Vehicle System Dynamics*, 2017. **55**(10): pp. 1498–1571.
47. Spiryagin, M., P. Wolfs, C. Cole, V. Spiryagin, Y.Q. Sun, and T. McSweeney, Design and simulation of heavy haul locomotives. Ground Vehicle Engineering Series. 2017, Boca Raton, FL: CRC Press.
48. Spiryagin, M., Q. Wu, C. Bosomworth, et al., *Understanding the Impact of High Traction Hybrid Locomotive Designs on Heavy Haul Train Performance*, 2022. CQUniversity.
49. Parbst, B., E. Dolmans, and T.F. Rasmussen, *Digital Twin and Railway System Safety*, 2020. University of Twente.
50. Ahmad, S., M. Spiryagin, C. Bosomworth, et al., Digital twin study to investigate the impact of heavy haul train driving strategies on rail wear, in *Conference on Railway Excellence*, Melbourne, 2023.
51. Krishna, V.V., Q. Wu, S. Hossein-Nia, et al., Long freight trains & long-term rail surface damage – a systems perspective. *Vehicle System Dynamics*, 2022. **61**: pp. 1500–1523.
52. Bernal, E., M. Spiryagin, E. Vollebregt, et al., Prediction of rail surface damage in locomotive traction operations using laboratory-field measured and calibrated data. *Engineering Failure Analysis*, 2022. **135**: p. 106165.
53. Liu, Z., S.N. Araghi, A. Sarkar, et al., Demystifying the buzzword behind Digital Twin: a novel generic evaluation model. CoRR, 2023. https://arxiv.org/abs/2311.12961.
54. Ahmad, S., C. Bosomworth, E. Bernal, M. Spiryagin, M. Rahaman, Q. Wu, Y. Sun, and C. Cole. Prediction of Rolling Contact Fatigue Caused by a Full Train Operation by Using a Digital Twin Method. in The Third International Conference of Rail Transportation (ICRT). Shanghai. 2024.
55. Pillai, N., J.-Y. Shih and C. Roberts, Enabling data-driven predictive maintenance for S&C through Digital Twin models and condition monitoring systems. *The PWI Journal,* 2021. **139**: p. 14–19.
56. Bustos, A., H. Rubio, E. Soriano-Heras, et al., Methodology for the integration of a high-speed train in Maintenance 4.0. *Journal of Computational Design and Engineering*, 2021. **8**(6): pp. 1605–1621.
57. Meng, Z., T. Tang, G. Wei, et al., Digital twin based comfort scenario modeling of ATO controlled train. *Journal of Physics: Conference Series (JPCS)*, 2020. **1654**: p. 012071.

58. Moya, B., A. Badias, I. Alfaro, et al., Digital twins that learn and correct themselves. *International Journal for Numerical Methods in Engineering*, 2022. **123**(13): p. 3034–3044.
59. Wagg, D.J., K. Worden, R.J. Barthorpe, et al., Digital twins: State-of-the-art and future directions for modeling and simulation in engineering dynamics applications. ASCE-*ASME Journal of Risk and Uncertainty in Engineering Systems, Part B: Mechanical Engineering*, 2020. **6**(3): p. 030901.
60. Kapteyn, M.G., D.J. Knezevic, D.B.P. Huynh, et al., Data-driven physics-based digital twins via a library of component-based reduced-order models. *International Journal for Numerical Methods in Engineering*, 2022. **123**(13): pp. 2986–3003.
61. Funfschilling, C. and G. Perrin, Uncertainty quantification in vehicle dynamics. *Vehicle System Dynamics*, 2019. **57**(7): pp. 1062–1086.
62. Bolton, A., L. Butler, I. Dabson, M. Enzer, M. Evans, T. Fenemore, F. Harradence, E. Keaney, A. Kemp, and A. Luck, Gemini principles. https://doi.org/10.17863/CAM.32260, 2018, Centre for Digital Built Britain.
63. An, A., S. Blakeney, R. Bowden, J. Burgess, et al., Digital twin TOWARDS A MEANINGFUL FRAMEWORK. 2019, ARUP.
64. Fjeld, T.M.B., *Digital Twin-towards a Joint Understanding within the AEC/FM Sector*, 2020. NTNU.
65. Ye, C., L. Butler, C. Bartek, M. Iangurazov, et al. A digital twin of bridges for structural health monitoring. in *12th International Workshop on Structural Health Monitoring 2019*. Stanford University, California, USA. 2019.
66. Grieves, M., Digital twin certified: Employing virtual testing of digital twins in manufacturing to ensure quality products. *Machines*, 2023. **11**(8): p. 808.
67. Chinesta, F., E. Cueto, E. Abisset-Chavanne, et al., Virtual, digital and hybrid twins: A new paradigm in data-based engineering and engineered data. *Archives of Computational Methods in Engineering*, 2020. **27**(1): pp. 105–134.
68. Bickford, J., D.L. Van Bossuyt, P. Beery, et al., Operationalizing digital twins through model-based systems engineering methods. *Systems Engineering*, 2020. **23**(6): pp. 724–750.

7 Development of Human–Digital Twin Interaction

Anjum Naweed

7.1 DESIGN REQUIREMENTS IN HUMAN-DRIVEN SYSTEMS

7.1.1 Digital Twins and Human-Driven Systems

Historic problems with defining "digital twin" (DT) (see Chapter 1) impact how we understand the human–DT relationship. Proposed definitions vary in their points of emphasis and areas of application. A description for DT given by Wang et al. [1] simply states that it is "a digital replica of a living or nonliving physical entity," whereas the definition formulated within the ISO/IEC 30173:2023 "Digital twin – concept and terminology" defines it as: "a digital representation of a target entity with data connections that enable convergence between the physical and digital states at an appropriate rate of synchronization" [2]. The use of the word "entity" in both these examples usefully resonates with the human–DT relationship because it carries distinctions to describe and characterize what is being "twinned" (i.e., digitized). Here, the term "human–DT" refers to the interaction and communication between humans and DTs and, more broadly, a system relationship where the function of a human operator and a DT are integrated. An "entity" is defined within the ISO/IEC 30173:2023 as a "thing (physical or non-physical) having a distinct existence." To extend on this, "entity" might also be qualified by drawing on philosopher Heidegger's (1889–1976) understanding of metaphysics, which, in its simplest form asks, "what entities are (or what an entity is)," which means "asking about the being of those entities" [3]. In the context of DTs, this postulates fundamental questions about what an "entity" *is* and *does*, as much as exploring the *being* of a DT if a human is at the center of design, because considerations of cooperation, exchange, and trust, also apply to DT of both human and non-human physical entities.

Human involvement in cyber-physical systems often treats humans as experts imbued with the job of performing operations and making decisions based on the acquisition of knowledge from cyberspace [4,5]. However, the breadth of knowledge produced in the human factors field shows that human potential is contextualized by human limitations and creates a need for human-system optimization recognizing the need for both human wellbeing and productivity. Therefore, while cyber-physical system integration creates improvement, overall system performance degradation and reduced productivity are inevitable when human resourcing and depletion start to misalign with task demand. Ariansyah et al. [4] view a human DT as a valuable participant in a cyber-physical system not only because an understanding of cognitive and physical capacity facilitates decision-making and promotes sustainability but also because sharing locally generated information and individual system capabilities means all humans participating within a cyber-physical system become "intelligently enriched and functionally empowered" to perform. The cyber-physical nature of humans and technology has prominent application in rail, where similar theoretical approaches like "joint-cognitive systems" [6] have been used to characterize the relationships between the human driver and rail vehicle operation [7–9].

The design requirements for human-driven systems depend on the stance designers hold around human involvement and error. The trend for human integration in future industries is predicted to

DOI: 10.1201/9781003492146-7

fall somewhere between two scenarios of human agency in relation to the diffusion of cyber-physical systems [4,10]. In a "technocentric" scenario, humans are seen to have oversight and play a supportive role, intervening in systems governed by information technology (IT) only if there are anomalies or erroneous processing in the system state. In an "anthropo-centric" scenario, IT enacts the role of a decision-support system to assist human-governed systems. Despite the varying levels of centricity, humantechnology relationship dynamics in both these scenarios place humans at the core; however, the most common scenario is the first, reflecting transference of most (if not all) decision-making tasks to technical systems with humans having a supportive role for process execution [11]. This carries the potential to ostensibly devalue the work of humans and makes it less meaningful, which may introduce challenges when looking at developing human integration in DT and gaining acceptance by the workers [12]. It is therefore critical that human dimensions be approached and treated as an invaluable and meaningful element of the system, rather than one that is only error-prone and introduces reliability concerns. As Bécue et al. [11] note, human workers add flexibility and problem-solving competence to all processes.

The modeling and simulation of human performance remains inexact because of a tendency to lean toward two extremes: (1) assuming human actions are perfect or flawless; or (2) treating humans as entirely unpredictable and random [13]. Human variability is constrained by physiological, cognitive, and psychosocial factors that influence performance—which renders human decision-making and actions imperfect. However, as Golightly et al. [14] note, human behavior in operational contexts is generally rational and often adapted to meet local demands and conditions. This makes it highly predictable and neglecting this aspect while modeling human performance runs the risk of creating models and simulations that have low verisimilitude to real-world rail environments. Accurately understanding and capturing this variability is therefore an important design requirement.

For example, human performance modeling for (human) train drivers may assume that all drivers will operate perfectly, meaning that they will always adhere strictly to speed limits and signaling protocols as laid out in Rule Books, with the likelihood of error related to external factors. Alternatively, the model might assume that performance is highly random and unpredictable, leading to scenarios where drivers make seemingly erratic decisions, resulting in a failure to stop at signals. However, a train driver's performance reflects the demands of the task and their own human factors limitations, which can be physiological (e.g., fatigue) or cognitive (e.g., distraction), as well as interactions with other broader cultural and environmental factors and constraints. Therefore, a passenger train driver who routinely pushes their train to its speed limit instead of slowing it down at critical points may be acting entirely rationally within a system if that system prioritizes on-time performance and productivity, even where these actions increase safety risks. Similarly, a freight train driver of a priority service who maintains their speed instead of slowing down after encountering a caution signal may be acting rationally if they are routinely given proceed signals aspects prioritizing them over other traffic.

Rail, like many transportation sectors, often depends on the actions of one or occasionally two individuals [15], but these actions only make sense when viewed as a broad and distributed system. The technical functions of rail interact with others in similarly complex non-technical systems within the broader environment [16]. The complexities and uncertainties of human integration in complex DTs mean that the digital representation of human entities in DT form is a burgeoning area with a range of philosophies around the human role in error production with implications for design requirements in human-driven systems.

7.1.2 The Socio-Technical Context

The rail system has observed a rapid pace of change and it is now characterized as a complex and dynamic system with opaque features in its operations [17]. This necessitates proactive anticipation and optimization of new developments to achieve performance gains while maintaining—and ideally enhancing—safety [13]. It is now standard practice across industry and academia to characterize the human factors and safety criticality inherent in complex systems through a socio-technical systems framework [18]. Regardless of the position that a designer takes with human centricity in design, the social context surrounding human work is apparent and a socio-technical framework describes the conceptual boundaries that exist across various levels of a system. These levels show how actions and decisions within and across levels interact with each other to shape behavior and create implications for design. One example of this is Rasmussen's risk management framework [19], which has found prevalent application in rail human factors and safety science.

The risk management framework [19] depicts a system of "vertical integration," meaning levels of the system exude downward from the top, starting at the government level which creates laws and regulations. The next level is the regulatory level which creates industry standards based on laws and regulations, followed by the company level, which creates policies and procedures based on the industry standards developed to govern work processes. The next level down is the management level which implements the policies and procedures, followed by the staff level, representing the activities and characteristics of people performing the processes. Finally, the bottom work level captures the equipment and environment within the work context.

For rail, a socio-technical systems approach recognizes the interconnected nature of railway functions, acknowledges the existence of competing demands in need of continuous trade-offs, and understands that successful outcomes depend as much on human factors and processes as they do on engineering [20]. Rail operations are guided by targets and performance measures, typically encompassing service delivery (i.e., performance) and safety indicators [18]—all within financial constraints. This introduces hurdles and enablers that influence and shape the design requirements and impose its broader context. For DTs, considerations around the socio-technical system allow the modeling of human–DT interaction to occur in a way that recognizes the design requirements and constraints that exist at different levels.

Figure 7.1 depicts the key elements of a human–DT conceptual framework. Key design requirements, also included in Figure 7.1, include: (1) real-time responsivity and data integration; (2) the user interface (UI) and user experience (UX); (3) trust and transparency; (4) safety and security; and (5) ethical and privacy considerations. These design requirements are each discussed in the next sections.

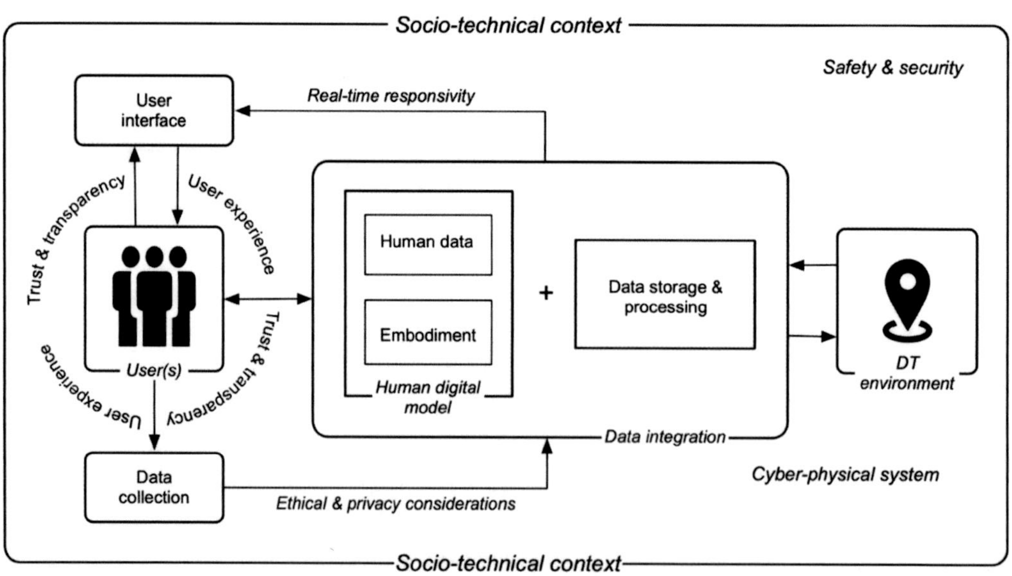

FIGURE 7.1 Conceptual human–digital twin framework showing key elements and design requirements.

7.1.3 Real-Time Responsivity and Data Integration

Real-time responsivity is an important parameter to consider for human-driven systems but is especially important for human–DT interaction, and a critical requirement for an accurate DT replica. Operating in real-time means that a DT can process data, simulate scenarios, and deliver feedback instantly, allowing quicker and more informed decision-making. This is particularly relevant in dynamic and safety-critical environments like rail transportation where near-misses may introduce risks or inefficiencies in operations [21]. As covered in other parts of this handbook, achieving real-time responsiveness means that systems must handle large volumes of data and process them efficiently and without latency. Updating and feedback in UIs must invariably also be "live" and seamless. Data for DTs can be collected from a range of sources (e.g., sensors, IoT devices), and consideration must be given to how the human interacts with these data.

The requirements for DTs of humans are more complex. Much like the twinning of non-human physical entities, technical challenges include transmission and storage of data, but twining a human DT involves (psycho)physiological measurement and collecting data about the human operator's functional state. A large body of research shows a correlation between human work and physiological activity. One example relevant to rail is that an increase in heart rate variability accompanies a demanding task [22], with this change ostensibly reflecting a stress response. Like DTs of non-human entities, consideration for how humans interact with sensor devices and data sources must also be given. This is especially important as some of the associated sensors and devices can collect human data with high accuracy and reliability only if they are attached to them or encroaching on their workspace. A range of (psycho)physiological measures are given in the next part of this chapter, focusing on specific examples from design concepts published in academic literature.

7.1.4 UI and UX

UI and UX design play a large part in the development of human–DT interaction and human-driven systems, in general. UI refers to visual ergonomic and design components that users interact with when using a system or technology. This includes elements like buttons, icons, menus, and layouts, essentially the overall aesthetic design [23]. When well-designed, the UI allows users to intuitively navigate the environment of an interface, and in the case of a DT interface, facilitates access to its data, models, and insights with minimal effort. The design, format, and presentation of data associated with DTs should therefore recognize the importance of human factors and visual ergonomics in terms of readability and digestibility of information so that the cognitive load of its users may be optimized with decision-making efficiency. This has links with points made earlier in the context of real-time responsivity in the rail system where rapid and accurate decisions can be necessary and have high stakes. Poor design in UIs not only places additional cognitive demands on humans but can also create "error traps," meaning scenarios where poor human factors considerations in systems design lead to unavoidable and high rates of error.

While UI is about the look and feel of a system or technology, UX design represents the overall interaction experience [23]. UX focuses on usability, functionality, and the ease with which goals can be achieved by users. This extends from understanding user needs to ensuring that their interactions are smooth, efficient, and satisfactory. In the context of DTs, well-executed UI and UX design is a requirement that may ultimately bridge the gap between complex DTs and human users, producing seamless interaction, enhanced productivity, reduced frustration (and error rates), and ensuring that the system can be engaged with effectively for the user to meet their goals. To achieve this, designers may consider different user profiles, from experts who need advanced features to non-specialists who require simpler access.

In the context of human interaction design and DTs, poor UI design considerations are likely to make the interface more challenging to understand and engage with, with the result that they will require more human effort and cognitive demands and introduce errors in usability. In comparison, poor UX design will garner a less positive experience and decreased effectiveness in interaction, from "clunky" interaction to reduced overall satisfaction. This may also influence perceptions of trust and transparency, leading to graver issues.

7.1.5 Trust and Transparency

The widely applied and extensively validated technology acceptance model [24] focuses on perceived usefulness and perceived ease of use as factors to explain how users come to accept and use a technology. The former refers to a user's belief that technology will enhance their performance or efficiency, while the latter is the degree to which the user believes the technology will be easy to operate without effort. Both factors influence user attitude toward technology use, intention to use, actual use, and trust. Resistance to technology is an area that has received coverage in rail [25–27]. Given unionized considerations around human work—an important part of the system in rail—lack of success in technology implementation born from low acceptance is not always a reflection of poor design or problematic concepts. Rather, it may simply be because of a lack of consultation with end users, which strongly influences their acceptance and smoothness of implementation, regardless of how robust and reliable the technology might be. Issues of trust and transparency ultimately intersect with other kinds of design requirements. Good UI and UX design are often achieved through end-user consultation, a process which is often labeled as "co-design" in current participatory rhetoric. This means that interaction with a DT will feel more intuitive and actionable, and the information transmitted may not adversely impact cognitive load. The enhanced decision-making brought on by real-time responsivity may also build and establish trust between humans and the system. This may ensure that the DT is perceived and interpreted as a reliable extension of real-world processes rather than a detached and lagging simulation that makes work harder. This is even more important in the context of human DT where a lack of congruence between the human and its digital replica will result in perceptions of poor verisimilitude.

7.1.6 Safety and Security

Safety and security contextualize the design requirements in human-driven systems when they are safety critical. The substantive factor for a DT is that it should accurately represent the conditions, processes, and information about the entities it simulates (i.e., rail infrastructure, rolling stock, operations); otherwise, decisions and actions made based on the data received may not only be uninformed but also introduce safety risks (e.g., derailments, signal failures). Human–DT interaction would therefore require comprehensive risk assessments on many levels. For example, in the context of rail driving and network control, both tasks are cognitively demanding and prone to distraction and inattention from task- and non-task-related intrusion [28,29]. Sensor equipment and data collection devices placed within the workplace of a rail driver (train cab) or network controller (control room) or attached to the human in the case of an accurate DT replica (e.g., wearables), may introduce unintended risks and safety-related consequences in the real world.

Security is a design requirement focusing on protecting the DT and associated data from misuse, cyberattacks, or other forms of unauthorized access. DTs typically collect and process large quantities of sensitive information. This includes real-time data from IoT devices and operational systems, thus strong encryption, access controls, and regular security updates are essential. Safety and security influence many of the other design requirements, such as trust, insofar as interaction with DT and its infrastructure in a controlled and protected environment will build trust in its reliability while safeguarding people and data.

7.1.7 Ethical and Privacy Considerations

Noting the role humans occupy in design frameworks, ethics and privacy concerns are a key design requirement, thus human-driven systems need to apply these and take carriage of the responsibility for them in design. Emerging technologies often raise considerations around ethics and privacy [30] based on interaction data collected, and as various personal and human functional state data are collected in the context of human DTs, privacy and ethics requirements alongside others already mentioned (e.g., security, trust, acceptance) are critical to address. He et al. [31] specifically raise privacy-preservation and user consent as requirements for human DTs as they influence user acceptance of data acquisition methods, noting that acceptance may also be influenced by its intrusiveness with daily usage. They also indicate that centralized processing of user data by third parties may potentially add to privacy concerns for fear of data leaks and cyberattacks, which are now very much part of the zeitgeist [32]. In the case of driver-facing cameras, for example, privacy and concerns around third-party "human oversight" have been shared for the heavy vehicle road sector [33].

In rail, ethics and privacy concerns for any technologies that collect human personal and sensitive data, like in-vehicle cameras, have long been a concern for workers. Paradigm-shifting technology like artificial intelligence (AI) has created a call to understand the implications [34]. For DT integration, real-time data collection from sensors, users, and various connected systems, raises questions about who controls these data and how it is used. There are clear requirements for policy compliance, and clear consent mechanisms about how a user's data impacts the function of a DT, potentially with an ethical obligation to avoid bias in algorithms underpinning DTs, ensuring fair treatment across different user groups. Balancing DT development and integration with ethical practices may ensure DTs operate in a framework that considers trust and accountability and respects individual privacy, promoting responsible use.

7.2 HUMAN—DT INTERACTION DESIGN METHODOLOGY

7.2.1 Concepts and Approaches for Digitalized Human Behavior and Performance

The advent of Industry 5.0 and ongoing technological development (particularly advancements in AI) invites futures where potentialities of concepts such as DT intelligent systems may be realized, with scope for twinning entire rail network operations (including the human entities). Advancement toward this is hampered by technical and non-technical challenges, especially the requirements around data resolution and volume, but while this imposes limits on what can and cannot be captured and simulated as a DT, it has not impeded the conceptualization of approaches for modeling and simulating human performance. Kumar et al. [35] for example, propose a DT-centric approach that includes edge computing, machine learning, and 5G communication to predict driver intention, and Chen et al. [36] propose a Digital Behavior Twin framework to share models of drivers among connected vehicles is proposed to predict future actions of other vehicles. However, both are suggested to have somewhat simplistic conceptualizations of DT and uncertainty in the involvement of IoT and lack systematic integration of DT in cloud/architecture design and clarity in data structures and workflows [1].

In relation to human performance modeling with systems models and large-scale analysis, Lawton and Gauthier [37] question the feasibility of integrating detailed cognitive models capturing finer points of human perception, decision-making, and response. Rather than approaching simulation of cognitive models or viewing humans as a system, they define humans as one "element of a system" supporting its functions, such as mobility, communications, and sensing, and model probability-based human performance within a time-step stochastic simulation tool designed for multi-echelon operation and support activities. Probability-based human performance modeling means quantifying tasks and parameterizing "human error" in task completion failure, with this considered to occur predominantly via mistakes and absences (i.e., reliability errors). Thus, Lawton and Gauthier [37] conceptualize human performance variability and differentiation based on performance-shaping human factors that are time in/dependent. These include: (1) training; (2) level of experience; (3) perceptual skills; and (4) fatigue. All but fatigue are time-independent factors, meaning that people with greater training, experience, and skills are less likely to experience reliability errors. Fatigue level is modeled as a time-dependent factor to illustrate time-on-task effects and increased probability of error rate with time in a linear relationship. An example drawing on fatigue is given in the context of a military combat unit to demonstrate the utility of the model where each human is assigned a basic error rate distribution. The model derives six measures of performance along with 20 human elements (i.e., individuals). While some consideration is given to the broader context, this approach is an example of a person-centric focus in the development of systems models. As mentioned in the earlier section, the social environment is also a key factor in design.

The importance of the social environment and context around human–DT integration is articulated in the context of a "parallel driving/control" by Wang [38], where both social complexity and engineering dimensions are argued as important considerations for a data-driven approach to modeling, analysis, and decision-making in the transportation management. Here, the physical, mental, and artificial worlds are modeled in parallel, with interactions considered among (connected) vehicles, and human drivers.

Several DT studies have been conducted in the realm of transportation [e.g., 35,36,39]. An important observation given by Wang et al. [1] in relation to these studies is the general paucity of holistic frameworks that have connected and/or integrated the range of disparate entities, such as the vehicle and the human. This has started to change. Recent research, review, and commentary on the topic, including the work by Wang et al. [1], have proposed human interaction design concepts and methodologies to fill this gap, with many of these adopting the concept of multiple planes to drive human interaction design methodology. Conceptualizing multiple planes or levels of integration existing in relation to the physical objects/spaces/realms and their digitally represented replicas/counterparts was provided by [40].

Frameworks integrating human factors and digital techniques to deal with real-world complexity and dynamism have been proposed. For example, a unified framework incorporating human body models with body movement and cognitive status is conceptualized by He et al. [31]. Data sources include physiological signals emitted electrically by skeletal muscles, neural and circulatory systems, and other vital biometrics from wearable sensors. Kinematics and measurement of real body movement, postural, and motion data is also represented. Demographical data (age, sex, body mass index) is used to create personalized profiles. Notably, the framework also proposes that data about the social and contextual environment may be drawn from various social media platforms. All these data sources are inputs into a digital engine modeling human health, biomechanics, and cognitions, driven with the support of AI inference, and realized in an interface that includes functional, visualization, and multimodal interaction components [31].

Another human interaction design methodology is proposed by Chen et al. [36] with the integration of both physical human and non-human entities proposed in the form of a mobility DT framework for mobility services. It is defined as an AI-based data-driven cloud-edge framework with the emphasis on "talking" occurring between the multiple planes, which here comprise the: (1) physical space, which includes human beings, vehicle, and other substantive infrastructure (e.g., traffic); and (2) the digital space containing their replicas; with (3) the "talking" mediated by a communication place between the spaces. The human side of DT is realized through user management and driver-type classification.

In earlier work on a DT for advanced driver assistance systems in connected vehicles, Wang et al. [41] proposed replication of both human and non-human entities (i.e., vehicles, drivers, road infrastructure) through cloud-based data and cellular-based vehicle to cloud (V2C) communication. In their subsequent mobility DT concept [1], a more fleshed-out version is given, with the three entities classified into building blocks, and greater emphasis placed on the importance of maintaining synchronization and a communication plane facilitating off-load of computing work into the digital space. As an end-to-end framework, sampling of information between the physical space and the actuation process is conducted by drivers. In terms of the human, data sources include in-cab status sensing (via cameras, seat sensors, etc.), wellness monitoring (via wearables, physiological data, etc.), and behavior preference via active (e.g., driver manual selection of control input) and passive means (e.g., vehicle camera recording pedestrian navigating a crosswalk). Data sampling can also be obtained actively through the human-machine interface. The human DT has a data lake that stores samples data, with separate personal databases to differentiate between humans, and drivers are algorithmically classified into specific types. The framework is purported to consider, not only the drivers, but every human being substantive to the system (e.g., passengers, pedestrians, cyclists). The stated utility of human integration is multifaceted [1]; for example, the human DT can use data from other building blocks (e.g., the vehicle) to predict future driver behavior, and if the human driver in the physical space of the framework is provided data from the digital space, their actuation would be more accurate. Lane changing intent [42] and adjacent-vehicle lane change possibility [43] are given as an example of one scenario that demonstrates the benefits of the human–DT relationship.

The human-driven context of industrial and manufacturing work has played a large part in mobilizing much of the research in human–DT design. A DT framework integrating a human DT to predict human cognitive and physical status, perception, and behaviors in an industrial setting was proposed by Ariansyah et al. [4], along with a sample case of an instrumented human and "cobot" (a portmanteau of "collaborative" and "robot") carrying a heavy load. Like the concept proposed by Wang et al. [1], the framework draws on a physical and virtual plane, mediated by a communication plane. Given the context, human real-time state integration within the DT is intended to promote deeper insight into available resources and human status for task allocation and to mitigate dangerous and distressing situations. Descriptions of the human-technology dynamic therefore characterize a collaborative relationship based on realizing a "partnership." Such a view is unsurprising, given the nature of industrial work and cobots, but as mentioned in Section 7.1.1, such cyber-physical system concepts have strong applications to human work in rail service delivery. The integrated models for the physical plane proposed by Ariansyah et al. [4] draw on production, data perception, and feedback interface elements, with the digital plane integrating both geometrical, physical, behavior, and rule-based models (for equipment), with psycho/physiological models (for humans). AI-based processing support is described in conjunction with an "insight element" for optimization, decision-making, and visualization. In terms of human data measurement, this includes electromyography for muscle fatigue and electrocardiography for mental workload.

Another, relatively well-theorized concept for human DTs, again driven by future needs in industry and manufacturing, is proposed by Bécue et al. [11]. Here, human behavioral modeling is approached predominantly via real-time monitoring and defining of AI-based user profiles (and model training and validation). This is used to understand whether human behavior conforms with usual performance—meaning that "unusual activities" can be discerned as such, followed by a predictive impact on the process, costs, and corrective actions. Interactions between human and non-human entities (e.g., machines, equipment, peripherals, software applications) are also monitored and modeled in addition to the human condition to enable inference of the human state of mind. As seen in other concepts, the emphasis here is human mental state (fatigue, sleepiness, inattention) and human performance optimization, for example, detecting and preventing cognitive overload and underload. Bécue et al. [11] also reference affective computing and interacting emotionally with humans, both as a means of understanding human behavior and for recognizing human factors such as fatigue and stress. Non-intrusive integration of mental chronometry, for example, via reaction time testing, is also used to trigger go/no go tests to infer anomalous states. Many of the foregoing human DT design concepts draw on the measurement of the human functional state. Table 7.1 shows different types of measurements and measures that can be used to

TABLE 7.1
Overview of Example Measures to Capture Information on Human State

Type of Measurement(s)	Measure(s)
Physical	Muscular and skeletal positioning/posturing
	Muscle response/tension
	Accelerometry
	Gait analytics
Psychophysiological	Electrodermal activity
	Cardiovascular signals
	Respiration/breathing rate
	Body/skin temperature
	Brain activity
	Eye-gaze/tracking
	Pupillometry
	Blink rate
	Emotional speech perception
	Facial expression recognition
	Mental chronometry
Other	Video analytics
	Proximity
	Anthropometrics
	Demographics
	Self-reported/survey-based data
	Social media

obtain these data, though accurately twinning human cognition is perhaps the most challenging, and as mentioned in the previous section, a range of other requirements apply.

By way of a summary, Figure 7.2 illustrates the key factors and components of human–DT interaction design. The three spaces (physical, digital, communication) are central to current thinking, with performance-shaping human factors, human functional state measures, the social and contextual environment, and demographical data underpinning key human-centric components. Approaches that integrate multiple planes, AI inference and cloud-edge computing, real-time state, and personalized profiles are among some of the most technical factors. Importantly, human–DT concepts are underpinned by holistic/unified frameworks that depict collaborative relationships. Human-driven approaches for design are the next key element and are discussed next.

7.2.2 Human-Driven Approaches for Design

One of the most important measures for understanding the extent to which the design of a human-driven system meets usage requirements is its usability. Human interaction with a technology or system happens through an interface focusing on various indicators (e.g., usefulness, efficiency, effectiveness, learning curve) and broader system factors which shape how humans interact. This falls into the remit of various multidisciplinary approaches or frameworks that emphasize a human perspective in design. Human-centered design (HCD) is one of the main approaches here, which includes human factors and ergonomics knowledge, techniques, and elements to ensure systems

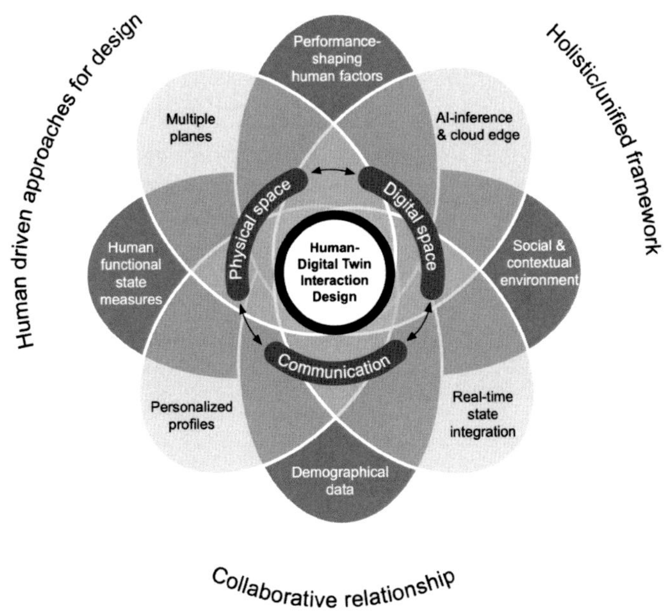

FIGURE 7.2 Key factors and components for human–digital twin interaction design.

are usable [44]. HCD addresses the Industry 4.0 challenge of complex and dynamic work by placing the human at the center of smart design and recognizing that design complexity must consider aspects of human-system interaction in terms of cooperation, exchange, and trust [45]. Applying the HCD approach leads to increased productivity, usability, and UX. While the context of technology and system use is emphasized, HCD also encourages the involvement of others throughout design and development, and the impact on stakeholders other than direct users, which corresponds with accommodation of the broader social context.

Participatory ergonomics is a human-driven design approach that includes the user, researcher, and designer, defined formally by Wilson [46] as a methodology that involves "people in planning and controlling a significant amount of their own work activities, with sufficient knowledge and power to influence both processes and outcomes to achieve desirable goals." Participatory ergonomics places even a great emphasis on stakeholders than HCD and on mutual learning. It harnesses a range of tools and techniques to expand the capacity for stakeholders to implement knowledge and procedures and create improvement in the working environment in terms of safety, quality, productivity, comfort, and morale. It includes the user, researcher, and the designer. The approach has been shown to improve engagement and enhance collaboration in a variety of settings and is perhaps most synonymous with "co-design" in the current rhetoric. As noted by Niemelä et al. [30], participatory design helps users garner a better understanding of the expected impacts of large-scale socio-technological changes and empowers them with influence. Participatory design has broad application in rail, for example by the example of rail simulator design provided by Naweed et al. [47], which bridges mutual learning with underlying values. The implications of not involving both users and those with substantive subject matter expertise, such as human factors, in the process of design are well-reported in the rail context [48].

The role of values is emphasized further in approaches like value-sensitive design (VSD) [49], which seeks to integrate and operationalize conceptually defined values considered important in life (e.g., trust, wellbeing, freedom from bias, accountability, privacy, data handling) in the systematic design of technology and systems. Importantly for DT, VSD has ethical imports, meaning that ethics in the design of technology is a key constituent across all phases. The process starts with a conceptual phase that identifies all stakeholder types who will have in/direct interaction with the designed technology or system, along with their values. Follow-up phases include empirical, technical, and research phases.

The importance of values also underpins life-based design [50], a multidimensional and holistic approach where an understanding of human life is a design foundation in and of itself, conceptualized from psychological, sociocultural, and biological human life sciences perspectives. The concepts are explored in ways that can, for instance, examine the ethics, benefits, and meaningfulness received from developed systems, and how they can be improved for a better quality of life [30].

Finally, Niemelä et al. [30] assimilate key principles from a number of human-driven frameworks to propose a human-driven design, an approach which aims to be holistic, and foster collaboration and ethical reflection by adopting a human and social view of technology, empowering both users and stakeholders, and ensuring awareness of responsibility for human societal phenomena (e.g., sustainability).

7.3 BROADER CONSIDERATIONS FOR DEVELOPMENT OF HUMAN–DT INTERACTION

The necessity for human integration in DTs creates a concomitant need to understand human roles in different future contexts and how advancements in technology and AI change the way humans interact and exchange information. This handbook is a comprehensive reference guide for this in the context of rail applications; however, human DTs still have a long way to go, with key areas of development, research, and context converging around theoretical and technological advancement, and implementation with the public. However, the importance and multifaceted role of the human cannot be evaded. This point is particularly pertinent because while humans remain at the core of DT design due to the complexity of many processes, integration between DT models and human factors (including ergonomics) still tends to be neglected [51,52]. This concluding section extends the previous sections with broader considerations substantive to the development of human–DT interaction.

7.3.1 Human Factors and Task Analysis

Human factors and ergonomics principles can guide the design of human interaction design methodologies for DTs and ensure that they are optimized for user efficiency, safety, and satisfaction. This blend of engineering and psychological science focuses on understanding the capabilities, limitations, and behaviors of users when interacting with complex systems—much like DTs—which may minimize errors and reduce cognitive load. One well-established methodology for doing this is the family of task analysis techniques, such as hierarchical task analysis, and applied cognitive task analysis, which have prominent applications in rail human-systems contexts already [7,15,53,54]. These approaches will enable designers to break down specific tasks that users will perform with the DT, identifying each step, its complexity, and potential challenges. Such methods help designers create workflows and interfaces that align with how users naturally operate, which may ensure that DTs enhance, rather than complicate, their tasks.

7.3.2 (Re)conceptualizing "Human Error" as "System Error"

A "human error" approach is conceptualized in many ways in DT modeling and simulation for the optimization of system performance. "Human error" has a tendency to be seen in terms of being accidental (e.g., a mistake), malicious [11], and unreliable [37], which is one reason why comparison with "normal" or routine behavior patterns often underpins DT design philosophies. Here, the context of the socio-technical system for DT may be enhanced by applications of "systems thinking," which reposition notions of error in relation to safety. A fundamental aspect of this is that systems are inherently complex and unsafe, thus safety is an emergent property [55] and created by humans at all levels, meaning that errors are produced by multiple interacting system factors rather than individual actions alone [56]. A key concern with adopting HCD that interprets "malicious intent" and parameterizes "human error" in task failure is that it encourages the assignment of blame. In the context of human–DT interaction, this is likely to lead to maladaptive design. Problems with the conceptualization of "human error" [57] have been shared, and have particular relevance with mainstream commentary in rail [8]. As the focus intensifies on technological performance and efficiency in the utility of DTs, a parallel need for more mature and robust safety cultures grows more apparent.

7.3.3 Iterative Design and Evaluation

Achieving real-time performance in DTs often involves leveraging advanced techniques, such as edge computing, distributed networks, and cloud-based infrastructures, to minimize data transfer delays. As we head toward the development and realization of seamless updates, live feedback will allow users to interact naturally with the DT, making the UI of any DT interaction methodology an important area. In terms of rail applications, the safety-critical nature of the rail system also necessitates robust validation of DT data and simulations to avoid errors or malfunctions. Evaluation frameworks will enable iterative assessment of how well a DT meets user needs and supports task completion, as well as how it reduces inefficiencies and supports risk management and decision-making under pressure. Usability testing, feedback loops, and iterative design are essential for refining the experience. The guidance be Bearman et al. [27] may offer a valuable starting point in the context of rail. Simulation and augmented reality (AR) / virtual reality (VR) approaches may be used to design, test, and refine human–DT interaction methods.

7.3.4 Privacy and Ethical Concerns

As mentioned in Section 7.1.7, privacy concerns center on the potential misuse of information collected, including unauthorized access to sensitive data, regardless of whether it is personal or operational. Compliance with relevant data protection regulations will help safeguard user privacy, and edge computing and the emerging field of federated learning may help process sensitive data more securely. Ethical concerns go hand in hand with privacy and must address transparency around data collection, analysis, and use in decision-making. This may also require compliance with regulations and a proactive, transparent, and responsible approach involving user consent. While robust data anonymization and encryption methods will invariably help safeguard sensitive information, it is important to ensure that ethical design practices are used (as part of HCD) so DT algorithms and user profiles are free from bias. Frequent audits and assessments may be used to evaluate the ethical implications of system decisions and help maintain accountability.

7.3.5 Embedding DTs into Safety Culture

Finally, dealing with many of the points and interaction considerations raised in this chapter may benefit from the inclusion of stakeholders from concept design through to implementation. For rail applications, this means that, like most new systems and technologies, DTs must become part of the safety culture. This means good communication about how data use, access, and purpose allows users to make informed decisions about their participation, fostering a culture of responsibility by providing training for designers, developers, and users on the potential challenges to promote a good culture of trust, fairness, and accountability throughout the DT lifecycle. This also means developing an inclusive approach involving all stakeholders, particularly unions, to ensure that political and non-technical roadblocks are accommodated. As mentioned earlier, participatory design offers one holistic framework and research strategy to foster collaboration and co-design through the involvement of non-academic stakeholders. This process may also engender "transdisciplinarity" [47], meaning that all underlying values and philosophies are brought to the surface and every individual involved gains disciplinary literacies and knowledge. This also resonates with the broader context of interaction for the development of human-driven systems, which is, of course, the foundational argument around buy-in and procurement of DTs.

REFERENCES

1. Wang, Z., R. Gupta, K. Han, et al., Mobility digital twin: Concept, architecture, case study, and future challenges. *IEEE Internet of Things Journal*, 2022. 9(18): 17452–17467.
2. International Standard, BS ISO/IEC 30173:2023. *Digital Twin – Concepts and Terminology*, 2023. Edition 1.0. ISBN 978-2-8322-7680-8.
3. Thomson, I., *Heidegger on Ontotheology: Technology and the Politics of Education*, 2005, Cambridge University Press, New York.
4. Ariansyah, D., A. Buerkle, A. Yacoub, et al., Towards a digital human representation in an industrial digital twin, in *9th International Conference on Through-life Engineering Service*, 2020, Cranfield, UK.
5. Lee, J., B. Bagheri, and H.-A. Kao, A Cyber-Physical Systems architecture for Industry 4.0-based manufacturing systems. *Manufacturing Letters*, 2015. 3: 18–23.
6. Hollnagel, E. and D.D. Woods, *Joint Cognitive Systems: Foundations of Cognitive Systems Engineering*, 2005, CRC Press, Boca Raton.
7. Naweed, A., Investigations into the skills of modern and traditional train driving. *Applied Ergonomics*, 2014. 45(3): 462–470.
8. Naweed, A., Getting mixed signals: Connotations of teamwork as performance shaping factors in network controller and rail driver relationship dynamics. *Applied Ergonomics*, 2020. 82: 102976.
9. Naweed, A., L. Bowditch, J. Trigg, et al., Injury by design: A thematic networks and system dynamics analysis of work-related musculoskeletal disorders in tram drivers. *Applied Ergonomics*, 2022. 100: 103644.
10. Dworschak, B. and H. Zaiser, Competences for cyber-physical systems in manufacturing – First findings and scenarios. *Procedia CIRP*, 2014. 25: 345–350.
11. Bécue, A., E. Maia, L. Feeken, et al. A new concept of digital twin supporting optimization and resilience of factories of the future. *Applied Sciences*, 2020. 10, doi:10.3390/app10134482.
12. Müller, J.M., Assessing the barriers to Industry 4.0 implementation from a workers' perspective. *IFAC-PapersOnLine*, 2019. 52(13): 2189–2194.
13. Naweed, A. and D. Golightly, Advancement in rail transportation and safety: Key themes for human factors and systems thinking, in *Policy Messages for Planning and Implementing High-Speed Rail in Asia*, N. Bugalia, et al., Editors, 2022, Asian Development Bank Institute, Japan, pp. 89–97.
14. Golightly, D., K. Pierce, R. Palacin, et al., A feasibility assessment of multi-modelling approaches for rail decarbonisation systems simulation. *Proceedings of the Institution of Mechanical Engineers, Part F: Journal of Rail and Rapid Transit*, 2021. doi:10.1177/09544097211039395.
15. Naweed, A., G. Balakrishnan, and J. Dorrian, Going solo: Hierarchical task analysis of the second driver in "two-up" (multi-person) freight rail operations. *Applied Ergonomics*, 2018. 70: 202–231.
16. Larue, G.S., A.J. Filtness, J.M. Wood, et al., Is it safe to cross? Identification of trains and their approach speed at level crossings. *Safety Science*, 2018. 103: 33–42.
17. Naweed, A., G.R.J. Hockey, and S.D. Clarke, Designing simulator tools for rail research: The case study of a train driving microworld. *Applied Ergonomics*, 2013. 44(3): 445–454.
18. Read, G.J., A. Naweed, and P.M. Salmon, Complexity on the rails: A systems-based approach to understanding safety management in rail transport. *Reliability Engineering & System Safety*, 2019. 188: 352–365.
19. Rasmussen, J., Risk management in a dynamic society: A modelling problem. *Safety Science*, 1997. 27(2): 183–213.
20. Wilson, J.R., Fundamentals of systems ergonomics/human factors. *Applied Ergonomics*, 2014. 45(1): 5–13.

21. Naweed, A., J. Trigg, S. Cloete, et al., Throwing good money after SPAD? Exploring the cost of signal passed at danger (SPAD) incidents to Australasian rail organisations. *Safety Science*, 2018. 109: 157–164.
22. Young, M.S., K.A. Brookhuis, C.D. Wickens, et al., State of science: Mental workload in ergonomics. *Ergonomics*, 2015. 58(1): 1–17.
23. Norman, D., *The Design of Everyday Things*. Revised and expanded edition, 2013, Basic Books, New York.
24. Davis, F.D. and A. Granić, *The Technology Acceptance Model: 30 Years of TAM*. Human–Computer Interaction Series, 2024, Springer, Switzerland.
25. Naweed, A. and J. Rose, Assessing technology acceptance for skills development and real-world decision-making in the context of train driving. *Transportation Research Part F: Traffic Psychology and Behaviour*, 2018. 52: 86–100.
26. Naweed, A., Simulator integration in the rail industry: The Robocop problem. *Proceedings of the Institution of Mechanical Engineers, Part F: Journal of Rail and Rapid Transit*, 2013. 227(5): 407–418.
27. Bearman, C., et al., *Evaluation of Rail Technology: A Practical Human Factors Guide*, 2017, CRC Press, Surrey, UK.
28. Naweed, A., Psychological factors for driver distraction and inattention in the Australian and New Zealand rail industry. *Accident Analysis & Prevention*, 2013. 60: 193–204.
29. Naweed, A. and P. Murphy, One-track mind: Investigating positive and negative applications of non-technical skills in rail network control. *Applied Ergonomics*, 2023. 106: 103840.
30. Niemelä, M., V. Ikonen, J. Leikas, et al. Human-driven design: A human-driven approach to the design of technology, in *ICT and Society*, 2014, Berlin, Heidelberg, Springer, Berlin Heidelberg.
31. He, Q., L. Li, D. Li, et al., From digital human modeling to human digital twin: Framework and perspectives in human factors. *Chinese Journal of Mechanical Engineering*, 2024. 37(1): 9.
32. Marassi, L. and S. Marrone, What would happen if hackers attacked the railways? Consideration of the need for ethical codes in the railway transport systems, in *Applications of Artificial Intelligence and Neural Systems to Data Science*, A. Esposito, et al., Editors, 2023, Springer Nature Singapore, Singapore, pp. 289–296.
33. Higginson, A., D. Dawson, and M. Sprajcer, *Fatigue Distraction Detection Technology Use in the Australian Road Freight Transport Sector: Research Report Phase 2*, Queensland, AU, 2019, NHVR.
34. Bešinović, N., L. De Donato, F. Flammini, et al., Artificial intelligence in railway transport: Taxonomy, regulations, and applications. *IEEE Transactions on Intelligent Transportation Systems*, 2022. 23(9): 14011–14024.
35. Kumar, S.A.P., R. Madhumathi, P.R. Chelliah, et al., A novel digital twin-centric approach for driver intention prediction and traffic congestion avoidance. *Journal of Reliable Intelligent Environments*, 2018. 4(4): 199–209.
36. Chen, X., E. Kang, S. Shiraishi, et al., Digital behavioral twins for safe connected cars, in *Proceedings of the 21th ACM/IEEE International Conference on Model Driven Engineering Languages and Systems*, 2018, Association for Computing Machinery, Copenhagen, Denmark, pp. 144–153.
37. Lawton, C.R. and J.H. Gauthier. Human performance modeling in system of systems analytics, in *2012 7th International Conference on System of Systems Engineering (SoSE)*, Genoa, Italy, 2012.
38. Wang, F.Y., Parallel control and management for intelligent transportation systems: Concepts, architectures, and applications. *IEEE Transactions on Intelligent Transportation Systems*, 2010. 11(3): 630–638.
39. Alam, K.M. and A.E. Saddik, C2PS: A digital twin architecture reference model for the cloud-based cyber-physical systems. *IEEE Access*, 2017. 5: 2050–2062.
40. Kritzinger, W., M. Karner, G. Traar, et al., Digital Twin in manufacturing: A categorical literature review and classification. *IFAC-PapersOnLine*, 2018. 51(11): 1016–1022.
41. Wang, Z., X. Liao, X. Zhao, et al., A digital twin paradigm: Vehicle-to-cloud based advanced driver assistance systems, in *2020 IEEE 91st Vehicular Technology Conference (VTC2020-Spring)*, Antwerp, Belgium, 2020.
42. Shou, Z., et al. Long-term prediction of lane change maneuver through a multilayer perceptron, in *2020 IEEE Intelligent Vehicles Symposium (IV)*, 2020.
43. Liao, X., Z. Wang, X. Zhao, et al. Online prediction of lane change with a hierarchical learning-based approach, in *2022 International Conference on Robotics and Automation (ICRA)*, Philadelphia (PA), USA, 2022.
44. International Organization for Standardization, *Ergonomics of Human-System Interaction—Part 210: Human-Centred Design for Interactive Systems (ISO Standard No. 9241-210:2019)*, 2nd ed., ISO, Switzerland, 2019.

45. Nguyen Ngoc, H., G. Lasa, and I. Iriarte, Human-centred design in industry 4.0: Case study review and opportunities for future research. *Journal of Intelligent Manufacturing*, 2022. 33(1): 35–76.
46. Wilson, J. and N. Corlett, Ergonomics and participation, in *Evaluation of Human Work: A Practical Ergonomics Methodology*, J. R. Wilson and N. Corlett, Editors, 2005, CRC Press, UK, pp. 933–962.
47. Naweed, A., D. Ward, C. Gourlay, et al., Can participatory ergonomics process tactics improve simulator fidelity and give rise to transdisciplinarity in stakeholders? A before–after case study. *International Journal of Industrial Ergonomics*, 2018. 65: 139–152.
48. Naweed, A., L. Bowditch, J. Trigg, et al., Out on a limb: Applying the person-environment-occupation-performance model to examine injury-linked factors among light rail drivers. *Safety Science*, 2020. 127: 104696.
49. Friedman, B., P.H. Kahn Jr., A. Borning, et al., Value sensitive design and information systems, in *Early Engagement and New Technologies: Opening up the Laboratory*, N. Doorn, et al., Editors, 2013, Springer Netherlands, Dordrecht, pp. 55–95.
50. Leikas, J., P. Saariluoma, J. Heinilä, et al., A methodological model for life-based design. *International Review of Social Sciences and Humanities*, 2013. 4(2): 118–136.
51. Berti, N. and S. Finco, Digital twin and human factors in manufacturing and logistics systems: State of the art and future research directions. *IFAC-PapersOnLine*, 2022. 55(10): 1893–1898.
52. Sgarbossa, F., E.H. Grosse, W. Patrick Neumann, et al., Human factors in production and logistics systems of the future. *Annual Reviews in Control*, 2020. 49: 295–305.
53. Naweed, A., J. Chapman, C. Vandelanotte, et al., 'Just Right' job design: A conceptual framework for sustainable work in rail driving using the Goldilocks Work Paradigm. *Applied Ergonomics*, 2022. 105: 103806.
54. Rose, J., C. Bearman, and A. Naweed, Using task analysis to inform the development and evaluation of new technologies, in *Evaluation of Rail Technology: A Practical Human Factors Guide*, C. Bearman, et al., Editors, 2013, Ashgate, Surrey, UK, pp. 129–169.
55. Leveson, N., A new accident model for engineering safer systems. *Safety Science*, 2004. 42(4): 237–270.
56. Knobel, P. and A. Naweed, How does the regulatory context influence systems thinking in work health and safety (WHS) inspectors? *Safety Science*, 2023. 166: 106237.
57. Read, G.J.M., S. Shorrock, G.H. Walker, et al., State of science: Evolving perspectives on 'human error'. *Ergonomics*, 2021. 64(9): 1091–1114.

8 Digital Twins for Rail Vehicle Designs

Nicola Bosso, Matteo Magelli, and Nicolò Zampieri

The design of railway vehicles has historically been one of the first activities affected by the development of numerical models capable of reliably reproducing the physical reality of the vehicle, which can be considered the precursors of the modern concept of the digital twin. The use of numerical modeling techniques, initially adopted for the main structural elements and for the analysis of vehicle dynamics, has been extended over time both methodologically and in terms of specific applications to an increasing number of vehicle subsystems, because of the increasing complexity and technology employed in modern railway vehicles. This was due to the continuously increasing complexity and technology used in modern railway vehicles.

To better clarify the typology of digital twin techniques used in railway vehicle design, in accordance with the classification given by Kritzinger [1], three classes will be considered based on the level of integration:

- *Class 1*: Digital model: a digital twin of a physical system where the data flow is not automated. The model reproduces the physical reality but there is no communication between the two entities. The numerical model is simply used to perform simulations aimed at predicting the behavior of the physical system.
- *Class 2*: Digital shadow: the models are duplicates of the physical reality and a one-way communication (from the physical system to the model) is implemented.
- *Class 3*: Digital twin: this type of approach is the most integrated and involves a two-way data flow. In this sense, the digital twin can eventually control its physical counterpart. This level of integration allows a synergy of the physical and digital systems, where they can work together.

In the case of vehicle design, Class 1 digital twins are frequently used for the design and pre-design phases, Class 2 twins are developed in the advanced design phase, when preliminary experimental data is available, and finally Class 3 twins are limited to some high-tech subcomponents that require the integration of active control elements during operation.

Since the application of digital twin techniques is strongly differentiated based on the type of design, this application will be analyzed separately below for geometric design, structural design, dynamic analysis, and the design of subsystems.

8.1 GEOMETRICAL DESIGN

The design of every mechanical system in modern industry, including the railway vehicle, is based at first on geometric modeling techniques developed using 3D computer-aided design (CAD) software packages that allow the creation of a geometric model faithful to the real system that will be subsequently produced. The 3D CAD geometric model can obviously be considered a Class 1 digital twin according to the previous classification and constitutes the minimum requirement of modern design, ensuring the repeatable replication of components with identical characteristics within the limits of the defined tolerances.

DOI: 10.1201/9781003492146-8

The development of 3D CAD systems already used in other industrial fields (since the late 70s) has begun to spread also in the railway field [2,3], where at least the large companies in the sector made a complete transition by the early 90s, thus leading the path toward the V-shaped design model sketched in Figure 8.1.

The use of 3D CAD systems initially had a benefit and an impact on the production system [4], allowing for the optimization of the production process and better management of components, these outcomes also being thanks to the integration of geometric models in numerically controlled machine tools (CAM: computer-aided manufacturing) and in automated geometric control systems (CAT: computer-aided testing). These new design tools were essential to realizing new types of structures for vehicle car bodies, no longer based on a metal frame structure but on a tubular structure realized through metal sheet bending to obtain complex profiles, ensuring the optimization of the stiffness–weight ratio. Such structures, especially those made of light alloys for high-speed vehicles, would not have been possible without the use of geometric modeling and computerized production control techniques.

The availability of an accurate three-dimensional model of the structure has also made it possible to adopt calculation techniques based on geometric models generated from the geometry of the system, such as finite element (FE) methods, used for structural calculation, which will be analyzed in the following Section 8.2, and multibody (MB) methods used for the calculation of vehicle dynamic performance, which will be analyzed in Section 8.3. These numerical modeling techniques are part of the computer-aided engineering (CAE) paradigm and allow the creation of digital models of the mechanical system capable of predicting its performance in terms of dynamics and structural resistance. These models can be integrated with other models in a multiphysics environment to better predict the loads applied to the structure (fluid-dynamic codes) or with codes capable of simulating the behavior of electrical or active components that can be installed on the vehicle (traction, braking system, etc.).

As shown in Figure 8.1, the vehicle simulation and production cycle is completed during the testing and operation phase with tools capable of managing series production and component procurement (CAP: computer-aided planning), verifying the reliability of the system in operation (CAQ: computer-aided quality) and managing maintenance (CAS: computer-aided service). These stages provide feedback on the requirements agreed with the customer and may lead to changes and improvements to the project if the initial requirements are not met. Figure 8.1 represents the vehicle

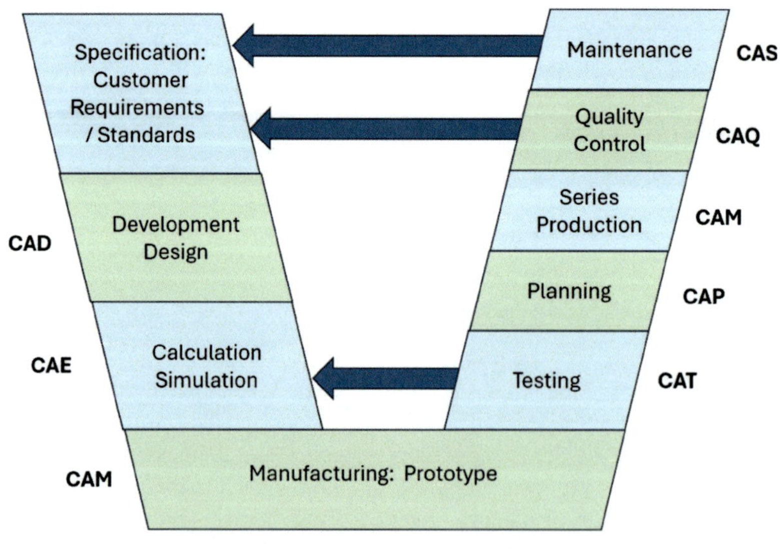

FIGURE 8.1 Typical design cycle for a railway vehicle.

design cycle at the system level and can normally be broken down by considering subsystems and components.

Initially, the design model illustrated in Figure 8.1 was achieved using different numerical tools (e.g., a CAD code, one or more codes for CAE, a code for testing, etc.). Recently, a trend has been growing toward the development of integrated platforms capable of ensuring data exchange in the correct formats as well as compatibility between different software environments, hence improving the overall project management capability.

These integrated platforms are sometimes referred to as "digital mockups" and their application in the railway sector is growing, especially in large companies. Siemens was one of the first companies to introduce this approach to train system design, but other companies have also moved in this direction. The digital mockup can be considered a digital twin of the railway vehicle at a more extensive level than a single component, system, or function of the vehicle, since it includes the production process, testing, and operating activity.

Normally a digital mockup [5] must at least include the 3D CAD, FE, and MB models for structural and dynamic calculations, the management of virtual prototyping systems to perform intermediate checks, and the management of the production process according to the concepts of the "digital factory" envisaged by the Industry 4.0 production systems.

8.2 STRUCTURAL DESIGN

The availability of a 3D model of the vehicle was the prerequisite for the creation of numerical models developed with FE codes for the structural analysis of the vehicle components.

These models can be considered as digital twins of the real vehicle component under investigation. In fact, from a structural point of view, the FE model is intrinsically the physical representation of an elementary portion of the analyzed structure and responds through functions that reproduce the behavior of the adopted material. Eventually, the FE model allows us to evaluate stresses and strains in the component starting from the displacements calculated on the nodes, defined during the geometry discretization process.

The accurate geometric representation, the correct representation of the stress–strain relationships, and an adequate discretization of the geometry are the parameters that determine the compliance of the model with respect to the real behavior of the physical component. The constitutive laws of the material, relating stresses and strains, can be linear or non-linear with different levels of complexity, thus substantially determining the type of FE element to be selected. The use of more complex laws and geometries normally leads to a greater fidelity of the model but increases the computation time required for its solution, and it is therefore normally necessary to adopt compromises between the complexity of the model and the computational efficiency. For this reason, although the approach guarantees to a certain extent an "intrinsic" correspondence of the results to physical reality, the models normally also require experimental validation to certify the results and the accuracy obtained.

Figure 8.2 schematically illustrates the approach that can be used to create FEM models in the railway sector for the analysis of the main structural components of the vehicle. In the field of vehicle structural design, historically these obviously started with models developed according to the Class 1 digital models, starting from the 80s, for the most critical components such as axles and wheels, and this type of approach is still widely adopted at least for the initial design phase. However, given the importance of the structural strength requirements for the safety and durability of the railway vehicle, design using a digital model is substantially always supported by experimental verifications. These experimental verifications are defined based on mandatory regulations that differ according to the railway administration where the vehicle is intended to operate. In the past, experiments were not necessarily related to the calculation models used for design and this often led to the need for several iterations to comply with the project requirements. Recently, however, given the growing interest in weight reduction and optimization of the manufacturing process, the design

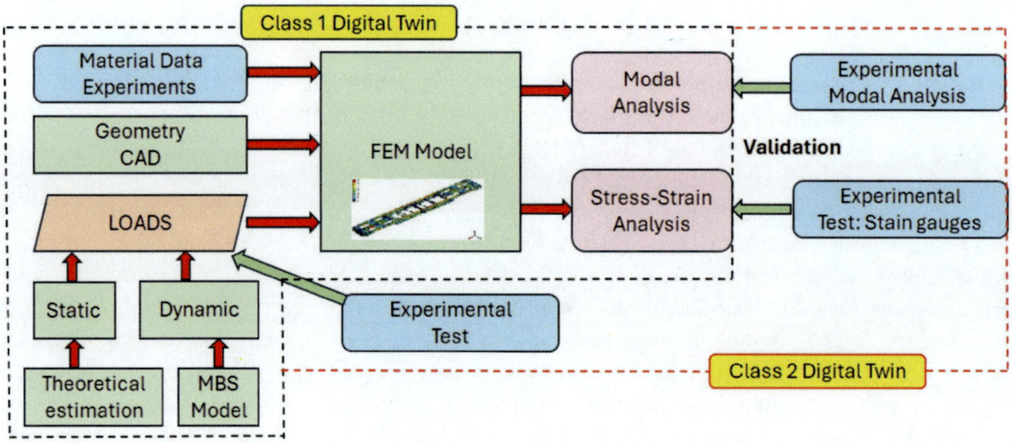

FIGURE 8.2 FEM model common design process.

with digital models and the subsequent experimental verification have been aligned, thus obtaining a digital twin according to Class 2 of the previous list. In fact, experimental validation cannot be performed on each physical point of the structure or all of the potential operating conditions, so specific locations and working scenarios are selected for a comparison between experimental measurements and model results. Validation by measuring strains using strain gauges applied to the vehicle structure in predefined positions is used for validation in static conditions, while dynamic validation can be carried out by comparing natural frequencies calculated with the FE model with the results of experimental modal analyses. Validation of the dynamic structural behavior, for example for fatigue life, is much more complex and normally cannot be carried out directly given the large statistical dispersion of the fatigue phenomenon. In this case, regulations normally provide for a subsequent verification of the most critical components (wheels [6], bogies [7], axles [8]) by means of 1:1 scale fatigue tests which must demonstrate resistance within the same cycles also considered as design specifications for the numerical model. In this case, however, the results of the numerical model and the experimental test are not directly comparable. Some studies are carried out any way to develop methods to validate the fatigue tests [9].

In the structural evaluation of the model, the determination of dynamic load has primary importance. The applied loads can be prescribed by standards and regulations, which often indicate the maximum design loads [10]. Nonetheless, to consider more realistic loading conditions, loads can be derived from dynamic models, for example, MB models of the vehicle and the infrastructure, which will be analyzed in the next section. The determination of stresses sometimes requires the use of codes capable of operating in a multiphysics environment, for example, to estimate the aerodynamic load [11] or the thermal load on the components of the braking system [12]. In this case, the digital twin can assume a more complex architecture if the simulations require a bidirectional interaction between two or more codes that can be achieved with co-simulation techniques [13].

The use of FE numerical models for structural calculation has been of fundamental importance for the development of modern vehicle structures that are increasingly constrained by weight reduction requirements, e.g., for the structure of the car bodies and the bogie frame. Recent structures have been weight-optimized using numerical topological optimization techniques that involve running multiple simulations by varying the component geometry, determining each time the stress and strain field, and iterating until reaching the lightest solution capable of satisfying the structural requirements [14]. This process is configured with a "V" cycle repeated iteratively until the minimization of an objective function (the weight in this case) is achieved while maintaining the initial requirements unchanged (see Figure 8.3a).

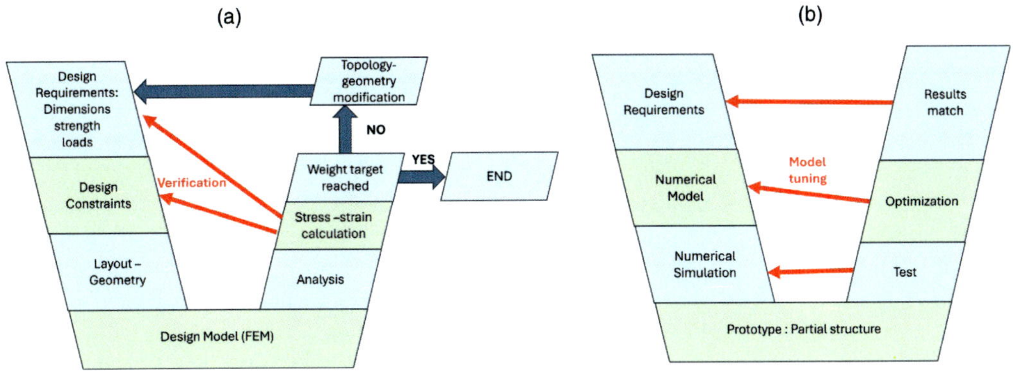

FIGURE 8.3 FEM process in the case of topological optimization (a) and in the case of manufacturing process tuning (b).

The structural design of railway vehicles presents aspects of significant complexity related to welding processes, especially when considering that innovative processes such as friction welding, stir welding, and spot welding, which are of particular relevance with the use of innovative or high-resistance materials being introduced. Given that the weld can be the critical part of the component and that it is necessary to evaluate in detail the calibration of the process to achieve the required performance, innovative simulation methods of the process are being developed with FE codes dedicated to the study of dynamic and highly non-linear phenomena that however require advanced validation techniques [15–17]. It is often necessary to calibrate the model based on welding tests performed on the real component and this configures a process with bidirectional data exchange between the digital twin (model of the welding process) and the test performed on the component in the initial stages of production. This type of approach can be considered a Class 3 digital twin because the fine-tuning of the manufacturing process occurs through continuous interaction with the numerical model (see Figure 8.3b).

A field where structural analysis has become crucial for the development of new technologies is that related to crash analysis [18,19]. The need to ensure suitable crash resistance of railway vehicles is increasingly relevant in design and has been incorporated into the most recent regulations. However, pursuing the development of crash-boxes and crash-resistant structures through a complete test on the vehicle is not experimentally feasible. The same regulations [20], therefore, allow for integrated design between FE numerical models and tests limited to scale prototypes or portions of the structure to be tested, combining the results of the experimental test with the results of the numerical model to ensure adequate validation [21]. This approach configures a bidirectional data flow between the numerical model of the vehicle (digital twin) and the experimental analysis and can be classified at least partially as a Class 3 system. Since the numerical model of the vehicle used for the crash can also be used considering non-destructive loads, it is also possible to carry out further experimental validations of the model through static and dynamic tests of a non-destructive nature, achieving a strengthening of the destructive tests carried out on specific portions of the structure that mostly dissipate the deformation energy. The destructive tests, even if partial, are still essential to validate the characterization of the material in a non-linear context.

8.3 DESIGN OF VEHICLE DYNAMICS

In railway vehicle design, MB codes are widely used to predict and optimize the dynamic behavior of the vehicle. The development of MB codes, specifically for railway applications (GENSYS, MEDYNA, NUCARS, VAMPIRE, and VOCO), can be traced back to the turn of the 70s and 80s. Subsequently, some general-purpose MB codes, such as Simpack and Adams, were enhanced with specific libraries for railway applications.

MB codes can be used to model an entire vehicle or specific vehicle substructures and the numerical model represents a virtual copy of the real object to be designed. For this reason, the MB model can be considered a Class 1 digital twin. In this case, there is no data exchange between the physical world and the numerical model, which in this case is only used to carry out simulations to assess how the mechanical system will behave. Although the numerical model can be built with a high level of detail, it always includes uncertainties with respect to the physical system, such as input data, irregularities, the actual shape of wheel and rail profiles, damping, component tolerances, etc. For this reason, the vehicle approval process requires experimental tests to assess that the vehicle guarantees suitable dynamic performance.

Annex T of the EN14363:2016 Standard [10] proposes, for certain cases, to replace the standard approval process with a virtual approval process, thus using the numerical model as a digital twin of the vehicle. In general, the requirements to apply the method proposed in Annex T are as follows:

- Extension of tests under conditions where the full test program has not been completed;
- Approval of already approved vehicles following modifications;
- Approval of new vehicles by comparison with an already approved reference vehicle;
- Analysis of the dynamic behavior of an already approved vehicle in the case of failure.

For each of the cases listed above, the standard requires compliance with specific and severe conditions to be able to apply the virtual approval process. For all cases, it is required that the numerical model, which will be used to carry out the simulations, is validated against experimental data. The numerical model of the vehicle can therefore be categorized as a Class 2 digital twin, as the data flow is unidirectional.

Annex T of the EN14363 Standard provides two possible methods for the validation of the numerical model (see Figure 8.4):

- Method 1 is used when data from standard approval tests is not available. In this case, the Standard allows validation of the numerical model against experimental results obtained on the real vehicle, when it is subjected to various inputs (described in the Standard).
- Method 2 is used if the experimental results recorded during a standard approval test of the vehicle are available. Experimental data is then used to validate and calibrate the numerical model.

In both methods, once the model is validated, a Class 2 digital twin of the vehicle is available. This, in the cases described by the Standard, replaces the real vehicle to conduct on-track standard approval tests. Australian Standard AS7509:2017 [22] also contemplates the possibility of replacing standard on-track approval tests with numerical calculation methods.

Figure 8.5 highlights the differences between the traditional vehicle design and approval strategy and the new paradigm of virtual homologation. In the past, the numerical MB model was built and adopted as a support to the early design stages (e.g., calibration of primary and secondary suspension stiffness/damping, location of lateral bump stops, calibration of traction and trailing system), but then the manufacturing of the prototype and the execution of experimental homologation tests were mandatory. Nowadays, virtual homologation is leading the path toward the possibility of assessing the vehicle dynamic performances using a validated numerical model, which acts as a Class 2 digital twin and replaces the real physical vehicle.

8.3.1 Active Systems

Active systems mounted on railway vehicles to improve their dynamic performance have been studied for several decades. This section is not intended to provide a comprehensive state-of-the-art; rather, it aims to present the literature works that best fit within the digital twin paradigm.

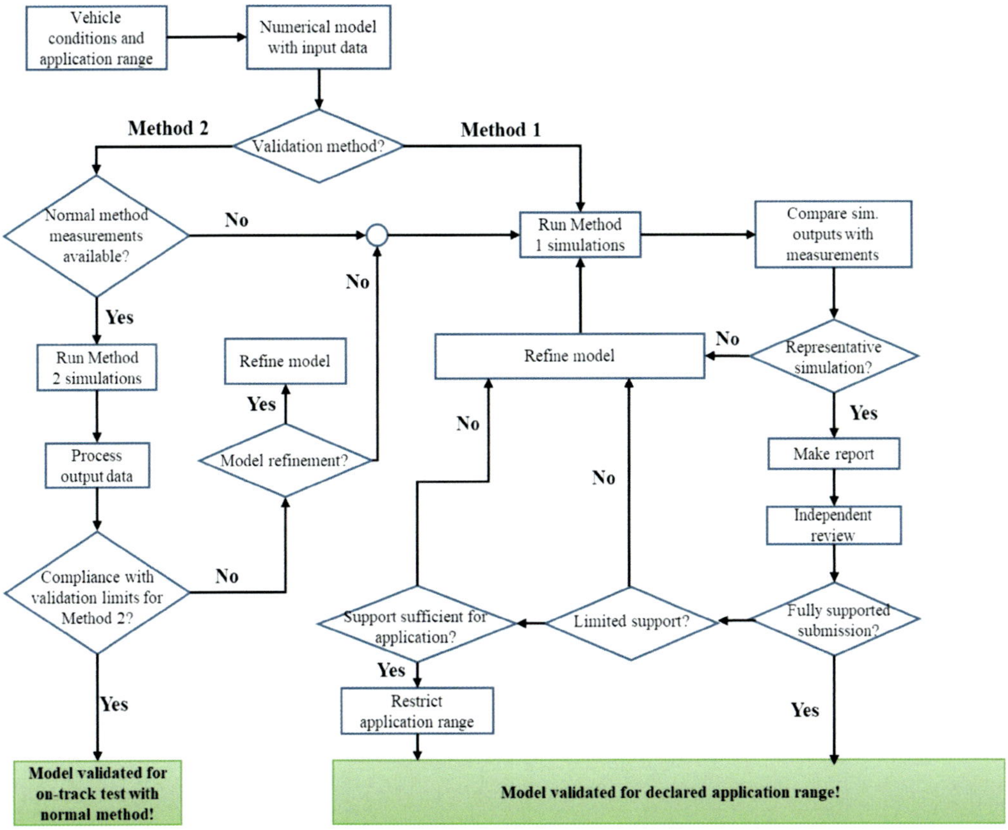

FIGURE 8.4 Virtual homologation according to methods 1 and 2 of the EN14363 Standard.

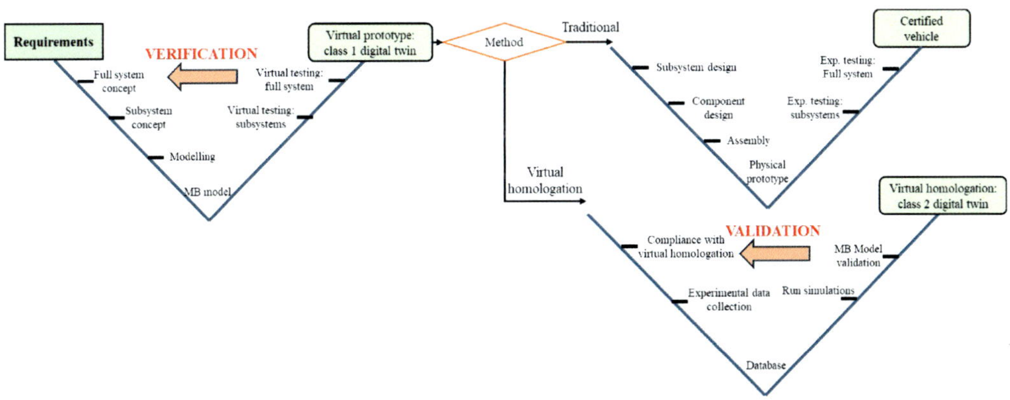

FIGURE 8.5 Railway vehicle design and acceptance: traditional and virtual approval.

The use of digital twins in the design and testing phase of vehicle dynamics is particularly used for the study of active suspensions. Commonly, during the design phase, the algorithm controlling the behavior of the active suspension is calibrated and tested on a laboratory test bench, which simulates the entire rail vehicle or subsystem. In this case, the algorithm controlling the behavior of the suspension is a digital twin of the vehicle, which calculates and generates the command signals for the actuators controlling the physical system, based on the input data it receives from the physical counterpart (test bench). In this case, the control algorithm can be considered a Class 3 digital twin since there is bidirectional communication between the digital twin and the physical world. An example can be found in [23], where the authors evaluate the possibility of using an active air suspension, together with an active lateral suspension to control body tilting and improve passenger comfort during curve negotiation. The test bench consists of a full-scale bolster supported by two air springs, which are loaded by means of vertical hydraulic actuators simulating the static and quasi-static loads generated when the vehicle is running on a curve. The control is realized by a combination of a feedforward control, which sets the deformation rate of the suspension to ensure rapid response of the system, and a position feedback control, which controls in a closed loop the deflection of the suspension. Obviously, the control system for the evaluation of spring deflections and deformation rates includes the mathematical model of the air suspension, which is indeed the digital twin of the vehicle's secondary suspension.

Another example of the use of a digital twin is proposed in [24], where the authors describe the design of an active vertical primary damper for the reduction of coach vertical vibrations. The work describes the development of the control algorithm that can determine in real time the force to be generated by the vertical damper to reduce the vertical oscillations of the coach. The linear-quadratic-Gaussian control law uses a Kalman filter to estimate the force that must be exerted by the damper to reduce the vertical bounce of the coach. The Kalman filter, based on a simplified 5 degrees of freedom (DOFs) model of the vehicle, allows the estimation of the dynamic state of the vehicle from the vertical accelerations measured on the bogie frames. The control system was then verified on a test vehicle and a real track to assess its efficiency. In this case, the Kalman filter can be considered a Class 3 digital twin of the vehicle since it interacts in real time with the physical system.

In [25], the authors use a hardware-in-the-loop (HIL) test rig to design a semi-active secondary suspension to reduce the lateral vibrations of the coach. The control of the suspension is realized by means of a magneto-rheological fluid damper. The HIL system consists of a dSPACE system that runs the vehicle numerical model in real-time, a magneto-rheological damper, and a ball screw drive, which imposes the required deformation on the damper. The force generated by the damper is measured using a load cell. The numerical model of the vehicle, running in real time, evaluates the relative displacement between the bogie and the car body. This is then used as the reference position for the ball screw drive, which applies the load on the damper. The force generated by the damper is measured by the load cell and used as input by the real-time numerical model of the vehicle. This architecture allows the damper control to be studied and developed in a laboratory environment without the need for expensive on-track experimental campaigns. Also, in this case, the vehicle model represents an example of a Class 3 digital twin.

Another application of the use of digital twins can be found in [26], where the authors propose a HIL simulation for the study of the control system of the lateral position of a tramway vehicle with independent wheels. The control is based on providing different torque to the two wheels of the wheelset to control the bogie's lateral position. The system is composed of the driver's lever, which regulates the traction level, the traction control which determines the traction torque to the motors and the centering torque, the speed control which determines the resistant torque that the load motor must supply to maintain the required speed profile, and the real-time model of the vehicle which runs on a dSPACE system. The traction control communicates in real time with the dSPACE system, and, based on the lateral position and yaw angle of the wheelset calculated by the real-time model, it determines the difference in torque to be applied to the left and right wheels to keep the wheelset centered with respect to the track. The vehicle model has 26 DOFs and is the digital twin of the real vehicle. Again, the digital twin is Class 3 because there is a real time, bidirectional data exchange between the digital model (numerical model running in real time) and the physical system (traction motors).

Further examples of digital twins in the dynamic design of railway vehicles concern the study of pantograph–catenary interaction. HIL test benches are often used to study active control systems that maintain contact between pantograph and wire in the presence of track irregularities and high speeds. For example, in [27], the authors propose a HIL system in which the physical pantograph interacts with the numerical model of the catenary. The numerical model of the catenary, which is the digital twin of the real catenary, runs in real time and exchanges data bidirectionally with the actuators acting on the pantograph (physical system).

8.4 SUBSYSTEMS AND COMPONENTS DESIGN

8.4.1 Brake Components

Braking is one of the aspects that most influence the safety of rail vehicle operations. Today, most rail vehicles rely on an air brake system, with a working principle that strongly depends on the adhesion conditions at the wheel–rail contact [28]. The latter can vary greatly due to environmental conditions and the possible presence of contaminants at the interface, acting as a third-body layer that can lead to a drop of adhesion with respect to dry conditions [29,30]. Common examples of degraded adhesion include wet conditions, as well as the presence of snow or leaves at the interface. Under these circumstances, braking torques can lead to a saturation of the available adhesion with possible wheel locking, thus generating undesirable consequences such as increased stopping distances, damage to the rail surface, and generation of wheel flats. All these issues result in reduced operation safety and comfort, with an increase in the costs required for wheel and track maintenance.

To overcome the aforementioned drawbacks, modern railway vehicles, especially passenger vehicles, are equipped with wheel slide protection (WSP) systems, which can cut or limit the brake cylinder pressure when full sliding conditions are identified on one or more wheelsets. The operation of WSP systems is based on the integration of data measured by the onboard sensors (wheelset rotation speed, brake cylinder pressure), physical models (wheelset/vehicle dynamics, valve fluid dynamics, degraded adhesion models), and proper control logic for slide detection and the implementation of the required countermeasures through the actuation of additional electro-pneumatic components.

Therefore, it can be stated that the operation of the WSP system requires the development of a digital twin with the railway vehicle as its physical counterpart. The architecture of the digital twin is shown in Figure 8.6a. During a braking operation, the digital twin receives information from its physical counterpart via sensors, which provide parameters measured onboard the vehicle, such as brake cylinder pressure and speed. In addition, the digital twin includes a physical model, which predicts the dynamics of the wheelset or the entire vehicle. The model can estimate the available adhesion level and detect sliding on the wheelsets. At the same time, in future upgrades, the digital twin could also use experimental data recorded in previous braking operations, allowing for the identification of sliding by means of data-driven approaches [31]. The control logic implemented in the digital twin is then able to estimate possible sliding on one or more wheelsets and make decisions to maintain constant braking pressure (hold) or even temporarily release it (release). Consequently, the WSP system can be considered as a Class 3 digital twin, since it is able to actively control the braking operations based on measured quantities collected from the real vehicle that are processed by a physical model and dedicated control logic. To ensure better performances and enhanced safety, modern WSP systems are potentially able to evolve and self-configure according to the actual operating conditions of the vehicle. This is the case of the adaptive WSP system described by Imbert et al. [32] from Wabtec, which can perform an ad-hoc tuning of the controller parameters to minimize errors thanks to adaptive least mean square filters.

The architecture of the operating logic shown in Figure 8.6a is similar to the one proposed by Alturbeh et al. [33] in the frame of the LABRADOR project, where several physical models implemented for each wheelset and each vehicle concur to predict vehicle dynamics and implement the appropriate correction strategies. This is achieved by actively controlling not only the WSP system but the electrodynamic braking and sanding systems as well.

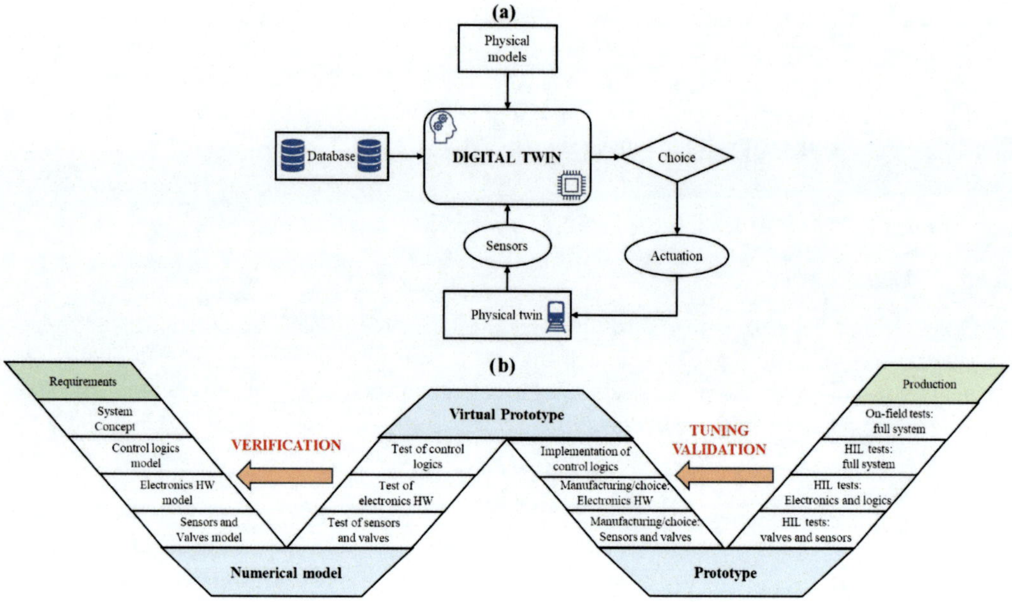

FIGURE 8.6 WSP system: (a) working principle and (b) W-shaped design model.

The digital twin at the foundation of the WSP system must be developed and conceived at the earliest stages of design, which nowadays can be carried out according to a W-shaped model, as shown in Figure 8.6b. The system design starts with the definition of the product requirements, which involve both customer needs and prescriptions from international standards such as UIC 541-05 [34] and EN 15595 [35]. In the descending branch of the first V phase, a system concept is designed which complies with the requirements, with a dedicated control logic. The latter requires a corresponding concept of the electronic hardware, the pneumatic system, and sensors. The result of the first stage is hence the development of a numerical model for each subsystem, which is then tested in a virtual environment in the ascending branch of the first V phase. This leads to a virtual prototype of the system, which can drive the prototyping of the electronics, pneumatics, and sensors, thus obtaining a final prototype of the full system. In the last branch of the W, the subsystem prototypes are tested, initially with HIL benches as allowed by international rules and standards [35] to ensure faster times in the early prototyping stages. In fact, HIL benches make it easier to test control logic and pneumatic/electronic components by reproducing vehicle dynamics using numerical models running in real time [36–38]. This also makes it possible to simulate failure or undesired working conditions in controlled environments, which would not be easily reproducible on track. The experimental testing phase can then be concluded with online testing of the entire WSP system, which then leads to the marketing and series production of the final system.

8.4.2 Traction Components

Like braking systems, traction systems for rail vehicles also rely on adhesion at the wheel–rail contact and play a key role in the safety, punctuality, and efficiency of the entire rail transport system. Although the evolution of modern passenger transportation networks is increasingly leading to distributed traction configurations, especially in the case of high-speed trains, in freight trains wagons are trailed and traction is provided by a small number of locomotives, which provide the tractive effort for the entire composition.

Spiryagin et al. [39] from the Centre for Railway Engineering (CRE) at Central Queensland University have recently highlighted how locomotive design is becoming increasingly complex since it has to consider different requirements from different stakeholders:(1) prescriptive rules and standards, (2) performance demands from several customers who will operate the locomotives, and (3) the needs of the railway infrastructure managers who require that new solutions do not produce excessive damage to the track. For this reason, the same authors have proposed a three-stage virtual V-shaped design model with the development of a digital twin of the system as the end point (see Figure 8.7). In the first stage, an MB model of the locomotive is developed and virtually verified to simulate running dynamics. At this stage, the main input data can be derived from CAD models of the carbody, bogies, power unit, etc. In the second stage, a mechatronic model of the traction system is implemented and virtually tested. Finally, in the bottom Stage 3 of the V, the previous models are combined, thus enabling the integration of additional models and modules, such as models for wear and rolling contact fatigue prediction and virtual driver models, etc. This leads the path toward a digital twin of the entire system. Please note that, for models developed in Stages 1 and 2, CRE researchers proposed a reproducible method for virtual verification.

The design strategy proposed by CRE researchers is extremely versatile and can be applied to various traction solutions, such as diesel-electric locomotives [39], battery-electric locomotives [40], and hybrid hydrogen-battery locomotives [41], by only changing the model developed in Stage 2. Obviously, the digital twin conceived according to Figure 8.7 only derives from virtual models of the locomotive and traction system and is therefore not based on experimental data measured on real physical prototypes. In the actual realization of the locomotive, a second triple V could therefore be added. The second triple V, not shown in Figure 8.7, could include the experimental tuning and validation of the developed model, with reinforcement of the control logic thanks to experimental data. More in detail, if the traction system also includes an antiskid system, the latter could have its own dedicated digital twin to be developed according to the strategy shown in the previous section for the WSP system.

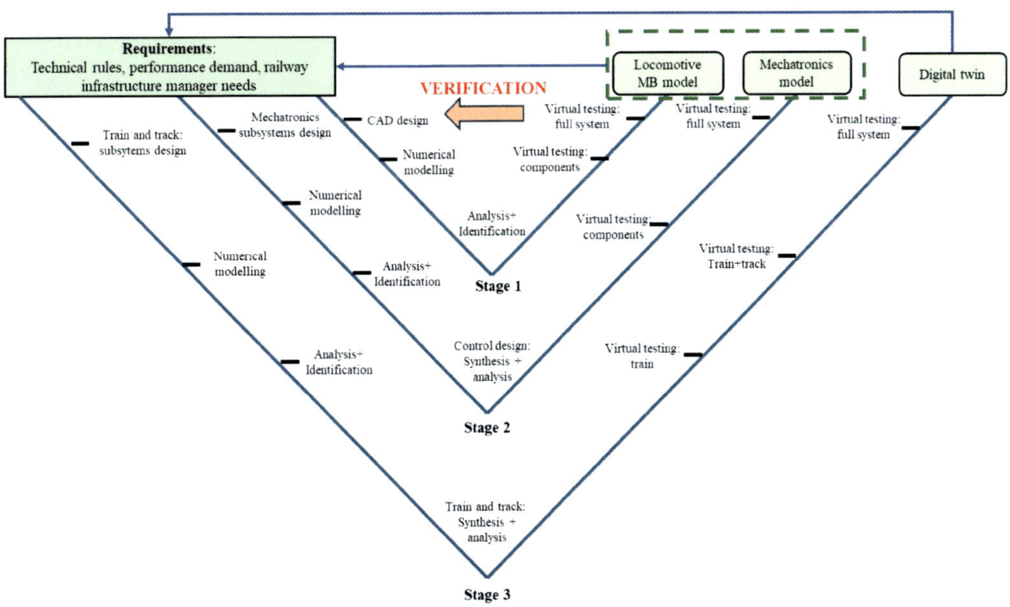

FIGURE 8.7 Triple V-shaped virtual design model for locomotives proposed by CRE researchers. (Redrawn from Spiryagin M, Wu Q, Polach O, et al. Problems, assumptions and solutions in locomotive design, traction and operational studies. *Railway Engineering Science*. 2022;30(3):265–288.)

8.4.3 Train Control Systems

Modern railway vehicles are equipped with automatic train protection (ATP) systems, which support the train engineer during railway operations. As a remarkable example, the ERTMS/ECTS system, which can now be considered a global "de facto" standard, allows information on speed limits and line occupancy conditions to be sent to the train driver in the locomotive. Based on information provided by the driver, including train composition and braked weight, the system can estimate the limit speed profile for the train. If the speed of the train exceeds the permissible limit value, the system automatically activates emergency braking, thus ensuring high levels of safety for the rail transport network. Although the estimation of the limit braking curve is carried out according to simple kinematic models, based on the inertial characteristics of the train and its braking capacity, this type of system can be considered a Class 3 digital twin, as it is capable of automatically commanding braking if the limits of the vehicle or line are not respected.

As highlighted by Flammini et al. [42], the integration of ATP and automatic train operation systems makes it possible to achieve grades of automation equal to 2 or higher, resulting in an automatic train control system. Although these systems are automatic, at the current state, they cannot be considered autonomous, as autonomy requires intelligence, i.e., the capability of making choices based on previous experience. For this reason, the same authors have defined the paradigms of intelligent train protection, operation, and control as future upgrades of current automatic systems. Intelligent systems should be able to rely on artificial intelligence (AI) and machine learning (ML) to combine the physical models of train dynamics and kinematics with measured data and historical data. However, as these systems must guarantee high levels of safety, the greatest limitation for the effective implementation of intelligent trains is still represented by regulatory bodies, since intelligent solutions require international standards to explicitly allow for the implementation of AI and ML techniques.

REFERENCES

1. Kritzinger W, Karner M, Traar G, et al. Digital twin in manufacturing: A categorical literature review and classification. *IFAC-PapersOnline*. 2018;51:1016–1022.
2. Taylor DR, Scholes A. Computer aided design of railway vehicles. *Railway Engineering Journal*. 1975;4(6):63–71.
3. Schneider JB. A review of a decade of applications of computer graphics software in the transportation field. *Computers, Environment and Urban Systems*. 1984;9(1):1–20.
4. Lashin G, Feldhusen J. A CAD-based tool for development of large layouts. *Research in Engineering Design*. 1996;8:217–228.
5. Ihme J. *Rail Vehicle Technology*. Springer, Wiesbaden, 2022.
6. EN13262:2004: EN13261:2009: Railway applications – Wheelsets and bogies – Wheels – Product requirements. European Standard.
7. EN13749:2011: Railway applications – Wheelsets and bogies – Method of specifying the structural requirements of bogie frames. European Standard.
8. EN13261:2009: Railway applications – Wheelsets and bogies – Axles – Product requirements. European Standard.
9. Kraft S, Blum H, Alves C. Calibration and validation of fatigue design models for railway car bodies considering uncertainty. *Fatigue & Fracture of Engineering Materials & Structures*. 2023;46. doi:10.1111/ffe.14160.
10. EN14363:2019 Railway applications – Testing and Simulation for the acceptance of running characteristics of railway vehicles – Running Behaviour and stationary tests. European Standard.
11. Lu Y, Zhang D, Zheng H, et al. Analysis of the aerodynamic pressure effect on the fatigue strength of the carbody of high-speed trains passing by each other in a tunnel. *Proceedings of the Institution of Mechanical Engineers, Part F: Journal of Rail and Rapid Transit*. 2019;233(8):783–801. doi:10.1177/0954409718809469.

12. Grzes P, Oliferuk W, Adamowicz A, et al. The numerical–experimental scheme for the analysis of temperature field in a pad-disc braking system of a railway vehicle at single braking. *International Communications in Heat and Mass Transfer.* 2016;75:1–6. doi:10.1016/j.icheatmasstransfer.2016.03.017.
13. Li T, Zhang J-Y, Zhang W-H. Co-simulation of high-speed train fluid-structure interaction dynamics in crosswinds. *Zhendong Gongcheng Xuebao/Journal of Vibration Engineering.* 2012;25:138–145.
14. Cascino A, Meli E, Rindi, A. A strategy for lightweight designing of a railway vehicle car body including composite material and dynamic structural optimization. *Railway Engineering Science.* 2023;31:340–350. doi:10.1007/s40534-023-00312-6.
15. Afazov S, Mistry K, Uzunov K. Welding simulation of railway bogie frame side beam: Analyses of residual stresses, clamping forces, distortion and prediction of fatigue S-N curves. *Proceedings of the Institution of Mechanical Engineers, Part F: Journal of Rail and Rapid Transit.* 2023;237(1):33–40. doi:10.1177/09544097221094986.
16. Ya-na L, Cheng-tao L, Bin Y, et al. The application of welding numerical simulation on two typical welded structures in railway vehicles. *Interaction and Multiscale Mechanics.* 2012;5(2):145–155. doi:10.12989/imm.2012.5.2.145.
17. Chen FF, Xiang J, Thomas DG, et al. Model-based parameter optimization for arc welding process simulation. *Applied Mathematical Modelling.* 2020;81:386–400. doi:10.1016/j.apm.2019.12.014.
18. Gao G, Wang S. Crashworthiness of passenger rail vehicles: A review. *International Journal of Crashworthiness.* 2019;24(6):664–676. doi:10.1080/13588265.2018.1511233.
19. Nie Y, Tang Z, Liu F, et al. A data-driven dynamics simulation framework for railway vehicles. *Vehicle System Dynamics.* 2017;56(3):406–427. doi:10.1080/00423114.2017.1381981.
20. EN15227:2020: Railway applications – Crashworthiness requirements for railway vehicle bodies. European Standard.
21. Kirkpatrick SW, Schroeder M, Simons JW. Evaluation of passenger rail vehicle crashworthiness. *International Journal of Crashworthiness.* 2001;6(1):95–106. doi:10.1533/cras.2001.0165.
22. AS7509:2017: – Rolling Stock – Dynamic Behaviour. Australian standard. Rail Industry Safety and Standards Board, Brisbane, Australia. Published June 2017. https://www.rissb.com.au/products/as-7509-rolling-stock-dynamic-behaviour/
23. Alfi S, Bruni S, Diana G, et al. Active control of airspring secondary suspension to improve ride quality and safety against crosswinds. *Proceedings of the Institution of Mechanical Engineers, Part F: Journal of Rail and Rapid Transit.* 2011;225(1):84–98.
24. Sugahara Y, Kojima T. Suppression of vertical vibration in railway vehicle carbodies through control of damping force in primary suspension: presentation of results from running tests with meter-gauge car on a secondary line. *WIT Transactions on the Built Environment.* 2018;181:329–337.
25. Kwak MK, Lee JH, Yang DH, et al. Hardware-in the-loop simulation experiment for semi-active vibration control of lateral vibrations of railway vehicle by magneto-rheological fluid damper. *Vehicle Systems Dynamics.* 2014;52(7):891–908.
26. Oh YJ, Lee JK, Liu HC, et al. Hardware in-the-loop simulation for active control of tramcars with independently rotating wheels. *IEEE Access.* 2019;7:71252–7126.
27. Facchinetti A, Bruni S. Hardware-in-the-loop hybrid simulation of pantograph-catenary interaction. *Journal of Sound and Vibration.* 2012;331(12):2783–2797.
28. Günay M, Korkmaz ME, Özmen R. An investigation on braking systems used in railway vehicles. *Engineering Science and Technology, an International Journal.* 2020;23(2):421–431.
29. Lewis SR, Dwyer-Joyce RS. Effect of contaminants on wear, fatigue and traction. In: Lewis R, Olofsson U, editors. *Wheel-Rail Interface Handbook.* Cambridge, UK: Woodhead Publishing Limited; 2009. pp. 437–455.
30. Fulford CR. *Review of Low Adhesion Research.* Rail Safety & Standard Boards; 2004.
31. Namoano B, Ruiz-Carcel C, Emmanouilidis C, et al. Data-driven wheel slip diagnostics for improved railway operations. *IFAC-PapersOnLine.* 2022;55(19):103–108.
32. Imbert L, Frea M, Filoni M. The adaptive wheel slide protection to increase performance and lower LCC. *13th World Congress on Railway Research*, June 6–10, 2022, Birmingham, UK.
33. Alturbeh H, Santos J, Stow J, editors. Towards a valid model of train braking system at low adhesion condition. *The Fifth International Conference on Railway Technology: Research, Development and Maintenance (RAILWAYS2022)*, 2022, Montpellier, France. Edinburgh, UK: Civil-Comp Press.
34. UIC 541–05:2020. Brakes – Specifications for the construction of various brake parts – Wheel Slide Protection device (WSP).
35. EN 15595:2018. Railway applications – Braking – Wheel slide protection.

36. Kim H-Y, Lee N-J, Lee D-C, et al. Hardware-in-the-loop simulation for a wheel slide protection system of a railway train. *IFAC Proceedings*. 2014;47(3):12134–12139.
37. Lee N-J, Kang C-G. Wheel slide protection control using a command map and Smith predictor for the pneumatic brake system of a railway vehicle. *Vehicle System Dynamics*. 2016;54(10):1491–1510.
38. Pugi L, Paolieri G, Giorgetti M, et al. HIL testing of wheel slide protection systems: Criteria for continuous updating and validation. *Railway Engineering Science*. 2023;31(2):108–121.
39. Spiryagin M, Wu Q, Polach O, et al. Problems, assumptions and solutions in locomotive design, traction and operational studies. *Railway Engineering Science*. 2022;30(3):265–288.
40. Spiryagin M, Cole C, Wolfs P, et al., editors. Heavy haul digital twin – Proof of concept for 8-axle high adhesion 10 MWh battery electric locomotive. *Proceedings of the 12th International Heavy Haul Association Conference 2023 (IHHA Rio 2023)*, 2023, Rio De Janeiro, Brasil. Virginia Beach, VA: International Heavy Haul Association (IHHA).
41. Spiryagin M, Wright G, Wolfs P, et al., editors. Preliminary investigation of an 8-axle freight locomotive design as a possible net-zero-emission motive power solution for Australian rail operational tasks. *CORE2023: Conference on Railway Excellence Conference Proceedings*, June 19–21, 2023, Melbourne, Australia. Online: Railway Technical Society of Australasia (RTSA).
42. Flammini F, De Donato L, Fantechi A, et al. A vision of intelligent train control. reliability, safety, and security of railway systems. In *Reliability, Safety, and Security of Railway Systems. Modelling, Analysis, Verification, and Certification*. RSSRail 2022. Lecture Notes in Computer Science, vol 13294, Collart-Dutilleul, S., Haxthausen, A.E., Lecomte, T., Editors, 2022. Cham: Springer International Publishing. https://doi.org/10.1007/978-3-031-05814-1_14

9 Digital Twins for Railway Infrastructure Asset Management and Maintenance

Marco Macchi, Adalberto Polenghi, and Irene Roda

9.1 INTRODUCTION

The digital twin (DT) technology has recently gained momentum in the railway industry. The expectation is that this technology will contribute to the fulfillment of the usual goals to ensure safety, reliability, and service performance standards in the operations of the transportation system; this will occur through an innovative and cost-effective approach that is potentially exploitable across the different phases of the lifecycle of the infrastructure assets.

The DT technology is concretized into a physical-digital convergence that is developed by leveraging different underlying technologies (such as the Internet of Things (IoT) and computer-based modeling), ultimately allowing the virtualization of, and the connection with, the physical entities and systems in the built environment.

Each underlying technology has been around for a while as can be proven by various projects related to the digital transformation of the railway infrastructures. Nevertheless, when it comes to DTs being adopted in railway industry applications, with a particular focus on railway infrastructures, DTs have gained more attention only in very recent years.

In their literature survey, Ghaboura et al. [1] offer one of the most up-to-date and comprehensive overviews of DTs in the railway sector, with a particular focus on advanced railway systems. Their findings indicate that DT technology for railways remains in an early adoption phase. Therefore, current challenges relate to the identification of the best technologies that can be combined to fit the purpose of DTs for railway application cases (referred to as DTRs in the remainder of this chapter), creating value for the management of railway infrastructure assets and their maintenance.

The challenge is not far from similar research discussions in other application contexts. A DT is generally understood as an aggregation of technologies like the IoT, big data, and artificial intelligence (AI) and simulation as some of the fundamental technologies for DT development in different fields of application such as manufacturing, health care, and transportation [2–5]. Despite the growing interest in these fields, no agreement has been reached so far on the best blend of technologies that may allow DTs to be deployed and used for decision-making with evident benefits. Also, when it comes to the management of railway infrastructures and their maintenance, the DTR can leverage various promising technologies, but the methods needed to exploit such technologies seem under-explored, and further research progress is required [1].

The trend toward the integration of computer-based systems with advanced information and communication technologies is generally observed in the architecture, engineering, and construction industries [6]. Within this trend, the first step toward the virtualization of civil infrastructure systems consists of the creation of digital models of the physical entities and systems [7]. This step is required to enable further development and integration steps for the full deployment of DTs to operate in connection with the built environment, finally realizing the physical-digital convergence.

As an end result, the DT development promises to result in a framework that provides advanced decision support built on the monitoring and prediction, and the optimization and improvement

DOI: 10.1201/9781003492146-9

tasks, for an engineering asset along "cradle-to-cradle" processes [6], from the beginning of life to end of life of the railway infrastructure assets [2]. This attractive perspective leads to envisioning the DTs as a means to deal with the engineering and management complexity of advanced railway systems. This is setting a potential for future applications and benefits.

Considering such expectations, the objective of the present chapter is to outline both the current state and the future direction of the research and innovation that puts the DT at the center as a transformative concept to move from conventional computer-based systems to digital platforms fed by varied data sources. As a methodological approach, the literature findings are adopted to provide a collection of observations from published works, finally leading to insights into the current state. Possible directions for the future state are then deduced, resulting in foresight of the DT adoption for infrastructure asset management (AM) and maintenance. The insights at the current state are organized in three application areas of DTs: condition assessment and predictive maintenance, operations and maintenance management, and railway infrastructure lifecycle management; therefore, the foresight at the end of this chapter aims at condensing the expectations from the three application areas into a set of conjectures providing a vision for the future development of a DTR.

This chapter is correspondingly organized. Section 9.2 presents a railway asset taxonomy as the base ground for the management of assets in the railway infrastructures, which also opens the possibility of integrating advanced concepts due to digitalization, such as the DTs. Then, Section 9.3 provides an overview of the opportunities and challenges currently perceived by addressing the DT development in the railway transportation industry, with a particular emphasis on railway infrastructures. Section 9.4 discusses the DTR application areas, pinpointing the more relevant insights for each area at the current state-of-the-art. This chapter ends with Section 9.5 whose objective is to outline an integrated view of DTR development stemming from the current advances and, subsequently, to hypothesize the way the DTRs shall be adopted for managing railway infrastructures and their maintenance.

9.2 ASSET TAXONOMY AS A BASIS FOR THE RAILWAY INFRASTRUCTURE AM AND MAINTENANCE

A railway infrastructure network consists of a large collection of diverse types of systems and structures operating with high interdependencies. Given this complexity, it is advisable that the DTR development resolves around the ability to decompose the railway system into sub-systems and elementary components to reach a level of discretization and indenture that fits the purpose. Indeed, the DTs should have specific assets targeted and then support certain asset-related decisions [8]; the effectiveness of such decisions will then depend on the ability to describe the assets within the railway system so that data management is exploited at best [9].

To this end, it is worth remembering that a railway infrastructure can be defined as a network or portfolio since it is composed of many assets whose types and functions are different [10]. If compared with other similar networks like the one of electricity, gas, and water, a classification is valid also for a railway infrastructure, which leads to distinguishing assets in terms of distributed and discrete assets:

- Distributed assets are assets whose dimensions are not comparable to each other, and one of them could be much larger than the others to the point that they cannot be identified and located as points on a geographic map.
- Discrete assets are assets whose dimensions are comparable so that they can be located and identified appropriately, as point, on a geographic map.

The classification between distributed and discrete assets allows the listing of the main railway infrastructure assets, as depicted in Figure 9.1.

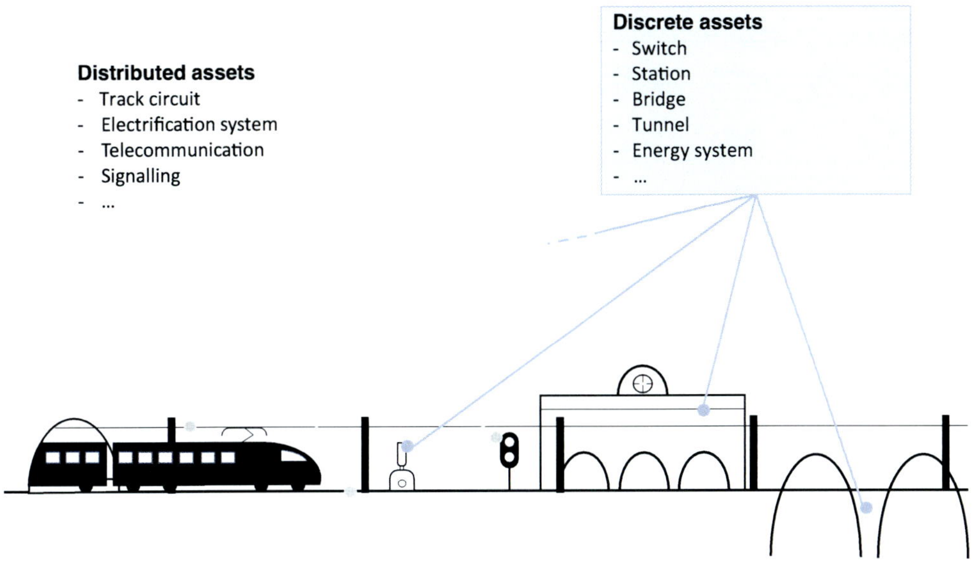

FIGURE 9.1 Classification in terms of distributed and discrete assets, with examples.

The logical interconnections between distributed and discrete assets generate systems and structures at next-higher levels such as the line of route on which the service is carried out. Different levels of aggregation of assets in the railway infrastructure are in fact required in light of the levels of decision-making. According to the UIC Asset Management Working Group [11], this comprises five levels: asset component, asset class, route section, line of route, and transport network. While the first two levels imply a focus on a specific asset, either a distributed asset or a discrete asset, the last three levels relate to the multiple interactions between different types of assets as well as the connection with auxiliary assets or third-party assets.

Moving from specific asset components to the transport network, the complexity increases as it results from decision-making that should consider a multi-stakeholder perspective. Each level of aggregation has its own related decision-makers, with influencing stakeholders, both internal and external; as a matter of fact, seamless integration between the various aggregation levels is deemed necessary within the management system. Therefore, the potential of DTs for railway systems should be investigated to this end, considering their role at the different levels of aggregation of the railway infrastructure and their support for multi-stakeholder decision-making.

Figure 9.2 shows the multiple levels of aggregation, built upon the aforementioned UIC taxonomy, remarking on the influential role of the human decision-makers with respect to which it is advisable to conceive the DT-based decision support.

From the modeling perspective implied by the DTs, a few general requirements can be correspondingly identified:

- The ability to properly describe the railway infrastructure is an essential basis for the DTR development and needs to deal with the discretization problem in the scope of a wide geographical distribution; particularly, proper segmentation is then required, especially for the distributed assets [12].
- The DTR development must also cope with the time dimension: a railway infrastructure is not only dispersed geographically in space but unevenly evolving over time, hence proper time modeling is fundamental for the management supported by the DTs of the different assets, systems, and structures [13].

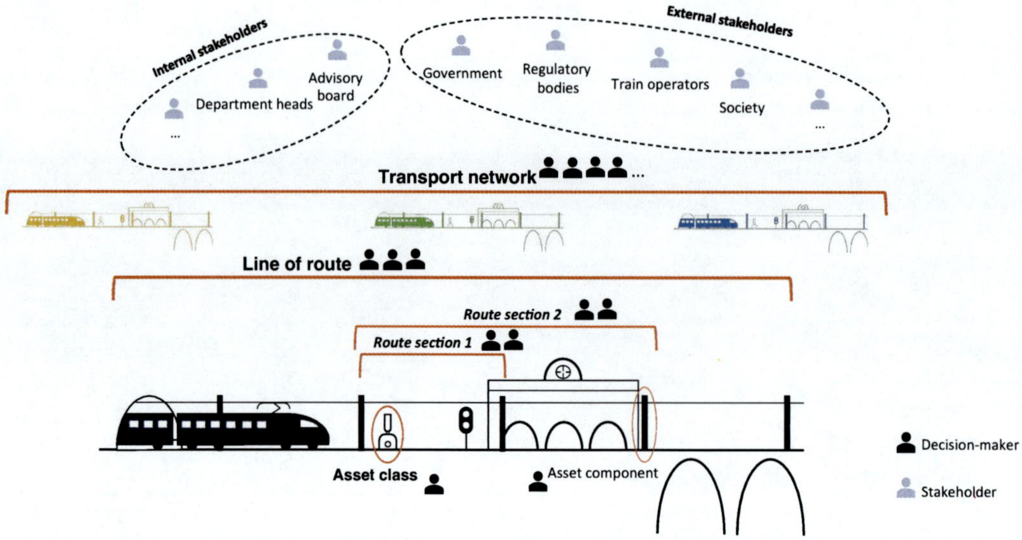

FIGURE 9.2 Multiple levels of aggregation, from asset component to transport network with several decision-makers and stakeholders.

In the end, a good taxonomy of distributed and discrete assets within the railway system, able to cope with space and time information, is the first enabler to allow proper management of the heterogeneity of the properties in the railway infrastructure. This should be considered as a basis from which to start to support the transition from the current information systems toward the integration of digital platforms where DTs are developed and hosted [14].

9.3 OPPORTUNITIES AND CHALLENGES FROM DTs FOR RAILWAY INFRASTRUCTURE AM AND MAINTENANCE

9.3.1 Railway Infrastructure AM and Maintenance

In the railway sector, infrastructure AM is to be considered as a strategic approach that aims to deliver value to customers, funders, and stakeholders by managing infrastructure assets sustainably and cost-effectively throughout their lifecycle. AM goes beyond traditional management practices in three key ways [15]:

- *Sustainable value delivery*: AM aligns maintenance, renewal, and enhancement activities with stakeholder priorities. Rather than focusing solely on asset condition, it prioritizes actions that deliver the most valued outcomes—such as safety, reliability, and efficiency—at the lowest whole-life cost.
- *Cross-functional integration*: AM provides a unified framework that spans organizational functions and asset disciplines. It integrates different maintenance and operational roles, and often extends coordination between infrastructure managers and contractors, ensuring that all efforts contribute to a common set of objectives.
- *Evidence-based decision-making*: By leveraging data on how assets degrade and fail, AM informs maintenance decisions, optimizing when and how interventions should occur. This approach enhances reliability and helps to extend the useful life of railway assets.

Maintenance, as an essential component of AM, focuses on the actions necessary to keep assets in optimal condition. Effective maintenance involves not only corrective actions but also preventive and predictive strategies that prevent asset failures before they occur. In the railway context, maintenance activities must be closely aligned with AM principles to ensure sustainable asset performance.

Despite the potential benefits, railway AM faces several challenges. These include the need for high-quality data to support evidence-based decisions, difficulty in coordinating across functions and with external partners, and the pressure to maintain performance levels while minimizing costs. For all these reasons, the DTR offers promising solutions to these challenges. By creating a virtual representation of physical railway assets, DTs may provide real-time data to enhance both AM and maintenance activities. For AM, DTs facilitate a holistic view of asset conditions and performance, improving strategic planning. For maintenance, DTs enable predictive insights, allowing us to identify and address issues before they impact operations, thus optimizing resources and extending asset life.

9.3.2 Overview of DT-Related Technologies and Techniques for Railways Systems

Management systems in the railway industry typically aim to incorporate an integrated view built on asset and maintenance management practices [16]. This is a working method that has been apparent in railway organizations for some years. It is motivated by the purpose of addressing the socio-political complexity due to diversified stakeholders' requirements, and the need to leverage multiple disciplines to approach the technical complexity of the railway systems as a networked infrastructure type.

DT-related technologies and techniques are promising in this context. They bring about multifaceted and multidisciplinary characteristics in the applications adopted to manage the assets' portfolio, thus leading to advanced support with a high potential to cope with the complexity of the railway infrastructures [17].

However, the current state of DT adoption in railway management systems seems to be driven more by the vision and long-term expectations surrounding these technologies than by practical benefits that have been directly observed or fully realized in this field.

The evidence collected by extant literature reviews is helpful to provide a first insight into the relevance of DT development in the railway transportation industry. Particularly, a broad overview is hereinafter provided to remark on the most prominent issues at the state-of-the-art and practice of the research and innovation actions aimed at embedding DTs in the management systems concerned with railway infrastructure AM and maintenance.

First, it is worth noting the high variety of DT-related technologies and techniques. These provide meaningful expectations for the opportunities brought about by their usage in railway systems.

- The most popular technologies and techniques for DTs are the IoT and computer-based modeling techniques including building information modeling (BIM), simulation, and AI. Virtual reality (VR), augmented reality (AR), and extended reality (XR) are also cited in recent works [1]. The most discussed ones are evidently BIM and IoT; AI is observed with a steep growth of interest; VR/AR/XR models and technologies are perceived to be in an incubation phase. Among the entire set of computer-based modeling techniques, BIM has been a key topic for a relatively long period [18].
- The maintenance management system appears to be the major area that is generally addressed by the DTR for building opportunities. This is coherent with specific insights on maintenance types and techniques applicable within the maintenance management system, such as structural health monitoring (SHM), defect/damage detection, and, more generally, condition monitoring and predictive maintenance [1,19].
- The DTR is also envisioned as a promising approach to optimize system performance and explore any improvements regarding the efficiency of the railway systems. In particular, various domains can be involved within the optimization or improvement enabled by the DTs; it is leading to consider multiple viewpoints, concurrently related to industrial,

financial, structural, planning, and/or operational aspects [1]. It finally confirms the multifaceted and multidisciplinary nature envisioned by the DT-based applications in railway systems.

Based on this first insight, it is notable that the computer-based modeling techniques in the realm of a DTR demonstrate a high potential for application in the railway infrastructure AM and maintenance field, finally leading to different types of models, i.e., BIM and simulation, AI, and VR/AR/XR models. This conforms with the general evidence of the DT research where the digital representations of the physical entities and systems are an important lever for the (high-fidelity) modeling of the real world [20]. Particularly looking at the railway systems, modeling can then take different shapes not only because of the different digital representations of the railway infrastructure but also as a consequence of the different types of data that are adopted. Indeed, data should be taken into account as they are a determinant factor for the variety of choices when developing the DTR; they lead to enriching the DTs from a diversity of data sources, which is a key feature when talking about the developments that can be expected by digitalization. Given this broad sense of DTR development, the opportunities and challenges currently perceived are elaborated in the next sub-sections.

9.3.3 Opportunities from Modeling Techniques in DTs for Railway Systems

Computer-based modeling techniques are a key part of the research and innovation with DTs in general. Correspondingly, they are frequently discussed in the applications related to the DTR. In particular, different modeling types can be exploited as discussed in the literature; among others, BIM and 3D/6D modeling, physics-based/physics-informed modeling and simulation, and AI models are hereinafter presented as they are claimed for their high potential in a DTR. Their description is enriched with some examples showing the possible applications in railway systems; these are presented as concrete, although limited, proofs of the related opportunities.

- BIM is popularly used by relying on its hybrid modeling type which makes it flexible for the integration of different data sources, such as geometrical, spatial, and non-spatial data [21]. This allows us to understand the BIM as a key means for the visualization of asset-related information to support the asset and maintenance managers in their decision-making [22]. For example, a BIM model can help to visualize and analyze asset-related information such as the location and condition of the assets [23]. Therefore, a BIM model can allow the asset and maintenance managers of railway infrastructure to judge informed decisions based on data and information from the monitored assets.
- 3D modeling is a physical modeling type and it is used to represent the physical dimensions (width, length, depth) of railway infrastructure assets. This may be an aid to support the optimization/improvement of, e.g., track geometry, track maintenance and inspection, and even the design of a new infrastructure [24]. 6D modeling adds to the physical dimensions other types of information. In the case of a construction project, these may regard the construction sequencing (4D), cost information (5D), and project lifecycle management (6D) [25]; in the case of maintenance operations, these may be concretized in maintenance costs and time schedules (4D and 5D) together with the maintenance impacts (6D) [26]. In general terms, having a 6D modeling type allows to address the diversity of asset-related information to be managed according to the requirements of the different activities along the asset lifecycle; this extends the physical (geometrical) dimensions, to be considered as a baseline, with other key dimensions such as cost, time, and impact dimensions. To extend the asset-related information, some scholars have mentioned even more information layers (e.g., 7D) to incorporate environmental impacts and related measures (e.g., carbon footprint, toxicity in case of hazardous materials). It is worth finally noting that BIM, by

its native flexibility, is an elective "tool" to develop 6D or more dimensional perspectives, being potentially applicable along the lifecycle of the assets.

- Physics-based modeling is a modeling type that leverages the numerical simulation technique to model and analyze complex systems [27]. Physics-based modeling is also defined as an approach that relies on a partial understanding of the physical phenomenon and then transferred into a set of mathematical equations representative of physics; therefore, the numerical modeling technique allows the mathematical equations to be solved by means of numerical methods on computers. Physics-based modeling is an opportunity that can be used to predict the behavior of complex assets within the railway system, even without the need to wait for the occurrence of failures or anomalies, e.g., to predict how a bridge will perform in different environmental conditions, to apply in advance the maintenance efforts to the purpose [28]. This modeling approach can exploit numerous numerical methods, such as the finite element method (FEM), finite difference method, and finite volume method. Moreover, it can leverage open source and/or commercial multi-physics simulation packages; this becomes an opportunity for the DT implementation. The advantage of physics-based modeling is that the models are ruled by the laws of nature, which allows the avoidance of potential biases due to data-driven modeling.
- Simulation is a key modeling technique in the realm of DTs, as originally conceived in general for DTs and also envisioned in applications from diverse sectors [29]. Particularly, simulation leverages the data integration provided by the DTs, and the possibility of a continuous collection of operational data that allows to envision, as a potential, even its (close to) real-time adoption to support decision-making. Considering the specific viewpoint of railway operations, a real-time simulation could be used to model and simulate the behavior of a railway system, inclusive of different physical assets such as trains, tracks, electrification systems, and signal and control systems. Particularly, Adeagbo et al. [6] embed simulation in their conceptual framework for the SHM-focused advanced train transit systems DTs. In this context, they remark on different functionalities of the simulation required within DTs, such as (1) training data-driven models in different cases, from unobserved characteristics or unmeasurable data to extreme events; (2) prediction; (3) inference and re-adaptation; (4) visualization. Other authors expect that future applications for advanced decision-making could include real-time simulation for different possible use cases, e.g., (1) for adjusting the schedules based on real-world conditions due to trains' delays or changes in track conditions and/or other assets represented by the DTs [1]; or (2) for assessing specific properties of the transport network during its operational runtime, for example, by means of a load-flow analysis with different solutions for the substations sharing the power between different tracks [30]. These types of models are then informed by physics (i.e., physics-informed), even though they require various simplifications of the physical phenomenon that is modeled, especially building on discretization and segmentation of the railway systems into more manageable models for a (close to) real-time usage. This may lead to hybridized simulation with data-driven methods.
- AI is a key technological lever that can be embedded in DTs. Common opportunities can be achieved by incorporating AI such as the followings: on the one hand, the DTs could be used to generate synthetic data related to the assets' failures by means of fault injection techniques [31], which is an aid to configure and improve the performance of AI models in their detection and prediction capabilities for defects and faults. On the other hand, AI could be exploited to develop intelligent DTs that, besides their prediction capabilities, could take advantage of the recent trend toward the so-called cognitive approaches and, in particular, the cognitive DT concept [32]. An AI technique that is popularly cited in DTs for railways is machine learning, with an emphasis on its adoption in predictive maintenance as the target application [33]. Nevertheless, this is not the only technique. In their survey, Ghaboura et al. [1] collect different applications of interest to AI, and three

primary domains of techniques are indicated as the most evident ones to be employed for the DTR, such as machine learning, deep learning, and computer vision. Overall, it can be asserted that various types of AI techniques are employed at the basis of railway applications, across a wide range of different types of assets such as railway stations, bridges, tracks, overhead contact systems, etc., and for different use cases in terms of tasks related to the railway AM and control, planning optimization, prediction of maintenance needs, prediction of delays, and defect inspection [6].

As a concluding remark of the present section, it can be asserted that BIM is evidently a key computer-based technique upon which developing the foundations of the DTR. Indeed, the representation via BIM is a key step as it provides the main lens that brings the DTR to life [6]. On the contrary, different kinds of simulation models and AI models should be adopted in light of the target railway assets, use cases and tasks that have to be supported in the decisional process. Opportunities generated by such modeling techniques can be identified and exploited in this wide scope of application.

9.3.4 Opportunities from Data, Information, and Knowledge Management in DTs for Railway Systems

The exploitation of data is a key aspect leading to integrated data-driven modeling within the DTs. It corresponds to incorporating AI and any other data-driven modeling technique within the DTR, with the ultimate effect of increasing the capabilities to predict, optimize, and improve the system performance in (close to) real-time operations.

Different types of data are managed through the digital representations provided by DTs. Data flowing through a DTR may include the following types: engineering data (i.e., from railway design), operational and maintenance data (i.e., from the activities performed in railway operations and maintenance), environmental data (influential for the assets in their effective behavior, see for example the weather conditions), and any other recorded data stored building up historical datasets. This variety is not only an opportunity but also a necessity to deal with the multifaceted and multidisciplinary characteristics of the assets' portfolio resulting from the heterogeneity of multiple assets in the railway infrastructure.

The adoption of a DTR is expected to take advantage of the data already available in the legacy systems for the data and information management of each railway organization, such as the enterprise resource planning and computerized maintenance management systems as extant solutions. The DT development may also induce a step forward to manage the railway infrastructure, by leveraging novel solutions from the advanced engineering informatics.

Among such advances, sensors and monitoring systems evidently play a crucial role in real-time data collection for decision-making and optimization, even though they are not enough to fully exploit the value generated from data. Therefore, different systems, tools, and devices for data and information management, also extending to knowledge management, could be adopted to facilitate effective data flowing through the digital representations resulting from DTR development. An outline of the major systems, tools, and devices is hereinafter provided, spanning from data collection to data handling and management [1,6]; the opportunities along the entire process are evoked through some exemplary cases.

First of all, the rise of new sensing techniques and technologies is today a remarkable trend, which is promising for DTR development, besides the adoption of traditional sensors. Some techniques/technologies are herein presented, with their advantages.

- The optical fibers [34–37] allow to build sensing networks capable of measuring various responses from the target structures; this is especially relevant for the SHM applications, where the measures lead to different features gathered along multiple directions and with high frequencies.

- The scanning techniques and sensors are today demonstrating their validity for the reverse engineering of any extant system, and, in particular, they could be exploited for faster modeling of the extant railway infrastructures [38]. Among others, see some examples: (1) light detection and ranging technologies can measure distances through laser light and can thereafter create 3D maps of the physical entities and the environment [39]; they can be used to model tracks, tunnels, bridges, and other structures in the railway infrastructures; (2) point cloud scanning is a 3D scanning technique capable to capture shape and position of the physical entities through the collection of large sets of data points designated as the point cloud data; this is another way to build infrastructure models for DTRs [40].
- Accelerometer sensors should be also cited, as a type of more "traditional" sensors applicable to moving entities to detect changes in motion and orientation but also to non-moving entities to detect unexpected behaviors. They can be applied to specific target physical entities and systems such as railway crossings or for a track line assessment based on the rail stress–strain evaluation [41,42].

Besides the sensing techniques/technologies, the emergence of DTs is facilitated by the combined adoption of other technological enablers required by data-driven modeling [6]. These are required throughout the entire process from the data collection to the data handling and management: (1) IoT is adopted for communication with sensors, actuators, and other devices (such as personal computers and mobile devices), with the ultimate goal of supporting the data collection from field operations as it is required by the DTs; (2) communication technologies, such as 5G and WiFi, are used to facilitate proper data exchange between the physical entities and their DTs, ensuring the requirements in terms of target performances such as speed and latency; (3) cloud computing offers the support of multiple services for computation, storage, and retrieval of massive amounts of data, which is finally allowing the DTs to be continuously updated, as well as data to be further processed to achieve a thorough information and knowledge management. In the end, with the data flowing through these technologies up to the digital representations provided by the DTs within cyberspace (i.e., typically within the cloud), some key opportunities can be considered in general; these are also relevant to railway infrastructures. A particular focus is now provided in regard to the seamless and secure collection of data from the field and to the data handling and management up to knowledge integration in railway infrastructures.

- IoT is expected to provide a platform to facilitate a seamless collection to move data toward data processing and analysis within BIM tools, with the final aim of monitoring the condition of railway infrastructure assets [43,44]. Moreover, the integration under the umbrella of the IoT paradigm should consider data security; thus, new wired and wireless networking protocols are required to incorporate data encryption [45] as a functionality to mitigate data security concerns.
- Once a seamless and secure data collection is achieved, different data handling and management methods and systems have to deal with the huge and heterogeneous amount of data from the real physical entities along their entire lifecycles. Among them, it is worth mentioning the role of semantic technologies and ontologies therein [32] and standards for product model data exchange [46]. In particular, conceptual, logical, and ontological models should ease the integration of knowledge into the data-driven modeling; it leads to a promising mechanism for understanding the complex relationships among the different railway characteristics, which then emerges as an aid for decision-making and optimization [47]. More generally, semantic modeling approaches are expected to provide effective support in the complexity management of DTs, which is a key property of the cognitive DT concept [32].

Overall, when DTs are being developed, a general opportunity emerges in relationship to the management of data, information, and knowledge related to the infrastructure assets: data, information, and knowledge are in fact handled and integrated by using DTs with an approach that inherently leads to a railway asset-centricity. This is an essential aspect of the philosophy of AM in general [48] and has evidently a validity also for the railway infrastructure case.

9.3.5 Challenges in the Development and Usage of DTs for Railway Infrastructure AM and Maintenance

Given their interesting opportunities, it is now worth remarking also how DT-related technologies and techniques are subject to different challenges for their usage within railway systems. Hereinafter, a short list of selected challenges is outlined, inspired by different extant publications as [1,6,33].

- Interoperability is an essential requisite for DTR development. This is motivated by the fact that current railway infrastructures are typically managed by different systems, tools, and devices. It leads to challenges in data integration, eventually impacting decision-making, especially for the tasks expected at middle-to-top management levels. Therefore, interoperability is a key to facilitating the DTR's actual applicability and usage.
- Railway infrastructure is a complex multi-asset and multi-stakeholder system. It is evidently not reasonable to consider the creation of a single DT for the entire railway infrastructure. This can be motivated by the different purposes determined by the expected applications and the heterogeneous characteristics of the multiple assets within the railway infrastructure. The asset taxonomy (see Section 9.2) could be used as a key guiding concept to help manage such complexity and heterogeneity.
- Guidelines are required for the design and development of DTs for railways. These should allow to focus on promising applications of DTs for railways, helping to move from requirements specification and the subsequent digital representations, through data collection, handling, management, and analysis, up to the decisional processes supported and their validation when put in place.
- The computer-based modeling techniques may have practical limits, and this could lead to some specific choices for the digital representations in a DTR. First of all, some limitations may exist due to the knowledge and expertise required to build an accurate model for the physical system. Also, the computational intensiveness required for the DT runtime may impose some practical constraints. In this regard, it is worth citing the simulation modeling. Today, the computational efficiency may not habilitate simulation exploitation in a full DT sense, in (close to) real-time modality. Simulating the heterogeneous assets such as catenaries, rails, trains, power supply systems, etc., is in fact a demanding activity due to the large scale of the transport network. Therefore, simplifications and modeling approaches should be properly introduced to enable meeting the requirements, including the time constraints related to the horizon when the decisions should be supported.
- Understanding of the full potential of the DTs in the railway domain has not yet been achieved as a lack of exploration of the underlying technologies applied in the railway systems is still evident. A comprehensive framework of DT-related technologies and techniques could be useful, to provide a deeper understanding of opportunities and challenges brought by the DTR with the lens of specific experiences and/or even expert judgments in possible railway use cases. The framework might include both technologies that have been discussed for some years (such as BIM and IoT), partially explored (such as AI), and those technologies that are still in their incubation phase (such as VR/AR/XR).

9.4 MAJOR APPLICATION AREAS FOR DTs IN RAILWAY INFRASTRUCTURE AM AND MAINTENANCE

9.4.1 Condition Assessment and Predictive Maintenance

Condition assessment and predictive maintenance are emerging as the relevant research focus for the application of a DTR. Indeed, it can be observed that DTR models are primarily proposed for the health monitoring of complex systems and for the identification of failure risks before their occurrence. Coherently with this trend, SHM is one of the predominant cases when discussing the application of DTs to the management of railway infrastructures. In terms of target asset classes, the application of a DT for SHM to improve condition assessment and predictive maintenance of railway infrastructure mainly refers to bridges (particularly steel bridges) and switches, followed by tunnels, catenaries, and tracks.

SHM entails four main steps [49]: defect identification, damage localization, damage assessment, and structure life prediction. Despite the controversial definitions of DTs in railway systems and the lack of an intensive adoption (see Section 9.1), the use of DT for SHM may have a tremendous impact as it will support all four phases of SHM [6]. In the same work, the authors argue that a DT is at least composed of a digital model, that many times collapses into the BIM model, joined with sensors, whose data are elaborated by employing additional discipline-specific models such as FEM or crack propagation, which are the core part of a DTR. However, different forms of DT may be found, all aimed at supporting SHM, as can be exemplified by the following cases, starting from the most targeted asset class, namely the bridges.

For DT-based SHM, in [50], the DT is built upon a fatigue analysis system (FAS) joint with BIM for a steel bridge. The former represents the key system, integrating various models, whose results are displayed to the railway infrastructure manager via BIM. Indeed, the FAS relies on the adoption of a linear damage accumulation method (LDAM) combined with a FEM model developed in Ansys. The data that feed the proposed DT-based SHM are internally generated by both static and dynamic analyses provided by the LDAM-FEM model, and also externally gathered, specifically referring to real traffic, which is manually introduced for simulation purposes. A more automatic data collection from the physical to the digital entities leverages vibration data; these are sent to the cloud for a real-time fast Fourier transform whose outputs are analyzed by a supervised machine learning model that describes normal vibration clusters [51]. Differently, Chacón et al. [52] focused on the bridge structure test data, and, in particular, the integration of load tests inside the DT, focusing mainly on the need for seamless information integration to favor the digital birth of the twin; this DT will be later used for condition assessment during the operational life of the represented assets. In addition, continuous monitoring of the strain is implemented by [34], whose goal is to guarantee strain sensing in real-time and to analyze the best trade-off between sensitization and numerical FEM modeling. The data are gathered automatically via fiber optics embedded in the bridge structure and elaborated by statFEM [53]. A similar application applied to an old, riveted steel bridge is shown by [54], which leverages as well on physics-based modeling for operational and environment structural response.

Concerning railway switches, a recent example by Jung et al. [55] shows the DT for switches and crossings as a supervised machine learning model able to detect faults. The authors also developed a numerical model to describe the behavior of the system and generate related synthetic data; these data are necessary as most of the time in practice the available datasets are imbalanced, with few or nil recorded faults. The fault injection activity allowed the authors to train the supervised learning models without the switch having experienced those faults in reality, to be ready to detect them upon occurrence. A physics-based model-driven DT has also been developed to assess fatigue at the switch crossing nose [56]. This was designed specifically for the discontinuity in the wheel-rail contact. Similarly to that, in Zhang et al. [57], a DT-based condition monitoring and root cause analysis for switch movement has been developed. The final solution entails a monitoring system for electric current value joint with a numerical simulation model able to generate the baseline of functioning, which is essential to bring forward the identification of anomalies and perform possible root cause analyses.

Similarly to bridges, DTs for tunnels encompass BIM as a core technology for the representation and inspection of data mapping with respect to the physical entities. In the work by [22], the DT is intended as the representation through BIM of data coming from inspection campaigns that are reported in a grid shape for the matter of maintenance. The authors in [58] employed the tunnel serviceability index to measure and predict future states based on current available information, although Bellini Machado and Futai [22] reviewed related literature and spotted far less exploration of railway tunnels if compared to subway or road ones. Despite the less explored application cases, image-based defect detection has received increased attention from researchers as it may offer the possibility to automatically record and elaborate images of tunnels and inform about possible defects [59,60]. This has certain practical implications and constraints to be considered, such as space availability and lighting conditions.

Assuming a certain degree of DT-related knowledge transferability between the subway and road tunnels to the railway ones, it is interesting to see how semantic technologies are gaining momentum, like ontologies and knowledge graphs, enabling logical inference [61–63]. This phenomenon may be justified by the need to integrate multiple information, from different actors, in a shared knowledge base that enables reasoning, but also semantic integration so that decision-making processes are favored.

The same issues due to imbalanced data sets have already emerged in other cases and, in particular, issues of limited fault data are also suffered by the DT for catenary, in which fault injections are well exploited, backed by numerical models [64]. In this same work, the authors proposed an edge-based AI to perform catenary geometric diagnostics, with emphasis on the interpretability of the results as well as guaranteeing improved interoperability with other railway systems.

When considering a DT for rail tracks, many solutions include the wheel-track interaction as the main factor to be considered, being both directly included in the physics-based modeling and considered relevant for data-driven modeling of fault identification [65,66]. Specifically, in-service or diagnostic vehicles operated by the infrastructure manager provide relevant data sources to collect data about track health, being expressed as the track geometry assessment or the fault detection, so as to identify possible anomalies [67].

9.4.2 Operations and Maintenance Management

When dealing with operations and maintenance management, DTRs can be grouped into two main categories based on the targeted system level of the railway infrastructure and the managerial objective.

- On the one hand, when dealing with operations management, DTRs are mainly focused on combining traffic management with automation of assets like rail level crossing and switches; another stream of work is recently growing in regard to the operations management of railway stations.
- On the other hand, maintenance management implies the use of DTs to increase the effectiveness of inspections and maintenance planning activities.

To better manage level crossing operations to optimize their usage, but also improve rail-road traffic management, Djordjevic et al. [68] propose a DT able to first monitor the equipment to detect anomalies in their behavior; then possible failures are pointed out. Therefore, the DT solution leverages real-time traffic data and a mixed-integer linear programming model to eventually modify the operation of the level crossing itself and, as a future scenario, of the level crossing-train system, e.g., considering the train speed as an additional decision variable. To support decision-making for operations management, Chandaluri and Nelakuditi [69] developed a DT representing the interlocking system, modeled from an electrical point of view. The DT is firstly used for verification, but it can also be employed to analyze the effects of specific failures on system behavior and modify the operations accordingly. Concerning interlocking systems, Salierno et al. [70] developed a big data architecture for DTs to support the argument that DTs can be used for synthetic data generation. The final aim is to be able to accurately reproduce anomalous behavior via a generative adversarial network of switches to influence railway operations and maintenance management.

Concerning the operations management of railway stations, specific works are presented in the literature. Chen et al. [20] proposed a service-oriented DT for improved station management integrating data on the equipment, operation data, and fault data. A sustainability-oriented study regarding station operations management refers to the evaluation of carbon emissions given current processes and geometrical construction [24]. Further, despite not being directly related to the management of railway assets, it is worth mentioning the recent trends in developing DT solutions for passenger flows in station operations management, especially for safety reasons. In this regard, an interesting work is proposed by Padovano et al. [71], which developed a DT to improve pedestrian flow prediction, early congestion warning system as well as intelligent gate management. The proposed artifact is a simulation-based DT, run in Anylogic, that leverages data from multiple, already-in-place solutions for traffic management or surveillance systems. Assuming a broader perspective, the proposed DT may also support evacuation response planning and layout optimization. Wang et al. [72] adopt a similar approach to Padovano et al. [71] but with the aim of simulating responses with respect to emergency disposal action in case of fire, natural disaster, social security, or public health events. In both works, the main challenge referred to how to model system state changes given the complexity. Finally, Wang et al. [73] also proposed a thorough solution for passenger flow management; this is built by integrating various data sources and by merging the physical entities with virtual ones, looking for suitable business, personnel, and equipment functions based on station area to mix 3D views with related insightful information.

On the side of maintenance management, DT for railway assets is mainly developed and employed to improve inspection management and make maintenance management more effective. Regarding the former goal, a degradation-informed inspection management problem is faced by a work in Swedish railway AM that leverages multiple data sources to better manage preventive maintenance strategies and inspections [74]. To look for a way to better plan preventive maintenance, Cavieres-Lagos et al. [75] exploited BIM as the visualization interface to allow decision-makers to better plan the next maintenance actions and deadlines based on always-up-to-date information. A similar approach was followed by [76] in which current data from bridge sources can be combined with physics-based modeling to offer a thorough view of the health of the bridge to plan eventual maintenance actions.

9.4.3 Railway Infrastructure Asset Lifecycle Management

Considering the AM principles, the use of a DTR for infrastructure asset lifecycle management should extend from a local and detailed modeling of specific assets toward a more global view of the whole infrastructure in the long term, addressing performance, cost, and risk. Indeed, the risk is especially argued by Alibrandi [77] as a relevant dimension to be included when considering the lifecycle rather than a single-point-in-time analysis for civil structure.

Starting from this general mindset, the adoption of advanced technologies for railway infrastructure AM should support different decisions along the railway infrastructure lifecycle, which can be categorized as lifecycle delivery decisions and AM support decisions [78,79]. The former are the ones that are directly related to the infrastructure and for which the advanced technologies are deemed paramount for improved decision-making, including the asset creation and acquisition phases as well as the defining of operations and maintenance strategies; the latter decisions refer to the supporting processes that are vital to make the lifecycle delivery decisions effective, such as financing and human resource management.

According to the review by Liu [80], BIM and LCA are currently the most promising methods that can span the railway infrastructure lifecycle to support decision-making with long-term effects. Indeed, BIM is currently perceived as one of the main techniques that is promoting this evolution toward an integrated management of railway infrastructures. This is supported by the intrinsic, by-design ability of the BIM to follow the infrastructure lifecycle not only with concern to the virtual geometrical representation of the infrastructure assets but also including other dimensions of interest to the lifecycle activities (see also Section 9.3, for the discussion on the opportunities from BIM and 3D/6D modeling). The experience from practice, provided by Bentley and reported in a white paper, demonstrates that BIM is effectively a core part of the DT and becomes even more powerful when integrated with data from multiple sources and adopted for cross-departmental and multi-stakeholders data access and collaboration processes (see their article "Deliver better rail asset lifecycle outcomes with digital twins" on *Global Railway Review* [81]). In turn, BIM enables a combined use with LCA to support the lifecycle assessment of alternative design options [82]. In particular, BIM with multiple dimensions of information allows the creation of a DT framework where it is possible to perform the lifecycle assessments for environmental impacts, such as carbon emission, and energy consumption, and even for costs according to Life Cycle Costing; and this can even be updated by maintenance and repair activities as they are carried out.

Based on the analysis above, it can be concluded that current through-life DTR implementations primarily extend from virtual representations, such as BIM, which serve as a strong foundation for data integration and visualization. The methodologies currently applied throughout the infrastructure lifecycle mostly focus on sustainability-related performance metrics and lifecycle assessments, which also consider cost factors. However, comprehensive, fully realized examples of DTR applications for asset lifecycle management remain relatively limited compared to other application areas.

9.5 EVOLUTION TOWARD THE ADOPTION OF DTs IN ADVANCED RAILWAY SYSTEMS

The importance of DTs in railway infrastructure AM and maintenance has been increasing in recent years as is evident within three main application areas, namely the condition assessment and predictive maintenance, the operations and maintenance management, and the railway infrastructure asset lifecycle management. Each area demonstrates a different level of maturity regarding the DTR development. As the state-of-the-art is evolving at different paces in different areas, it is reasonable to assert that even the meaning of the DTR cannot yet be considered consolidated according to a transversal consensus; indeed, to the authors' understanding, the definition of a DTR could be subject to further insight that will emerge from the research works and practical experiences across the application areas, to be capable to embrace all the areas' requirements.

Building on the current status, the present section concludes this chapter by providing an outlook for the future. This is achieved by foreseeing a possible path for the evolution of DTs applied in railway infrastructures. The path is based on two main aspects:

i. An integrated view of DT development in advanced railway systems, defined as a scenario expected based on the current advances in the three application areas
ii. A foresight consisting of three conjectures for future development, proposed as a vision to drive future studies on the way the DTRs will be adopted for managing railway infrastructures and their maintenance

The integrated view of DT development in advanced railway systems consists of three levels. The level definitions are intended to correspond to the results of the progressive adoption of DTs. It is also assumed that the first two levels are centered around condition assessment and predictive maintenance, while the third and last level is concerned with operations and maintenance, and railway infrastructure asset lifecycle management.

The first level, subsequently referred to as DTR1, corresponds to the development of the first DT-based solutions for decision-making built on design, construction, and traffic data. These data can be integrated and elaborated via physics-based modeling, and this may especially be beneficial during the design phase of the infrastructure assets. Examples of such models may rely on different techniques, such as FEM and fracture mechanics. A signal-processing approach could be added to complement the physics-based models with field information. The output of physics-based models and signal processing is displayed visually to the decision-makers through the BIM so that it is possible to relate the results with geospatial references.

It is notable that this first level deals with an initial modeling approach to develop technical and engineering decisions that can prepare the ground for further applications, such as the identification of criticalities and the first steps to move into an SHM application. Moreover, this level figures out a more static view of the DTR as the physics-based models are built considering design and construction data while traffic data are fed on demand. This way, the traffic data do not refer to specific real load and traffic conditions, rather they are intended as data fed to gain a scenario analysis for the decision-maker to assess and judge the effects of their possible decisions. Comprehensive decisions should also rely on the operational data and maintenance data, but at the DTR1 level a seamless integration with extant systems is not considered; it is in fact replaced by the tasks of the analyst/decision-maker, to manually manage data on-demand to carry out the scenario analysis.

The second level of DTR, subsequently referred to as DTR2, integrates multiple data in the development of the same DT solutions, which opens the potential to leverage both the physics-based models and the data-driven models, all integrated via the BIM before being displayed to the decision-makers. Particularly, the DTR2 level can adopt all the data from the railway infrastructure, ranging from design, construction, and traffic data to operation data and maintenance data. This should be relevant for an advancement step of the SHM application.

The advanced solutions at the DTR2 level should involve the elaboration of the aforementioned data by synergically adopting the physics-based and data-driven models. Specifically, the use of physics-based modeling is common to the DTR1, but it is now extended at this second level, especially for the synthetic data generation, e.g., by fault injection technique. It is, in fact, difficult to have real datasets including a comprehensive set of data that can represent both normal, anomalous, and fault conditions and, for the latter, all possible failure events; therefore, it is an option to leverage physics-based modeling to generate highly probable fault conditions of interest with the end purpose to generate the synthetic data. These data are then used to train AI/ML models to recognize anomalous conditions and then to alert decision-makers. As another possible view of the synergistic use, it is foreseeable that hybrid simulation models, built on the data-driven methods but also informed by the physics, could be developed to improve the data-driven model performance with physics-informed properties that are tractable in the railway runtime.

At the DTR2 level, there could be a potential development that is pulled by the heterogeneity of data from design, construction, operation, and maintenance. In addition to AI/ML, it is indeed notable that the recent scientific literature outlines the utility of knowledge-based AI methods, like ontologies and knowledge graphs; these can be relevant to harmonize the data fusion from multiple sources and to habilitate automatic inference in reasoning processes. This might be relevant to obtain a benefit as augmented decision-making from the wide set of different data available at this level.

Overall, BIM still remains a vital part of integrating the results from the variety of models now available at this level, as well as displaying the interesting information that may result in the utility of the decision-makers.

The third level, subsequently referred to as DTR3, extends toward a management perspective that is oriented to a systemic view and the improvement or optimization of the railway systems. To this end, this level is particularly oriented to capture various stakeholders' expectations, often expressed in terms of performance, cost, and risk related to the railway infrastructure AM and maintenance. Such a management perspective is today less mature; however, it is promising for the future to gain an effective capability to manage the planning and control of maintenance strategies, even in the larger scope due to the relationship with the operations in the railway networks (i.e., traffic management) and the service offered toward the end-users of the transportation activities (i.e., passengers operations). Furthermore, a comprehensive view along the entire lifecycle of the railway infrastructure is also expected, to make cost-effective and value-based management decisions through the life of infrastructure assets, thus fully exploiting the potential of DTRs throughout life.

This third level is expected to require the adoption of a variety of innovative methods, such as AI-based methods, simulation, and others, in conjunction with more "traditional" ones, especially for the lifecycle-oriented assessment tools, seeking to support industrial management decisions over the long term (e.g., life cycle analysis (LCA) and life cycle costing (LCC)). The adoption of these methods is generally oriented to the decision-making support over a mid- and/or long-term horizon. BIM continues to be a cornerstone as it makes easy, also at this level, the interfacing with the human decision-makers, while providing multiple dimensions of information and knowledge required for asset-related decision-making. In particular, as mid- and long-term planning typically requires multiple criteria for decision-making, it is relevant to integrate a 6D/7D modeling logic through the BIM. Eventually, it is also foreseeable that this level will facilitate the exploration of the potential of knowledge-based AI methods, like ontologies and knowledge graphs.

Figure 9.3 summarizes the view of DT levels, relating DTR1, DTR2, and DTR3. As it is possible to see, BIM is vital for DTs at all levels as it represents the interface between the outputs of the models and the decision-makers, finally allowing them to have an easier visualization of results.

The foresight hereinafter proposed is meant as a vision to drive future studies on the way the DTRs will be adopted for managing railway infrastructures and their maintenance. The three conjectures for the future development are now expressed having the integrated view of Figure 9.3 in mind as a background scenario.

- *First conjecture* "Network of DTs of the railway infrastructure as viable solution": It is assumed that the creation of a network of DTs is more viable rather than building an individual digital representation of the entire railway infrastructure. Each individual DT should aim at providing its capabilities to support specific analysis and decisional tasks (i.e., mirroring physical entities and processes, predicting their behavior, and improving

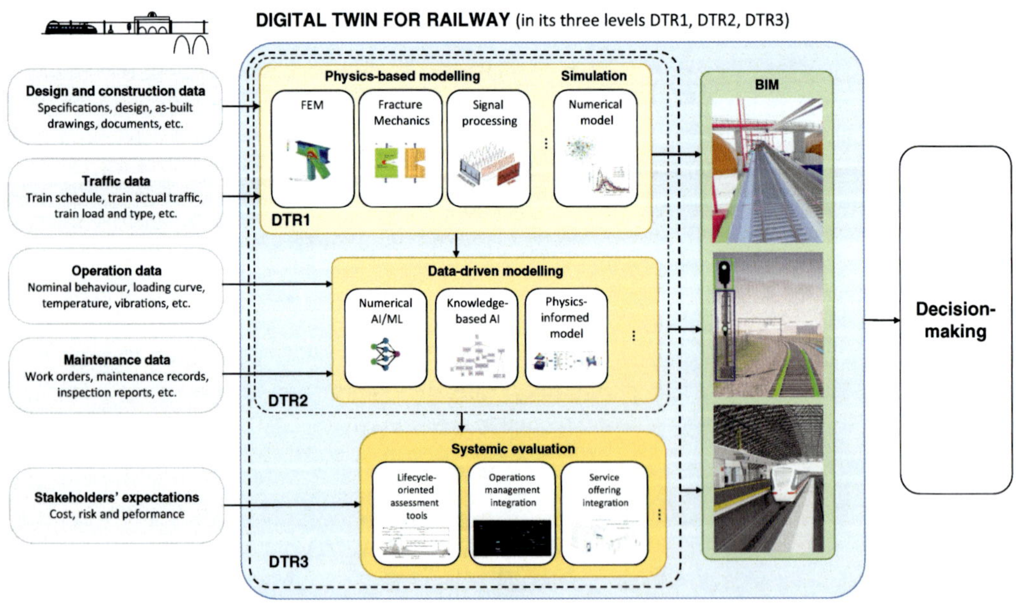

FIGURE 9.3 Digital twin levels for railway systems.

and/or optimizing the decisions). At the same time, the DTs related to the individual assets and components should be considered as a local view that has to interact with other DTs at the higher levels, i.e., route section, line of route, transport network, to reach a more global view up to a more systemic orientation.

- *Second conjecture* "Progressive and scalable deployment of the network of DTs": The network of DTs allows to provide a digital ecosystem where modeling methods, techniques and technologies are deployed according to the business and operational needs and also to the achieved maturities of the selected technologies. This leads to outlining the importance of an integrated architecture to embed the different DTRs in a seamless way; standing on this architecture, the different DTs – providing both local and global views – could be progressively incorporated, finally reaching a scalable deployment. At the current state-of-the-art, the architecture could leverage the BIM with all its potential.
- *Third conjecture* "Smooth integration of the Human-In-the-Loop in the DT-based decision-making": Human decision-makers shall interact with the ecosystem of DTs in a transparent way, which requires the design of a proper interface. This is based on the assumption that humans will continue to keep their central role; therefore, DT-based decision-making is intended toward the augmentation of decision-making tasks. The interface should take advantage of BIM's potential but also requires an understanding of the role of humans within the DT-based decisional processes. Organizational roles, and also knowledge and skills of humans, should be studied in light of the new DT-based management approach to ensure an effective design of the integration of the human-in-the-loop.

Figure 9.4 is a pictorial representation of this vision. This is a final outline that underlines the collaboration between different DTs in the network of DTs and human decision-makers to augment the decision-making capabilities within the complexity and heterogeneity of railway infrastructure assets.

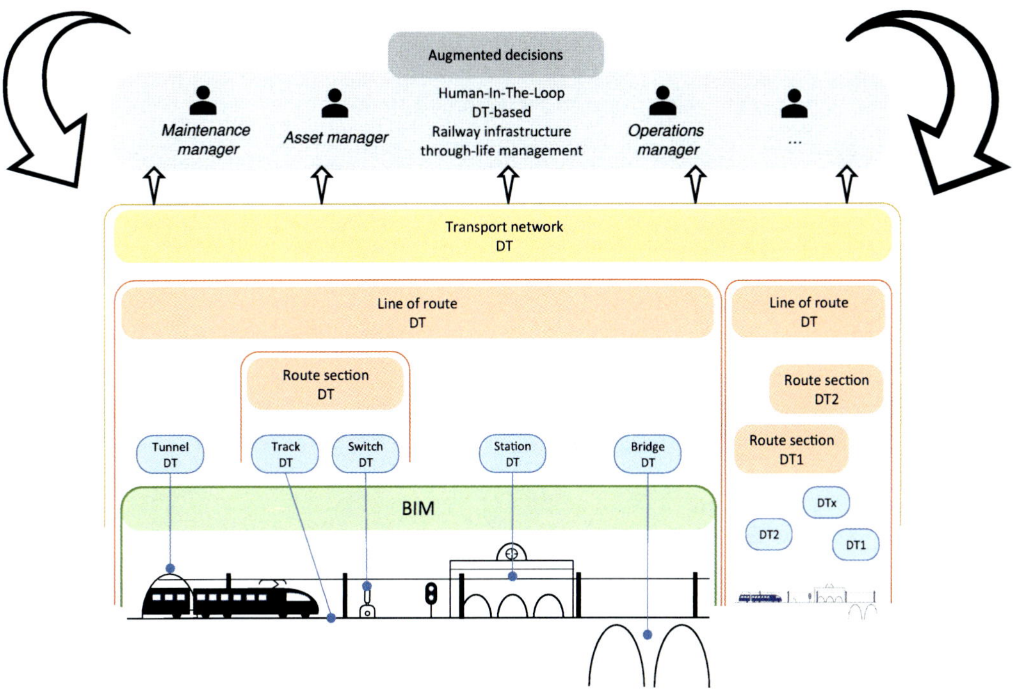

FIGURE 9.4 Future vision for human-in-the-loop, DT-based railway infrastructure through life management.

REFERENCES

1. S. Ghaboura, R. Ferdousi, F. Laamarti, C. Yang, and A. E. Saddik, 'Digital twin for railway: A comprehensive survey', *IEEE Access*, vol. 11, pp. 120237–120257, 2023, doi:10.1109/ACCESS.2023.3327042.
2. M. Jafari, A. Kavousi-Fard, T. Chen, and M. Karimi, 'A review on digital twin technology in smart grid, transportation system and smart city: Challenges and future', *IEEE Access*, vol. 11, pp. 17471–17484, 2023, doi:10.1109/ACCESS.2023.3241588.
3. M. Liu, S. Fang, H. Dong, and C. Xu, 'Review of digital twin about concepts, technologies, and industrial applications', *Journal of Manufacturing Systems*, vol. 58, pp. 346–361, 2021, doi:10.1016/j.jmsy.2020.06.017.
4. C. Semeraro, M. Lezoche, H. Panetto, and M. Dassisti, 'Digital twin paradigm: A systematic literature review', *Computers in Industry*, vol. 130, p. 103469, 2021, doi:10.1016/j.compind.2021.103469.
5. F. Tao, B. Xiao, Q. Qi, J. Cheng, and P. Ji, 'Digital twin modeling', *Journal of Manufacturing Systems*, vol. 64, pp. 372–389, 2022, doi:10.1016/j.jmsy.2022.06.015.
6. M. O. Adeagbo, S.-M. Wang, and Y.-Q. Ni, 'Revamping structural health monitoring of advanced rail transit systems: A paradigmatic shift from digital shadows to digital twins', *Advanced Engineering Informatics*, vol. 61, p. 102450, 2024, doi:10.1016/j.aei.2024.102450.
7. M. Callcut, J.-P. Cerceau Agliozzo, L. Varga, and L. McMillan, 'Digital twins in civil infrastructure systems', *Sustainability*, vol. 13, no. 20, Art. no. 20, 2021, doi:10.3390/su132011549.
8. G. C. Doubell, A. H. Basson, K. Kruger, and P. D. F. Conradie, 'A digital twin system for railway infrastructure', *R&D Journal*, vol. 39, pp. 23–34, 2023, doi:10.17159/2309–8988/2023/v39a3.
9. M. Jafari, A. Parlikad, L. Robazetti-Concho, and B. Jafari, 'Review of asset hierarchy criticality assessment and risk analysis practices', 2014, [Online]. Available: https://rosap.ntl.bts.gov/view/dot/32347.
10. S. Petchrompo and A. K. Parlikad, 'A review of asset management literature on multi-asset systems', *Reliability Engineering and System Safety*, vol. 181, no. March 2018, pp. 181–201, 2019, doi:10.1016/j.ress.2018.09.009.
11. UIC, 'A framework for whole system decision-making', 2024, [Online]. Available: https://shop.uic.org/en/other-reports/14727-a-framework-for-whole-system-decision-making.html.
12. M. R. Hernández, A. Crespo Márquez, A. G. López, and E. C. Fernandez, 'Hierarchy definition for digital assets. Railway application', in *16th WCEAM Proceedings*, A. Crespo Márquez, J. F. Gómez Fernández, V. González-Prida Díaz, and J. Amadi-Echendu, Eds., in *Lecture Notes in Mechanical Engineering*, Cham: Springer International Publishing, 2023, pp. 416–427, doi:10.1007/978-3-031-25448-2_40.
13. Y. Guo, Q. Zhu, Y. Ding, Y. Li, H. Wu, Y. He, et al., 'Efficient distributed association management method of data, model, and knowledge for digital twin railway', *International Journal of Digital Earth*, vol. 17, no. 1, p. 2340089, 2024, doi:10.1080/17538947.2024.2340089.
14. S. Zhou, A. Meierhofer, O. Kugu, Y. Xia, and M. Grafinger, 'A machine-learning-based surrogate modeling methodology for submodel integration in the holistic railway digital twin platform', *Procedia CIRP*, vol. 119, pp. 345–350, 2023, doi:10.1016/j.procir.2023.02.141.
15. UIC, 'Guidelines for the application of asset management in railway infrastructure organisations', International Union of Railways, Union Internationale des Chemins de fer (UIC), Paris, France, September 2010. isbn 978-2-7461-1878-2.
16. Y. Shang, M. Nogal, H. Wang, and A. R. M. (Rogier) Wolfert, 'Systems thinking approach for improving maintenance management of discrete rail assets: A review and future perspectives', *Structure and Infrastructure Engineering*, vol. 19, no. 2, pp. 197–215, 2023, doi:10.1080/15732479.2021.1936569.
17. S. Zhou, S. Dumss, R. Nowak, R. Riegler, O. Kugu, M. Krammer, et al., 'A conceptual model-based digital twin platform for holistic large-scale railway infrastructure systems', *Procedia CIRP*, vol. 109, pp. 362–367, 2022, doi:10.1016/j.procir.2022.05.263.
18. Y. Gao, S. Qian, Z. Li, P. Wang, F. Wang, and Q. He, 'Digital twin and its application in transportation infrastructure', in *2021 IEEE 1st International Conference on Digital Twins and Parallel Intelligence (DTPI)*, Beijing, China, 2021, pp. 298–301, doi:10.1109/DTPI52967.2021.9540108.
19. E. Dimitrova and S. Tomov, 'Digital twins: An advanced technology for railways maintenance transformation', in *2021 13th Electrical Engineering Faculty Conference (BulEF)*, Bulgaria, Varna town, 2021, pp. 1–5, doi:10.1109/BulEF53491.2021.9690822.
20. R. Chen, C. Jin, Y. Zhang, J. Dai, and X. Lv, 'Digital twin for equipment management of intelligent railway station', in *2021 IEEE 1st International Conference on Digital Twins and Parallel Intelligence (DTPI)*, 2021, pp. 374–377, doi: 10.1109/DTPI52967.2021.9540153.

21. C. Shim and H. Moon, 'Recent BIM developments and applications on railway infrastructure', in *Digital Railway Infrastructure Beijing*, China, D. Ribeiro, P. A. Montenegro, A. Andersson, and M. D. Martínez-Rodrigo, Eds., Cham: Springer Nature Switzerland, 2024, pp. 211–227, doi:10.1007/978-3-031-49589-2_11.
22. L. Bellini Machado and M. Massao Futai, 'Tunnel performance prediction through degradation inspection and digital twin construction', *Tunnelling and Underground Space Technology*, vol. 144, p. 105544, 2024, doi:10.1016/j.tust.2023.105544.
23. S. Kaewunruen and Q. Lian, 'Digital twin aided sustainability-based lifecycle management for railway turnout systems', *Journal of Cleaner Production*, vol. 228, pp. 1537–1551, 2019, doi:10.1016/j.jclepro.2019.04.156.
24. S. Kaewunruen and N. Xu, 'Digital twin for sustainability evaluation of railway station buildings', *Frontiers in Built Environment*, vol. 4, 2018, doi:10.3389/fbuil.2018.00077.
25. R. Charef, H. Alaka, and S. Emmitt, 'Beyond the third dimension of BIM: A systematic review of literature and assessment of professional views', *Journal of Building Engineering*, vol. 19, pp. 242–257, 2018, doi 10.1016/j.jobe.2018.04.028.
26. J. Sresakoolchai and S. Kaewunruen, 'Integration of building information modeling and machine learning for railway defect localization', *IEEE Access*, vol. 9, pp. 166039–166047, 2021, doi:10.1109/ACCESS.2021.3135451.
27. A. Rasheed, O. San, and T. Kvamsdal, 'Digital twin: Values, challenges and enablers from a modeling perspective', *IEEE Access*, vol. 8, pp. 21980–22012, 2020, doi:10.1109/ACCESS.2020.2970143.
28. U. Alvarado, P. Ciáurriz, A. S. Emeterio, I. Errandonea, and J. Rodriguez, 'MAINRAIL: Maintenance optimization of railway infrastructure through physical modeling and advanced analytics', in *Proceedings of the 31st European Safety and Reliability Conference, ESREL 2021*, Angers, France, 2021, pp. 200–207, doi:10.3850/978-981-18-2016-8_424-CD.
29. R. K. Phanden, P. Sharma, and A. Dubey, 'A review on simulation in digital twin for aerospace, manufacturing and robotics', *Materials Today:* Proceedings, vol. 38, pp. 174–178, 2021, doi:10.1016/j.matpr.2020.06.446.
30. S. Milovanović, S. Strobl, P. Ladoux, and D. Dujić, 'Hardware-in-the-loop modeling of an actively fed MVDC railway systems of the future', *IEEE Access*, vol. 9, pp. 151493–151506, 2021, doi:10.1109/ACCESS.2021.3125050.
31. D. Orive, N. Iriondo, A. Burgos, I. Saráchaga, M. L. Álvarez, and M. Marcos, 'Fault injection in Digital Twin as a means to test the response to process faults at virtual commissioning', in *2019 24th IEEE International Conference on Emerging Technologies and Factory Automation (ETFA)*, Zaragoza, Spain, 2019, pp. 1230–1234, doi:10.1109/ETFA.2019.8869334.
32. J. Lu, X. Zheng, L. Schweiger, and D. Kiritsis, 'A cognitive approach to manage the complexity of digital twin systems', in *Smart Services Summit*, S. West, J. Meierhofer, and C. Ganz, Eds., Cham: Springer International Publishing, 2021, pp. 105–115, doi:10.1007/978-3-030-72090-2_10.
33. R. Dirnfeld, L. De Donato, F. Flammini, M. S. Azari, and V. Vittorini, 'Railway digital twins and artificial intelligence: Challenges and design guidelines', in *Dependable Computing – EDCC 2022 Workshops*, S. Marrone, M. De Sanctis, I. Kocsis, R. Adler, R. Hawkins, P. Schleiß, S. Marrone, R. Nardone, F. Flammini, and V. Vittorini, Eds., in *Communications in Computer and Information Science*, vol. 1656., Cham: Springer International Publishing, 2022, pp. 102–113. doi:10.1007/978-3-031-16245-9_8.
34. E. Febrianto, L. Butler, M. Girolami, and F. Cirak, 'Digital twinning of self-sensing structures using the statistical finite element method', *Data-Centric Engineering*, vol. 3, p. e31, 2022, doi:10.1017/dce.2022.28.
35. P. Ferdinand, 'The evolution of optical fiber sensors technologies during the 35 last years and their applications in structure health monitoring', *Presented at the EWSHM – 7th European Workshop on Structural Health Monitoring, INRIA*, Nantes, France, July 2014, p. 914. Accessed: August 05, 2024. [Online]. Available: https://inria.hal.science/hal-01021251.
36. H.-N. Li, D.-S. Li, and G.-B. Song, 'Recent applications of fiber optic sensors to health monitoring in civil engineering', *Engineering Structures*, vol. 26, no. 11, pp. 1647–1657, 2004, doi:10.1016/j.engstruct.2004.05.018.
37. R. Sieńko, M. Zych, Ł. Bednarski, and T. Howiacki, 'Strain and crack analysis within concrete members using distributed fibre optic sensors', *Structural Health Monitoring*, vol. 18, no. 5–6, pp. 1510–1526, 2019, doi:10.1177/1475921718804466.

38. C.-S. Shim, N.-S. Dang, S. Lon, and C.-H. Jeon, 'Development of a bridge maintenance system for prestressed concrete bridges using 3D digital twin model', *Structure and Infrastructure Engineering*, vol. 15, no. 10, pp. 1319–1332, 2019, doi:10.1080/15732479.2019.1620789.
39. A. Patwardhan, A. Thaduri, R. Karim, and M. Castano, 'Federated learning for enablement of digital twin', *IFAC-PapersOnLine*, vol. 55, no. 2, pp. 114–119, 2022, doi:10.1016/j.ifacol.2022.04.179.
40. B. Ton, F. Ahmed, and J. Linssen, 'Semantic segmentation of terrestrial laser scans of railway catenary arches: A use case perspective', *Sensors*, vol. 23, no. 1, Art. no. 1, 2023, doi:10.3390/s23010222.
41. A. Avsievich, V. Avsievich, N. Avsievich, D. Ovchinnikov, and A. Ivaschenko, 'Railway track stress–strain analysis using high-precision accelerometers', *Applied Sciences*, vol. 11, no. 24, Art. no. 24, 2021, doi:10.3390/app112411908.
42. M. D. G. Milošević, B. A. Pålsson, A. Nissen, J. C. O. Nielsen, and H. Johansson, 'Demonstration of a digital twin framework for model-based operational condition monitoring of crossing panels', in *Advances in Dynamics of Vehicles on Roads and Tracks II*, A. Orlova and D. Cole, Eds., Cham: Springer International Publishing, 2022, pp. 95–105, doi:10.1007/978-3-031-07305-2_11.
43. C. Ramonell Cazador and R. A. Chacón Flores, 'Towards automated pipelines for processing load test data on a HS railway bridge in Spain using a digital twin', *Presented at the Proceedings of the 39th International Symposium on Automation and Robotics in Construction*, Bogotá, Colombia, 2022, pp. 231–237. Accessed: August 06, 2024. [Online]. Available: https://upcommons.upc.edu/handle/2117/371392.
44. J. Yang, Y. Sun, Y. Cao, and X. Hu, 'Predictive maintenance for switch machine based on digital twins', *Information*, vol. 12, no. 11, Art. no. 11, 2021, doi:10.3390/info12110485.
45. M. Mashaly, 'Connecting the twins: A review on digital twin technology & its networking requirements', *Procedia Computer Science*, vol. 184, pp. 299–305, 2021, doi:10.1016/j.procs.2021.03.039.
46. M. J. Pratt, 'Introduction to ISO 10303—The STEP standard for product data exchange', *Journal of Computing and Information in Science and Engineering*, vol. 1, no. 1, pp. 102–103, 2001, doi:10.1115/1.1354995.
47. H. Li, Q. Zhu, L. Zhang, Y. Ding, Y. Guo, H. Wu, et al., 'Integrated representation of geospatial data, model, and knowledge for digital twin railway', *International Journal of Digital Earth*, vol. 15, no. 1, pp. 1657–1675, 2022, doi:10.1080/17538947.2022.2127949.
48. I. Roda, M. Macchi, and S. Albanese, 'Building a total cost of ownership model to support manufacturing asset lifecycle management', *Production Planning & Control*, vol. 31, no. 1, pp. 19–37, 2020, doi:10.1080/09537287.2019.1625079.
49. V. R. Gharehbaghi, E. N. Farsangi, M. Noori, T. Y. Yang, S. Li, A. Nguyen, et al., 'A critical review on structural health monitoring: Definitions, methods, and perspectives', *Archives of Computational Methods in Engineering*, vol. 29, no. 4, pp. 2209–2235, 2022, doi:10.1007/s11831-021-09665-9.
50. I. A. Nhamage, N.-S. Dang, C. S. Horas, J. Poças Martins, J. A. Matos, and R. Calçada, 'Performing fatigue state characterization in railway steel bridges using digital twin models', *Applied Sciences*, vol. 13, no. 11, Art. no. 11, 2023, doi:10.3390/app13116741.
51. A. Armijo and D. Zamora-Sánchez, 'Integration of railway bridge structural health monitoring into the Internet of Things with a digital twin: A case study', *Sensors*, vol. 24, no. 7, Art. no. 7, 2024, doi:10.3390/s24072115.
52. R. Chacón, C. Ramonell, H. Posada, P. Sierra, R. Tomar, C. Martínez de la Rosa, et al., 'Digital twinning during load tests of railway bridges – Case study: The high-speed railway network, Extremadura, Spain', *Structure and Infrastructure Engineering*, vol. 20, no. 7–8, pp. 1105–1119, 2024, doi:10.1080/15732479.2023.2264840.
53. M. Girolami, E. Febrianto, G. Yin, and F. Cirak, 'The statistical finite element method (statFEM) for coherent synthesis of observation data and model predictions', *Computer Methods in Applied Mechanics and Engineering*, vol. 375, p. 113533, 2021, doi:10.1016/j.cma.2020.113533.
54. A. Rageh, S. E. Azam, Q. Alomari, D. Linzell, and R. Wood, 'Model updating and parameter identification for developing digital twins for riveted steel railway bridges', in *Recent Developments in Structural Health Monitoring and Assessment ? Opportunities and Challenges*, Achintya Haldar, Abdullah Al-Husseinis, Editors, WSPC, London, UK, 2021, pp. 285–318, doi:10.1142/9789811243011_0010.
55. C. Jung, A. K. A. Toguyeni, and B. O. Bouamama, 'Supervised machine learning from digital twin data for railway switch fault diagnosis', in *2023 European Control Conference (ECC)*, Bucharest, Romania, June 2023, pp. 1–7, doi:10.23919/ECC57647.2023.10178257.
56. M. Hamarat, M. Papaelias, and S. Kaewunruen, 'Fatigue damage assessment of complex railway turnout crossings via peridynamics-based digital twin', *Science Reports*, vol. 12, no. 1, p. 14377, 2022, doi:10.1038/s41598-022-18452-w.

57. S. Zhang, H. Dong, U. Maschek, and H. Song, 'A digital-twin-assisted fault diagnosis of railway point machine', in *2021 IEEE 1st International Conference on Digital Twins and Parallel Intelligence (DTPI)*, Beijing, China, July 2021, pp. 430–433, doi:10.1109/DTPI52967.2021.9540118.
58. X. Li, X. Lin, H. Zhu, X. Wang, and Z. Liu, 'Condition assessment of shield tunnel using a new indicator: The tunnel serviceability index', *Tunnelling and Underground Space Technology*, vol. 67, pp. 98–106, 2017, doi:10.1016/j.tust.2017.05.007.
59. T. Dawood, E. Elwakil, T. Zayed, and Z. Zhu, 'Data fusion of multiple machine intelligent systems for the condition assessment of subway structures', *Tunnelling and Underground Space Technology*, vol. 126, p. 104512, 2022, doi:10.1016/j.tust.2022.104512.
60. B. V. Farahani, F. Barros, P. J. Sousa, P. J. Tavares, and P. M. G. P. Moreira, 'A railway tunnel structural monitoring methodology proposal for predictive maintenance', *Structural Control and Health Monitoring*, vol. 27, no. 8, p. e2587, 2020, doi:10.1002/stc.2587.
61. H. J. Pang, S. Y. Li, L. F. Dai, J. T. Kong, M. K. Liu, F. Jia, et al., 'Intelligent metro shield tunnel structure assessment based on knowledge graph', in *Trends on Construction in the Digital Era*, A. Gomes Correia, M. Azenha, P. J. S. Cruz, P. Novais, and P. Pereira, Eds., Cham: Springer International Publishing, 2023, pp. 476–489, doi:10.1007/978-3-031-20241-4_33.
62. G. Yu, Y. Wang, Z. Mao, M. Hu, V. Sugumaran, and Y. K. Wang, 'A digital twin-based decision analysis framework for operation and maintenance of tunnels', *Tunnelling and Underground Space Technology*, vol. 116, p. 104125, 2021, doi:10.1016/j.tust.2021.104125.
63. G. Yu, D. Lin, Y. Wang, M. Hu, V. Sugumaran, and J. Chen, 'Digital Twin-enabled and knowledge-driven decision support for tunnel electromechanical equipment maintenance', *Tunnelling and Underground Space Technology*, vol. 140, p. 105318, 2023, doi:10.1016/j.tust.2023.105318.
64. I. Errandonea, P. Ciáurriz, U. Alvarado, S. Beltrán, and S. Arrizabalaga, 'Edge intelligence-based proposal for onboard catenary stagger amplitude diagnosis', *Computers in Industry*, vol. 144, p. 103781, 2023, doi:10.1016/j.compind.2022.103781.
65. M. Karakose, O. Yaman, K. Murat, and E. Akin, 'A new approach for condition monitoring and detection of rail components and rail track in railway', *International Journal of Computational Intelligence Systems*, vol. 11, no. 1, pp. 830–845, 2018, doi:10.2991/ijcis.11.1.63.
66. C. Shen, P. Zhang, R. Dollevoet, A. Zoeteman, and Z. Li, 'Evaluating railway track stiffness using axle box accelerations: A digital twin approach', *Mechanical Systems and Signal Processing*, vol. 204, p. 110730, 2023, doi:10.1016/j.ymssp.2023.110730.
67. P. Weston, C. Roberts, G. Yeo, and E. Stewart, 'Perspectives on railway track geometry condition monitoring from in-service railway vehicles', *Vehicle System Dynamics*, vol. 53, no. 7, pp. 1063–1091, 2015, doi:10.1080/00423114.2015.1034730.
68. B. Djordjević, E. Krmac, C.-Y. Lin, O. Fröidh, and B. Kordnejad, 'An optimisation-based digital twin for automated operation of rail level crossings', *Expert Systems with Applications*, vol. 239, p. 122422, 2024, doi:10.1016/j.eswa.2023.122422.
69. R. Chandaluri and U. R. Nelakuditi, 'Performance evaluation of electro-mechanical railway interlocking system for digital twin application', *Computers and Electrical Engineering*, vol. 116, p. 109225, 2024, doi:10.1016/j.compeleceng.2024.109225.
70. G. Salierno, L. Leonardi, and G. Cabri, 'A big data architecture for digital twin creation of railway signals based on synthetic data', *IEEE Open Journal of Intelligent Transportation Systems*, vol. 5, pp. 1–18, 2024, doi:10.1109/OJITS.2024.3412820.
71. A. Padovano, F. Longo, L. Manca, and R. Grugni, 'Improving safety management in railway stations through a simulation-based digital twin approach', *Computers & Industrial Engineering*, vol. 187, p. 109839, 2024, doi:10.1016/j.cie.2023.109839.
72. X. Wang, W. Bai, Y. Su, G. Yang, C. Li, X. Lv, et al., 'Digital twin for multi-scenario emergency of railway passenger stations', *Frontiers in Physics*, vol. 11, 2024, doi:10.3389/fphy.2023.1291785.
73. X. Wang, T. Shi, X. Lv, C. Li, G. Yang, and D. Dong, 'Digital twins for daily operation management of railway station', in *2023 IEEE 3rd International Conference on Digital Twins and Parallel Intelligence (DTPI)*, Orlando, Florida, USA, November 2023, pp. 1–6, doi:10.1109/DTPI59677.2023.10365475.
74. Y. K. Al-Douri, P. Tretten, and R. Karim, 'Improvement of railway performance: A study of Swedish railway infrastructure', *Journal of Modern Transportation*, vol. 24, no. 1, pp. 22–37, 2016, doi:10.1007/s40534-015-0092-0.
75. S. Cavieres-Lagos, F. Muñoz La Rivera, E. Atencio, and R. F. Herrera, 'Integration of BIM tools for the facility management of railway bridges', *Applied Sciences*, vol. 14, no. 14, Art. no. 14, 2024, doi:10.3390/app14146209.

76. A. Najafi, Z. Amir, B. Salman, P. Sanaei, E. Lojano-Quispe, A. Maher, et al., 'A digital twin framework for bridges', in *ASCE International Conference on Computing in Civil Engineering*, Corvallis, OR, 2024, pp. 433–441, doi:10.1061/9780784485231.052.
77. U. Alibrandi, 'Risk-informed digital twin of buildings and infrastructures for sustainable and resilient urban communities', *ASCE-ASME Journal of Risk and Uncertainty in Engineering Systems, Part A: Civil Engineering*, vol. 8, no. 3, p. 04022032, 2022, doi:10.1061/AJRUA6.0001238.
78. UIC, 'Maximising the benefits of big data for asset management', March 2022. [Online]. Available: https://shop.uic.org/en/oher-guidelines/14264-maximising-the-benefits-of-big-data-for-asset-management.html.
79. I. Roda, A. Polenghi, and V. Männistö, 'Big data adoption in strategic decision-making for railway infrastructure asset management', in *16th WCEAM Proceedings*, A. Crespo Márquez, J. F. Gómez Fernández, V. González-Prida Díaz, and J. Amadi-Echendu, Eds., Cham: Springer International Publishing, 2023, pp. 428–438, doi:10.1007/978-3-031-25448-2_41.
80. C. Liu, P. Zhang, and X. Xu, 'Literature review of digital twin technologies for civil infrastructure', *Journal of Infrastructure Intelligence and Resilience*, vol. 2, no. 3, p. 100050, 2023, doi:10.1016/j.iintel.2023.100050.
81. Bentley, 'Deliver better rail asset lifecycle outcomes with digital twins on *Global Railway Review*', in *Global Railway Review*, 22 November 2023. [Online]. Available: https://www.globalrailwayreview.com/webinar/146309/how-digital-twins-drive-better-outcomes-in-rail/.
82. A. O. Borjigin, J. Sresakoolchai, S. Kaewunruen, and J. Hammond, 'Digital twin aided sustainability assessment of modern light rail infrastructures', *Frontiers in Built Environment*, vol. 8, 2022, doi:10.3389/fbuil.2022.796388.

10 Digital Twins for Train Operation Optimizations

Colin Cole

10.1 INTRODUCTION

The practice of train simulation for evaluating trains and providing train operator training is well developed [1–4]. In the late 1980s, both the measurement and simulation of coupler forces in large heavy-haul and freight trains were reported in Australia [5,6] and in South Africa [7]. The need to model inter-wagon forces is less important in shorter passenger trains and short freight trains, so for operational optimizations, these were often considered as single vehicles for the purpose of understanding and controlling acceleration, speed, and energy. One such example is the energy management system MetroMiser, first used in 1990 [8]. Freight trains, however, are more complex to control because the external input forces vary along the length of the train due to the locations of grades and curves, speed restrictions, etc. These factors, combined with the complexity of accelerating and braking the large inertia of the train mass, lead to the need for both the development of and training in train handling. The limitations of the friction that can be supported at the wheel–rail contact require carefully controlled locomotive traction and braking, and the system is constrained to relatively long stopping distances even when utilizing all vehicles within the train to apply brakes. Full-trip simulations of large heavy-haul trains or high-speed passenger trains are routinely completed with the simulation models controlled by emulations of human drivers or automated control systems providing controls of the "virtual train" [9]. This virtual control, of course, mirrors real human, cruise, or automated controls in operating trains. By combining detailed longitudinal train dynamics modeling, track topography and curvature, speed limits and restrictions, stops and starts, and signaling conditions, it is possible to assemble a digital twin of a train-track freight haulage or train-track passenger system, similar to approaches developed in [10]. The resulting "virtual twin" can then be used for applications such as scheduling and routing of trains and network capacity planning, energy optimization and decarbonization, and analysis of coupler forces in relation to operational risks and rollingstock fatigue.

If field data is available from a condition monitoring system, the virtual twin for each operating train can be fine-tuned and updated to maintain accuracy and retain confidence in the outputs of the virtual twin and allow its use as a "predictive twin".

10.2 A DIGITAL TWIN FOR TRAIN OPTIMIZATION

It is useful to first discuss what a digital twin is in simple terms. A common question is that of how it differs from simulation with validation. A digital twin differs in both use and scope compared to traditional simulation although it can look similar. Recalling that traditional simulation has had applications primarily in analysis, for design or commissioning purposes, or for training purposes, it can be concluded that the modeling is primarily generic for train type. In other words, it was important that the modeling was representative of an "average" train type or fleet. The user of the results understands the inherent uncertainties with the expectation that results may vary in certain conditions and that the equipment itself would have a range of conditions and responses. It can be said that traditional simulations deliver a model that is sufficiently generalized to represent an operating fleet. Another way the risk of uncertainty is managed is that the use case is near to operating

DOI: 10.1201/9781003492146-10

conditions as those where the validation had been performed. If a use case involving very different conditions is required another validation is requested. Of course, it is also a matter of judgment when further validations are needed.

Conversely, the intention of a digital twin is to have modeling that is as precise and comprehensive as possible. This means that the model is not a fleet generalization of a train type but a specific train made up of specific vehicles and components. Furthermore, to ensure as broad a range of validity as possible, measurable, testable, "white box" physics-based models should be developed and used wherever possible. The modeling begins by producing a virtual twin, but it also can continually evolve into a more accurate digital twin as data is provided by condition monitoring systems.

A systematic approach for developing a digital twin for train operation will allow the effects of operation and train-track system decisions to be evaluated. A simple digital twin model of a short train system for draft gear impact analysis as shown in Figure 10.1 and introduced in [11] is illustrative of a train-track system without the introduction of traction, braking, and control.

For train operation optimization, the model grows, as shown in Figure 10.3.

10.2.1 Introducing a Laboratory Size Digital Twin

Before focusing on a digital twin for train optimization, it is useful to consider a laboratory-sized digital twin to illustrate the importance of physics-based models. Operational data is often not of an adequate range to allow the development of robust models. Data from normal operations can be expected to have a large population of data points close to normal conditions and therefore don't provide a validation set for the actual components. This problem mirrors the difficulty that manufacturers face in providing a product that is adequate for the full range of operating conditions that are possible. The manufacturing solution is proof or destructive tests. For coupler draft gears (the vital component of train coupling systems), one such proof test is impact testing as per AAR M-921 [12]. The approach is published in [11]. To develop the digital twin, the proof test system is first modeled as a longitudinal train model to allow detailed train simulation as per [1–3]. The modeling, in this case, can be regarded as a virtual twin (see Figure 10.1). Using test data, validation can be ensured by comparing the twin with the test data. The virtual twin provides a platform for the

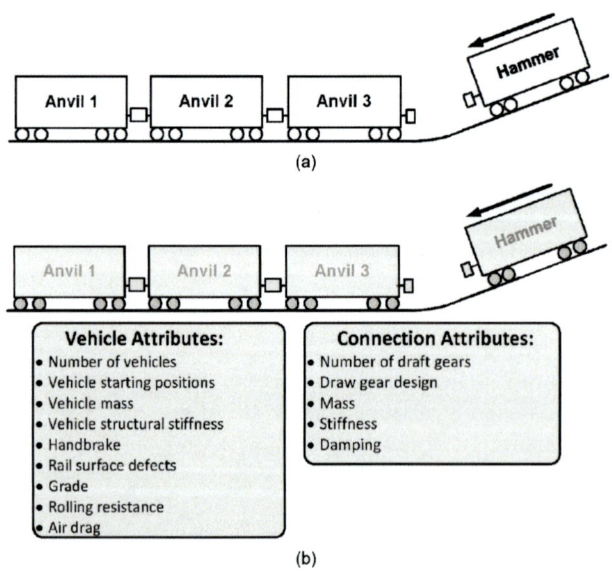

FIGURE 10.1 Impact testing arrangement. (a) Impact testing arrangement for the actual train system in the real world environment. (b) Virtual twin of the impact testing vehicles.

detailed study of alternate draft gears both for manufacturing and as a building block for the larger digital twin for train optimization.

The longitudinal train modeling captures all the parameters of the vehicles, including for this case:

1. Track topography, including the downhill grade of the Hammer car
2. Any incidental vertical track defects
3. The handbrake application of the lead Anvil car
4. Rolling resistance and air resistance modeling
5. Hammer car release modeling
6. Wagon connection modeling
7. Open wagon connection modeling of the Hammer car

By comprehensively capturing the surrounding situation of the draft gear tests, including any changes to the track, the Anvil cars, or the brake performance, the aerodynamics of the Hammer car can be captured in the model and validated (see Figure 10.1b). The digital twin, in this case, offers the possibility of many cycles of virtual testing of new designs prior to physical tests or train optimization tests.

10.2.2 Train-Track Operation Digital Twin

A systems engineering approach ensures a logical approach to the modeling and simulation of several systems. A good introduction to railway applications was provided by Spiryagin et al. [13]. A similar methodology for the train-track-operation digital twin is introduced in Figure 10.2 and the virtual twin model is illustrated in Figure 10.3.

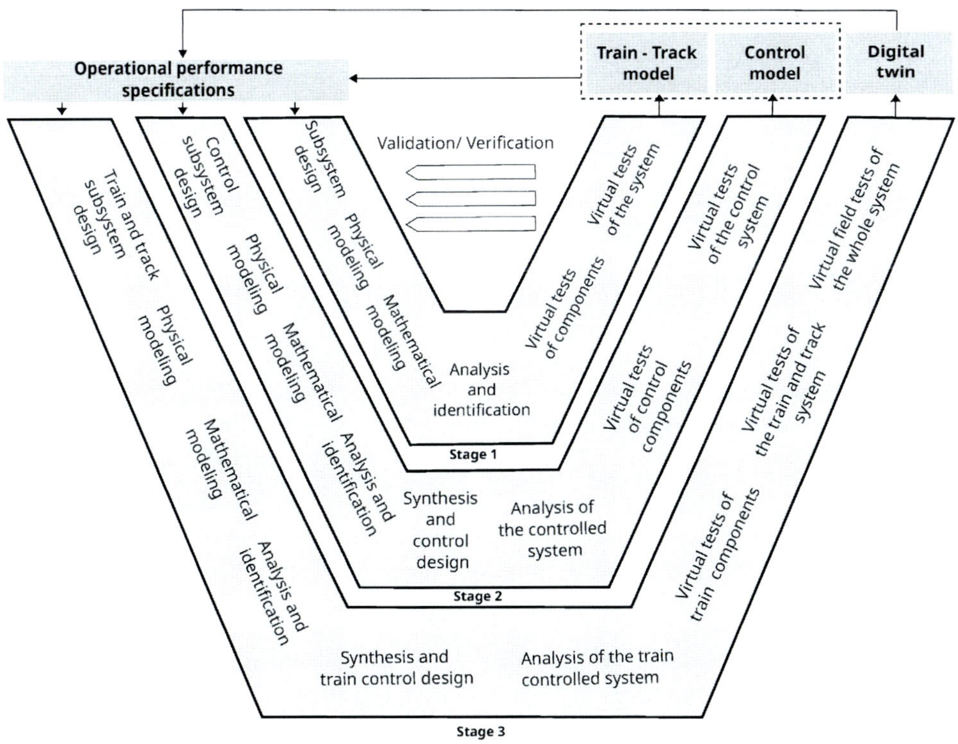

FIGURE 10.2 The V-model with several system integration stages.

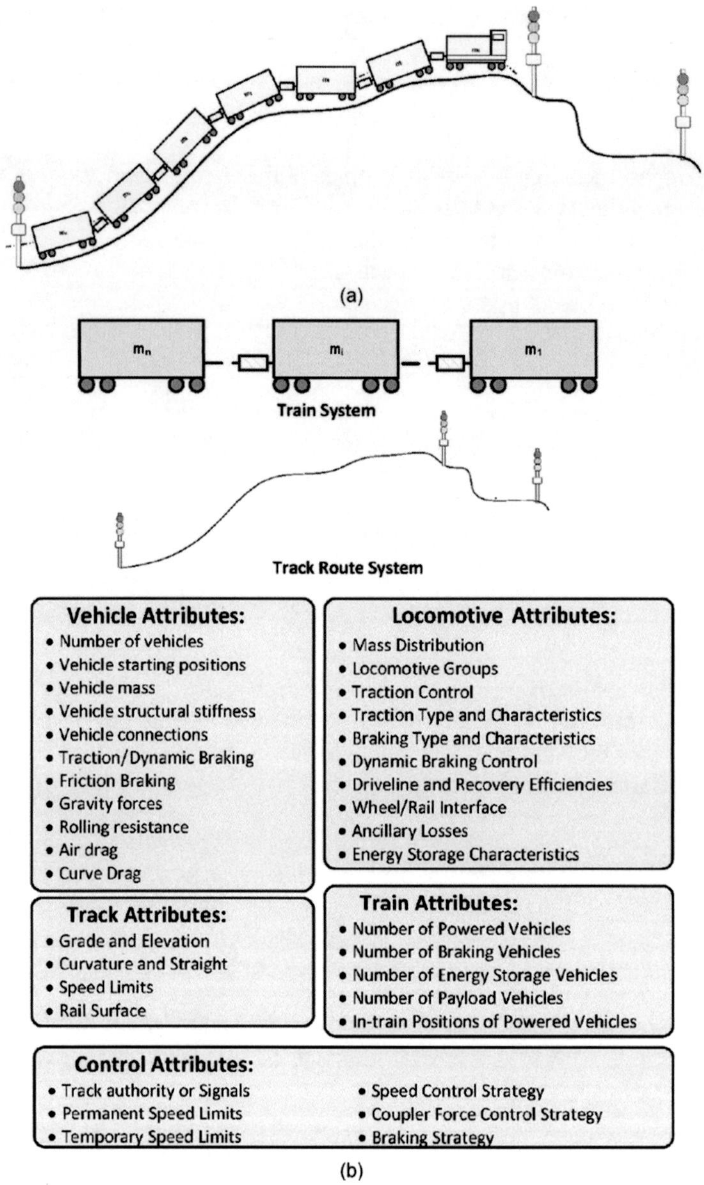

FIGURE 10.3 Train optimization model. (a) Real-world train-track operation system. (b) Comprehensive virtual twin of the train-track operation system.

The methodology is based on the V-model system approach [14,15]. In this application, the digital twin approach is used to combine train operation parameters with control options and train route operation. The digital twin should include four important parts [16–18]:

- A model of the object built based on an integrated multiphysics, multiscale, probabilistic simulation approach
- A method of obtaining and evolving sets of data relating to the object
- A means of dynamically updating or adjusting the model in accordance with the data
- The ability to mirror and predict activities/performance over the life of its corresponding physical twin

Stage 1 of the V-model represents the development of a longitudinal train model to allow all train operational parameters to be simulated and key performance parameters derived. Building on the detailed inter-vehicle connected modeling and validation as presented in Section 10.2.1, complete longitudinal train models can be developed as established in [1–3]. The established approach includes modeling of each vehicle as a single mass along with applied forces of traction, braking, connections, rolling resistance, air resistance curving resistance, and the effects of grades. Single mass models of vehicles are easily represented as a train by a set of differential equations, but the complexity comes from the non-linear nature of wagon connections and draft gears and the extensive detail required for the force inputs. The train model has a high level of co-dimensionality, requiring traction and braking systems to be modeled in detail along with several resistances to motion, namely, rolling resistance, air resistance, curve resistance, and the effects of grades. With the exception of grades, all these subsystems require non-linear modeling to ensure the required realism and accuracy of the twin.

Stage 2 represents the development of the controls and constraints that are input to the train system. This includes the operating environment of the track route, the signaling conditions, and the need to stop-start or comply with speed restrictions. It also includes the methodology of train control, whether it be human or semi-automated (such as cruised control) or fully automated (such as Metro systems or AutoHaul [19]). The selection of the train handling approach (human) or algorithm (software control) has a significant influence on average train speed and energy use [20].

Stage 3 concerns the whole train/track/control/ system that includes modeling of the track topography and curves; operational limits such as speeds, accelerations, maximum coupler forces, and control inputs to the locomotive; and the option of either emulating a human driver or providing an automated control system such as in [9].

Analysis in Stage 3 is achieved by implementing the comprehensive modeling as a virtual train operation using simulation on either a computer workstation or a high-performance computing (HPC) cluster. Larger-scale analysis can be achieved using multiple HPC nodes for parallel completion of train operation simulations as developed by Wu [21].

10.3 STAGE 1 – DEVELOPMENT OF THE FULL TRAIN-TRACK MODEL

This section presents the development of the longitudinal train dynamics model that is to be used for operational performance evaluations. Train operation is a multi-object problem. Typically, there are three main objectives: production rate, safety, and costs. These translate into parameters such as production: average running speed, train reliability; safety: dynamic stability, coupler forces; and costs: electric power or fuel usage, maintenance, etc.

10.3.1 Longitudinal Train Dynamics Components

A summary of the modeling approach is included briefly. A fully detailed longitudinal model can be used to evaluate all three objectives in the multi-objective problem. Modeling and simulation of longitudinal train dynamics are presented in detail in [1,2]. The dynamic model is based on Newton's second law, which gives a set of ordinary differential equations. Using a numerical integrator solution allows non-linear representative physics models of the components to be used. The connected mass model has only three forms, as shown in Equations 10.1 to 10.3.

For the lead vehicle:

$$m_1 a_1 + f_{we}\left(v_1, v_2, x_1, x_2\right) = F_{t/db1} - F_{r1} - F_{g1} \tag{10.1}$$

For the *i*th vehicle:

$$m_i a_i + f_{we}\left(v_i, v_{i-1}, x_i, x_{i-1}\right) + f_{we}\left(v_i, v_{i+1}, x_i, x_{i+1}\right) = F_{t/dbi} - F_{ri} - F_{gi} \tag{10.2}$$

For the *n*th or last vehicle:

$$m_n a_n + f_{wc}(v_n, v_{n-1}, x_n, x_{n-1}) = F_{t/dbn} - F_m - F_{gn} \tag{10.3}$$

where f_{wc} is the non-linear function describing the full characteristics of the vehicle connection. A schematic of the generalized train dynamics system for use in optimization is shown in Figure 10.4.

The mathematical model has one differential equation for each vehicle mass, but considerable complexity is added by the forcing functions. These include parameters as shown in Table 10.1.

The solution to the system of differential equations gives the speeds and accelerations of each vehicle along with the inter-wagon forces from the coupler connection model. As the many subsystems have to be evaluated in each integration step, full details of the train system can be reported and further analyses completed, as shown in Table 10.2 and Figure 10.5.

10.3.2 Energy Components

The train work–energy balance can be defined as:

$$\sum W = \Delta E_{PE} + \Delta E_{KE} \tag{10.4}$$

where W is the work done on the train system and ΔE_{PE} and ΔE_{KE} are changes in gravitational potential energy and kinetic energy, respectively.

As the initial energy balance is defined at the wheel via the operational train dynamics, driveline efficiencies (Figure 10.6) can be added according to manufacturers' data or measured data and

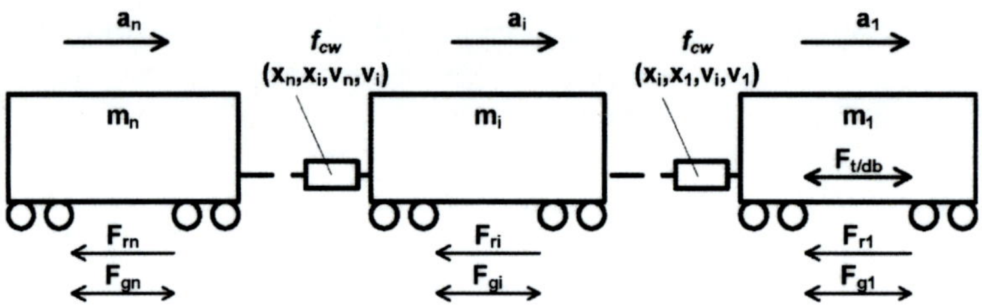

FIGURE 10.4 Generalized longitudinal train dynamics model. (Courtesy of M. Spiryagin, P. Wolfs, C. Cole, V. Spiryagin, Y. Sun, T. McSweeney, *Design and Simulation of Heavy Haul Locomotives and Trains.* Ground Vehicle Engineering Series, CRC Press, Boca Raton, FL, 2017; and C. Cole, M. Spiryagin, Q. Wu, Y.Q. Sun, Modelling, simulation and applications of longitudinal train dynamics. *Vehicle System Dynamics* 55(10): 1498–1571, 2017.)

TABLE 10.1
Dynamic System Parameters

System Parameters	Symbol	Source
Locomotive traction and dynamic braking forces	$F_{t/db}$	Equations modeling manufacturers' specifications
Grade force	F_g	Grade angle and wagon weight, $mg \sin\theta$
Resistance forces	F_r	$F_r = F_b + F_{pr} + F_{cv}$
Air braking force	F_b	Simulated values from a brake pipe and cylinder system model
Propulsion resistance	F_{pr}	Empirical equations for various rolling stock, includes rolling resistance and air resistance, $F_{pr} = F_{rolling} + F_{air}$
Curving resistance	F_{cv}	Empirical equation for the increase in resistance on curves
Connection model	$f_{wc}(v_i, v_{i-1}, x_i, x_{i-1})$	Coupler draft gear model incorporating inter-wagon slack and draft gear providing both stiffness and damping

TABLE 10.2
Typical Reporting Parameters

Reporting Parameters	Symbol	Source
	Production rate	
Average trip speed	V_{ave}	Simulation output time series
Payload mass	m_{net}	Train system data
Section time	T	Distance travelled time series
	Energy inputs	
Traction	$W_{TE} = N_L \int F_{TE}\ dx$	Where $F_{\text{TE}} = F_{t/\text{db}}$, for $F_{t/\text{db}} > 0$ $F_{t/\text{db}}$ is the traction/braking force distance series for each powered vehicle
Gravity	$W_{PE} = N_v \int_1^{N_v} F_g dx$	F_g grade force distance series for each vehicle (as an energy input to the system)
	Energy consumed	
Dynamic braking	$W_{DB} = N_L \int F_{DB}\ dx$	Where $F_{DB} = F_{t/db}$, for $F_{t/db} < 0$ $F_{t/db}$ is the traction/braking force distance series for each powered vehicle
Air braking	$W_{BP} = N_v \int F_b\ dx$	F_g grade force distance series for each vehicle
Propulsion resistance	$W_{Drag} = N_v \sum_{i=1}^{N_v} \int F_{pr}\ dx$	F_{pr} propulsion resistance force distance series for each vehicle
Curve resistance	$W_{Curves} = N_v \sum_{i=1}^{N_v} \int F_{cv}\ dx$	F_{cv} curving resistance force distance series for each vehicle
Gravity	$W_{PE} = N_v \int_1^{N_v} F_g dx$	F_g grade force distance series for each vehicle (as an energy consumed the system)
	Risks to equipment	
Coupler forces	F_c	Simulation output time series $F_c = f_{wc}(v_i, v_{i-1}, x_i, x_{i-1})$ modeling as per [1,2]
Coupler angles	Φ	Simple equations can be added to the simulation process to give coupler angles as per [1,2]
Lateral coupler forces	$F_c \Phi$	
Locomotive forces	$F_{t/db}$	Equations modeling manufacturers' specifications
Resistance forces	F_r	$F_r = F_b + F_{pr} + F_{cv}$
Air braking force	F_b	Simulated values from a brake pipe and cylinder system model
Propulsion resistance	F_{pr}	Empirical equations for various rolling stock, includes rolling resistance and air resistance, $F_{pr} = F_{rolling} + F_{air}$
Curving resistance	F_{cv}	Empirical equation for the increase in resistance on curves
Inter-wagon forces	$f_{wc}(v_i, v_{i-1}, x_i, x_{i-1})$	Coupler draft gear model time series

simply changed when different driveline systems or technologies are considered such as fully electric traction, diesel-electric traction, and zero emissions alternatives and hybrids including batteries, hydrogen, and gas (see Figure 10.6). Importantly, there are differences in the energy needs of ancillaries depending on the type of locomotive power that is in use. Ancillary energy use is significant, up to 10% on existing diesel-electric locomotives, and includes the power needed for locomotive systems, cabin systems, and the train brake air compressor. Zero-emission locomotives as they are

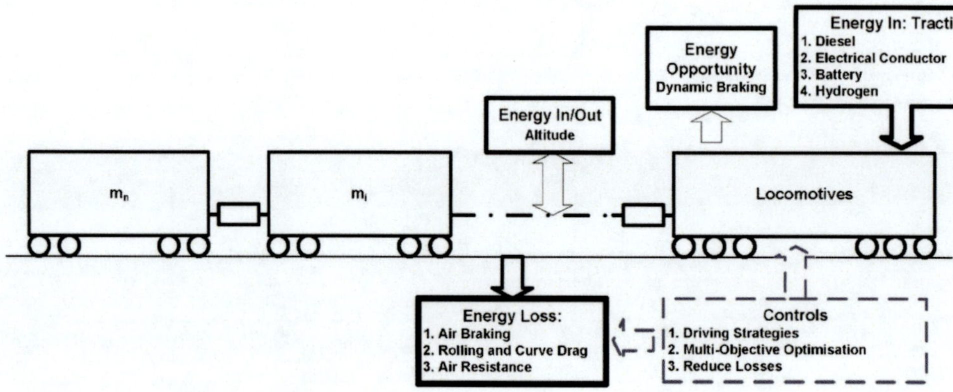

FIGURE 10.5 Train energy components.

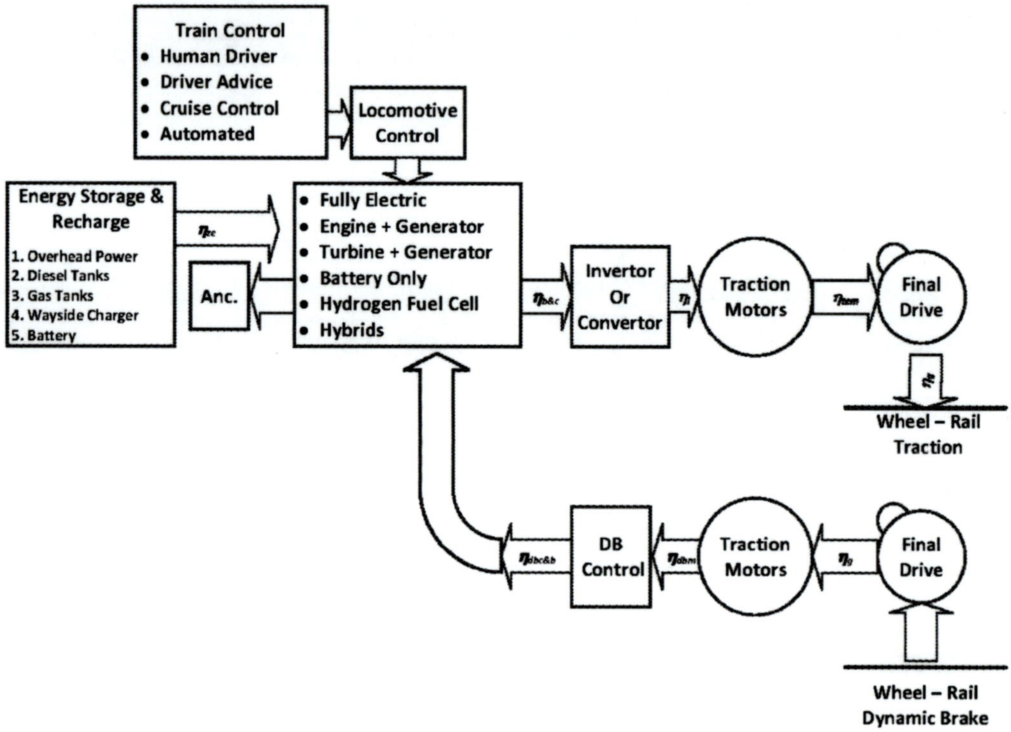

FIGURE 10.6 Train energy driveline components.

emerging will have different ancillary energy needs depending on the required operating needs of the new technologies (i.e., batteries and fuel cells).

10.3.3 Validation

Confidence in the modeling should be established at an early stage. As noted previously, this discussion was introduced with inter-wagon draft gear connection modeling in Section 10.2.1. It is important that this difficult area of modeling is carefully managed as different draft gear types can give very different results in peak forces, particularly if the maximum draft gear displacements are changed. If the draft gear characteristics are unknown, physical testing should be completed.

Other parts of the train-track system can also present difficulty. The traditional train brake pipe systems require knowledge of the brake application propagation speed. This can be achieved via time-dependent empirical functions or lumped models of computational fluid dynamic analysis, but for train dynamics, only two things are important: (1) the delays along the train of brake cylinder applications; and (2) the brake cylinder fill time characteristic. The brake cylinder pressure is translated to a brake retardation force according to the brake design. Modeling is simplified by manufacturers' data, which usually provides the maximum cylinder force levels supplied to the brake shoes or brake pads as maximum cylinder pressures. Railway standards also sometimes specify these levels. Brake shoe or disk pad friction and the friction dependence on speed also significantly affect stopping distance. If a digital twin is going to be used for stopping distance investigations, the modeling should be validated with field test data. The remaining challenges in validation are easier to manage. Locomotive manufacturers generally provide performance curves for the locomotives supplied. Empirical estimates of rolling, curving, and air resistance are well established [1,2] and are sufficiently accurate as initial values. These can be tuned by comparisons with field performance using whole trip simulations based on locomotive logs and comparing with logged train performance data (i.e., traction forces, speed, etc.). The use of fuel consumption assists in ensuring that energy values are representative, preventing an incorrect locomotive traction value from being effectively concealed by incorrectly tuning the rolling resistance to match it. Further errors can, of course, be introduced by other factors such as incorrect track survey or train mass.

Further validation can be achieved by considering the energy balance of the whole trip if the locomotive logs include reporting of traction and dynamic braking forces. If the starting and ending points of the trip are known, then:

$$\sum W = W_{TE} - W_{DB} - W_{Losses} = \Delta E_{PE} + \Delta E_{KE} \tag{10.5}$$

where the rolling, air, curving, and braking losses are now deduced as W_{Losses}, all other terms in the equation being known. While this gives certainty to the overall energy losses, details of each loss are difficult to extract. Some validation can be achieved by examining responses on track sections without braking and without curves and then comparing them with other track sections.

It will be noticed that validation approaches will still leave some uncertainty in the details of the unrecoverable losses, but the locomotive and gravitational energy flows can be determined with reasonable accuracy. Of course, very expensive instrumentation programs could be used to get better validation, but such programs are rare. Given the amount of signaling systems that are already in use, it can be argued that adequately accurate validation is being achieved by the use of physics-based models, manufacturers' data, and established empirical estimates, albeit with some adjustments.

10.4 STAGE 2 – CONTROLS AND CONSTRAINTS

The train operation is constrained by requirements for observing the authority to proceed as given by the signaling system, the safe speed on curves, speed restrictions for operational reasons, the grade forces from the track, the drag forces on the train, and the operational traction and braking limits. A typical track is shown in Figure 10.7. The train control system must first be shown to be able to stop the train and comply with speed restrictions in all expected conditions at any position on the track route. Second, the train must have sufficient traction to negotiate the worst-case train mass and track grade scenario. Ideally, there should be enough tractive effort to start the train from any track position. It is possible, however, to operate successfully with less than this requirement if the steep sections are sufficiently short for the train to utilize rolling momentum to negotiate the grade. It also means that, on such sites if a mistake is made, a train might need the assistance of more locomotives to be pushed clear if a stall occurs.

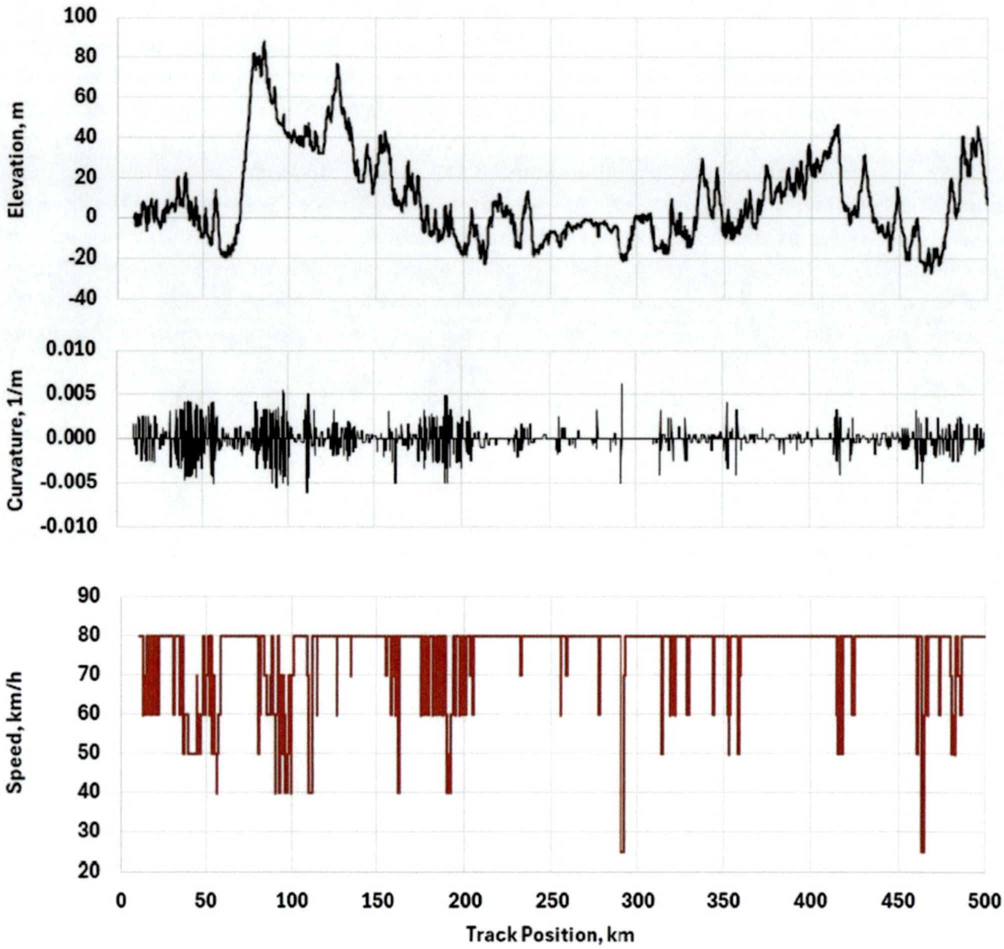

FIGURE 10.7 Track elevation, curvature, and speed restriction data.

Third, the actual methodology of driving or locomotive control has a significant effect on energy use [20] and on coupler forces [22].

Irrespective of whether the control of the train is only by human agency, human plus driver advice system assistance, cruise control, or automation, very different results will be obtained from:

- Changes to maximum running speed
- Changes in the amount of coasting
- Changes to limits in coupler forces

In the paper [22], two control strategies are presented, namely, minimum-time train control and energy-efficient train control [23]. The controller provides a live solution to the multi-objective problem of (1) on-time running; (2) energy conservation; and (3) coupler force management.

10.5 STAGE 3 – COMPREHENSIVE DIGITAL TWIN OF TRAIN-TRACK CONTROL SYSTEM

The final digital twin can be used to examine the consequences of design and operational decisions and optimize either or both design and operation. Common application areas are:

- Scheduling, routing, and network capacity planning
- Energy and decarbonization
- Train safety, stability, and rollingstock fatigue

10.5.1 Scheduling, Routing, and Network Capacity Planning

Scheduling and routing can benefit from the more precise modeling of accelerations and braking. This is true of both short suburban and very long heavy-haul trains. Importantly, the digital twin allows the variability of train operation to be modeled under different control strategies, rolling stock types, braking capabilities, etc. A series of simulations of the digital twin allow statistical expectations to be built up. These results should be validated against field measurements in the early stages of using the digital twin. Similarly, network capacity planning can be enhanced by accurate section times and the ability to calculate realistic operation limits. Useful studies for capacity planning include:

1. *Continuous running, fastest train path*: All green signals with speeds only restricted below the section speed limit by curves, level crossings, community requirements, and locomotive power limitations;
2. *Continuous running, slowest train path*: Speeds further restricted by the train taking all crossing loop options via speed-restricted turnout rails and observing slower speeds if applicable on the turnout loop;
3. *Continuous running, train path variants*: Variants between the fastest and slowest train paths can be investigated with multiple simulations of the many variants of train paths through fast and slow sections on the route;
4. *Stop-start running, fastest train path*: Adding just the expected minimum number of stop signals, otherwise green signals with speeds restricted as per continuous running, fastest train path;
5. *Stop-start running, slowest train path*: Adding the expected maximum number of stop signals, with speeds restricted as per continuous running, slowest train path;
6. *Stop-start running, train path variants*: Variants between the fastest and slowest train stop-start paths can be investigated with multiple simulations as per item 3 above.

10.5.2 Energy and Decarbonization

The digital twin gives a comprehensive picture of the energy consumed in the system. Investigation into energy use and costs can be obtained with an understanding of the effect of driving controls and speed policies. There is an important trade-off between energy use and speed. Faster operation increases aerodynamic drag. Slower trains increase the time to delivery but may not necessarily reduce the overall rate of tons delivered if the train traffic system is moving more slowly on the day. Examples of three different running times on a route are shown in Figure 10.8. Compared to the "typical" strategy the "slow" strategy not only increases section time substantially but also decreases energy use substantially. Conversely the "fast" strategy does not reduce running time much when compared to the "typical" strategy, but it will consume considerably more energy due to the need for excessive braking as the train attempts to hold speed at the highest possible levels for the longest time. The "typical" strategy is a suitable balance between "fast" and "slow", achieved by adjusting coasting and maximum speed. A digital twin enables the exercise of obtaining the best compromise of speed and energy costs with precise matching to the train operation and route. The solution chosen will depend on the value of the freight delivery rate versus energy costs. The understanding of possible trip run times may also lead to more efforts being directed toward traffic scheduling, yard operations, train provisioning, and crew changes.

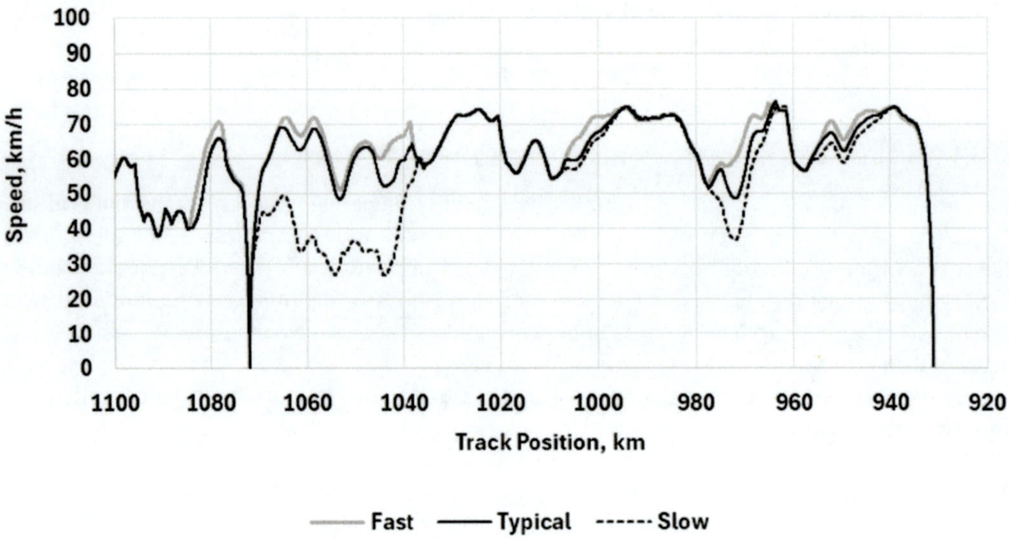

FIGURE 10.8 Examples of speed profiles from different driving or locomotive control strategies.

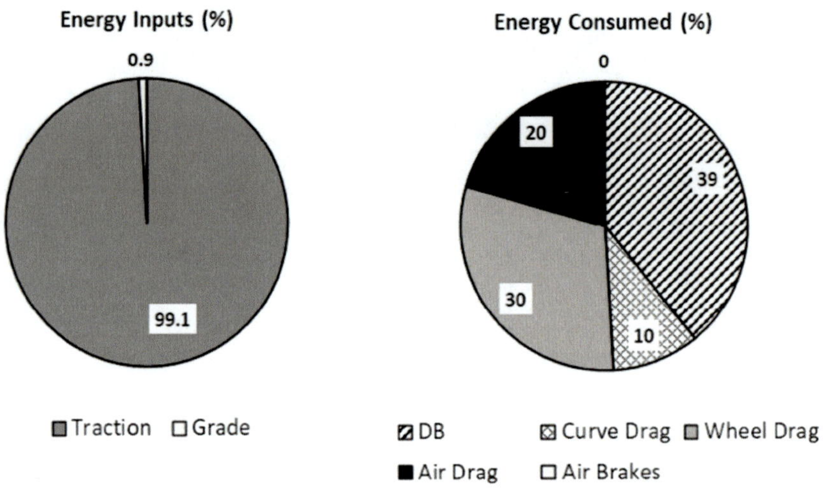

FIGURE 10.9 Typical freight train energy balance taken at the wheel.

Decarbonization options can also be evaluated by first understanding where energy is being consumed. A typical freight train operation energy balance is shown in Figure 10.9.

10.5.3 Train Safety, Stability, and Rollingstock Fatigue

The digital twin gives a comprehensive picture of the coupler forces in the system. The controls chosen for either speed or energy considerations might require adjustment to ensure that coupler forces remain within limits. High accelerations can result in unnecessarily high coupler forces or impacts. Trains longer in length can exhibit larger coupler forces, which increase risks of drawgear failure, vehicle instability, or shortened vehicle life. An example of coupler forces in a large freight train is given in Figure 10.10. A consequence of larger coupler forces in longer heavier trains are larger lateral force components and increased roll-over and derailment risks on curves as per [24].

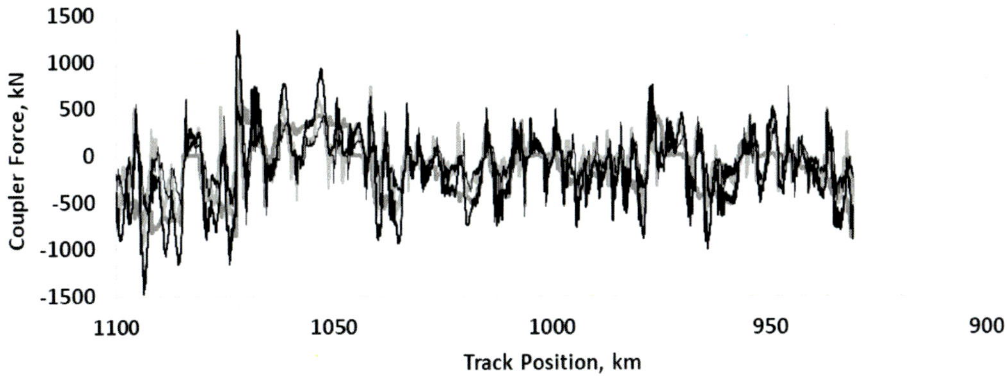

FIGURE 10.10 Example of freight train coupler forces.

Worst cases of vehicle instability caused by longitudinal train dynamics occur for empty wagons on sharper curves. The situation can be made worse when coupler forces are larger on grades, particularly when trains start on grades.

10.6 SUMMARY

Digital twins for train operation optimization are an easily realizable step from the existing simulation capabilities of train operation, coupler forces, energy use, and running times. These past analyses have been forced to consider train-track system-wide issues, and full-trip monitoring and simulation systems are well developed. Digital twins enable further precision and evolution of the modeling to match each operating train and so predict more precise train operation production rates, energy use, and equipment risks as conditions change for the train-track system being monitored. Putting all aspects together in a more comprehensive single digital twin model will allow the various railway management groups to coordinate, understand, and manage changes in conditions or improve operating statistics. A summary flow chart of a typical model is given in Figure 10.11. Essential to the functionality of the model for analysis is the ability to study new scenarios using different driving strategies provided by either a human operator or an algorithm-based control scheme. Note that measured operational logs only have the function of validating and tuning the

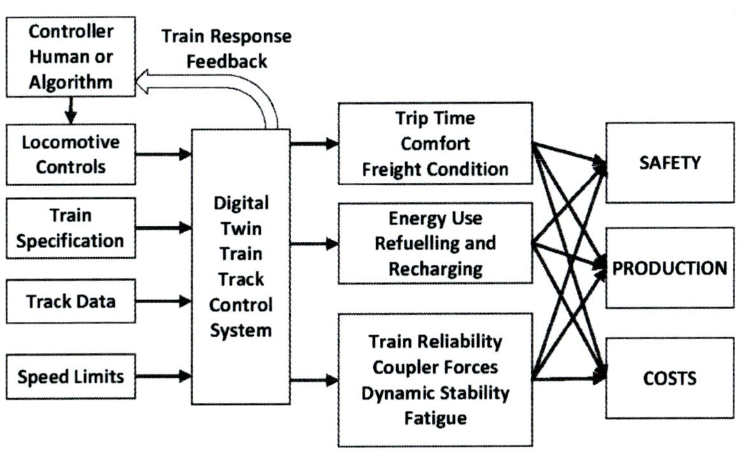

FIGURE 10.11 Simplified flowchart of the whole digital twin system for train optimization.

model. A single change to speed, train size, etc. requires a full-trip simulation with new control levels developing interactively with the train response to allow examination of its new trajectory.

Although various studies are typically done focused on single aspects of train dynamic performance, be that speed, equipment fatigue, etc., a fully detailed digital twin provides a single source of information and analysis to de-risk management decisions, facilitate change management, and monitor changing conditions of the equipment and consequences for operation and efficiency.

REFERENCES

1. M. Spiryagin, P. Wolfs, C. Cole, V. Spiryagin, Y. Sun, T. McSweeney, *Design and Simulation of Heavy Haul Locomotives and Trains*. Ground Vehicle Engineering Series, CRC Press, Boca Raton, FL, 2017.
2. C. Cole, M. Spiryagin, Q. Wu, Y.Q. Sun, Modelling, simulation and applications of longitudinal train dynamics. *Vehicle System Dynamics* 55(10): 1498–1571, 2017.
3. V.K. Garg, R.V. Dukkipati, *Dynamics of Railway Vehicle Systems*. Academic Press, New York, 1984.
4. S. Iwnicki, M. Spiryagin, C. Cole, T. McSweeney (Eds.), *Handbook of Railway Vehicle Dynamics*, 2nd ed. CRC Press, 2019. https://doi.org/10.1201/9780429469398.
5. I.B. Duncan, P.A. Webb, The longitudinal behaviour of heavy haul trains using remote locomotives. *Proceedings of the 4th International Heavy Haul Conference*, Brisbane, Australia, pp. 587–590, 1989.
6. B.J. Jolly, B.G. Sismey, Doubling the length of coals trains in the Hunter Valley. *Proceedings of the 4th International Heavy Haul Conference*, Brisbane, Australia, pp. 579–583, 1989.
7. R.D. Van Der Meulen, Development of train handling techniques for 200 car trains on the Ermelo-Richards Bay line. *Proceedings of the 4th International Heavy Haul Conference*, Brisbane, Australia, pp. 574–578, 1989.
8. P.G. Howlett, I.P. Murray, P.J. Pudney, Energy-efficient train control. *Control Engineering Practice* 2(2): 193–200, 1994.
9. C. Cole, Control System for Operating Long Vehicles, United States Patent No. 7,359,770, 2003.
10. M. Spiryagim, J. Edelmann, F. Klinger, C. Cole, Vehicle system dynamics in digital twin studies in rail and road domains. *Vehicle System Dynamics* 61(7): 1735–1784, 2023. https://doi.org/10.1080/00423114.2023.2188228.
11. C. Cole, Q. Wu, M. Spiryagin, Testing the impact performance of a long pocket dual draft gear using digital twin technique. In: Huang, W., Ahmadian, M. (eds) *Advances in Dynamics of Vehicles on Roads and Tracks III*. IAVSD 2023. Lecture Notes in Mechanical Engineering. Springer, Cham, 1180–1188, 2025. https://doi.org/10.1007/978-3-031-66971-2_122.
12. AAR Manual of Standards and Recommended Practices freight Car Draft Components – Section B, Cushioning Devices, End-of-Car, Active Draft End of Car, Specification M-921 -F, *Association of American Railroads*, 107–120, 2019.
13. M. Spiryagin, Q. Wu, O. Polach, J. Thorburn, W. Chua, V. Spiryagin, et al., Problems, assumptions and solutions in locomotive design, traction and operational studies. *Railway Engineering Science* 30: 265–288, 2022. https://doi.org/10.1007/s40534-021-00263-w
14. I. Graessler, J. Hentze, T. Bruckmann, V-models for interdisciplinary systems engineering. In: Marjanović, D., Štorga, M., Škec, S., Bojčetić, N., Pavković, N. (eds) *Proceedings of the DESIGN 2018-15th International Design Conference*. University of Zagreb, Dubrovnik, Croatia, pp. 747–756, 2018.
15. J. Gausemeier, S. Moehringer, New guideline VDI 2206 – A flexible procedure model for the design of mechatronic systems. In: Folkeson, A., Gralen, K., Norell, M., Sellgren, U. (eds) *Proceedings of 14th International Conference on Engineering Design (ICED 03)*. The Design Society, UK, Stockholm, Sweden, 2003, available at https://www.designsociety.org/publication/38/DS+31%3A+Proceedings+of+ICED+03%2C+the+14th+International+Conference+on+Engineering+Design%2C+Stockholm.
16. L. Wright, S. Davidson, How to tell the difference between a model and a digital twin. *Advanced Modelling and Simulation in Engineering Sciences* 7(13): 1–13, 2020.
17. T. West, M. Blackburn, Is digital thread/digital twin affordable? A systemic assessment of the cost of DoD's latest Manhattan project. *Procedia Computer Science* 114: 47–56, 2017.
18. Defense Acquisition University. Digital Twin. DAU Glossary of Defense Acquisition Acronyms and Terms. Available at https://www.dau.edu/glossary. Accessed on June 8, 2021.
19. K. Smith, Rise of the machines: Rio Tinto breaks new ground with AutoHaul. *International Railway Journal*, 2019. [Online]. Available at https://www.railjournal.com/in_depth/rise-machines-rio-tinto-autohaul.

20. A. Albrecht, P. Howlett, P. Pudney, X. Vu, P. Zhou The key principles of optimal train control—Part 1: Formulation of the model, strategies of optimal type, evolutionary lines, location of optimal switching points. *Transportation Research Part B: Methodological* 94, 482–508, 2016. https://doi.org/10.1016/j.trb.2015.07.023
21. Q. Wu, M. Spiryagin, C. Cole, T. McSweeney, Parallel computing in railway research. *International Journal of Rail Transportation* 8(2): 111–134, 2020.
22. M. Hayman, C. Bosomworth, C. Cole, M. Spiryagin, Next generation train control – beyond 40000 tonnes. *Proceedings of the 12th International Heavy Haul Conference*, Rio de Janeiro, Brazill de Je, 2023.
23. G.M. Scheepmaker, H.Y. Willeboordse, J.H. Hoogenraad, R.S. Luijt, R.M.P. Goverde, Comparing train driving strategies on multiple key performance indicators. *Journal of Rail Transport Planning & Management* 13: 100163, 2020. https://doi.org/10.1016/j.jrtpm.2019.100163
24. C. Cole, M. Spiryagin, Y.Q. Sun, Assessing wagon stability in complex train systems. *International Journal of Rail Transportation* 1(4): 193–217, 2013.

11 Digital Twins for Train Operational Safety and Reliability

Qing Wu, Xiaohua Ge, Shengyang Zhu, and Shuai Su

11.1 INTRODUCTION

Efficient railway operations are crucial for ensuring the smooth and reliable transportation of passengers and goods, directly impacting economic growth and sustainability. They minimize delays, reduce operational costs, and enhance service reliability, contributing to a better user experience. Additionally, efficient operations help decrease energy consumption and lower carbon emissions, making railways a more environmentally friendly mode of transportation. Overall, efficient railway operations are essential for maintaining competitiveness, meeting rising demand, and supporting broader environmental and economic goals.

Several factors influence railway train operational efficiency, including train scheduling, track utilization, and maintenance practices. Advanced signaling systems and communication technologies play a crucial role in optimizing train movements and reducing delays. Additionally, the condition and design of infrastructure, such as tracks and stations, impact the smooth operation of trains. External factors like weather conditions and unexpected incidents can also affect efficiency. Effective management of these elements, along with the adoption of innovative technologies, is essential for enhancing the overall efficiency of railway operations.

Railway virtual coupling (VC) (Wu et al. 2023b) as shown in Figure 11.1 is an advanced technology that enables trains to operate in closer proximity by maintaining a dynamic, real-time connection, rather than relying on traditional fixed block systems. This system uses continuous communication, sensors, and automated control mechanisms to adjust the spacing between trains, allowing them to run closer together safely. VC enhances the flexibility and efficiency of train operations by optimizing track capacity, reducing travel times, and improving overall service reliability. By leveraging real-time data and advanced algorithms, VC ensures that trains can respond swiftly to changes in speed and conditions, thereby maintaining a high level of safety and performance.

While railway VC technologies offer significant benefits in terms of efficiency and capacity, they also present safety and reliability challenges that need careful management. The primary concern is ensuring consistent and reliable real-time communication between trains, as any disruption could lead to unsafe conditions. Additionally, the accuracy and robustness of the sensors and control systems are critical; any malfunction or delay in response could pose serious safety risks. Cybersecurity is another major issue, as VC systems are vulnerable to hacking and other cyber threats, potentially compromising the safety and reliability of train operations. Therefore, rigorous testing, robust fail-safe mechanisms, and continuous monitoring are essential to address these challenges and ensure the safe and reliable operation of virtually coupled trains.

Digital twin technologies can significantly enhance the safety and reliability of VC in railway systems by providing a comprehensive virtual replica of the physical trains and their operating environment. These digital twins allow for real-time monitoring and analysis of train performance, enabling the detection and diagnosis of potential issues before they escalate into safety hazards. By simulating various operational scenarios and stress-testing the VC system, digital twins help in

 DOI: 10.1201/9781003492146-11

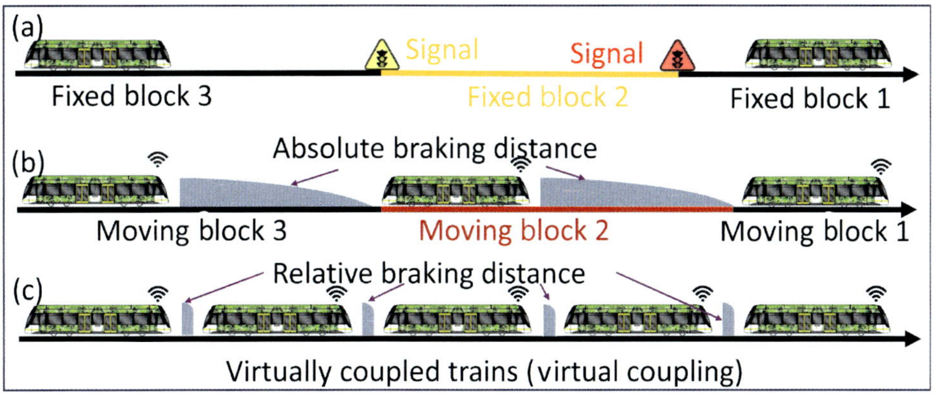

FIGURE 11.1 Railway signaling systems.

identifying vulnerabilities and optimizing response strategies. Additionally, they facilitate predictive maintenance by forecasting potential failures and scheduling proactive repairs, thereby reducing unexpected downtimes. With continuous data integration and advanced analytics, digital twins ensure that VC systems operate safely and reliably, adapting dynamically to changing conditions and preventing accidents.

In a futuristic picture, researchers should be able to create a comprehensive digital twin in the virtual world to replicate railway track systems, train systems, signaling control systems, and overhead powering systems. This chapter discusses the modeling of individual components and paints an overall picture of a holistic digital twin that can be used to enhance train operational safety and reliability. Various challenges that are currently faced by researchers are also discussed.

11.2 CONCEPT OF HIGH-FIDELITY DIGITAL TWINS FOR RAILWAY OPERATIONAL SYSTEMS

The concept for high-fidelity digital twins is shown in Figure 11.2, which integrates various computational models and systems using a Message Passing Interface (MPI) to achieve a comprehensive and synchronized simulation environment. This approach leverages parallel computing to model the complex interactions within a railway system, enabling detailed analysis and optimization.

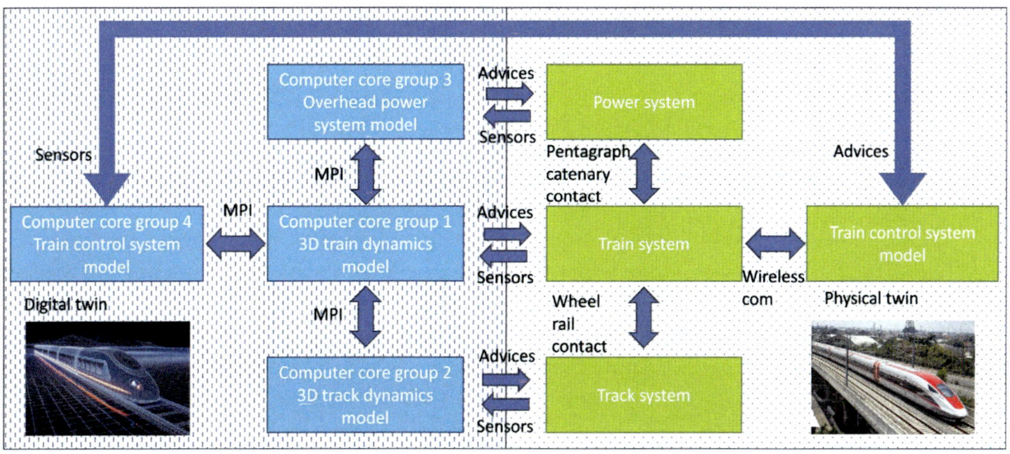

FIGURE 11.2 Concept for high-fidelity digital twins.

The concept uses a bidirectional flow of information between the physical and digital twins of the train system. On the one hand, for the physical twin, we primarily consider the actual train, track, power systems, and train control systems from the train dynamics and control point of view. These components are equipped with sensors that collect real-time data about their operations. The train system includes locomotives and carriages, which interact with the track system through wheel–rail contact. The track system itself consists of the rails, sleepers, ballast, and underlying substructure, all of which affect the train's motion and stability. The power system includes the overhead catenary system and the pantograph mounted on the train, ensuring a continuous supply of electricity. However, the pantograph parts are in the train models while the catenary is located in the power system model. They interact with each other via pantograph-catenary contact forces. In the meantime, for the train control system, real-time data gathered from the train system is transmitted wirelessly to the train control system model. And the train control system generates new commands for the train systems.

On the other hand, the digital twin encompasses several computer core groups, each dedicated to modeling different aspects of the railway system. The 3D train dynamics model simulates the dynamic behavior of the train, including its motion, forces, and especially the wheel–rail contact forces and pantograph centenary contact forces. The 3D track dynamics model focuses on modeling the track system, accounting for factors like track geometry, stiffness, and the dynamic response to train loads. The overhead power system model simulates the reactions of the catenary wire to the pantograph forces. The train control system model interacts with the train model to generate new control commands according to the current train status and future predictions. The MPI technique is used to facilitate communication and data exchange between these core groups.

The digital twin receives real-time sensor data from the physical twin, which is then used to update the models and provide accurate, up-to-date simulations. These simulations generate insights and optimization strategies that are fed back to the physical system as advice, helping to enhance performance, safety, and reliability. The integration of detailed models of the train, track, and power systems with real-time data from the physical system enables precise control and optimization of railway operations. This approach not only enhances the efficiency and safety of the railway system but also provides a powerful tool for predictive maintenance and strategic planning. The continuous feedback loop between the physical and digital twins ensures that the system adapts to changing conditions and maintains optimal performance at all times.

11.3 HOW TO DEVELOP THE COMPONENTS

As discussed in the previous section, this chapter focuses primarily on four parts of the twins: train system, track system, power system, and control system. This section will then introduce the modeling methods to obtain these high-fidelity models.

11.3.1 Train Mechanical Dynamics

Train system dynamics (TSD), specifically mechanical system dynamics, is a comprehensive modeling approach that incorporates wheel/rail contact, suspension systems, multibody coupler systems, and air brake systems. Each of these components plays a crucial role in accurately simulating the dynamics of train operations. Key modeling techniques for TSD include the modeling approaches of longitudinal train dynamics (LTD) and vehicle system dynamics (VSD).

LTD modeling assumes only longitudinal movements, focusing on elements like draft gear forces, air brake forces, traction forces, dynamic brake forces, propulsion resistance, gravitational components, and curving resistance. Draft gears, which dissipate energy and act as dampers, are integral to coupler systems. Air brake systems, essential for train safety, consist of brake pipes, brake valves, brake cylinders, etc. These systems are modeled using empirical, fluid dynamics, and combined empirical-fluid dynamics approaches, each offering different levels of detail and accuracy.

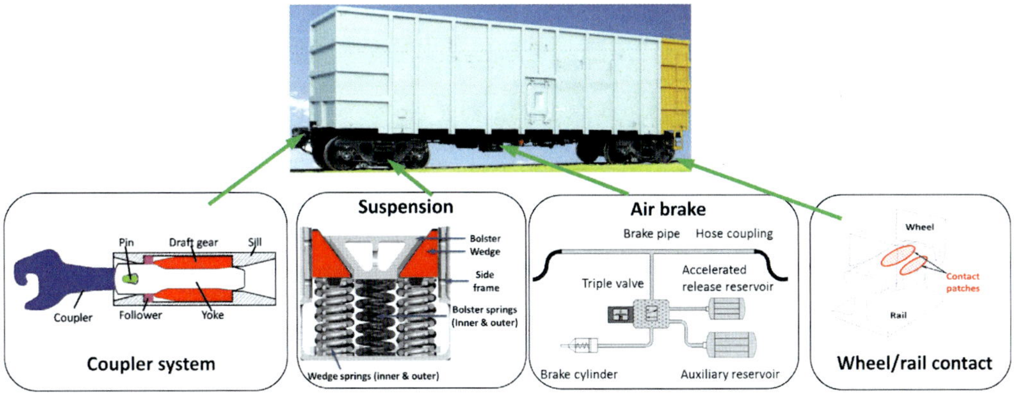

FIGURE 11.3 Key subsystems for wagon dynamics model.

VSD considers separately modeling each individual vehicle in a train, including both wagons and locomotives, with a particular emphasis on the wheel/rail contact and suspension systems. The wheel/rail contact model involves determining contact points, normal force distributions, relative motions, and tangential forces. Suspension models for freight wagons, particularly friction wedge suspensions, are more complex due to the use of friction dampers. These models include detailed representations of the spring nest, wedge, and friction components. An illustration of the key subsystems of a freight wagon is shown in Figure 11.3 according to (Wu et al. 2017).

The multibody coupler system includes components like the coupler, pin, yoke, follower, and draft gear. These systems must simulate both compressive and tensile conditions. Detailed multibody coupler models with alignment controls, such as friction pairs or bump-stops, help prevent excessive rotations and jackknifing. These models are crucial for accurately simulating the dynamic interactions between coupled vehicles.

For locomotives or electric multiple unit trains, a multibody pantograph model is also needed. The development of such models begins with defining the mechanical properties of each part, such as the arms, springs, and joints, which are essential for simulating the pantograph's motion and response to forces. Using principles of rigid body dynamics, each component is modeled to capture its mass, inertia, and connectivity to other parts. The interactions between these components are represented by constraints and force elements, including springs and dampers that mimic the pantograph's behavior in real-world conditions. The model also includes the contact interface between the pantograph and the overhead catenary wire, accounting for the forces and displacements resulting from this interaction.

TSD models are developed by connecting individual VSD models via multibody coupler system models. Compared to traditional LTD models, TSD models offer several advantages, such as eliminating the need to separately model rolling and flanging resistances, as these are inherently included in the wheel/rail contact models. Additionally, TSD models can incorporate advanced traction control systems and mechatronic models, providing a more comprehensive and accurate representation of train dynamics. Overall, the integration of LTD and VSD modeling techniques into TSD provides a robust framework for simulating the complex dynamics of train operations. This comprehensive approach allows for more accurate predictions of train behavior, leading to improved safety, reliability, and efficiency in railway operations.

11.3.2 Track Structural Dynamics

The track model discussed in this chapter is of the ballasted type as shown in Figure 11.4 according to (Wu et al. 2023a). It is structured around four main layers: rails, sleepers, ballast, and subballast, with the rails modeled using finite element methods (FEMs). The rails are represented by

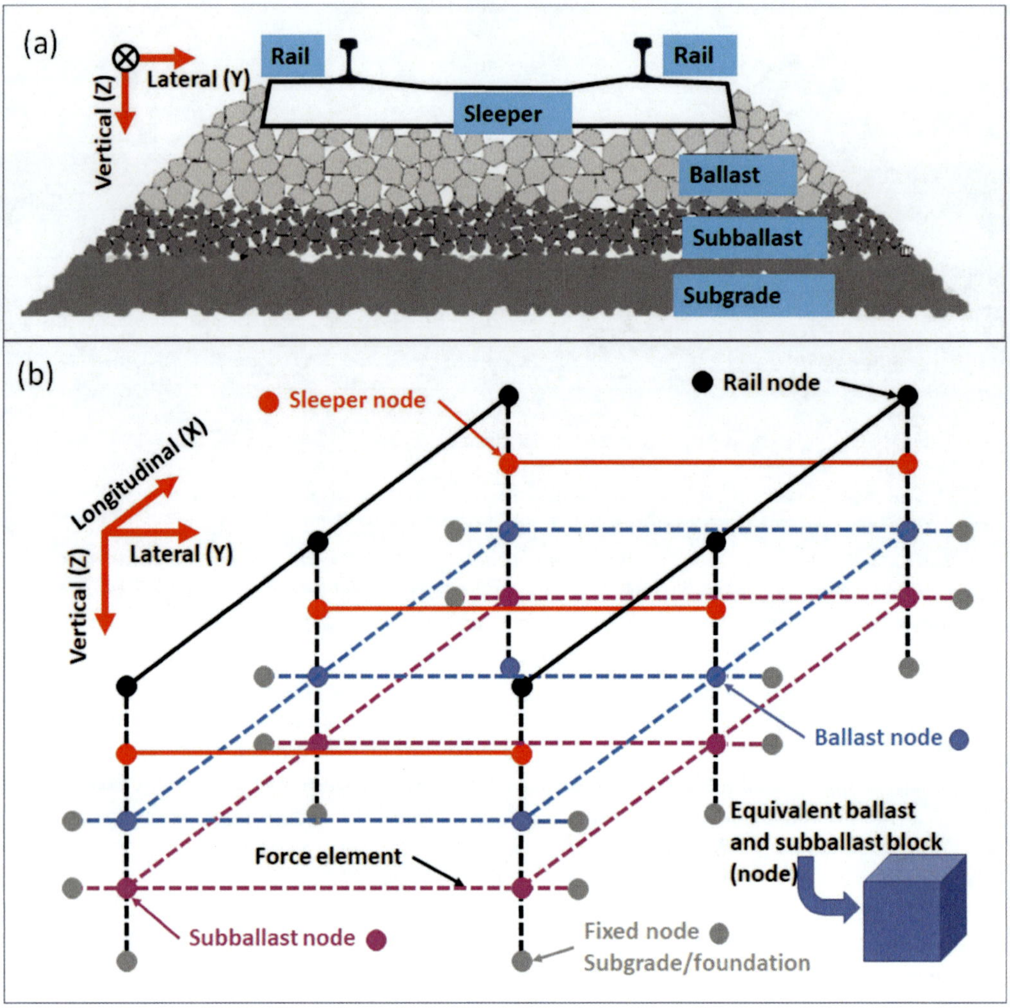

FIGURE 11.4 Ballasted railway track structure and model.

Euler-Bernoulli beams, chosen for their numerical simplicity and suitability for heavy haul railway simulations. Each node in the rail model has six degrees of freedom (DoFs), with nodes spaced according to sleeper positions.

The sleepers are modeled as rigid bodies with six DoFs, connected to the rails through fastening models that provide three translational and three torsional stiffnesses. The ballast and subballast layers are represented by block models with three translational DoFs each. These blocks are interconnected, with ballast blocks connected to adjacent ballast and subballast blocks, and subballast blocks also connected to the subgrade. The connections between these blocks include translational stiffness and damping properties. Frictional behavior is considered in the longitudinal and lateral directions of the sleeper-ballast interface, as well as in the longitudinal direction of the fastening.

For the rail model, basic element matrices for lateral and vertical movements are provided, incorporating mass and stiffness parameters. Longitudinal movement is modeled using torsional and axial vibration rod theories. Damping is incorporated using Rayleigh damping coefficients to form the damping matrix, which includes contributions from both mass and stiffness matrices. Shape functions for interpolation are detailed, differentiating between lateral/vertical and longitudinal movements.

The ballast and subballast blocks' equivalent masses and stiffness properties are determined through specific equations. These equations account for parameters such as block density, sleeper dimensions, force distribution angles, and block height. The stiffness and damping properties of the subgrade are also considered, with values approximated based on the Poisson ratio and damping ratio.

Frictional behavior within the track model is addressed, recognizing macroscopic friction in longitudinal and lateral movements between sleepers and ballasts as well as in the longitudinal direction of the fastenings. The model differentiates between static and kinetic friction, with static friction represented as a linear spring-damping combination and kinetic friction characterized by a hysteresis effect. This friction model ensures that the dynamic interactions between track components are accurately represented.

11.3.3 Power Systems and Overhead Lines

Developing a mechanics-based model to examine the forces and vibrations in the overhead powering systems, particularly the catenary-pantograph systems, involves a comprehensive understanding of the mechanical properties and interactions of various components. The goal is to create a model that accurately represents the physical behavior of the system under various operational conditions, allowing for the analysis of forces, displacements, and vibrations.

The first step in this process is to define the system components. The key elements of an overhead catenary system include the messenger wire, contact wire, dropper, supports, etc., as shown in Figure 11.5a. The pantograph can have a base frame, arm assemblies, current collectors (contact strips), dampers, etc.

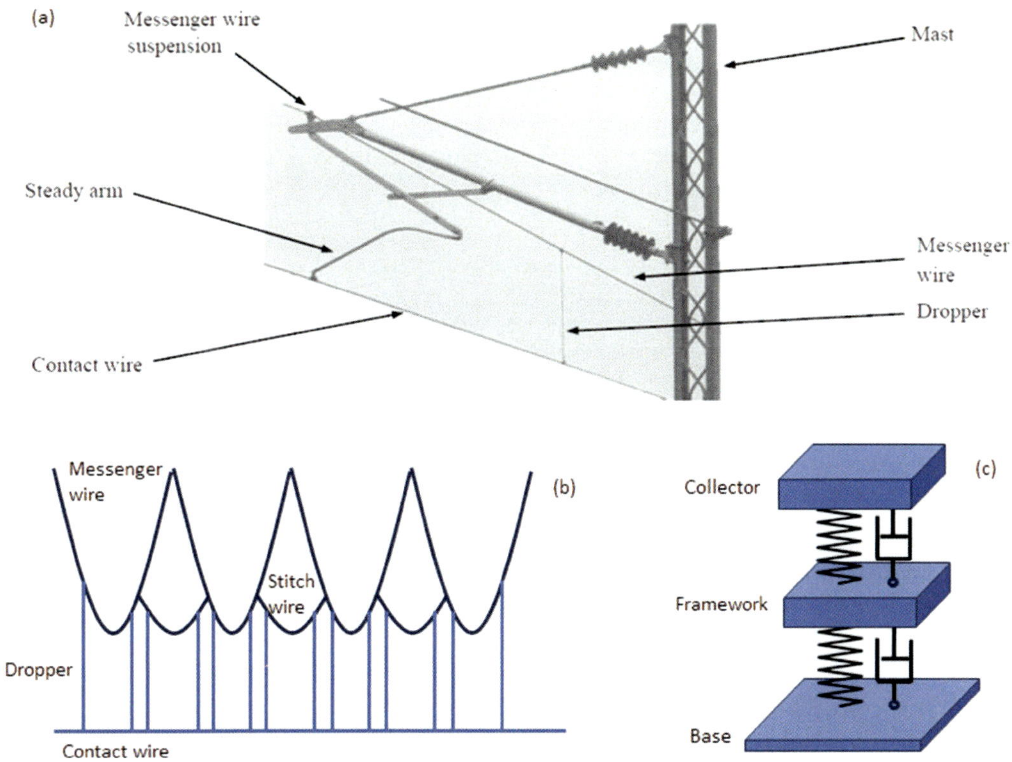

FIGURE 11.5 Pantograph-catenary system: (a) catenary system, (b) catenary finite element methods model, and (c) lumped mass pantograph model. (From Bruni S, Ambrosio J, Carnicero A, et al., 2015, The results of the pantograph–catenary interaction benchmark. *Vehicle System Dynamics*, 53 (3), 412–435, with permission.)

Next, it is essential to establish the mechanical properties of each component. For the catenary system, this includes defining material properties such as Young's modulus, density, cross-sectional area, and moments of inertia, as well as geometric properties like the lengths of the wires and the spacing between support structures. These properties form the basis for the subsequent formulation of the equations of motion. The equations of motion are derived using the principles of mechanics. This involves applying Newton's laws to develop differential equations that describe the static and dynamic behavior of the system. For the catenary wire, these equations account for the forces and displacements along the length of the wire, including tension forces and bending moments. Vibrational analysis is also necessary to understand the dynamic response of the system, which involves determining natural frequencies and mode shapes.

For a more detailed and accurate catenary model, the FEM is employed, and models, as exemplified in Figure 11.5b (Song et al. 2023), can be developed. This method involves discretizing the continuous system into a finite number of elements, each represented by its own set of equations. For the catenary system, the wires can be modeled as beam elements with appropriate stiffness and mass properties. The element stiffness and mass matrices are formulated based on the material and geometric properties, and these are assembled into global matrices that represent the entire system. Boundary conditions are applied to these matrices to account for the constraints imposed by supports and connections.

To model a pantograph using the rigid multibody dynamics simulation method, one begins by representing each component of the pantograph, such as the base frame, arms, and collector, as rigid bodies with defined mass and inertia properties. These components are connected through kinematic joints that allow for relative motion, typically modeled as revolute joints to replicate the actual movement mechanisms of the pantograph. Key forces include the spring forces providing upward motion, the damping forces controlling oscillations, and the contact force between the collector and the overhead catenary wire. The contact interaction is crucial and can be modeled using appropriate contact force models that capture the frictional and normal forces. Additionally, constraints are imposed to ensure that the pantograph's movements remain within realistic operational bounds. In addition to these, a simplified lumped mass model as shown in Figure 11.5c (Song et al. 2017) can be developed. In this model, the arm assembles can be represented by mass, damping, and stiffness. The rotational motions are simplified as un-and-down translational motions.

11.3.4 Communication Network and Train Controllers

The effective operation of digital twins for train operational safety and reliability hinges on the seamless integration of advanced communication networks and sophisticated train controllers. This section delves into the state-of-the-art technologies underpinning communication networks and train controllers, which are critical for enhancing safety, reliability, and efficiency in modern railway systems and enabling automated control and supervision of train movements.

11.3.4.1 Train-to-Train and Train-to-Ground Communication Technologies

Communication networks are pivotal in enabling real-time data exchange between trains, namely, train-to-train (T2T) communication, or between trains and railway control centers, namely, train-to-ground (T2G) communication, facilitating dynamic operational adjustments, and ensuring coordinated movements. Below are the primary technologies in use:

- *Global System for Mobile Communications-Railway (GSM-R)*: GSM-R is a widely adopted standard for railway communication, providing reliable voice and data services. It offers interoperability between different railway systems and countries, facilitating international railway operations. Furthermore, it operates on dedicated frequency bands and prioritizes critical communication, ensuring that emergency calls, shunting, and important operational data and messages are transmitted with high priority, even during network congestion. GSM-R is integral to modern train control systems like the European Train Control System (ETCS). It is primarily used to facilitate continuous T2G communication, enabling precise control of train movements and reducing the risk of accidents.

- *Long-Term Evolution for Railways (LTE-R)*: LTE-R is based on the long-term evolution (LTE) standard, commonly used in commercial mobile networks, but tailored to meet the specific needs of railway communication systems. While LTE-R can also support T2G communication, its primary focus is on enhancing the T2G communication infrastructure, offering better data rates, lower latency, and improved reliability compared to GSM-R. It also empowers advanced quality-of-service mechanisms, ensuring that critical communication, such as train control messages, is prioritized over less critical data, such as passenger internet use. LTE-R's scalability and flexibility cater to the increasing data demands of modern railways. It supports sophisticated train control systems like the ETCS and Communication-Based Train Control, which require reliable, low-latency communication. LTE-R allows for real-time monitoring of train systems and infrastructure, enabling predictive maintenance and reducing downtime.
- *Future Railway Mobile Communication System (FRMCS)*: FRMCS is poised to replace GSM-R, leveraging 5G and beyond technologies to provide ultra-reliable, low-latency communication. While specific implementations and use cases may vary, FRMCS aims to provide a comprehensive communication framework that enhances both T2T and T2G communication within the railway network. On the T2G side, it will primarily serve as a replacement for existing GSM-R and LTE-R, and facilitate essential functions such as signaling, control, operational data exchange, and maintenance communications between trains and ground-based control centers. On the T2T side, it enables trains to exchange information directly with each other, even over long distances, which can be vital for tasks such as automatic train protection, collision avoidance, and cooperative control. FRMCS will also support other critical applications, from automated train operation (ATO) to predictive maintenance. Its architecture is designed to accommodate future advancements, ensuring long-term relevance and adaptability.
- *Dedicated Short-Range Communications (DSRC) and Wi-Fi*: DSRC and Wi-Fi are both positioned as complementary technologies within the FRMCS ecosystem. Specifically, DSRC facilitates direct T2T communication over short distances, enabling rapid data exchange, critical for collision avoidance, traffic management, and real-time operational adjustments. Wi-Fi networks, often deployed in stations and along rail corridors, can be utilized for onboard passenger services such as internet access and infotainment, as well as for operational purposes like real-time video surveillance and data transfer when trains are in stations or depots. DSRC and Wi-Fi both can be integrated into FRMCS to enhance and support its core functionalities, providing a more reliable, efficient, and versatile communication system for railway operations.

11.3.4.2 VC Control Technologies

VC technology revolutionizes train operations by enabling trains to run in close proximity without physical coupling, akin to road vehicle platooning and truck platooning. This technology is integrated with the ATO system of each train and relies on reliable T2T/T2G data communication and precise control algorithms for each automated train to maintain its safety and efficiency more concretely.

Leveraging modern communication networking technologies, real-time data exchange of each train's motion state (position, speed, acceleration) and environmental data (track conditions, obstacles, weather) can be ensured by implementing key communication management protocols (e.g., QoS management, network slicing, redundancy mechanism, encryption, and fail-safe procedure). To model such T2T and T2G data exchange, the graph theory is typically employed to describe the information flow topology. For example, from an algebraic graph perspective, the information flow topology that characterizes the directional T2T/T2G data exchange can be modeled by a generic digraph $\mathcal{G} = \{\mathcal{V}_T, \mathcal{V}_G, \mathcal{A}_{T2T}, \mathcal{A}_{T2G}, \mathcal{E}_{T2T}, \mathcal{E}_{T2G}\}$, where $\mathcal{V}_T = \{1,2,\ldots,N\}$ is the set of train nodes (or vertices) with N being the number of total train nodes, $\mathcal{V}_G = \{1,2,\ldots,M\}$ is the set of M ground control center nodes, $\mathcal{A}_{T2T} = \left[w_{ij}\right]_{N\times N}$ and $\mathcal{A}_{T2G} = \left[b_{im}\right]_{N\times M}$ denote the adjacency matrices with nonnegative adjacency elements (or called coupling weights) $w_{ij} \geq 0$ and $b_{im} \geq 0$. For example, $w_{ij} > 0$, $i, j \in \mathcal{V}_T$ means that there is an information link from train node j to train node i; and $w_{ij} = 0$, otherwise. $\mathcal{E}_{T2T} \subseteq \mathcal{V}_T \times \mathcal{V}_T$ and $\mathcal{E}_{T2G} \subseteq \mathcal{V}_T \times \mathcal{V}_G$ are the edge sets of paired nodes with (i, j) denoting an edge rooted at train node j and ended at train node i and (i, m) denoting an edge rooted at control center node m and ended at train node i.

Advanced VC control strategies and algorithms then will be employed to process the exchanged data and make continuous and optimal adjustments of train spacing to ensure safe and efficient operations. We refer interested readers to Wu et al. (2023b) for a comprehensive survey of the latest control techniques used for railway VC, including consensus-based control, model prediction control, sliding mode control, machine learning-based control, and constraints-following control. To demonstrate the VC control strategy and algorithm development, the consensus-based control technique is adopted here, which can be formulated as:

$$a_{i,t}^{\text{des}} = K\left\{\sum_{j\in\mathcal{V}_T} w_{ij}\left(x_{i,t} - \tilde{x}_{j,t} - \tilde{d}_{ij,t}\right) + \sum_{m\in\mathcal{V}_G} b_{im}\left(x_{i,t} - s_{m,t}\right)\right\}, \; i \in \mathcal{V}_T$$

where $a_{i,t}^{\text{des}}$ stands for the desired high-level acceleration command input computed for each automated train $i \in \mathcal{V}_T$, $x_{i,t} = \left[p_{i,t}, v_{i,t}, a_{i,t}\right]^{\top}$ denotes the real-time train motion state of train i with $p_{i,t}, v_{i,t}, a_{i,t}$ being the longitudinal position, speed, and acceleration, respectively, $\tilde{x}_{j,t}$ denotes the received motion state from train j over the wireless communication link, K denotes the controller gain matrix, $\tilde{d}_{ij,t} = \left[d_{ij,t}, 0, 0\right]^{\top}$ represents the designated gap vector with $d_{ij,t}$ being the desired inter-train distance between train i and train j at time t, and $s_{m,t}$ denotes the environmental data received from the ground control center to regulate each train's longitudinal motion. It can be seen that the communication topology involving both T2T and T2G information flows is incorporated into the VC control strategy design, which eventually leads to the desired high-level acceleration command $a_{i,t}^{\text{des}}$ for each train. For scenarios where the T2T communication needs to be optimized, e.g., finding the optimal T2T communication paths with the lowest latency or highest reliability or minimizing the total communication cost, the desired acceleration control $a_{i,t}^{\text{des}}$ can suitably accommodate such communication requirements. Further, various network-induced constraints can be suitably modeled and analyzed under the above VC control strategy. For example, in the presence of a network-induced communication delay, the received motion state $\tilde{x}_{j,t}$ from train j can be formulated as $\tilde{x}_{j,t} = x_{j,t-\tau_t}$, where τ_t models the variable communication delay; when data losses or denial-of-service attacks occur, the received motion state $\tilde{x}_{j,t}$ may be partially dropped/lost, which can be modeled as $\tilde{x}_{j,t} = \theta_{ij,t} x_{j,t}$ with $\theta_{ij,t}$ being the random binary variable characterizing the arrival status of train j's state; when the wireless channel exhibits communication noises or suffers malicious deception attacks, the integrity of the received motion state $\tilde{x}_{j,t}$ may be violated, which can be modeled as $\tilde{x}_{j,t} = x_{j,t} + y_{j,t}$ with $y_{j,t}$ denoting the communication noise signal or the arbitrary deception attack signal.

11.3.4.3 A Digital Twin-Based VC Control Architecture

The integration of communication network technologies and VC control techniques into a cohesive digital twin framework represents a significant leap forward in railway operations. A representative digital twin architecture of the VC control system with T2T/T2G data exchange and control algorithms highlighted is illustrated in Figure 11.6.

Real-time train data from onboard sensors is transmitted via the communication network from the trains to the digital twin model. This data allows the digital twin model to continuously update

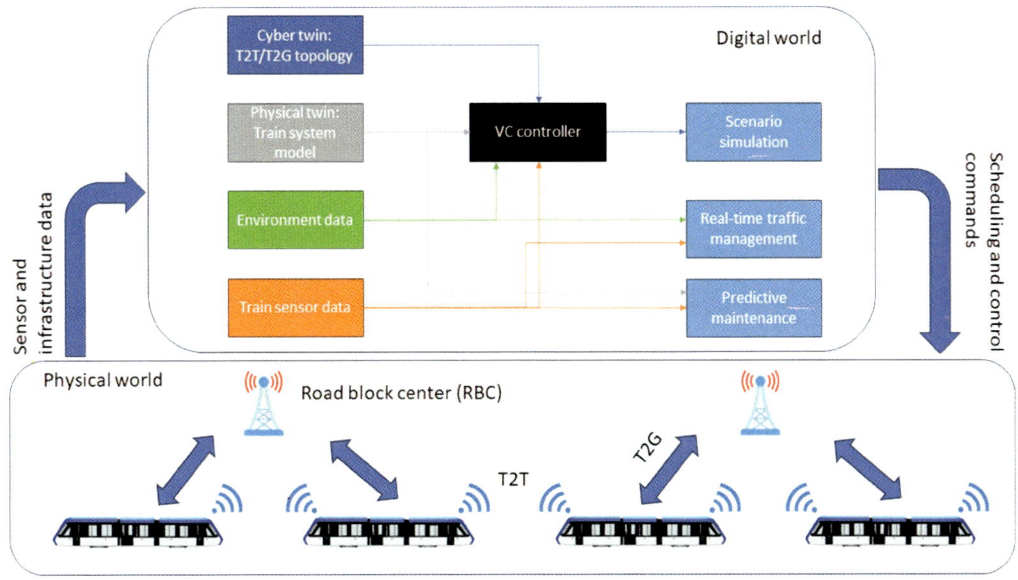

FIGURE 11.6 A digital twin-based virtual coupling control architecture.

its state, enabling (1) predictive maintenance, where early detection of potential equipment failures can be performed by analyzing sensor data from each train. Early detection of anomalies also allows for proactive actions to reduce the risk of sudden failures and enhance overall reliability. Fault detection systems within the VC framework ensure rapid identification and rectification of issues, minimizing operational disruptions; (2) real-time traffic management, where train schedules can be optimized and traveling delays can be minimized based on real-time position information; and (3) scenario simulation, where testing diverse scenarios and validating different control strategies can be carried out for virtually coupled trains within the safe and controlled environment of the digital twin.

In summary, the state-of-the-art communication networks and VC control technologies provide a robust foundation for the development of high-fidelity digital twins. These advancements promise to significantly enhance the safety, reliability, and efficiency of train operations, paving the way for the next generation of smart, interconnected railway systems.

11.4 THE COMPUTING CHALLENGES FROM LARGE-SCALE MODELS

Having been introduced to the modeling methods for previous subsystems, readers can comprehend the huge scale of computing tasks. To make the digital twin possible, two methods are introduced in this section: data-driven surrogate models and parallel computing.

11.4.1 Data-Driven Models

A data-driven surrogate model is a simplified representation of a complex system or process, created using machine learning and statistical methods. It approximates the behavior of the original model by learning from a large dataset generated through simulations or real-world experiments. These surrogate models are designed to be computationally efficient, enabling quick predictions and analyses without the need for extensive computational resources. They are particularly useful in scenarios where running detailed simulations is time-consuming or impractical, allowing for rapid evaluation and optimization of system performance based on the learned relationships within the data.

Developing a data-driven (AI-based) model for a 3D train track system involves integrating advanced machine learning techniques with traditional engineering principles to create a robust, accurate, and efficient simulation framework. The first step in creating a data-driven model is to collect comprehensive and high-quality data from various sources. In the context of a 3D train track model, this data can include sensor readings from the track and trains, historical maintenance records, weather conditions, and operational parameters such as train speed and load. Advanced sensors and IoT devices installed along the tracks and on the trains can continuously gather real-time data, providing a rich dataset that captures the dynamic interactions between the train and the track. This data serves as the foundation for building an accurate and reliable AI-based model. Another popular approach that is gradually being accepted by researchers is to use detailed high-fidelity models to generate the training data as shown in Figure 11.7 according to Bernal et al. (2024).

Once the data is collected, it must be preprocessed to ensure it is suitable for machine learning algorithms. Data preprocessing involves cleaning the data to remove any noise or inconsistencies, handling missing values, and normalizing or standardizing the data to a consistent scale. Additionally, feature engineering is performed to extract relevant features from the raw data, which can significantly enhance the model's ability to learn complex patterns. For instance, features such as track curvature, gradient, rail wear, and temperature variations can be derived from the raw sensor data.

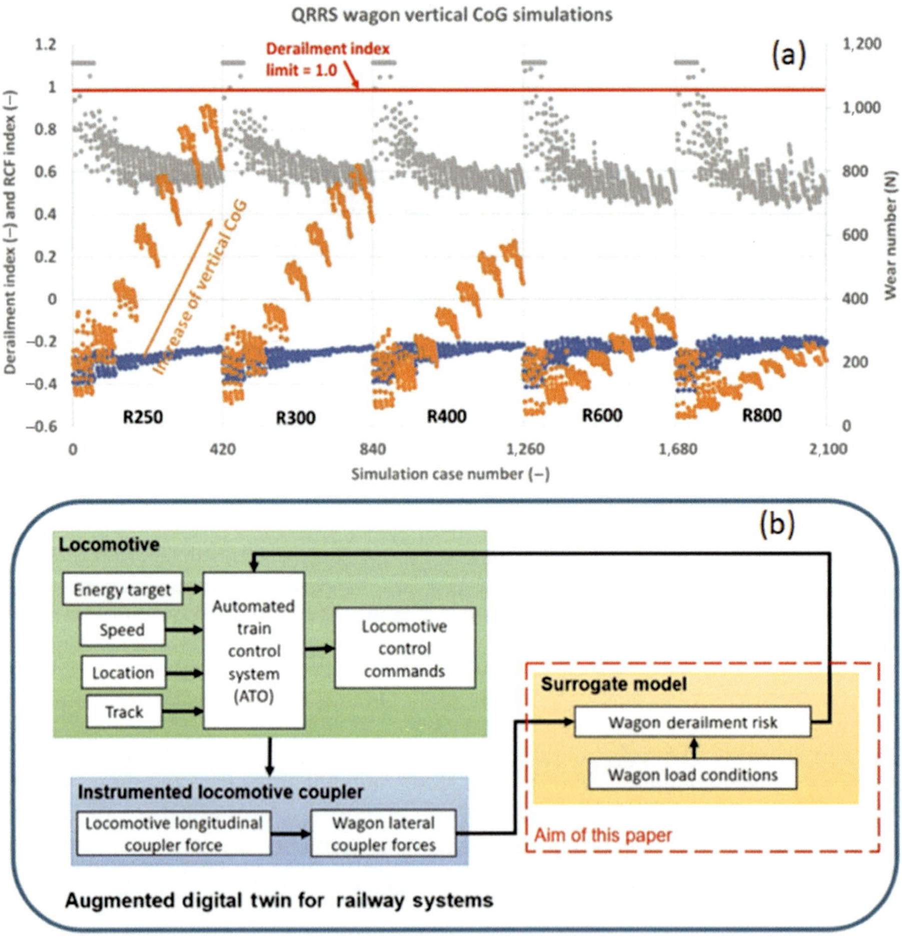

FIGURE 11.7 Surrogate model for wagon derailment risk assessment: (a) high-fidelity simulation results, and (b) data-driven surrogate mode.

With the preprocessed data ready, the next step is to select an appropriate machine learning model. Various AI techniques can be employed, depending on the specific requirements and characteristics of the train track system. For instance, supervised learning algorithms such as neural networks, decision trees, and support vector machines can be used to predict specific outcomes like track wear or failure. Deep learning models, particularly convolutional neural networks and recurrent neural networks, are particularly effective for handling spatial and temporal data, making them suitable for 3D train track modeling.

After selecting the model, it must be trained on the preprocessed data. The training process involves feeding the data into the model and adjusting the model parameters to minimize the prediction error. This is typically done using optimization techniques such as gradient descent. The training process is iterative, with the model continuously learning and improving its performance. It is essential to use a large and diverse dataset to ensure the model generalizes well to new, unseen data. During training, techniques like cross-validation can be used to prevent overfitting and ensure the model performs well on both the training data and new data.

Once the model is trained, it needs to be validated and tested to ensure its accuracy and reliability. This involves evaluating the model's performance on a separate validation dataset that was not used during training. Key performance metrics such as accuracy, precision, recall, and F1 score can be used to assess the model's effectiveness. If the model performs well on the validation data, it can be further tested on real-world data to ensure it can accurately predict and simulate the dynamics of the 3D train track system.

11.4.2 Parallel Computing

Parallel computing is a computational technique that involves dividing a large problem into smaller, independent tasks that can be processed simultaneously across multiple computing cores or processors. This approach significantly speeds up the overall computation by leveraging the power of modern multi-core and multi-processor systems, allowing for the efficient handling of complex and resource-intensive tasks. As shown in Figure 11.2, due to the large scale of the computing tasks for each individual subsystem (train dynamics, track dynamics, power system, and train control systems), computer core groups are required and then communication between computer groups is necessary to achieve the overall system simulations. This section will first introduce the parallel computing process within the individual groups and then the integration of the whole system.

11.4.2.1 Long Train Systems

Modeling a 3D train using parallel computing can be achieved as shown in Figure 11.8 according to Wu et al. (2017). The process begins with the distribution of tasks across a number of computing cores. Suppose the studied train consists of n vehicles; in this case, $n+2$ cores are required. These cores are numbered from Core 0 to Core $n+1$. Core 0 serves as the master core, responsible for coordination, while Core 1 to Core n function as slave cores dedicated to the single VSD simulations of individual train vehicles. An additional core, Core $n+1$, is assigned specifically for the simulation of the air brake system.

In this computing scheme, essential data such as track curvatures, gradients, and train driving controls (traction notches, dynamic brake notches, and air brake commands) are shared among all cores and synchronized by time. The master core initializes the process by sending out the necessary parameters to each slave core. Each slave core, except for Core 1 and the final core n, requires three sets of parameters: knuckle forces from the front and rear coupler systems and the corresponding brake pressure. The first and last vehicles only need knuckle forces from one coupler system. Once the knuckle forces and brake pressures are received, the slave cores can begin their VSD simulations in parallel. For locomotive simulations, the responsible cores must also search for traction and dynamic brake notches to determine the appropriate forces.

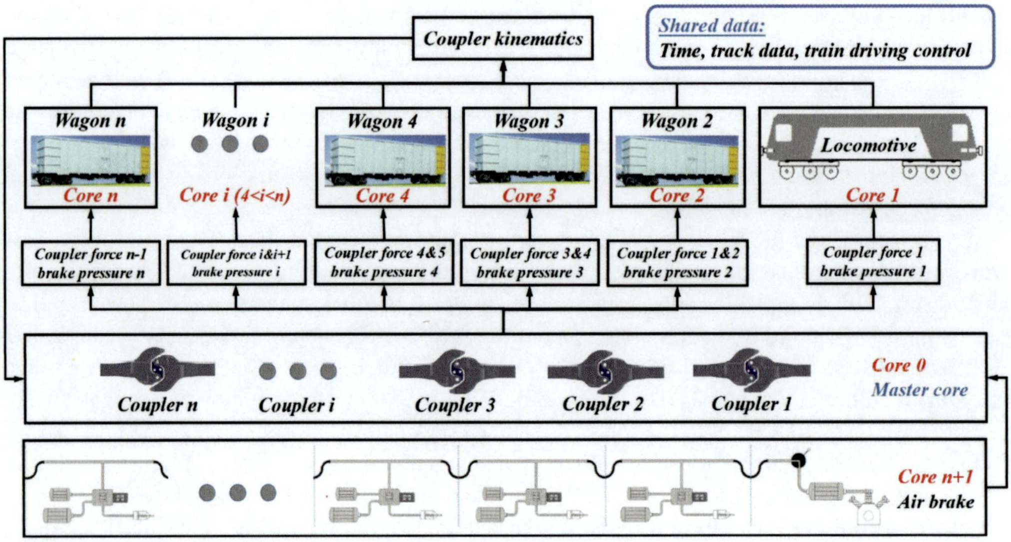

FIGURE 11.8 Parallel computing scheme for 3D long train dynamics.

As the VSD simulations proceed in parallel on the slave cores, Core $n+1$ simultaneously handles the simulation of the air brake system to determine the brake cylinder pressures along the train. Each time a slave core completes its simulation for a given time step, it sends the results back to the master core. Core $n+1$ transmits the brake cylinder pressures for each vehicle, while the other slave cores return the coupler kinematics results, which include displacements, velocities, angles, and angular velocities of the coupler pin.

With the kinematics data from the slave cores, the master core calculates the knuckle forces to be used in the subsequent time-step simulations. These newly determined knuckle forces, along with the updated brake cylinder pressures, are then distributed back to the slave cores. This cycle of sending and receiving data between the master core and the slave cores continues throughout the simulation, ensuring synchronized progress and accurate modeling of the train dynamics.

11.4.2.2 Long-Track Systems

Modeling a long (infinitely long) track using parallel computing involves a similar divide-and-conquer strategy as shown in Figure 11.9 according to Agustin et al. (2024). A long track is divided into several sections. Each section of the track is assigned to a specific core, forming a parallel track model where these sections are virtually connected through shared boundary conditions. The shared boundary sections ensure that adjacent track segments interact correctly, maintaining the physical continuity of the track. Forces and moments generated between the nodes at the boundaries of these sections are calculated using specific equations that account for displacements, velocities, rotational angles, and angular velocities of the nodes.

Similar to the 3D train model, one core is designated as the main communication core, referred to as Core 0. This core handles data transfer and communication between the other cores. Core 0 gathers position and velocity data from the first and last nodes of each track section, ensuring that boundary conditions are properly communicated and maintained throughout the simulation. This process is facilitated by the MPI framework, which manages the efficient exchange of data between cores.

The communication core employs algorithms for receiving and sending data. It collects data from the first and last node sets of each track section core, organizing and sorting this information to form the shared boundary. This core acts as a buffer zone, coordinating the flow of data while the track section cores perform calculations related to track variables. By distributing memory usage

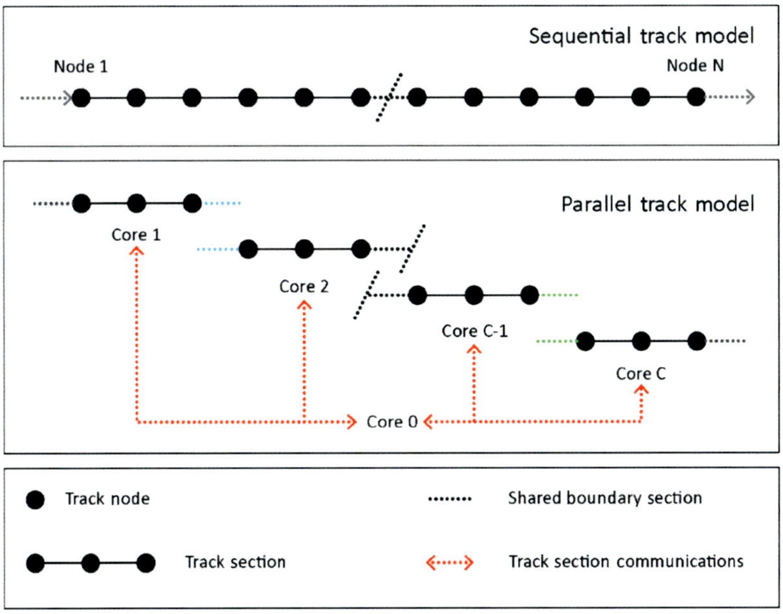

FIGURE 11.9 Parallel computing scheme for 3D long-track dynamics.

across multiple cores, the parallel computing approach ensures that the long-track model remains computationally manageable and efficient. By effectively managing the computational load and facilitating efficient data transfer, the model can simulate the complex interactions and forces within an infinitely long track.

11.4.2.3 Overhead Powering and Train Control Systems

Modeling of the overhead powering system is similar to the modeling of the long-track system considering its continuum feature. The whole system will be divided into shorter sections and processed by individual computer cores. The interactions between adjacent sections are then computed by simulating the boundary conditions. The details of the parallel computing scheme for the overhead powering system will then not be described future.

The train controller model in Figure 11.2 has the lowest computing loads. Its modeling can be handled by one single computer core or be distributed to the train model computer cores. It is noted that single-core simulations can also mimic the distributed train controllers as shown in Figure 11.10.

11.4.2.4 Integration of the Whole System

As shown in Figure 11.2, integrating the train model, track model, overhead powering model, and control system model into one comprehensive digital twin system involves creating a synchronized and interconnected simulation environment. This can be achieved by leveraging parallel computing techniques, specifically the MPI. To start it has to be assured that each model is developed with clear input and output parameters. For instance, the train model should output data regarding speed, position, and acceleration to the train control model and the track model should take wheel–rail contact forces and positions. The MPI is capable of facilitating the data exchange process.

Specific simulation starts with the synchronization of initial conditions and parameters across all models. The train control system model, acting as the central coordinator, receives data from sensors installed on the train system, power system and track system (if needed). These sensors provide real-time information about the train's operation, power condition and track conditions, and the control system then makes the decision about train operational actions. The train dynamics

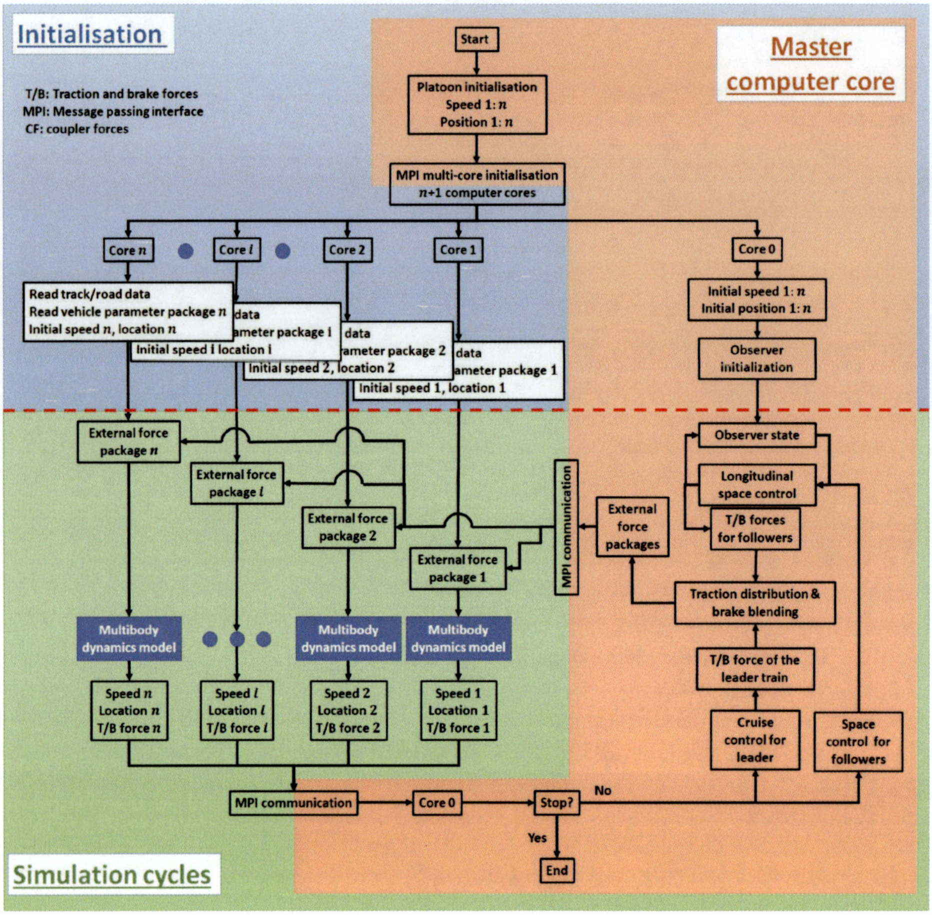

FIGURE 11.10 Linking the train controller to the train dynamics model.

model calculates the train's movement and forces, sharing the data with all other models. The track dynamics model uses this information to simulate track responses, including deformations and stresses, and sends rail status (postures and velocities) back to the train model. Simultaneously, the overhead power system model uses the train's positional data to simulate the interaction between the pantograph and catenary system. The power system model sends the contact forces back to the train model.

The integration of these models creates a high-fidelity digital twin that mirrors the physical railway system. By continuously updating with real-time data and providing actionable insights, the digital twin can optimize performance, predict maintenance needs, and improve overall safety and reliability. This interconnected system offers a powerful tool for railway operators, enabling them to make informed decisions and enhance the efficiency of their operations.

11.5 REMARKS

VC has great potential to improve railway operational efficiency but presents safety and reliability challenges. This chapter discusses an approach to developing a high-fidelity digital twin to address these challenges. Constructing a high-fidelity digital twin involves detailed modeling techniques for train dynamics, track dynamics, overhead powering systems, and train signaling control system. Modeling of the individual parts at various complexity levels is widely available in open literature.

The primary challenge is how to put them together and achieve acceptable computing speeds. This chapter proposes to address the computational challenges by using data-driven surrogate models and parallel computing. Data-driven models use machine learning techniques to approximate complex systems based on large datasets, enabling quick predictions and analyses. Parallel computing divides the problem into smaller tasks processed simultaneously across multiple cores, significantly speeding up computation. A digital twin system as such continuously updates with real-time data, optimizes performance, predicts maintenance needs, and enhances safety and reliability;, providing a powerful tool for railway operators to make informed decisions and improve operational efficiency.

REFERENCES

Agustin D, Wu Q, Zhu S, et al., 2024, Long railway track modelling – a parallel computing approach. *Proceedings of the Institution of Mechanical Engineers Part F Journal of Rail and Rapid Transit*, 239 (1), 3–16.

Bernal E, Wu Q, Spiryagin E, et al., 2024, Augmented digital twin for railway systems. *Vehicle System Dynamics*, 62 (1), 67–83.

Bruni S, Ambrosio J, Carnicero A, et al., 2015, The results of the pantograph–catenary interaction benchmark. *Vehicle System Dynamics*, 53 (3), 412–435.

Song Y, Ouyang H, Liu Z, et al., 2017, Active control of contact force for high-speed railway pantograph-catenary based on multi-body pantograph model. *Mechanism and Machine Theory*, 115, 35–59.

Song Y, Rønnquist A, Jiang T, et al., 2023, Identification of short-wavelength contact wire irregularities in electrified railway pantograph–catenary system, *Mechanism and Machine Theory*, 162, 104338.

Wu Q, Cole C, Spiryagin M, et al., 2023a, Loop track: an infinite long track model. *International Journal of Rail Transportation*, 11 (6), 872–885.

Wu Q, Ge X, Han Q-L, et al., 2023b, Railway virtual coupling: a survey of emerging control techniques. *IEEE Transactions on Intelligent Vehicles*, 8 (5), 3239–3255.

Wu Q, Spiryagin M, Cole C, 2017, Parallel computing scheme for three-dimensional long train system dynamics. *ASME Journal of Computational and Nonlinear Dynamics*, 12 (4), 044502.

12 Digital Twins for Rail Network Infrastructure

Farzaneh Tahmoorian, Travis March, and Maksym Spiryagin

12.1 DIGITAL TWINS IN ENGINEERING AND BUILDING PLANNING TASKS

Digital twins (DTs) are virtual representations of physical assets, systems, or processes connecting the real and digital worlds. The DT is a complex and evolving concept that is part of the Industry 4.0 revolution. It has emerged as a dynamic and fast-developing concept, finding its application across diverse industries with multi-faceted uses. The definition and framework for a DT are currently unregulated and still in development across the different industry sectors due to the highly customizable nature of DTs and their application and purpose.

However, what clearly distinguishes a DT from other digital models is its connection to its physical twin. The information received from the physical twin allows the DT end users to have a data-rich virtual representation of the real-world object, system, or process, thus allowing for insights, analysis, and simulations that can help improve planning, decision-making, and performance improvements to the real-world elements [1]. The concept of a DT goes beyond static 3D models as it involves real-time and current data integration, enabling the DT to mirror the behavior and status of its physical counterpart. Eleftheriou and Anagnostopoulos [2] discuss the differences in model definitions between digital models (DM), digital shadows (DS), and DTs by the degree to which dynamic information flows in real-time, and this is important when classifying a DT, as this is where misinterpretations can occur. These categories can be explained as follows:

- A DM is a 3D representation of a physical object, but it does not integrate any data flow, and if data is exchanged, it is completed manually.
- A DS builds on the DM concept. Data automatically updates from the physical object to the digital object, and the DS can represent the physical object's condition and status. The DS typically involves a one-way data flow from the physical object to the digital model and is one of the more prevalent DT concepts the industry publishes. In a DS, the flow of information or changes to the physical object is still controlled by human intervention and not automated.
- A DT builds on the DS concept and enables automatic data flow from the digital model to the physical object. This means that changes in the physical object are automatically reflected in the digital model and vice versa, allowing for real-time simulation, predictive modeling, and decision-making.

The Centre for Digital Built Brain developed a set of guidelines named *Gemini Principles* for effectively implementing the DT in the built environment, defining the DT as either dynamic or static models [3]:

DOI: 10.1201/9781003492146-12

- A dynamic model of an asset incorporates current performance data from the physical twin through live data flows from sensors and provides feedback to the physical twin via real-time control.
- A static strategic planning model of a system uses long-term condition data from the physical twin via corporate systems and provides feedback to the physical twin via the capital investment process.

The introduction of the ISO/IEC 30173 Standard by the International Electrotechnical Commission (IEC) provided a framework for defining the DT across various applications. This standard categorizes DTs based on their scope and functionality and defines different types of DTs to represent different levels of detail depending on the target entity, as summarized in Table 12.1.

In addition to the above levels of DT, as specified in ISO/IEC 30173, several organizations and standards (e.g., ISO/IEC 30173, IEC 62832, ISO 19650) have contributed to the development of another level of DT, the *Asset DT.* An asset DT is a detailed representation of a physical asset (such as a building, bridge, or any other infrastructure), including its properties, behaviors, and performance throughout its lifecycle. It generally comprises detailed component DTs integrated into system or process levels.

DTs can also be classified from the software developers' perspective into three categories, including Project DT to enhance collaboration and streamline the review process to maximize productivity, Construction DT for virtual planning, managing on-site construction and monitoring the progress, and Performance DT for optimizing maintenance based on the real-time data to enhance the lifespan and the performance [4].

The above discussions and definitions illustrate that the DT representation is not something that can be structured in the same way across all industries but is a concept that can be developed around the function and benefit that it provides.

Additionally, DTs can have different variations of digital ecosystems which provide variances in *purpose, scale, time, and approaches* [3]. In this context, a DT can have many *purposes*, including:

TABLE 12.1
Characterization of DT Levels

DT Type	Description	Application
Component DT	A vital component of an overall system. The component is defined as the primary element which substantially influences the performance of its associated entity.	Monitoring and modeling the performance of individual components, such as tracks or signaling systems in railway infrastructure.
System DT	A combination of components and digital counterparts that together perform a system or a network-wide function. It offers visibility into a network or system by highlighting the interdependent target entities.	Integrating the digital models of various components and their interaction to provide a comprehensive DT of an entire system. For example, integrating digital models of tracks, trains, signaling equipment, and other components in a railway network.
Process DT	A digital twin that provides insight into a set of activities or operations. While it may incorporate physical entities or system digital twins, its primary emphasis lies on the process itself rather than the physical components.	Modeling the operation process within a network, such as train routes, travel times, track maintenance requirements, etc.

- *Potential futures*: Strategy and planning support, running 'What if?' scenarios, predictive and preventive maintenance regimes
- *Current state*: Intervention management (operation and maintenance interventions or capital investment projects), real-time status monitoring and control, diagnostics, and prognostics to optimize the performance and safety of assets
- *History*: Record-keeping and learning from the past

Furthermore, a DT can have many *spatial* scales:

- Asset or building scale
- Network or neighborhood scale
- System, city, or regional scale
- National scale

A DT may also represent any point in the lifecycle of assets, processes, and systems and can be static or dynamic with different *temporal scales:*

- Operational timescale
- Reactive maintenance timescale
- Planned maintenance timescale
- Capital investment timescale

Moreover, a DT may use different modeling *approaches*, including:

- Geometric and geospatial modeling
- Computational/mathematical/numerical modeling
- Artificial intelligence and machine learning; 'learning systems'

Considering that DTs can vary in structure, purpose, and scale, it is essential to define each DT in relation to its proposed application. This may mean that many DTs may make up a connected digital ecosystem to cater to their specific purpose.

Further research into the DT structure and scheme is presented by Spiryagin et al. [5], which states that the DT needs 'a set of advanced models and respective data which comprehensively represents the properties and conditions of the physical object in its designated environment'. This representation allows the DT to be used for different purposes, such as lifecycle-specific simulation tasks, predictive maintenance, fault detection, and diagnostics. Figure 12.1 illustrates the basic scheme of a DT, which integrates information and data from various IT Systems and makes it accessible for all subsequent phases of the product's lifecycle [5].

In engineering and building planning, DTs offer a new and transformative approach to design, construction, and operational management. There are four key metrics commonly used to evaluate the state of a DT: autonomy, intelligence, learning, and fidelity [6]. These metrics have been used to define five levels of DTs based on their capabilities in construction projects:

- Level 1 Digital Model, with no autonomy, intelligence, learning capability, and low fidelity. This level incorporates a static 3D model representing the physical characteristics and geometry of a building or infrastructure. While this level of digital modeling can visually represent a project and help with initial design validations, it lacks real-time data integration and interactive capabilities (e.g., a basic 3D model of a building).
- Level 2 Digital Model, with limited autonomy (user-assisted), basic intelligence, minimal learning (through programming), and low to medium accuracy. This level integrates basic data such as material properties or construction progress into the model. While this

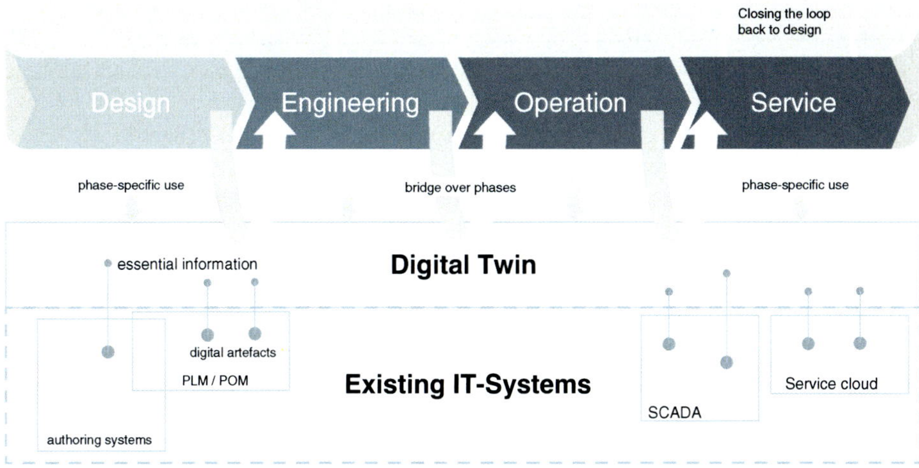

FIGURE 12.1 Basic digital twin schemes. (From Spiryagin, M., J. Edelmann, F. Klinger, et al., Vehicle system dynamics in digital twin studies in rail and road domains. *Vehicle System Dynamics*, 2023. **61**(7): pp. 1737–1786, with permission.)

level can provide feedback and control, it lacks real-time updates (e.g., house temperature sensor).

- Level 3 DT, with partial autonomy, intermediate intelligence (capable of learning from historical data), basic learning (trained using a supervised learning approach), and medium accuracy. This level allows for predictive maintenance and analytics through incorporating real-time data (e.g., monitoring systems for the life expectancy of railway machinery and components).
- Level 4 DT, with high autonomy (no/limited human assistance), high intelligence, intermediate learning (trained using an unsupervised learning approach), and high accuracy. This level can learn about the environment and supports predictive maintenance and autonomous decision-making for a particular task (e.g., real-time route recommendation).
- Level 5 DT, with full autonomy (no human intervention), high intelligence (human-like), advanced learning (continuous learning without pre-programming), and ultra-high accuracy. This level is an autonomous system that uses real-time data and is capable of reacting to unpredicted scenarios (e.g., an autonomous traffic management system of a smart city).

Although the construction sector has embraced the concept of the DT and different initiatives can be found to promote its development in engineering and the built environment, there are still limited demonstrations of DTs in practice. Further advances in modeling and simulation are needed to fully establish DTs in engineering applications [7].

12.2 COMPONENTS OF DTs FOR RAIL NETWORK INFRASTRUCTURE

The concept of the DT has been developed and successfully implemented across many industries over the past two decades, including but not limited to manufacturing, health care, building infrastructure, and smart cities, with the concept originating in the aerospace industry [8]. In engineering and construction, the conception of the DT begins with building information modeling (BIM) and continues with integrating modeling capability, big data, and Internet of Things (IoT) [9].

The term 'DT' needs to be clearly defined and understood for its specific application because, like many digital terminologies, it can vary slightly between implementations, audiences, and uses. This is evident from the variety of different applications of the DT.

Despite the various applications of DTs, the fundamental principles remain the same, and that is that the physical system is represented by a virtual system and that data flows between the two systems. The ideal DT is that the virtual system of the physical system will be continuously and dynamically updated over the life of the system from real-time data and the user's experience [10] so that it can be used to understand, simulate, predict, and optimize the physical world, as illustrated in Figure 12.2.

When developing and applying this DT approach to a rail network, it needs to be understood that a rail network is a 'system of systems' with some of the individual rail disciplines that make up a rail network as shown in Figure 12.3 [11] and include examples such as physical infrastructure (tracks, bridges, stations), rolling stock (trains), signaling and train control (ATP/European Train Control System (ETCS)), electrification (OHLE & traction power), communication and information systems (telecommunications), ticketing and fare collection, maintenance and asset management, safety and security, and passenger services and facilities. Most of these systems have the ability, if not now but in the near future, to capture real-time information from trains, signaling systems, infrastructure monitoring, weather information, passenger numbers, etc.

Generally, there are several key components of a DT that provide the foundation for the development and implementation of a DT. The frameworks for DTs have been widely researched and vary significantly across the different applications and developments. Boyes and Watson [12] suggest this is related to the different concepts closely related to the DT, such as 'simulations, IoT and cyber-physical systems (CPS)', all of which are components of a DT. Layered architecture models are the common approach to documenting the DT framework. Boyes and Watson [12] outlined

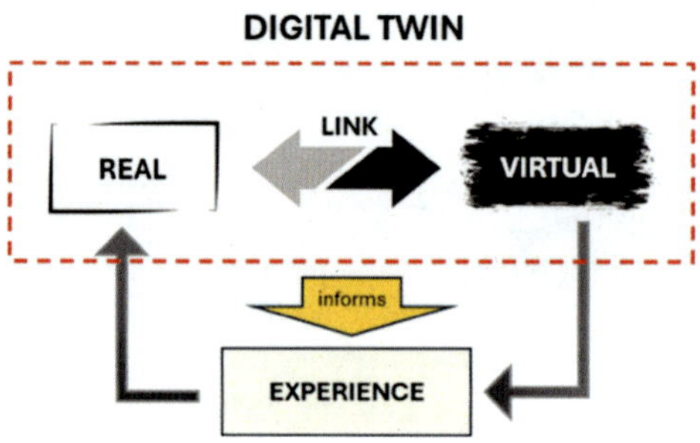

FIGURE 12.2 Typical digital twin components. (Based on Pregnolato, M., S. Gunner, E. Voyagaki, et al., Towards Civil Engineering 4.0: Concept, workflow and application of Digital Twins for existing infrastructure. *Automation in Construction*, 2022. **141**: p. 104421.)

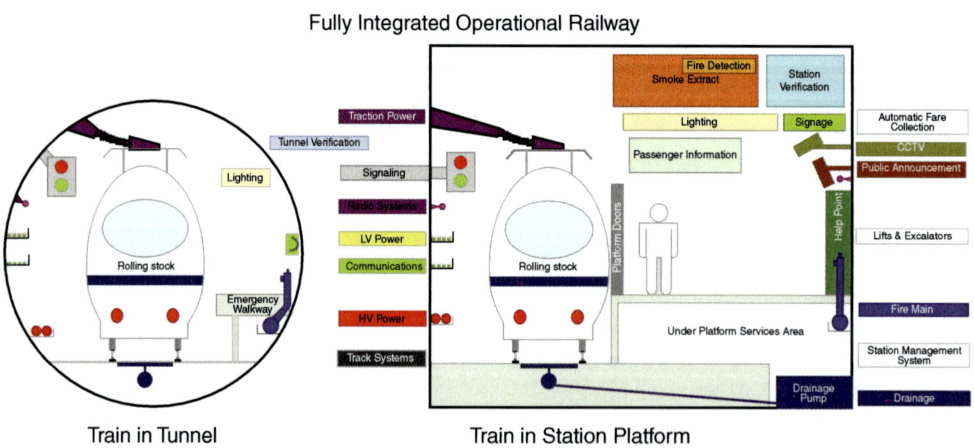

FIGURE 12.3 Operational railway systems of systems. (From Bhamra, S., G. Georgiou, and M. Georgaras, *Advanced Systems Integration in Major Rail Projects, Bechtel and Crossrail*, 2015, with permission.)

numerous literature examples of three, four, five, and six layered architectures, with each having subtle commonalities across the frameworks, which include:

- Three-layer models consist of four parts, including physical, digital, and cyber layers and communication for data exchange. The three-layer models are ideal for scenarios with limited complexity where the primary focus is on data exchange between physical and digital entities.
- Four-layer models consist of a physical layer, data extraction and consolidation layer, cyberspace layer, and the interaction layer. The additional layer of data extraction and consolidation makes the model suitable for systems requiring more detailed data processing and interaction with data from various sources.
- Five-layer models consist of data acquisition, transmission, digital modeling, model integration, and services layers. The five-layer models are beneficial in complex systems requiring advanced data management and model integration.
- Six-layer models consist of physical devices, data acquisition, local data repositories, IoT Gateway, cloud-based information repositories, and an emulation and simulation layer. The six-layer model is suitable for complex applications as it provides a detailed framework for extensive data acquisition, storage, and processing.

The fundamental framework for the DT revolves around five key categories outlined by Wang et al. [13], including physical entity (physical asset), virtual entity (DT), data (Data and Storage), connection (Integration), and services (Data Analytics). Figure 12.4 illustrates a proposed conceptual framework outlining the key components, connections, and interactions necessary to create a DT that can be adapted for a variety of railway infrastructure management activities.

The five key components presented in this figure also encapsulate the framework documented by Ferdousi et al. [14], which includes critical components such as data sensing (data), collection and storage (data storage), data pre-processing (integration), artificial intelligence (AI) inferencing (data analysis), and the actuation and control of the physical asset. This framework focuses explicitly on a DT that helps identify structural defects on rail wheels and is acknowledged as a conceptual framework. Throughout the literature review and case studies, it has been discovered that each DT developed follows the guiding principles from Dr Grieves's initial concept in 2002, which includes three components: 'real space, virtual space, and the link for data flow from real space to virtual space' and vice versa [15]. As technology advances, the fundamentals are still reflected in all DTs today.

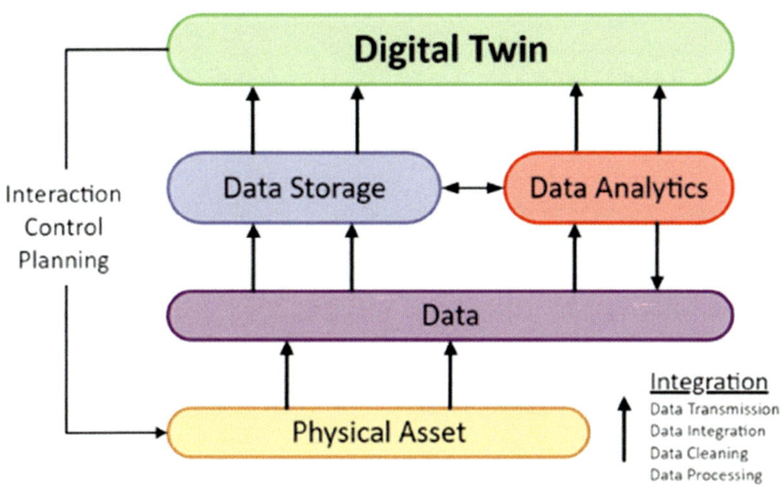

FIGURE 12.4 Conceptual framework for implementing a DT in a rail network.

Each component of the conceptual framework shown in Figure 12.4 is customized to achieve the defined purpose of a DT, which ultimately is the creation of a virtual representation of a physical world and the link of data transfer between the two. The key components in this framework include:

- *Physical asset*: The physical asset, process, or entity that will be virtually represented.
- *Data*: Data associated with the physical asset, which can be real-time (IoT devices) or static (enterprise asset information) and can be captured in many formats.
- *Integration*: Integration covers the communication and processing of data throughout the DT framework and varies according to the connection of key components and data formats.
- *Data storage*: Data storage comprises the platform(s) that stores real-time and static data sets and can consist of many cloud and land-based storage services.
- *Data analytics*: Data analytics is the component that defines the automation and machine learning facets and plays a pivotal role in empowering and advancing the DT's capabilities.
- *DT interaction*: The interaction with the DT represents the human connection with the virtual representation of the physical world, and its definition can vary across platforms depending on the purpose and function of the DT. This can take many forms, including but not limited to dashboards, 3D models, augmented reality (AR), and virtual reality (VR).
- *Connection to the physical asset*: The link or connection to the physical asset is the final feedback from the analysis of the virtual asset to the physical asset, and it can take the form of operation, maintenance, or planning changes or improvements.

As noted above, each component is to be defined or customized around the specific purpose and function of the DT [3]. The following sections summarize these components for managing railway infrastructure, particularly track maintenance.

12.2.1 Data

Various data sources contribute to the operational framework of a functioning railway network. The specific requirements for a DT will vary depending on its intended purpose. From a maintenance standpoint, these data sources might encompass:

- IoT data from selective track monitoring locations (e.g., at bridges or known soft spots), structural monitoring systems (SMAS), flood monitoring systems (FMS), and remote monitoring system data.
- Infrastructure monitoring data, including track data, contact wire data, and culverts.
- Rail corridor video footage, including photos.
- Rail corridor Light Detection and Ranging (LIDAR) and photogrammetry.
- Geospatial information.
- Existing Enterprise Asset Management Systems (EAMS) to provide fault notifications, lifecycle information, project data/work orders, SEQ scheduled corridor access system (SCAS) data, external data sources, environmental data, third-party data

12.2.2 Data Storage

Data storage for railway infrastructure information is essential to the DT architecture. Depending on the nature and complexity of the information, it is likely to involve a combination of different approaches and solutions. Railway systems produce extensive amounts of data from a variety of sources, including but not limited to IoT real-time devices, large spatial surveys (LIDAR, topographical, and photogrammetry), train video footage, and asset data.

To accommodate for the varying data sources, cloud-based platforms are often specified to allow for scalability and accessibility, enabling real-time monitoring, analysis, and simulation [16]. Additionally, edge computing solutions are increasingly deployed to process data closer to the source, reducing latency and enabling rapid response to critical events [17]. To support cloud and edge computing, data lakes and distributed storage architectures like ProjectWise are often used to manage the diverse data types and formats inherent in railway operations. Overall, the evolving landscape of DT storage solutions in railway infrastructure will likely reflect a combination of cloud, edge, and big data technologies tailored to meet the unique demands of the rail industry and is highly dependent on the existing systems that may already be utilized.

12.2.3 Integration

Integration is an all-encompassing term that covers the communication or connection of data and processes of a DT's architecture. The integration of data for a railway DT covers processes such as data cleaning, processing, conversions, standardization, labeling, augmentation, and transmission of various sources of information and, at its core, facilitates the combination of diverse data sources, ranging from real-time IoT devices, through to historical maintenance records and ultimately enables a comprehensive representation of the physical system [14].

Integration of data for a DT has many challenges, including but not limited to dealing with large sets of data, quality, and consistency of information, interoperability, and lifecycle management. Challenges specific to integrating railway infrastructure can include dealing with large linear and geographically sparse regions, so coordinate reference systems and defined standards need to be clearly defined, documented, and understood when managing and integrating data. Integrating and connecting different data sets and data analysis are key components in developing a DT. These activities will be customized specifically for the architecture and data sets of the rail network.

12.2.4 Data Analytics

Data Analytics is an essential component of the DT framework, as it aims to process vast amounts of data collected from the physical world, providing insight into trends and abnormalities in the data. The ability to handle complex and extensive data sets is becoming a critical part of the framework. It needs to be carefully understood and aligned with the purpose of the DT defined at the start of the development process. The output from the data analytics process helps provide asset managers, planners, engineers, and infrastructure workers with insights into the railway infrastructure, which is then used to change and improve the physical world through predictive and prescriptive maintenance strategies or capital works projects [18].

12.2.5 DT Interaction

DT user interaction and visualization can take many forms and are highly dependent on the intended purpose of the DT. Understanding and interacting with the maintenance of assets can be visualized in many ways, including 2D and 3D representation of the physical world, and can include maps, diagrams, animations, and 3D models, which could be interacted with using AR and VR tools [19]. Data dashboards are often one of the best platforms to present real-time and historical information related to asset conditions, performance metrics, and maintenance schedules, as these are readily available and can be scheduled to be updated regularly. These dashboards can be used to monitor trends, enable informed decision-making, and highlight predictive maintenance requirements of the network. If enabled, alerts and notifications can be set up to notify users of critical events such as bridge strikes, flood levels, and track levels outside maintenance tolerances. All these platforms can enhance communication across maintenance and planning teams to assess risk levels and priorities in terms of maintenance across the network.

In an era marked by rapid advancements in technology, the concept of DT has emerged as a transformative concept in various industries. As the adoption of DTs increases across diverse domains, there is a growing need for standardized practices and guidelines to ensure the consistency, interoperability, and reliability of DT implementations. It is important to note that DTs are composed of the collation of information from various systems and processes, which should all adhere to their own defined standards. However, it should be noted that additional standards need to be developed to allow for greater information sharing, consistency, and reliability, thus facilitating greater integration and interoperability. Organizations and consortia have been actively working to establish standards, providing guidelines and best practices for effectively creating and utilizing DTs. Some of these organizations are listed below:

- International Organization for Standardization (ISO), a standard-setting body for a variety of disciplines and processes, which includes a working group called ISO/IEC JTC 1/SC 41 – WG6, specifically focusing on DT concepts, terminology, IoT, use cases, and related technologies. ISO/IEC 30173 – DT – Concepts and Terminology has been recently released to help define and outline the DT terminology and concepts across various industries [20]. ISO/IEC 30172-DT – Use Cases is also a recent publication that documents various DT use cases in different stages of implementation [21].
- IEC that focuses on the 'electrotechnology' field, including power generation, transmission, semiconductors, and data models.
- International Telecommunication Union Telecommunication Standardization Sector (ITU), which focuses on standards for telecommunications and information communication technology.
- Institute of Electrical and Electronics Engineers Standards Association (IEEE), which focuses on digital representation related to DTs.

These bodies are developing a wide range of standards related to their respective disciplines and include elements such as data formats, communication protocols, security measures, and modeling methodologies. Wang et al. [13] aligned some of the existing standards across the five key architecture framework areas of a DT, as illustrated in Figure 12.5. By adhering to these standards, organizations can enhance collaboration, accelerate development, and drive innovation in the realm of DTs.

As DTs are developed, there are other more explicit examples of standards that capture the flow of information throughout the lifecycle of an asset. Some of these standards include but not limited to:

- AS ISO 19650 – Organization and digitization of information about buildings and civil engineering works including BIM – Information management using building information modeling [22].

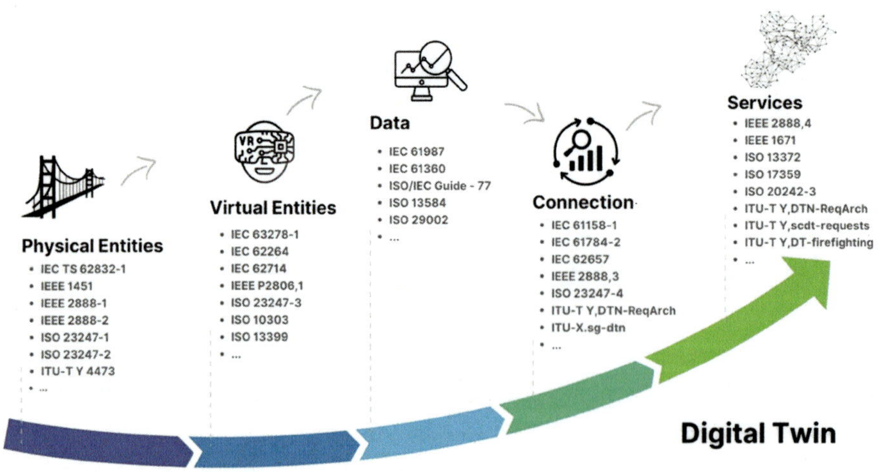

FIGURE 12.5 Framework of DT standards. (Based on Wang, K., Y. Wang, Y. Li, et al., A *Review of the Technology Standards for Enabling Digital Twin*. Version 2, Digital Twin, 2022.)

- IEC 62832 – Industrial-process measurement, control, and automation – digital factory framework [23].
- ISO 23247 – Automation systems and integration – DT framework for manufacturing [24].

12.3 BUILDING INFORMATION MODELING

BIM is a digital representation and collaborative process that facilitates the efficient planning, design, construction, and operation of buildings and infrastructure, including rail systems. BIM integrates geometric information (physical dimensions and appearance of the project) and non-geometric information (materials, cost, and performance criteria) and is categorized into different maturity levels based on the extent of collaboration and sophistication, as follows [10,25,26]:

- BIM Level 0, which is the most basic level for producing information in a 2D format using computer-aided design (CAD). This level can be used for simple drawings and documentation of rail projects. Collaboration at this level is either non-existent or minimal.
- BIM Level 1, in which a combination of 2D and 3D CAD drawings is used to develop models. Basic 3D models can be created for rail components such as tracks or stations at this level. There is partial collaboration at this level, and project stakeholders can share their information electronically using a common data environment.
- BIM Level 2 involves detailed 3D modeling to develop projects. This level allows for detailed modeling of various rail systems, such as track alignment, and goes beyond just 3D graphics. It also includes time management (4D) and project costs (5D). BIM Level 2 is a collaborative 3D BIM, where 3D models are shared using a common file format, allowing project stakeholders to access a unified BIM model to reduce clashes and improve planning.
- BIM Level 3 is a fully integrated BIM that enables the project stakeholders to work on a single unified model with real-time updates and full collaboration for improved decision-making. At this level, a new dimension is available for incorporating the lifecycle data to develop a 6D BIM model. This level enables comprehensive rail project modeling integrating design, construction, and operational data.

Another BIM level, BIM Level 4, was recently introduced. It focuses on achieving optimal performance, social outcomes, and sustainability, but it is not as commonly used as the first three levels [25]. This level can be applied to advanced rail projects incorporating sustainability, social impact, and performance optimization.

The concept of the DT builds on BIM and expands by integrating sensing capabilities, big data, and IoT. The synchronization between the real-world asset/process/system and its virtual representation is the defining feature of the DT, distinguishing it from other digital representation models like BIM. The methods used to achieve this synchronization, the frequency of virtual representation updates, and the utilization of the virtual representation further characterize the DT [9].

12.4 BIM VERSUS DT

In the design, construction, and operation of infrastructure projects, different advanced technologies are available to improve the functionalities and enhance the infrastructure system management through data integration, real-time monitoring, and predictive maintenance. These technologies include geographic information systems, the IoT, AR and VR, CAD, Asset Management Systems (AMS), and predictive analytics. However, BIM and DT technologies are at the forefront of these modern technologies, bringing unique capabilities to the field of infrastructure engineering.

The DT is clearly distinguished from other digital models by its connection to its physical twin. The information received from the physical twin allows the DT end users to have a data-rich virtual representation of the real-world object, system, or process, thus allowing for insights, analysis, and simulations that can help improve planning, decision-making, and performance improvements to the real-world elements [1]. The concept of a DT goes beyond static 3D models as it involves real-time and current data integration, enabling the DT to mirror the behavior and status of its physical counterpart. Furthermore, DTs can be used in all stages of a project, including planning, design, simulation, validation, construction, monitoring, and operation stages [27]. The use of live data in DTs enables real-time analysis and enhanced and informed decision-making.

Unlike a DT, BIM models are primarily static models, which are typically used during the design and construction phases to represent a data-rich 3D model integrating various types of data such as materials and cost. The BIM models are not continuously updated via real-time data from sensors and other IoT devices. Even a 6D BIM, which aims to incorporate data for the entire lifecycle of facilities management, is not optimized for real-time operational response, interoperability, and automation because it does not connect live data during the asset's lifecycle or utilize live calculation models [10].

12.5 HOW CAN BIM AND DTS WORK TOGETHER?

BIM models can work together with the concept of DTs by being part of the building blocks of a DT. BIM provides detailed 3D models and data-rich representations of physical structures, which are crucial for creating comprehensive DTs [28]. Combining real-time data metrics from IoT sensors with BIM models can provide users with a real-time virtual understanding of the physical world, allowing stakeholders to monitor, analyze, and optimize asset performance continuously [29,30]. The advantage of using 3D BIM models in a DT is that it allows users to virtually see the physical object, which can be used in Augmented and Virtual reality platforms. This interaction would be useful in design reviews, maintenance planning, and operational decision-making. If BIM models are not used in conjunction with a DT, the physical world data is often viewed through data metric platforms such as dashboards. While effective for analyzing trends and changes in large data sets, these platforms lack the interactive capabilities of a DT. Dashboards typically present information in charts and graphs, which, although useful, do not provide the spatial and contextual understanding offered by 3D BIM models. The integration of BIM and DTs has great potential for innovation across various sectors, as it can automate decision-making processes and enhance asset performance. Further development of open data standards and interoperability frameworks will enable the smooth integration of BIM and DT technologies and information across various platforms and applications.

12.6 CHALLENGES AND OPPORTUNITIES IN IMPLEMENTING DTs

There are several critical limitations to implementing a DT within the rail network which can be summarized into two essential categories. The first fundamental limitation concerns the DT framework and data availability and reliability. The framework needs to be flexible, robust, and structured so that large amounts of data can be captured, processed, and stored for the DT to absorb and utilize. Within a rail network, enterprise asset data is available for most, if not all, assets and this can be supplemented with near real-time information from IoT devices. Managing these types of data sets would require a well-documented and structured data processing and storage capability, which would require further research, development, and change within rail organizations. The availability and reliability of the data would also be significant limitations to implementing a rail DT. A DT needs to be trustworthy, and for it to be reliable, the data it relies on must be trusted. This is only done through validation and control checks, which, on a large scale, can be a massive limitation to the development of DTs. The second fundamental limitation is around the organizational barriers and regulatory requirements. Railway organizations are usually very large organizations with a long history of managing the rail network. The railway organizations are subject to many regulatory requirements to keep the operators and passengers of the train network safe. The requirements have been developed over the years of owning and managing the network, and implementing a DT into this type of organizational environment is a very considerable limitation. This limitation is probably one of the more difficult ones to overcome due to the controlled environment of a rail network, the strict regulatory requirements, and the historical-cultural barriers regarding change.

Furthermore, there are several key risks when implementing a DT, particularly an idealistic one, where the prescribed changes from the DT are implemented into the physical world on an automated basis. This type of DT is often used in manufacturing where the operational changes improve the manufacturing process on an ongoing real-time basis. This type of implementation is not viable within the railway network in the short term, considering the safety assurance processes that need to be considered when any changes are proposed within the railway corridor. This regulated process ensures that all rail network changes are managed, checked, and passed using a structured and assured approach.

Additionally, there are several risks related to the design of the DT, particularly when involving data analytics and machine learning, as this is a tool where AI is taught the specific rules of how to interpret and predict results based on the data sets it has access to. This entire process of machine learning requires a systematic approach and validation to ensure the predicted results are consistent with the expected output; in other words, the results need to be checked and understood outside of the AI 'black box'.

Operational risks exist when implementing any new technology into an operational railway; however, these risks are exacerbated when considering the idealistic or automated DTs, where the predictive changes from AI can be made to an operational railway without the control of the asset managers. In the short term, predictive changes to the rail network will likely revolve around maintenance planning and capital works programs to ensure that all changes are managed through the control change process and safety assurance processes that exist in the management of the railway network.

The anticipated benefits of implementing a DT for managing railway infrastructure include consolidating existing data sets and introducing access to new data through IoT devices, which can provide near real-time information on rail network assets. The DT will enable asset managers, planners, engineers, and infrastructure workers to improve visibility on the condition of their assets. This enhancement will facilitate processes such as predictive maintenance, leading to enhanced safety through targeted maintenance and reduced time in the danger zone. Additionally, cost savings are expected as reactive and preventive maintenance efforts are reduced in favor of targeted predictive maintenance. A key part of realizing these benefits is validating them through a pilot project on a selected section of the track to ensure the anticipated outcomes are achieved.

12.7 EMERGING TECHNOLOGIES AND FUTURE TRENDS

At the outset of defining a DT, it is important to stipulate how it interacts with the physical world. Bolton et al. [3] describe that a DT can be either dynamic or static, where a dynamic DT provides feedback into the physical world via real-time and often automated control, while a static DT provides feedback into the physical world via capital and/or planning works. Due to the high complexity of rail systems and the critical importance of reliability and safety, further validation and safety considerations would be required prior to implementing a dynamic DT, where the feedback into the rail systems is via real-time control. For the foreseeable future, DTs in the management of railway infrastructure would be categorized as static, to ensure the feedback to the physical world is conducted in a controlled, consistent, and safe manner.

In the short term, an immediate opportunity for advancing a Rail DT could be streamlining the existing track maintenance processes and identifying a pilot section of track to focus the development around. This would firstly involve examining the systems and processes already available within the rail business, such as EAMS and railway management systems (RMS), and any other data capture related to track maintenance. Once understood, the DT starting point should be to systematically step through the eight-step process outlined by Ariansyah et al. [31] which includes:

- *Step 1*: Define the Requirements/Purpose.
- *Step 2*: Develop the Information Flow.
- *Step 3*: Develop the Virtual Representation.
- *Step 4*: Develop the Data Exchange System.
- *Step 5*: Develop the Analysis Method.
- *Step 6*: Develop Action.
- *Step 7*: Verify.
- *Step 8*: Validate.

The start of this process should be done with the framework outlined in Figure 12.4. The foundation and methodological working through each of the key components of the workflow are highlighted in Section 12.2. The first step is to capture and collate existing enterprise asset information, which may involve connecting other systems into a more structured and centralized area. Additionally, further research and development would be required to automatically integrate the train scan data on a regular basis from a hosted data lake or edge computing, as suggested in the pilot study on the train wheel defect DT by Microsoft [17]. This process would likely involve capital investment for trackside equipment. Once the data has been captured and transferred into the hosted data lake, it will undergo a series of machine learning modules. This would involve training the data sets to improve and refine the results from the predictive analysis AI. Finally, before rolling out a DT to help manage track infrastructure, it is crucial to complete validation and verification to ensure that the purpose and function of the DT are achieved and can improve the existing functionality of the existing track maintenance reports and processes. As the data sets get richer, further development would be required to integrate the historical train data, which would be the next step toward comprehensive predictive and trend analysis.

Medium-term prospects, once successfully verified and validated across a pilot section of the rail corridor, would be to expand this across the wider rail network. Following this wider network adoption, diversifying into various DT applications, and deploying IoT devices to automate data collection on different rail assets should be planned. Examples of DTs to help manage railway infrastructure could be focused on other high-wear or high-maintenance infrastructure elements such as the train's pantograph and contact wire. The Pantograph Catenary Monitoring System defined by Intel [32] collects real-time information and analyzes the data collected for asset managers to investigate the wear and damage of the asset. Other examples could include monitoring and managing vegetation growth along the corridor. This would involve collecting the LIDAR data captured from the train or survey car in combination with high-resolution aerial imagery (drone or satellite imagery) and running through machine learning predictive models to identify and optimize vegetation maintenance schedules. This aims to improve safety by reducing vegetation-related incidents caused by high winds and storm events.

Other essential medium to long-term DT improvements need to include further advancements in data storage and accessibility of information, which would involve integrating project data, including 3D surveys, flooding information, detailed designs, and as-constructed data. To achieve these medium to long-term goals, the critical technical focus would be mapping and linking different data sources. The alignment of standards and protocols is crucial for expanding the representation of the rail physical world within the DT framework. It would involve standardizing many of the integration processes across the framework. Ultimately, the richer the data set of the rail corridor, the more DT opportunities will be available.

Future long-term safety benefits of developing a robust DT used in managing railway infrastructure would be through live or near real-time monitoring and AI, where every train would be fitted with monitoring equipment. The trains would feed the data captured through trackside huts, and

AI (machine learning) would be used to detect defects and implement speed restrictions through the use of the automated train operation and ETCS processes until the asset manager could clear it either through site inspection or maintenance [17]. This could include all high-wear railway infrastructure components, including train wheels, track structure, and contact/catenary wires. With this cyclic nature of data flowing from the physical system to the virtual system and having the predictive changes automatically incorporated into the physical system, this would then fit into the definition of an idealistic DT. Looking further ahead into the future, it is not unfeasible to consider 'smart' interconnected DTs, where data and predictive and trend analysis are collected and connected from various DTs, providing constant health checks and analyses of the entire rail network.

REFERENCES

1. Ben Lomax Thorpe, *Digital Twins: Where Physical Meets Digital*, Royal Haskoning DHV, 2020.
2. Eleftheriou, O.T. and C.-N. Anagnostopoulos, Digital twins: A brief overview of applications, challenges and enabling technologies in the last decade. *Digital Twin*, 2022. **2**: p. 2.
3. Bolton, A., M. Enzer, and J. Schooling, *The Gemini Principles: Guiding Values for the National Digital Twin and Information Management Framework*. Centre for Digital Built Britain and Digital Framework Task Group, 2018.
4. Bentley Systems, Digital Twins, 2024; Available from: https://www.bentley.com/software/digital-twins/.
5. Spiryagin, M., J. Edelmann, F. Klinger, et al., Vehicle system dynamics in digital twin studies in rail and road domains. *Vehicle System Dynamics*, 2023. **61**(7): pp. 1737–1786.
6. ARUP, *Digital Twin towards a Meaningful Framework*, ARUP, 2019.
7. Batty, M., Digital twins. *Environment and Planning B: Urban Analytics and City Science*, 2018. **45**(5): pp. 817–820.
8. Singh, M., E. Fuenmaor, E.P. Hinchy, et al., Digital twin: Origin to future. *Applied System Innovation*, 2021. **4**(2): p. 36.
9. Wright, L. and S. Davidson, How to tell the difference between a model and a digital twin. *Advanced Modeling and Simulation in Engineering Sciences*, 2020. **7**(1): p. 13.
10. Pregnolato, M., S. Gunner, E. Voyagaki, et al., Towards Civil Engineering 4.0: Concept, workflow and application of Digital Twins for existing infrastructure. *Automation in Construction*, 2022. **141**: p. 104421.
11. Bhamra, S., G. Georgiou, and M. Georgaras, *Advanced Systems Integration in Major Rail Projects*, Bechtel and Crossrail, 2015
12. Boyes, H. and T. Watson, Digital twins: An analysis framework and open issues. *Computers in Industry*, 2022. **143**: p. 103763.
13. Wang, K., Y. Wang, Y. Li, et al., *A Review of the Technology Standards for Enabling Digital Twin. Version* 2, Digital Twin, 2022.
14. Ferdousi, R., F. Laamarti, C. Yang, et al., Railtwin: A digital twin framework for railway, in *2022 IEEE 18th International Conference on Automation Science and Engineering (CASE)*, Mexico City, Mexico, 2022. IEEE.
15. Grieves, M. and J. Vickers, Origins of the digital twin concept. *Florida Institute of Technology*, 2016. **8**: pp. 3–20.
16. Microsoft, Introduction to Azure Storage, Microsoft Learn, 2023.
17. Microsoft, *IoT Edge Railroad Maintenance and Safety System*, Microsoft Learn, 2024.
18. Errandonea, I., S. Beltrán, and S. Arrizabalaga, Digital Twin for maintenance: A literature review. *Computers in Industry*, 2020. **123**: p. 103316.
19. Shotz, H. and J. Birdsall. A glimpse into the near future–Digital twins and the Internet of Things, in *International Road Federation World Meeting & Exhibition*, 2021. Springer, Dubai, UAE.
20. ISO, ISO/IEC AWI 30173 – Digital Twin – Concepts and terminology, in *International Standard*, ISO - International Organization for Standardization, 2023.
21. ISO, ISO/IEC TR 30172, *Technical Report – Internet of Things (IoT) – Digital Twin Use Cases*, IEC, 2023.
22. Standards Australia committee BD-104 , *AS Guide to ISO 19650 Organization and Digitization of Information about Buildings and Civil Engineering Works Including Building Information Modelling (BIM) – Information Management Using Building Information Modeling*, Standards Australia, 2019.

23. International Electrotechnical Commission, *IEC 62832-Industrial-Process Measurement, Control and Automation – Digital Factory Framework*, IEC, 2020.
24. ISO, *ISO 23247 – Automation Systems and Integration – Digital Twin Framework for Manufacturing*, International Standard, 2021.
25. Catenda, BIM Levels, 2024; Available from: https://catenda.com/glossary/bim-levels/.
26. Kotula, Y., BIM vs Digital Twin: When and How to Use Them, 2023; Available from: https://intelvision.pro/blog/bim-vs-digital-twin-when-and-how-to-use-them/.
27. ThoughtWire, Digital Twins vs. Building Information Modeling (BIM), 2023; Available from: https://www.iotforall.com/digital-twin-vs-bim.
28. Rausch, C., R. Lu, S. Talebi, et al., Deploying 3D scanning based geometric digital twins during fabrication and assembly in offsite manufacturing. *International Journal of Construction Management*, 2023. **23**(3): pp. 565–578.
29. Gurevich, U. and R. Sacks, Longitudinal study of BIM adoption by public construction clients. *Journal of Management in Engineering*, 2020. **36**(4): p. 05020008.
30. White, G., A. Zink, L. Codecá, et al., A digital twin smart city for citizen feedback. *Cities*, 2021. **110**: p. 103064.
31. Ariansyah, D., I. Fernàndez del Amo, J. A. Erkoyuncu, et al., *Digital Twin Development: A Step by Step Guideline*, 2020.
32. Intel, IoT in Railways, Intel Corporation, 2023; Available from: https://www.intel.com/content/www/us/en/transportation/iot-in-railways.html.

13 AI in Digital Twins for Rail Operations

Francesco Flammini, Elena Napoletano, and Stefano Ricci

13.1 TYPES OF ARTIFICIAL INTELLIGENCE USED FOR RAIL OPERATIONS

The adoption of artificial intelligence (AI) in rail operations has significantly increased in recent years, driven by the availability of data and advancements in processing technologies. Various types of AI are employed to enhance the efficiency, safety, and reliability of rail operations. Below, we explore the main types of AI used in the rail industry.

Machine learning (ML) represents a subset of techniques used within AI and can be divided into the following main subcategories (i.e., supervised vs unsupervised).

Supervised learning is a type of ML where the model is trained on a labeled dataset. This means that each training example is paired with an output label. The goal of supervised learning is to learn a mapping from inputs to outputs based on the provided examples. Within railway applications, this type of ML is mainly used for passenger demand forecasting and train scheduling. By analyzing historical data, ML algorithms can predict passenger flows and optimize train schedules to meet demand efficiently. For example, supervised learning models can analyze ticket sales, weather conditions, and special events to forecast peak travel times and adjust schedules accordingly (Besinovic et al., 2021).

Unsupervised learning differs from supervised learning because it deals with unlabeled data. The model tries to learn the underlying structure or patterns in the data without any explicit instructions on what to look for. Within railway applications, unsupervised learning is utilized, e.g., for anomaly detection in predictive maintenance. This approach helps identify unusual patterns in sensor data that may indicate potential equipment failures. By detecting anomalies early, maintenance teams can address issues before they lead to significant disruptions, thereby improving the reliability and safety of rail operations (Tang et al., 2022).

Within emerging developments of AI, we can mention *generative AI* (gen AI), which can be used to create realistic simulations of complex operational scenarios. These simulations allow operators to test and optimize strategies without disrupting actual services. For instance, generative models can simulate emergency situations, such as equipment failures or natural disasters, enabling operators to develop and refine response plans (Malavasi and Ricci, 2001).

Deep learning is a subset of ML that focuses on using neural networks with many layers (hence "deep") to model and understand complex patterns in data. Deep learning involves training artificial neural networks (ANNs) to perform tasks by learning from large amounts of data. These neural networks are inspired by the structure and function of the human brain, consisting of layers of interconnected nodes (neurons).

A special case of deep neural networks is used for applications involving image recognition: they are known as *convolutional neural networks* (CNNs) (Pappaterra et al., 2024). Within railways, CNNs are mainly employed for infrastructure monitoring and surveillance. For instance, these networks can analyze images of tracks, bridges, and other structures to detect defects and damages. Automated image recognition systems can identify issues such as cracks, wear, and obstructions, facilitating timely maintenance and reducing the risk of accidents (UIC, 2024).

Natural language processing (NLP) can be used within railways to enhance passenger interaction through chatbots and virtual assistants. These AI-powered systems provide real-time information

DOI: 10.1201/9781003492146-13

and answer frequently asked questions, improving the overall passenger experience. For example, NLP can be used to develop chatbots that assist passengers with ticket bookings, schedule inquiries, and travel updates (Ficzere, 2023).

Optimization algorithms are crucial for resource planning and management, including staff scheduling and rolling stock management. These algorithms help optimize the use of resources to improve operational efficiency. For example, optimization models can create efficient crew schedules that minimize downtime and ensure compliance with labor regulations. Similarly, they can optimize the allocation of trains to routes based on demand and maintenance requirements (Zhang and Zhang, 2023).

Expert systems are implemented to support decision-making in critical situations. These systems use a knowledge base to provide real-time recommendations to rail operators. For instance, an expert system can assist in managing traffic during peak hours or in coordinating responses to incidents on the tracks. By leveraging the expertise encoded in these systems, operators can make informed decisions quickly and effectively (Besinovic et al., 2021).

Reinforcement learning (RL) is used for dynamic scheduling and real-time decision-making. RL algorithms learn optimal strategies through trial and error, making them well-suited for complex, dynamic environments like rail operations. For example, RL can be used to develop adaptive train control systems that adjust speeds and routes in real time based on current conditions, such as traffic congestion or delays (Tang et al., 2022).

Fuzzy logic systems are employed for handling uncertainty and imprecision in rail operations. These systems can model complex, ambiguous scenarios where traditional binary logic falls short. For example, fuzzy logic can be used in automated braking systems to provide smooth and safe deceleration under varying track conditions (Malavasi and Ricci, 2001).

The integration of these AI types into rail operations not only enhances efficiency and safety but also contributes to creating a more sustainable and resilient transportation system. As AI technologies continue to evolve, their applications in the rail industry are expected to expand, offering new opportunities for innovation and improvement.

More specifically, the integration of AI within *digital twins* (DTs) can revolutionize rail operations by further supporting decision-making processes (De Donato et al., 2023). DTs are virtual replicas of physical assets, systems, or processes that are continuously updated with real-time data. When combined with AI, these digital models become powerful tools for predictive maintenance, operational optimization, and strategic planning.

Figure 13.1 presents a concept map illustrating the integration of DTs and AI in rail operations, along with its potential applications and advantages. One of the most significant applications of AI in DTs for rail operations is *predictive maintenance*. By analyzing data from sensors embedded in trains and infrastructure, AI algorithms can predict when components are likely to fail and schedule maintenance activities accordingly. This approach minimizes downtime and reduces maintenance costs while ensuring the safety and reliability of rail services. For instance, ML models can detect patterns in vibration data that indicate wear and tear on train wheels, allowing for timely interventions (Besinovic et al., 2021).

Within *operational optimization*, AI-enhanced DTs can optimize various aspects of rail operations, from scheduling and routing to energy management. By simulating different scenarios, operators can identify the most efficient strategies for managing train movements and resource allocation. For example, optimization algorithms can help in creating dynamic schedules that adapt to real-time conditions, such as delays or changes in passenger demand (Tang et al., 2022).

Cognitive DTs represent an advanced form of DTs that incorporate cognitive computing capabilities. These DTs not only simulate physical assets but also understand, reason, and learn from their environments. In rail operations, cognitive DTs can provide deeper insights and more accurate predictions by continuously learning from new data and adapting their models accordingly. They can assist in complex decision-making processes, such as optimizing traffic flow during peak hours or responding to unexpected disruptions (Malavasi and Ricci, 2001).

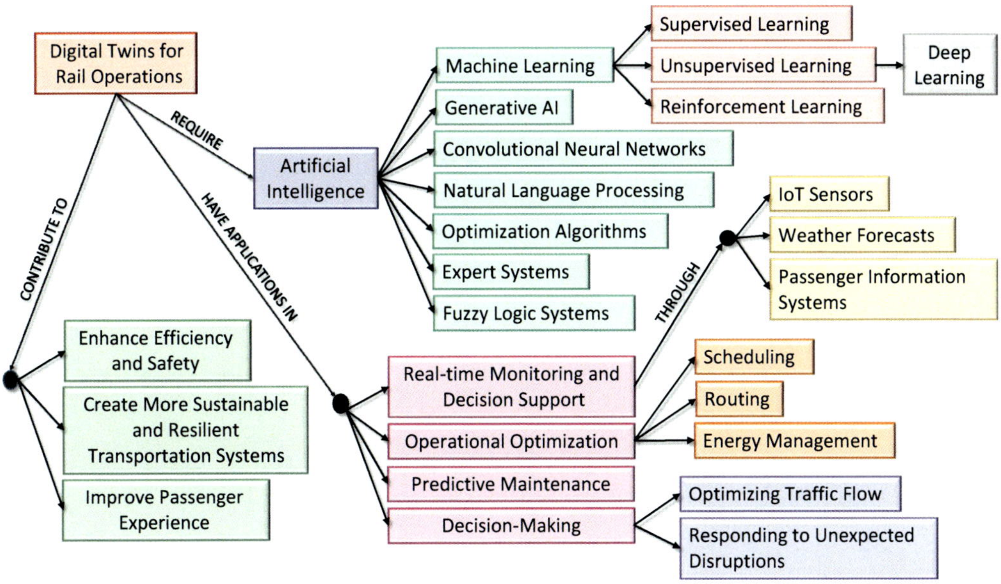

FIGURE 13.1 A concept map of the integration of digital twins in rail operations.

Within *real-time monitoring and decision support*, AI-powered DTs enable real-time monitoring of rail assets and operations. By integrating data from various sources, such as IoT sensors, weather forecasts, and passenger information systems, these twins provide a comprehensive view of the rail network. This holistic perspective allows operators to make informed decisions quickly and effectively. For instance, during an emergency, a cognitive DT can analyze the situation, predict potential outcomes, and recommend the best course of action (UIC, 2024).

AI in DTs also plays a crucial role in *improving the passenger experience*. By analyzing data on passenger behavior and preferences, AI algorithms can help design better services and facilities. For example, AI can optimize the layout of train stations to reduce congestion and improve passenger flow. Additionally, virtual assistants powered by NLP can provide real-time information and assistance to passengers, enhancing their overall travel experience (Ficzere, 2023).

The future of AI in DTs for rail operations is promising, with ongoing research and development aimed at further enhancing their capabilities (Dirnfeld et al., 2024). Emerging technologies such as edge computing, 5G connectivity, and advanced analytics are expected to play a significant role in the evolution of cognitive DTs. These advancements will enable even more sophisticated simulations, real-time analytics, and autonomous decision-making, paving the way for smarter and more resilient rail systems (Zhang and Zhang, 2023).

13.2 AI FOR TRAIN SCHEDULING AND DISPATCHING

Developing better solutions for train scheduling and re-scheduling problems has attracted the attention of researchers for several decades. The constraints forcing them to regulate rail services are the limitation of resources and the increasing competition among service providers. Planning train services, at both tactical and operational levels, has gained progressive relevance and, in both processes, inter-train conflicts need to be detected and solved to produce feasible schedules aligned with customer needs. The conflict resolution process normally causes delays and can bring trains to operate far away from their original schedules, causing deviations from timetables.

In such situations, the dispatchers are in charge of taking the necessary actions to solve inter-train conflicts. Because conflict resolution delays contribute to rail traffic regularity and trains' punctuality, the quality of dispatchers' re-scheduling actions requires monitoring, data gathering, data processing, forecasting, decision-making, and implementation. Therefore, as discussed by Besinovic et al. (2021) and Dirnfeld et al. (2022), traffic planning in general is a subdomain where AI can be largely and effectively applied to solve traffic timetabling and re-scheduling problems by using various approaches referred to as AI.

Indeed, dispatching and re-scheduling problems have generated extended and deep interest among scholars for more than 50 years. Schaefer and Pferdmenges (1970) proposed the first expert systems for real-time train dispatching. Assad (1980) published a comprehensive survey on railway transportation models including train scheduling models. Fukumori et al. (1987) presented one of the first algorithms for train scheduling based on AI. Cordeau et al. (1998) concentrated on optimization models for the commonly studied railway routing and scheduling problems. Fay (2000) presented a fuzzy logic-based algorithm to make dispatching decisions during a 10-hour period. Malavasi and Ricci (2001) proposed an approach to forecast delay generation in a real rail network using a neural network-based self-learning process. Törnquist (2005) produced possibly the most extensive and comprehensive review of 48 approaches published between 1973 and 2005 and provided a valuable classification framework with respect to problem type, solution mechanism, and type of evaluation. The analyzed approaches include priority-based heuristics and traditional operations research techniques such as optimization, simulation, and queuing theory. Corman et al. (2010) developed and tested the local optimization re-scheduling tool for train traffic under disturbed situations referred to as ROMA (railway traffic optimization by means of alternative graphs). In Yin et al. (2014), expert systems are used for intelligent train operation. More recently, the intensity of the applications presented in international literature has increased relevantly. Liu et al. (2018) designed a data analytics approach for train timetable performance measures with automatic train supervision. Kecman and Goverde (2015) analyzed the train delay patterns by using regressions and random forest techniques; meanwhile, Cerreto et al. (2018) analyzed them by data clustering techniques. Wen et al. (2019) provided a survey on the use of data-driven approaches for train dispatching management. Gorbachev et al. (2020) presented a method applicable to the decision support systems supporting the dispatching processes in complex railway networks.

Based on this framework, it is possible to conclude that the most consolidated approaches in the literature have focused on the use of evolutionary clustering, typical of *Genetic Algorithms*, and the applications of *ANNs*.

Genetic algorithms (GA) represent the genetic modeling of train schedules along a line with meeting, passing, or overtaking conflicts. Each conflict-free schedule alternative is encoded as binary strings (i.e., chromosomes) composed of bits (i.e., genes). The length of chromosomes is normally defined by the total number of conflicts. Each conflict is represented by a certain gene (bit in the binary string of the chromosome). The fitness value of each chromosome represents the total delay or the total weighted delay incurred in the system after the solution of all conflicts. The selection of individuals of the new generation is normally proportional (roulette wheel method). Since minimizing the re-scheduling delay is the objective of the algorithm, the fitness values obtained should be rescaled so that the individual having the lowest total (eventually weighted) delay would have the highest probability of being selected (total selection probability of individuals in a generation = 1). The termination criteria could be different, e.g., a convergence threshold and a maximum running time. After the termination, the algorithm stops running and does not initiate a new iteration, but the already-started iteration is normally completed until a solution is found, even beyond when the stopping criterion is fulfilled.

ANNs mimic the decision behavior of dispatchers in the conflict resolution process (Dündar and Sahin, 2013). The decision behavior is extracted from previous recorded conflict resolutions recorded. The ANN is an AI method developed on the human brain's basic operation principles, which can learn, remember, and memorize just as the human brain can since they were developed on the human brain's basic operation principles. The ANN is a black box model like the GA, and they both generate desired output(s) from supplied inputs. For this reason, the correctness of the output(s) is important rather than procedures generating outputs from input. The outputs are generated by joining artificial neurons to each other (totally or partially). Just like in the human brain, neurons process information gathered from outside the system and transmit it to other neurons or to the user as output(s). The ANN is normally structured in layers with one or more neurons: input layer, hidden layer(s), and output layer. Each neuron is connected to others in the previous and next layers. The learning process is by adjusting the values of synaptic connections (weights) just like brain cells. The differences (errors) between the outputs of the model and desired targets are propagated backward through the network by updating weights, thus the learning is accomplished in a supervised learning mechanism of feed-forward back-propagation. The ANN can be generalized by approximating target outcomes for the data not in the training set. To improve generalization performance, data is normally divided into two sets dedicated to training and testing. During the updating of weights, only the data in the training set is used; meanwhile, the data in the test set is used as a termination media to avoid memorization and to calculate the performance of the ANN itself, as well explained already in Mehrotra et al. (1996).

13.3 AI FOR SIGNALING SYSTEMS

The rapid development of railways toward high speed, high density, and heavy loads demands more and more that all the components achieve the highest levels of performance. In this context, the signaling systems are required to maintain and possibly increase the very high safety level of railway traffic, which makes more and more important the systematic and prompt diagnosis of any fault in safety critical systems and components that must comply with the standards defined by safety integrity level (SIL) 4, corresponding to a probability of dangerous failure per hour within the range 10^{-8} to 10^{-9}. In this context, the use of AI deep learning algorithms has been demonstrated as a highly effective support for these purposes. Therefore, most of the applications to the signaling systems focus on checking, verification, and detection of anomalies in normal operation. Recently documented demonstrations could be found in Liang et al. (2022), who studied the fault diagnosis model of turnout text data to synthesize a few category samples within the interlocking systems. At the same time, it can also provide new ideas for railway text classification and fault diagnosis.

Moreover, the open complex safety critical communications networks used for railways are potentially subject to security threats, which require the mitigation of the risk against malicious network attacks that disturb regular traffic deployment. The exceptional self-learning, environmental adaptability of AI technologies, and their massive data processing capabilities offer optimal performances for the detection of network anomalies in railway signaling as well, as described in Li and Wang (2023), where the potential network security threats of high-speed railway operation and comparative analysis of AI-based anomaly detection methods are considered. From this analysis the following options emerged, summarized in Table 13.1.

Classification algorithms are supervised learning algorithms involving training a classifier to predict unknown data, such as K-nearest neighbor, decision tree, naive Bayes, support vector machine, random forest, and attention mechanism. These methods demonstrate short training time and high accuracy, but it can be difficult to set parameters and detect unknown anomalies.

Clustering algorithms are unsupervised learning algorithms involving grouping similar data points into clusters while keeping different data points in separate clusters, such as K-means clustering, distance clustering, density clustering, and hierarchical clustering. These methods demonstrate the capability to detect unknown anomalies, but they often require a large amount of training data and time.

Deep learning algorithms are anomaly detection algorithms optimizing data analysis by ML. These methods demonstrate an ability to contribute to improving detection performance, but they often lack interpretability.

ANN *algorithms* are mathematical models simulating the information-processing mechanism of the nervous system in the human brain and operating in parallel and distributed processing structures. They demonstrate self-learning and generalization capabilities but can result in high model complexity.

TABLE 13.1
AI-Based Methods for Anomaly Detection and Fault Diagnosis in Railway Signaling Systems

AI-Based Method	Description	Pros	Cons
Classification algorithms	Supervised learning algorithms used to train a classifier to predict unknown data	Short training time and high accuracy	Difficulty in setting parameters and detecting unknown anomalies
Clustering algorithms	Unsupervised learning algorithms grouping similar data points into clusters while keeping different data points in separate clusters	High capability to detect unknown anomalies	Large amount of training data and time required
Deep learning algorithms	Anomaly detection algorithms that optimize data analysis through machine learning	Improved detection performances	Limited interpretability
Artificial neural network algorithms	Mathematical models simulating the information-processing mechanism of the human nervous system	Self-learning and generalization capabilities	High model complexity
Artificial immune algorithms	Inspired by the principles of the biological immune system	Self-learning ability, capability to detect unknown anomalies, fast convergence speed	Complex tuning required

Artificial immune algorithms are inspired by the principles of the biological immune system to detect anomalies.

It is possible to conclude that the most promising applications of these models are in the field of the safety assessment of signaling and control systems:

- In the design and implementation phases, to ensure that the systems and their components can perform the assigned operational tasks, fixed by rules and regulations, with the required SIL
- In the operational phases, to ensure that the systems and their components can maintain the required SIL, both under accidental disruptions and disturbances and under potential voluntary intrusions, that is, intentional, unauthorized interventions or disruptions that might be digital or physical, including cyberattacks

13.4 AI FOR ENERGY EFFICIENCY

In metro systems, unmanned operation technology has been a reality for more than 40 years, and it is now under rapid development for railway systems involving short line distance, low train speed, simple operation environment, and moderate crossline operation. Full automation is in place on over 70 metro lines in around 20 countries with a mileage largely exceeding 1,000 km.

However, current automatic train operation technology is based on traditional automata theory with objective profiles generated according to line and train conditions. The train is controlled by a control algorithm to follow the objective profile. Nevertheless, traditional automata theory cannot simultaneously consider operation efficiency and energy consumption.

To solve this problem, many research groups carried out studies on theories and applications of train control strategy. Albrecht et al. (2016) established the first energy consumption model of trains and proposed the optimization strategies, but Chang and Sim (1997) had previously proposed the use of GAs to optimize train movements by coasting control. Relevant studies on ML applications, such as neural networks for ATO, mostly focused on the control algorithm or expert experience (Yu et al., 2010), while Malavasi et al. (2011) later proposed to optimize the timetabling and Dominguez et al. (2012) to optimize regenerative energy recovery and the design of ATO speed profiles. Traditionally in control algorithm research, the studies went through the classic process of parameter adaptive control – intelligent control – integrated intelligent control, such as in Su et al. (2015), where proportional-integrative-derivative and fuzzy control are combined.

However, most of this research focused on the optimization of one aspect of the operation profile or control algorithm and is aimed at local optimization on energy consumption, passenger, comfort, or precise train stop while global considerations based on the key operation procedure being used are missing. According to the investigation by Migliorini et al. (2017), an experienced driver could achieve better performance in comparison with assisted driving and ATO, thanks to their broader perception capability and more comprehensive decision capability based on the richness of their experience. But a few research works consider only big data on operation conditions, train conditions, and driver experience.

AI algorithms and technologies, such as deep learning, reinforced learning, and imitative learning, have been widely applied in the traditional automata industry, and promising results in fields such as biomimetic robots, unmanned aerial vehicles, and driverless cars have shown huge potential and provided new ideas for the development of automatic train operation technology. An AI-based ATO can combine the advantages of both intelligent driving and automatic driving and improve the global train operation performance considering energy efficiency requirements integrated with safety and passenger comfort requirements. More recently, Zhang et al. (2018) proposed an AI-based automatic train operation (*iATO*) system to solve the automatic train operation as a multi-objective and multi-phase decision problem. They consider various typologies of operation-related data as the basis of train control decision generation and continuously optimize the problem according to a global *train running condition evaluation index*. It uses *deep reinforcement learning* (DRL), which combines the data mining ability of deep learning and motivates the decision ability of RL.

While in *traditional ATO*, according to the operation schedule, the speed profiles are calculated to ensure punctuality, and the energy cost is not considered in the calculation of the speed profiles; *iATO* uses the energy cost as an object to learn from train operation experiences with DRL and generate energy-efficient train operation profiles.

13.5 AI FOR PASSENGER INFORMATION AND TICKETING

Another important application field for AI concerns the acquisition, the elaboration, and the management of data on passenger flows onboard and to/from trains. Indeed, this application field is not completely railway-specific; it normally covers various public transport modes and can be roughly divided into the management of two main information streams:

A. Information about passengers
B. Information *for passengers*, which can include (A)

In stream (A), real-time monitoring can employ various typologies of algorithms and intelligent systems to monitor traffic patterns, predict the risk of congestion, and be utilized to dynamically adjust routes and schedules to ensure seamless traffic. The adaptability to evolving traffic conditions is the basis for optimizing public transport services, reducing travel times, and improving their reliability. Demand forecasting is paramount, both for real-time operation and for long-to-short-time monitoring. This makes the allocation of resources more efficient, adjusting the number and frequency of vehicles to specific routes based on demand. Optimizing service levels to match the expected passenger volume can effectively reduce waiting times, minimize overcrowding, and provide a more comfortable experience for passengers. A good example of this potential application is described by Gallo et al. (2019) with reference to a metro line in a densely populated area, which offers the possibility of continuous training of the selected ANN using counting of passengers at turnstiles. The general objective of this typology of applications is to estimate the number of passengers onboard in each section (between two stations). The numerical results show that, depending on an abundant amount of training datasets, the proposed approach can forecast the flows on metro sections with satisfactory precision.

In stream (B), by considering various relevant factors, such as traffic congestion, passenger demand, and weather conditions, AI systems can intelligently generate the most efficient routes for trains, as described by Petrovich Kozlov (2022). The results are more reliable at providing a time-effective transportation network, potentially favoring individuals to opt for public transport instead of private vehicles. AI-based systems can provide passengers with suggestions for the best routes and modes of transport and offer passengers personalized recommendations tailored to their individual preferences and travel history, constantly being updated with real-time information on travel times, disruptions, and expected delays.

Beyond operational enhancements, AI can also interact, with optimization purposes, with the ticketing systems, by integrating AI technologies into fare collection processes to make them more efficient and adherent to the scopes, keeping in mind that the implementation of smart ticketing systems improves the overall user experience and contributes to the overall efficiency of public transport operations.

In this context, an original application is the integration of business analytics and AI to assess the effects of variations in train ticket prices. The most effective example is provided by Huber et al. (2022) with regard to the effects of *Supersaver* discounts applied in Switzerland, assumed as *quasirandom*, due to the rich set of relevant parameters, using causal ML to assess the impact of the price reductions on re-scheduling of trips in favor of off-peak periods.

13.6 CHALLENGES FOR THE IMPLEMENTATION OF AI IN RAIL OPERATIONS

Challenging research directions have been identified by various studies and are well summarized by Besinovic et al. (2021) as future opportunities not yet investigated enough by scholars and practitioners to qualify them as promising future research and application fields. The most relevant of them, outlined in Table 13.2, are listed below.

TABLE 13.2

Key Challenges and Future Opportunities for Implementing AI-Based Digital Twins in Rail Operations

Challenges	Promising Research Directions
Trustworthiness of AI-based systems	Exploiting explainable AI for the development of regulations and standardized certification processes
Enhancing passenger satisfaction and personalized experiences	Advanced computer vision for crowd characterization and emotion recognition
User-friendly interfaces for traffic management	Visual support tools for real-time traffic planning and control
Informed decision-making based on human experience	Logic programming to incorporate expert knowledge
Ensuring operational safety	Operation-research-based models for safety and security optimization

1. Developing regulations and standardized certification processes required to quantify the trustworthiness of AI-based systems and thus assess their safety and dependable characteristics to provide performances, such as safe autonomous train operation. Therefore, it is necessary to finally exploit explainable AI approaches to make their processes more transparent and understandable.
2. Developing computer vision performances to provide advanced motion tracking at stations and onboard trains for passengers' crowd characterization and, finally, emotion recognition to monitor passenger satisfaction, driverless vehicles, personalized trip advisors, and experience for all, including visually impaired persons.
3. Developing visual support tools to help dispatchers with more user-friendly interfaces and provide the right information at the right time within an integrated computer vision system for traffic planning and management.
4. Developing logic programming for human-based decision-making to produce decisions based on experienced practitioners, e.g., service planners, timetable designers, dispatchers, maintenance planners, and workers.
5. Developing operation-research-based models to tackle safety and security checks. The increased awareness of AI usability for strategical decisions of policymakers can be expected, and new applications could arise for transport policy optimization and best resource use identification, such as revenue management and pattern recognition or autonomous systems and robotics and revenue management.

However, it is rather difficult to envision possible related applications in the future. The developments of AI and railway technologies could indeed generate new potential uses and users.

REFERENCES

Albrecht A., Howlett P., Pudney P., Vu X., Zhou P. (2016). The key principles of optimal train control-Part 2: Existence of an optimal strategy, the local energy minimization principle, uniqueness, computational techniques. *Transportation Research Part C*. DOI:10.1016/j.trb.2015.07.024.

Assad A.A. (1980). Models for rail transportation. *Transportation Research Part A*, 14 (3), 205–220.

Besinovic N., De Donato L., Flammini F., Goverde R.M.P., Lin Z., Liu R., Marrone S., Nardone R., Tang T., Vittorini V. (2021). Artificial intelligence in railway transport: taxonomy, regulations and applications. *IEEE Transactions on Intelligent Transportation Systems*, 4, 1–15. DOI:10.1109/TITS.2021.3131637.

Cerreto F., Nielsen B.F., Nielsen O.A., Harrod S.S. (2018). Application of data clustering to railway delay pattern recognition. *Journal of Advanced Transportation*. DOI:10.1155/2018/6164534.

Chang C., Sim S. (1997). Optimising train movements through coast control using genetic algorisms. *IEEE Proceedings-Electric Power Applications*, 144 (1), 65–73.

Cordeau J.F., Toth P., Vigo D. (1998). A survey of optimization models for routing and scheduling. *Transportation Science*, 32 (4), 380–404.

Corman F., D'Ariano A., Pacciarelli D., Pranzo M. (2010). Centralized versus distributed systems to reschedule trains in two dispatching areas. *Public Transport: Planning and Operations*, 2 (3), 219–247. DOI:10.1007/s12469-010-0032-7.

De Donato L., Dirnfeld R., Somma A. et al. (2023). Towards AI-assisted digital twins for smart railways: Preliminary guideline and reference architecture. *Journal of Reliable Intelligent Environments*, 9, 303–317. DOI:10.1007/s40860-023-00208-6.

Dirnfeld R., De Donato L., Flammini F., Azari MS., Vittorini V. (2022). Railway digital twins and artificial intelligence: Challenges and design guidelines. In: Marrone, S., et al. *Dependable Computing – EDCC 2022 Workshops. EDCC 2022. Communications in Computer and Information Science*, vol. 1656. Springer, Cham. DOI:10.1007/978-3-031-16245-9_8.

Dirnfeld R., De Donato L., Somma A., Azari M.S., Marrone S., Flammini F., Vittorini V. (2024). Integrating AI and DTs: challenges and opportunities in railway maintenance application and beyond. *Simulation*. DOI:10.1177/00375497241229756.

Dominguez M., Fernández Cardador A., Cucala A.P., Pecharroman R.R. (2012). Energy savings in metropolitan railway substations through regenerative energy recovery and optimal design of ATO speed profiles. *IEEE Transactions on Automation Science and Engineering*, 9 (3), 496–504. DOI:10.1109/TASE.2012.2201148.

Dündar S., Sahin I. (2013). Train re-scheduling with genetic algorithms and artificial neural networks for single-track railways. *Transportation Research Part C*, 27, 1–15. DOI:10.1016/j.trc.2012.11.001.

Fay A. (2000). A fuzzy knowledge-based system for railway traffic control. *Engineering Applications of Artificial Intelligence*, 13 (6), 719–729. DOI:10.1016/S0952-1976(00)00027-0.

Ficzere P. (2023). The role of artificial intelligence in the development of rail transport. *Home Archives*, 2, 4: Cognitive Sustainability. DOI:10.55343/CogSust.81.

Fukumori K., Sano H., Hasegawa T., Sakai T. (1987). Fundamental algorithm for train scheduling based on artificial intelligence. *Systems and Computers in Japan*. DOI:10.1002/scj.4690180306.

Gallo M., De Luca G., D'Acierno L., Botte M. (2019). Artificial neural networks for forecasting passenger flows on metro lines. *Sensors*, 19, 3424. DOI:10.3390/s19153424.

Gorbachev R., Novikov A., Kalinkin A., Cheranev A., Zakharova E. (2020). Applying virtual modelling to verify control systems decision with artificial intelligence in railway transport. *International Conference Engineering and Telecommunication En&T 2020, IEEE*, Dolgoprudny, Russia. DOI:10.1109/ENT50437.2020.9431256

Huber M., Meier J., Wallimann H. (2022). Business analytics meets artificial intelligence: Assessing the demand effects of discounts on Swiss train tickets. *Transportation Research Part B*, 163, 22–39. DOI:10.1016/j.trb.2022.06.006.

Kecman P., Goverde R.M.P. (2015). Predictive modelling of running and dwell times in railway traffic. *Public Transport*, 7 (3), 295–319. DOI:10.1007/s12469-015-0106-7.

Li, S., & Wang, J. (2023). Review of network anomaly detection in the high-speed railway signal system based on artificial intelligence. *2023 IEEE 3rd International Conference on Computer Communication and Artificial Intelligence (CCAI)*, Taiyuan, China (pp. 318–325). IEEE. DOI:10.1109/CCAI57533.2023.10201304.

Liang H., Wang X., Sharma A., Shah M.A. (2022). Application and study of artificial intelligence in railway signal interlocking fault. *Informatica*, 46, 343–354. DOI:10.31449/inf.v46i3.3961.

Liu F., Xu R., Fan W., Jiang Z. (2018). Data analytics approach for train timetable performance measures using automatic train supervision data. *IET Intelligent Transport Systems*, 12 (7), 568–577. DOI:10.1049/iet-its.2017.0287.

Malavasi G., Palleschi P., Ricci S. (2011). Driving and operation strategies for traction-energy saving in mass rapid transit systems. *Proceedings of the Institution of Mechanical Engineers, Part F: Journal of Rail and Rapid Transit*, 225 (5), 475–482. DOI:10.1177/2041301710395077.

Malavasi G., Ricci S. (2001). Simulation of stochastic elements in railway systems using self-learning processes. *European Journal of Operational Research*, 131 (2), 262–272. DOI:10.1016/S0377-2217(00)00126-0.

Mehrotra K., Mohan C.K., Ranka S. (1996). *Elements of Artificial Neural Networks*. The MIT Press.

Migliorini C., Ricci S., Tombesi E. (2017). Analysis and assessment of eco-driving strategies for train drivers training. *Ingegneria Ferroviaria*, 10, 727–758.

Pappaterra M.J., Pappaterra M.L., Flammini, F. (2024). A study on the application of convolutional neural networks for the maintenance of railway tracks. *Discover Artificial Intelligence*, 4, 30. DOI:10.1007/s44163-024-00127-2.

Petrovich Kozlov I. (2022). Optimizing public transport services using AI to reduce congestion in metropolitan areas. *International Journal of Intelligent Automation and Computing*, 5(2), 1–14.

Schaefer H., Pferdmenges S. (1970). An expert system for real-time train dispatching. *WIT Transactions on the Built Environment*, 7, 8.

Su S., Tang T., Chen L. (2015). Energy-efficient train control in urban rail transit systems. *Journal of Rail and Rapid Transit*, 229(4), 446–454. DOI:10.1177/0954409713515648.

Tang R., De Donato L., Besinovic N., Flammini F., Goverde R.M.P., Lin Z., Liu R., Tang T., Vittorini V., Wang Z. (2022). A literature review of artificial intelligence applications in railway systems. *Transportation Research Part C*, 140, 103679. DOI:10.1016/j.trc.2022.103679.

Törnquist J. (2005). Computer-based decision support for railway traffic scheduling and dispatching: A review of models and algorithms. *ATMOS 2005, 5th Workshop on Algorithmic Methods and Models for Optimization of Railways*, Palma de Mallorca, Spain. https://drops.dagstuhl.de/opus/volltexte/2006/659.

UIC. (2024). The journey toward AI-enabled railway companies, February 2024. https://uic.org/com/IMG/pdf/uic_layout_web_05032024.pdf.

Wen C., Huang P., Li Z., Lessan J., Fu L., Jiang C., Xu X. (2019). Train dispatching management with data-driven approaches: A comprehensive review and appraisal. *IEEE Access*, 7 (114), 547–571. DOI:10.1109/ACCESS.2019.2935106.

Yin J., Chen D., Li L. (2014). Intelligent train operation algorithms for subway by expert system and reinforcement learning. *IEEE Transactions on Intelligent Transportation Systems*, 15 (6), 2561–2571. DOI:10.1109/TITS.2014.2320757.

Yu J., Qian Q., He Z. (2010). Research on application of two-degree fuzzy neural network in ATO of high-speed train. *Journal of the China Railway Society*, 30 (5), 52–56.

Zhang J., Zhang J. (2023). Artificial intelligence applied on traffic planning and management for rail transport: A review and perspective. *Discrete Dynamics in Nature and Society*, 1832501, 17. DOI:10.1155/2023/1832501.

Zhang M., Zhang Q., Lv Y., Sun W., Wang H. (2018). An AI based high-speed railway automatic train operation system analysis and design. *International Conference on Intelligent Rail Transportation (ICIRT)*, Singapore, 2018, pp. 1–5. DOI:10.1109/ICIRT.2018.8641650.

14 AI for Rolling Stock

Sebastian Stichel, Wolfgang Birk, Carlos Casanueva, and Jonathan Leung

14.1 OVERVIEW

Artificial intelligence (AI) is a broad term interpreted differently depending on the context such as research, business, or legislative contexts. Due to its complexity, the EU commission [1] has proposed a common definition at three levels: an AI definition as such, a taxonomy of core and transversal AI domains, and a keyword list for each subdomain. In their study, common features in earlier definitions of AI were found:

- Perception of the environment and surrounding complexity
- Collecting, processing and interpreting data inputs
- Decision-making and consequent action taking, with a certain level of autonomy, which may include dynamically adapting to changes and learning
- Goal achievement, as the ultimate reason for AI systems

Without consideration of a unique definition for AI models, it is clear from these points that there is substantial overlap between AI-related technologies and adjacent digitalization-related technologies from areas like signal processing, fault detection and isolation, condition monitoring, maintenance intelligence, and business intelligence, all with wide application possibilities in the railway sector. Having a taxonomy and application area adapted definition of AI for railway transport is therefore important and was proposed by Bešinović et al. [2]. Besides defining AI, different subsystems and the applicability of AI-related technologies are given. Drawing from the known benefits of automation to increase the efficiency and performance of systems, a similar assumption is made for AI in railway transport.

For quite a long time AI has been a vague context and the benefits have remained unquantified, while rolling stock manufacturers have adopted it to try to improve the safety and operational efficiency of rail vehicles. Due to the lack of definition, large manufacturers and suppliers have endeavored to provide their own definitions and statements regarding AI and how they envision its use and impact. For example, Alstom has published a White Paper on their own approach to AI [3] highlighting the different types of AI needed in different contexts within the company, and acknowledging the potential and limitations of AI models in the railway sector. Further, CAF has a corporate information page [4] where efforts in "Advanced Analysis of Data and AI" are listed, coupling it explicitly to data analytics and digital twins (DTs), and targeting three different key goals: current performance optimization, design improvements, and detection of system malfunctions.

Looking for more specific application examples, Siemens has noted that AI techniques are used in obstacle detection associated with assisted and automated train operation [5,6], enabling the precise control and operation of the train using a combination of onboard sensor data and position data. They have also applied AI in the condition monitoring domain, where onboard and trackside sensor data have been analyzed to detect problematic conditions [7]. Hitachi Rail claims to have implemented AI-assisted collision detection and awareness systems in the Rio Tinto network [8] where the AI model's main focus is real-time computer vision-related tasks.

All rolling stock manufacturers and suppliers working with AI systems acknowledge in different ways that AI is not to be used in isolation and uncritically, but that it needs to be integrated into a

DOI: 10.1201/9781003492146-14

holistic engineering framework where the best tools are used for the best cases. However, it is not clear for any of these cases what is the actual AI method or tool used, so it is not possible to confirm that these are direct and actual applications of AI methods as defined in the literature or e.g. by the EU commission, or just advanced computational models that are published as AI. There is also a big overlap and confusion of terms when discussing different subtopics that fall under the broader "digitalization" umbrella due to the various definitions of AI across different sectors.

These examples cover only some of the rail vehicle companies, while many component manufacturers also make use of AI for their specific applications; and there are, as expected, many smaller AI-specialized companies that contribute to the railway sector with promises of higher safety, efficiency, or convenience. The objective of this handbook chapter is to clarify the different terms used in the extended context of AI systems and to showcase the guiding principles for choosing between different AI methods and their possibilities and limitations so that the readers can consider themselves informed on the actual use and application of AI in rolling stock.

14.2 WHAT IS AND IS NOT AI

AI "refers to systems that display intelligent behavior by analyzing their environment and taking action – with some degree of autonomy – to achieve specific goals", as defined by the European AI strategy [1]. AI is a component of digitalization where there are many overlapping concepts that need clarification.

The DT is a digital model of a singular product, system, or process that shares information with its real-world counterpart (the physical twin) and even affects and acts upon it. DT is a framework or umbrella where most of the digitalization technologies can be found, but AI is not necessarily a part of it, by definition. AI systems can also exist independently from DTs. Still, AI can be a powerful tool to integrate with DT systems as both align very well in their sensing, understanding, and decision-making characteristics.

Other terms that can be confounded in this area are "Industry 4.0" [9], which typically refers to the so-called "fourth industrial revolution" including the internet and interconnected cyber-physical systems in industrial, manufacturing, and technical contexts; "Internet of Things" (IoT), which specifically refers a network of physical components with digital connectivity; "big data", or the registration of all possible information in an orderly manner ("data management") for an increased understanding of the system ("data analytics"). It can clearly be seen that all these enable DTs or artificial intelligence in different ways, and how they can easily be mixed and tossed up in the same basket. The confluence of AI, data science, and IoT into Industry 4.0. can be visualized in Figure 14.1.

While Industry 4.0 is a pure technological concept, its extension to include humans, resilience, and sustainability into that context creates a more symbiotic view and where the technology goes hand in hand with the human worker, a companionship. This human-centric version is then denoted as Industry 5.0 [10].

14.3 APPLICATIONS OF AI IN ROLLING STOCK-ASSOCIATED TASKS

AI is used in rolling stock for the following objectives:

- Detection and/or prediction of system malfunctions
- Improvement or optimization of the performance of vehicles in operation
- Improvement or optimization of vehicle design

To describe the specific AI methods applied, this chapter adapts the taxonomy defined in [11]. "AI techniques" refers to methods and algorithms, while "AI research fields" represent research areas that would not exist without AI techniques.

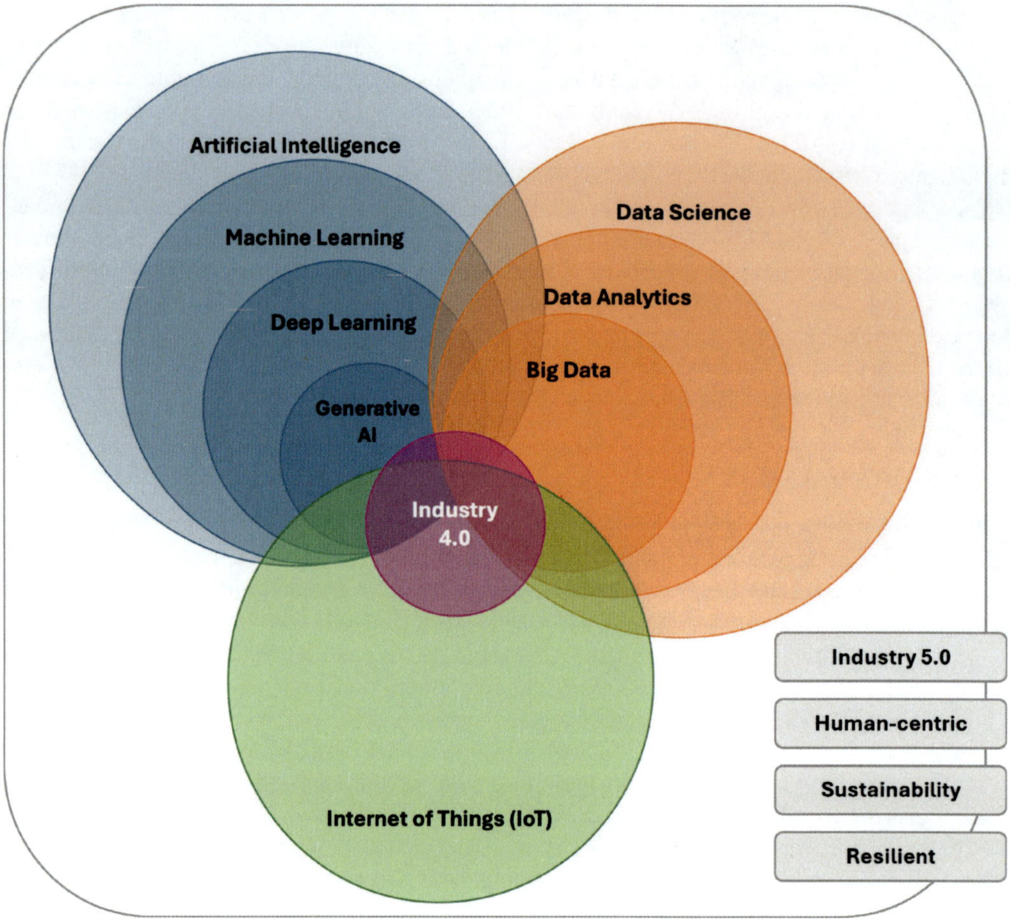

FIGURE 14.1 Confluence of AI, IoT, and data science that leads to "Industry 5.0".

- AI techniques
    - Evolutionary computing
    - Logic programming
    - Machine learning (ML), which contains:
        - Task
        - ML algorithm
        - Learning paradigm
        - Learning scenario
        - Training modality
- AI research fields
    - Expert systems
    - Data mining (DM)
    - Pattern recognition
    - Adversarial search

Sometimes two or more objectives and/or techniques are combined in a specific application, so the coming sections are divided into broader technical applications:

- Condition monitoring and diagnostics
- Predictive maintenance
- Traction drive systems control and automation
- Rail vehicle design

For condition monitoring and diagnostics, the main objectives are to follow the condition of an asset or component, detect faults and abnormal behaviors, and predict the future condition of said asset or component. The research fields of DM and pattern recognition are very important here, and the AI methods used belong to the ML group. Predictive maintenance is very similar, as this focuses on the later stages of the same processes after condition monitoring and diagnostics. It is very common to use ML techniques with Logic Trees.

For Traction drive systems control and automation, the objectives are typically focused on Performance and Design, such as minimizing energy consumption by finding the minima for very complex objective functions that include vehicle, track, and operational subsystems variables and models; reducing the complexity of the design space for novel battery and hydrogen traction systems; or examining vehicle traction and braking processes in operation for reducing the total risk in the railway system. Most AI techniques cluster around Evolutionary Computing and ML while, in the AI Research Fields group, only Expert Systems have been found as AI applications.

For Rail vehicle design, the objectives can be the reduction of the design space for creating a more manageable problem that requires less computational power; the reduction of model complexity, for the same reason; and the optimization of a desired performance attribute of the component or subsystem being designed. The AI techniques are clustered around evolutionary methods, support vector machines (SVMs), and artificial neural networks, with successful applications in all cases but unclear reasoning behind the selected methods.

In the coming chapters all these applications are explained in detail, including insights and practicalities about the AI techniques that would allow the reader to take informed decisions on the use of AI for their own applications.

14.4 AI FOR CONDITION MONITORING AND DIAGNOSTICS

Condition monitoring and diagnostics are tasks that benefit from automation, especially when the number of assets and components becomes large. For this automation to take place, data acquisition, ingestion, and harmonization from various sources can be seen as a pre-requisite but partial automation where the subsequent tasks for the condition assessment and the diagnostics are automated from pre-defined and structured data sources is also a good starting point. A detailed discussion of AI tools in this context is given in [12] where the emphasis is the use of the tools in the context of decision support for condition monitoring and diagnostics.

An important pre-requisite is a thorough understanding of the faults and failure modes that can be encountered, and which root causes are associated with them. This domain knowledge enables proper classification and ensures explainability of the AI outcomes to the user as the AI-based condition monitoring, and especially the diagnostics outcome, align with the user's understanding of the system behavior. If this understanding is lacking and not embedded in the AI, typical anomaly detection is usually the starting point.

We will therefore first discuss the failure modes and root cause analysis followed by monitoring schemes for typical failure modes.

14.4.1 FAILURE MODES AND ROOT CAUSE ANALYSIS

When considering failure modes of a component or system, root cause analysis is the logical starting point. Root cause analysis is a quality management tool which is used to determine the origins of a problem after an incident or issue is detected. The methods used are dependent on the type or severity of the faults. Failure mode and effect analysis (FMEA) is one of the most rigorous root causes analysis methods. With this approach, every possible failure mode is identified, and the impact of failure is assessed for each mode. Once the impact is assessed, the root cause of each failure is addressed to reduce the likelihood of failure. Fault tree analysis (FTA) follows a similar approach as FMEA. However, it utilizes visual mapping to understand the relationships and co-dependencies between faults and their causes. In this method, a logic tree is created starting at the fault and the dependencies throughout the system are mapped using Boolean logic as shown in Figure 14.2. Although the logic tree shown in Figure 14.2 has three levels, additional levels can be added depending on the complexity of the system. Pareto charts can also be used to identify faults and the common root causes. Unlike the FMEA and FTA, these charts rank identified faults in order of frequency to prioritize root cause analysis actions. Thus, this method requires the capability to identify multiple fault types and quantify the fault occurrence.

Fault tree analysis is particularly useful for supervised ML as the causality between input "events" and output "faults" is known and labeled, allowing for the algorithms to be trained along the preexisting mapped patterns.

Condition monitoring builds on root cause analysis techniques and uses fault detection and identification tools to identify faults and provide timely corrective actions based on the current operation conditions and recent operation history. Unlike root cause analysis where the mitigating actions are created after a fault has occurred, condition monitoring emphasizes the monitoring of features and event signals and correlating the evolution of these signals with respect to expected faults. In this way, condition monitoring takes a bottom-up approach to fault detection as opposed to a top-down approach as outlined in the FMEA and FTA methods.

14.4.2 From Measurements and Data to Diagnostics

While condition monitoring is targeted to assess the current condition of an asset or component, diagnostics need to map the condition value into the states "faulty" or "healthy". When a faulty state

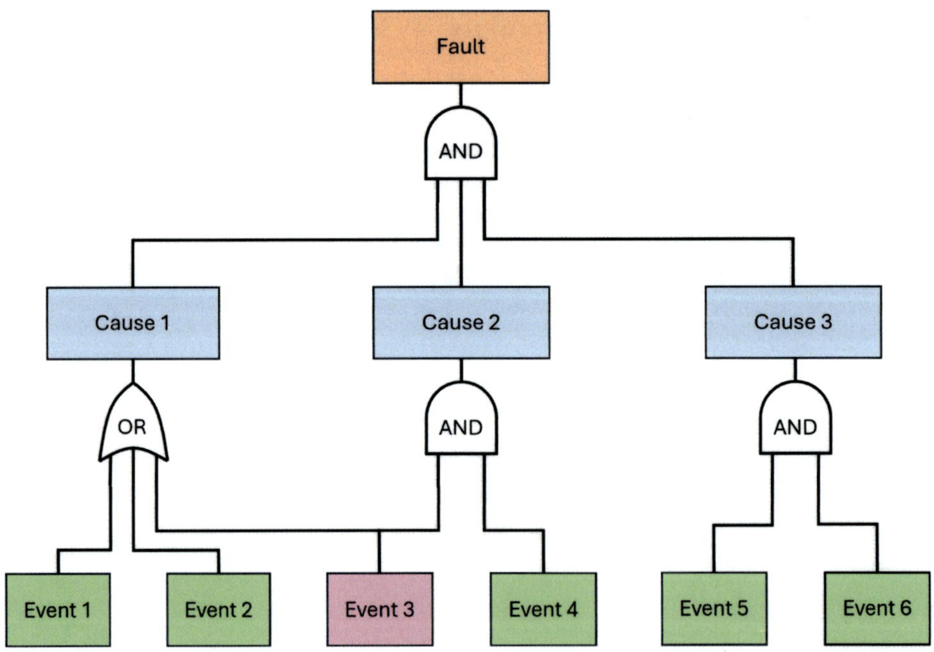

FIGURE 14.2 Example of a fault tree, which is used to systematically map the fault to event dependencies using Boolean logic to represent the dependencies between events, causes, and the fault.

is detected a root cause or failure mode needs to be determined for the faulty state, which usually depends on the features that are used to derive the condition indicator and to what degree the specific fault can be isolated using the available data and measurements.

Naively speaking, every combination of measurements and data at an instance should render the correct condition indicator value representing the condition of the asset. Due to the presence of measurement errors, such as noise, biases, drift, calibration issues, and device faults, the data quality needs to be considered, and proper handling needs to be in place. It also means that not every data point must be considered. This also extends into the correctness of meta data on the asset and the human assessment of the asset's true state (condition). There are possible errors and ambiguities that need to be considered. The AI should therefore make use of typical time series analysis principles as discussed in [13]. Essentially, it means that the condition monitoring outcomes should not be assessed in one instance in isolation.

When the outcome from the condition monitoring is mapped into a diagnostic outcome, the true state of the asset from a human inspection or assessment can be used to validate the performance of the condition monitoring and diagnostic system. Here, it is important to note that human errors and ambiguity in the judgment impact the assessment. Nevertheless, such an assessment builds trust in the AI solution.

In an ideal scenario there is no in-between and the faulty states are clearly defined, but in reality, this is not the case, and any mapping of the condition indicator value into the two categories, "healthy" and "faulty", will have a probability associated with it. Using a larger number of cases from the field will then enable the performance assessment of the diagnostic scheme using a confusion matrix [14] and thereof derived performance scores [15].

In Figure 14.3, the left-hand side shows the confusion matrix for binary classification in the two states "faulty" and "healthy". The diagonal contains then the number of cases that are correctly classified, and the off-diagonal is the cases that are falsely classified. In the multiclass cases where multiple failure modes are assigned, the diagonal still contains the correct assignments (classification) while the off-diagonal elements indicate the different combinations of the false assignments. Note that failure modes can be aggregated into groups if specific modes cannot be distinguished by the diagnostic scheme sufficiently well. This also enables the user to decide whether or not a diagnostic scheme is appropriate for the desired diagnostic task.

While this sounds straightforward, the assessment is not trivial and the expectations of the derived scores are often subjective [16]. The user should therefore consider that the cases should be balanced in numbers and that the ground truth (usually the human assessment of the true state) is correct and reflects the true state appropriately.

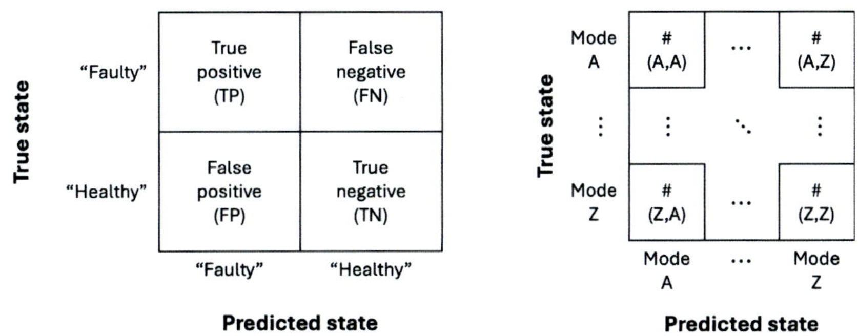

FIGURE 14.3 Confusion matrix for a general assessment of the categories "healthy" and "faulty" on the left and a detailed assessment for correct failure mode classification on the right. The elements of the matrix usually contain the number of cases for the specific combination predicted and true state. "#" indicates the number of combinations for the predicted and true state on the right.

A prescription of action is then the mere act of trusting the predicted state and selecting a pre-defined action for the failure mode. Having performed a detailed assessment of the diagnostic scheme using the confusion matrix enables a confident decision and line of action. It also helps in understanding what should not be done or decided about in an automated way due to desired performance requirements.

14.4.3 Wheel Wear and Wheel Damage

Monitoring the quality of wheel conditions using wayside and onboard detection methods allows for the early detection of problematic wheel properties prior to catastrophic failure. The types of damage considered include geometric degradation (wheel out of roundness, wheel polygonization, wheel wear, and wheel flats) and crack-based degradation (rolling contact fatigue (RCF) cracking).

Here, it is important to distinguish damages exhibiting growth patterns and the ones generated by instantaneous events. While all damages can be identified using detection and classification methods, their evolution and worsening during operation can also be tracked in time of use series or series depending on a different independent variable, like covered distance. Wheel flats are usually not good candidates to track as their occurrence is instantaneous and initiated by, for example, driver errors.

Damage detection methods are primarily focused on the damage to wheelsets and bogies, as the degradation of these components strongly influences the dynamic vehicle track interaction behavior and can influence the maintenance costs, both for the rolling stock and the track. Wheel damage detection methods can be classified into wayside and onboard detection methods. Wayside (also referred to as trackside) methods refer to detection method equipment located on the track infrastructure. Onboard detection methods refer to detectors placed on the vehicle as mentioned in Section 1.3.

AI techniques are applicable for condition monitoring as they provide an efficient method to process the measured data from onboard and trackside data to provide real-time estimation of the damage on the wheel, which is also exemplified in [17–19]. Shaikh et al. [20] note that AI can be used to improve the efficiency of condition monitoring systems, where machine vision techniques can be used to identify problematic wheels.

Recent examples of AI techniques being deployed in condition monitoring and wheel damage applications include the use of deep learning and neural networks to detect wheel flats and classify the severity of wheel flats [21]. This study compared different neural networks and found that deep neural networks were shown to have the highest accuracy in the identification and classification of wheel flats compared to other methods. Other studies have outlined an unsupervised learning framework to identify wheel damage based on onboard vehicle measurements [22,23]. The framework utilizes autoregressive clustering to determine the severity of wheel damage which can be deployed for vehicles at different speeds and for various sensor configurations.

14.4.4 Bearing Faults Rendering Hot Boxes or Acoustic Emission

Monitoring bearing conditions and determining bearing faults can be achieved through acoustics and vibrational signal processing from sensors mounted on the bearing housing. Determining the bearing fault type can be challenging as the drivers lead to faults that can occur over long- and short-time scales. For short-term damages, like bearing overloads, the change in signal is pronounced however, for long-term damages, like rolling contact fatigue, the change in measured signals can be imperceptible. Identifying the fault "signature" becomes challenging when there is a convolution of overlapping fault signals which can lead to challenges in predicting the bearing health, hot boxes, or acoustic emissions. These signal-processing approaches rely on expert knowledge of the range and frequencies of the signals where a fault signature occurs and employ the appropriate type of signal-processing filtering methods and signal analysis methods that should be most appropriate for the extracted signals. Nevertheless, Birk and Westerberg show in [24] how wayside data from acoustic measurements, hotboxes, and wheel impact load can be fused to determine a health index for bearings, where false positives are reduced by the use of wheel impact load data.

Recent studies have shown an increasing usage of ML and AI methods to identify bearing faults from measured signals [25]. Unlike conventional approaches, ML and AI methods can draw on more varied signal types and are more suitable for incomplete or sparse data sets. According to differences in data requirements and feature extraction, ML and AI methods for bearing fault identification can be categorized into shallow and deep learning methods. Shallow ML methods that have been studied include support vector machine (SVM) classification [26] and weighted K-nearest neighbor classifiers [27]. Both methods are used to classify the signal data into fault types. These methods require preexisting knowledge of the fault signature however these methods are beneficial and can use unlabeled data sets to classify the different fault types.

Deep ML and adjacent AI methods differ from shallow ML methods as they utilize a larger, more diverse dataset to obtain bearing fault diagnostics. To address the lack of training data sets for deep ML applications, unsupervised learning methods including auto-encoder and generative adversarial network models have been explored [28–30]. Sun et al. [28] have utilized transfer learning algorithms, which is a type of auto-encoder model, to translate domain data information from the training process directly to the new domain application. This is beneficial as it improves the quality of the fault resolution and has been shown to scale well for bearing fault diagnostics. Likewise, Liu et al. [29] have shown that multiscale residual generative adversarial networks and capsule networks can be used to realize bearing fault diagnosis.

To elucidate the bearing fault from ambiguous signals, supervised deep ML methods have been explored to classify data from noisy signals. Convoluted neural networks have been shown to effectively identify and extract the appropriate local features of the signal, which can be used to determine the fault signature [31]. These methods are useful as they enhance the resolution of the fault signatures and can be tuned to reduce noise effects that may lead to false diagnoses.

14.4.5 Brake Block Wear

Vision-based solutions located on the track wayside can monitor brakes and their wear. These techniques are similar to those used for wheel wear and wheel damage where pattern recognition and trained networks are used to identify the brake blocks and determine their fitness for continued operation [32]. There are also commercial solutions available to perform the task. Currently, brake blocks and associated undercarriage equipment are monitored using optical wayside sensors [33].

AI methods are used to automate the image processing of undercarriage images. AI has been shown to automate object recognition and brake block condition monitoring [34]. Multispectral vision solutions have also been developed to evaluate brake block thickness [35,36]. In these systems, visible and infrared data obtained from the wayside sensors is paired with brake disc degradation models to predict the brake condition and brake health. In this multi-step analysis, Automated AI image segmentation methods are used to analyze the image spectra at increasing levels of detail, starting with a coarse analysis of the overall underside to identify anomalies in the underside followed by a detailed analysis which focuses on the component anomaly detection and defect classification.

14.4.6 Failures in Other Components

Like failures in bearings, failures in other vehicle components are difficult to detect. They can, however, be safety critical or at least significantly deteriorate the running behavior of the vehicle. Failures like cracks in springs or bolts can be detected directly. Here, machine vision is an appropriate tool (see, e.g., [37]). With a track-based high-speed camera, for example, a passing train can be examined. Applying ML algorithms, the computer can learn to detect damaged components. Another example of using machine vision to detect loose screws or nuts on railway wheels is presented by Schmitz et al. [38]. Liu et al. propose machine vision to detect the presence/absence of a fastening bolt in a complex background [39]. The photos are classified with algorithms that have been trained with known correct and faulty screw arrangements.

Alstom proposes an acoustics-based diagnosis system to detect failures in the bogie. A pilot solution for trackside monitoring has been developed. The noise measurement can be performed with an ordinary smartphone [40]. From the acoustic data measured spectrograms are calculated. By comparing the measured acoustic data with known noise characteristics, the health of bogie components can be monitored. Failures in both rotating components like wheel flats and non-rotating components like loose yaw-damper connections can be detected. With the help of AI, the remaining life of the component could also be predicted.

14.4.6.1 Detection of Vehicle Running Instability

Another possibility is to detect failures indirectly by recognizing abnormal vehicle behavior. Running instability, for example, can be caused by worn wheels or rails, generally leading to increased wheel–rail conicity, or faulty suspension elements, e.g., deteriorated yaw-damper performance. Also for this task ML algorithms are the preferred tool. Two steps are necessary. First, the algorithm must decide whether the running behavior can be regarded as abnormal. To perform this task, features must be defined that characterize an operation condition with a significant periodic content in the vibration signal. To decide whether a hunting motion exists or not is not a trivial task, many different concepts have been proposed, e.g., [41–45].

In the next step the computer needs to determine which type of failure is causing the change in the running behavior (cf. Section 1.4.1). Regarding bogie hunting, this was tested, for example, in [46]. After training the algorithm with known data, it was – with different degrees of probability – able to detect a demounted damper, a damper with reduced capacity, and a bushing with reduced function on track both with high and, respectively, low wheel-rail conicity. The algorithms need to be trained with known data so that the algorithm can then be used to classify measured data and to decide whether and where there is a faulty component.

14.4.6.2 Fault Detection in Suspension Elements

Fault detection in suspension elements is also an important task that can even be safety critical. Therefore, many attempts have been made to detect faulty suspension components with the help of physical models. Physical models mean that advanced dynamic models are built up that can represent the dynamic behavior of the rail vehicle. Data needs to be collected from a real vehicle by sensors that interact with the models. Estimations of the vehicle states can then be gained by using various advanced versions of Kalman filters.

However, data-driven methods have also been explored in this area in recent years (cf. [47] or [48]). Classical statistics model-based or ML-based algorithms often struggle to determine which suspension component in a vehicle is deteriorated since the behavior of a rail vehicle is highly non-linear, and different failures can lead to similar changes in the behavior of the rail vehicle. The example in Ye et al. in [49] (Figure 14.4) shows that a failure of an anti-yaw-damper and a combined failure of an anti-yaw and a lateral damper led to the same acceleration pattern in the bogie.

In [49], therefore, so-called multiscale permutation entropy algorithms, together with linear local tangent space alignment methods to reduce the dimensionality of the feature matrix, are tested to detect suspension failures on a Chinese high-speed train. In [47] and [48], neural networks are applied to detect suspension faults. Chen et al. [47] use bidirectional neural networks with long- and short-term memory. By considering the temporal dependency of data in two directions, typically from past to future and from future to past, they can be more efficient than one-directional neural networks. In [48], a three-step methodology is applied. First Gaussian white noise is applied to the original signal to make it immune to the interference caused by track irregularities. In the next step, an edge sample training (EST) strategy is proposed to improve the robustness against the influence of wheel wear. Finally, a one-dimensional deep neural network-based fault diagnostic network of high-speed train suspension is built. The methodology is also tested on a Chinese high-speed train.

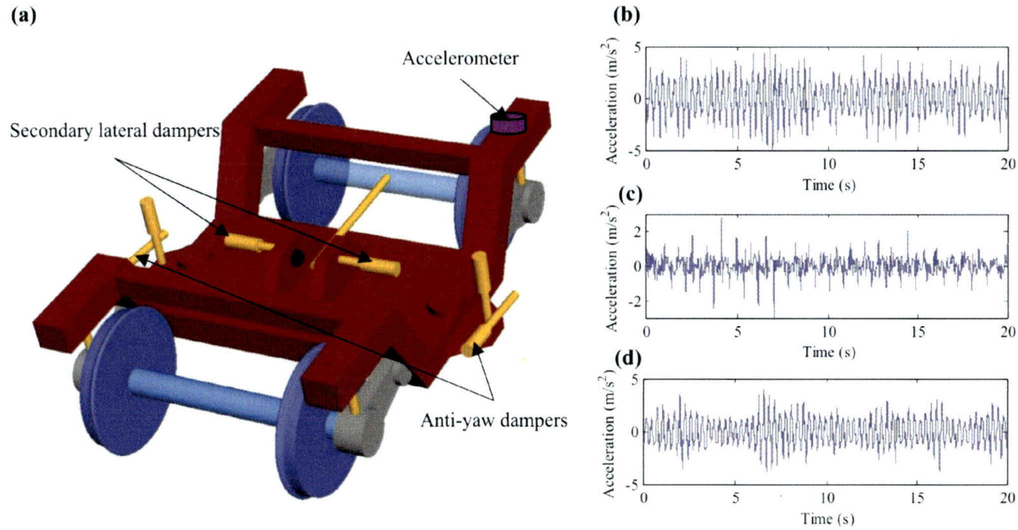

FIGURE 14.4 Bogie model: (a) bogie frame's lateral accelerations under different types of faults; (b) yaw-damper removal; (c) secondary lateral damper removal; (d) both yaw-damper and secondary lateral damper removal. (From Y. Ye, Y. Zhang, Q. Wang, Z. Wang, Z. Teng, and H. Zhang, 'Fault diagnosis of high-speed train suspension systems using multiscale permutation entropy and linear local tangent space alignment', *Mech. Syst. Signal Process.*, vol. 138, p. 106565, 2020, doi:10.1016/j.ymssp.2019.106565, with permission.)

14.5 AI FOR PREDICTIVE MAINTENANCE

Predictive maintenance aims at providing actionable insights for the decision-making on maintenance of components prior to the onset of a failure or fault based on an understanding of the future condition. If the degradation of the condition of a component, either on an individual level or a group level, is well understood and always follows the same pattern, the time of use in relation to the mean time between failure (MTBF) can be used as a representation of the current and predicted future condition. The maintenance regime would then follow the black arrows as shown in Figure 14.5. Notably, MTBF can also be given in terms of other operation or production variables as the independent variable.

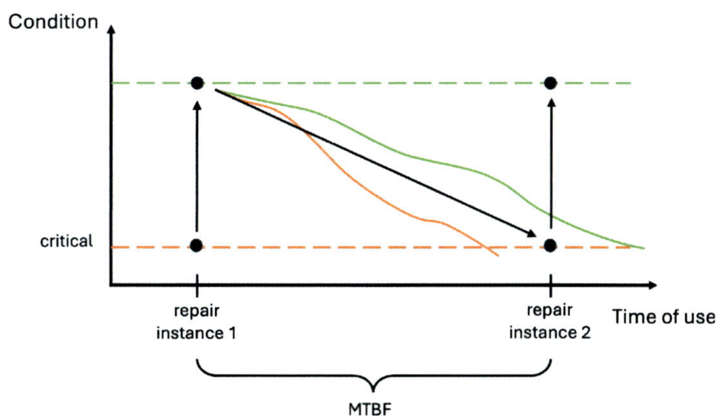

FIGURE 14.5 Time of use-based maintenance regime versus the real condition and its relation.

The issue with this approach is that the representation of the current and future condition does not use any insights into the actual condition; it merely uses the historic maintenance information in relation to the operational characteristics. If the actual degradation differs, the maintenance regime will be ineffective and either miss critical conditions rendering failures or over-maintenance, as indicated by the red and green traces in Figure 14.5. A more precise and effective maintenance regime would therefore consider condition monitoring of the components to understand the actual degradation.

14.5.1 Condition Forecasting

The degradation of an asset depends on the active failure modes and their combination. The condition forecasting model might therefore differ in line with the active failure modes, meaning that a model for one combination of failure modes might not be valid for others. Condition monitoring enables the diagnostic of the failure modes focusing on detection and classification. The outcome is then used in the selection and adaptation/training of the model for condition forecasting.

The forecasting can then be performed, either in time or a different independent variable like gross million tons or covered distance, which might be more of interest to the maintenance process. The maintenance process is usually organized such that there are certain desired windows for maintenance, e.g., every 6 months a rolling stock asset is taken out of service for inspection and maintenance. At the same time there are preferred or determined maintenance limits as well as safety limits. These limits are then represented by thresholds in the forecasting process. Using the forecasting model, the future condition and the remaining usage horizon can be determined as shown in Figure 14.6. Given that the quantified horizon directly depends on the determined windows and the threshold, maintenance effort and costs do not need to be fixed over time and can be adapted when these aspects are adjusted within the organization. The balancing of risk, cost, and asset availability become evident factors in such an optimization.

For the sake of clarity, Figure 14.6 only shows the nominal forecasting; the forecast itself is a statistical process with uncertainties. These uncertainties can be represented by confidence bounds leading to likelihood regions for the forecasted horizons. Namely, the critical horizon h_{crit} becomes a range with an associated confidence level that the maintenance threshold would be reached. Similarly, the minimum and maximum horizons, h_{min} and h_{max}, would be represented as a range.

14.5.2 From Diagnostics to Prognostics

The future evolution of the condition is also a statistical process which provides insights into the future condition. An additional complexity is that a forecasting or prediction horizon needs to be

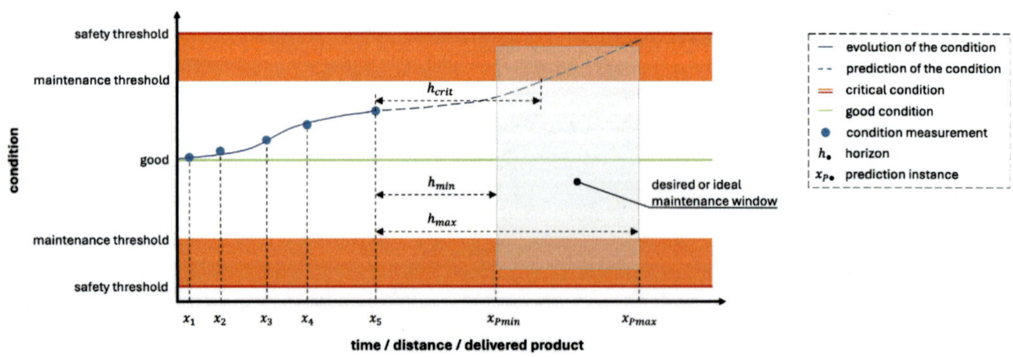

FIGURE 14.6 Condition forecasting based on the historic evolution of the asset condition and the physics governing the degradation for a combination of failure modes.

selected. As already shown in Figure 14.6 and discussed in Section 14.5.1, a desired maintenance window is a parameter used in the prognostic scheme. Obviously, the forecasted condition has not yet occurred and its occurrence in time or use has a probability associated with it. The predictions can also be used to estimate the remaining useful life (RUL) for an asset [50], where the critical level for the condition is seen as the end point and the time of use until this condition level is reached is the RUL. While RUL seems to be the ideal quantity to assess, it has a highly stochastic behavior making its interpretation difficult.

Any decision taken on that information means that a decision is made on a hypothetical condition. A prescription of an action has therefore a preventive character. Building trust in a prognostic scheme is difficult as the user needs to set up specific case studies to evaluate the performance, either by operating the asset until it reaches the forecasted condition in the future, which imposes high risks or by using domain experts to make a judgment about the current condition and its evolution.

Nevertheless, a prognostic scheme offers the advantage of reducing preventive maintenance costs, as only the assets or components that show a prognosed fault would be addressed by an action.

14.5.3 Rolling Stock Component Faults and Their Maintenance

Maintenance of rolling stock assets requires the shunting of specific units, usually wagons, vehicles, or locomotives. Each of these units contains multiple components that are subject to inspections and maintenance throughout their life cycle. Some of the actions are even based on regulations or contracts, e.g., the general contract of use for wagons [51].

Decisions on maintenance are therefore not only based on the current or future condition of a component within a unit but also on contractual and regulatory or even practical aspects. It is therefore important to note that the decision to act is based on a blend of multiple factors and condition indications. Understanding the current and future condition can therefore be an important ingredient to make appropriate and well-balanced decisions.

In Figure 14.7, a schematic view of a maintenance process with pre-defined windows for maintenance is shown. The windows can be the result of the current way of operating and include regulatory and contractual aspects in their definition. The orange ranges indicate the typical maintenance-based area for the condition of a component within a unit. The red line indicates when the condition becomes safety critical. The historic condition of a component is given by the solid traces and the bullets indicate the latest instance when the condition was established. The dashed traces indicate the forecasted condition.

Clearly, combining all aspects and the condition of multiple components requires a decision logic that decides when the units need to be taken out of operation and what actions need to be

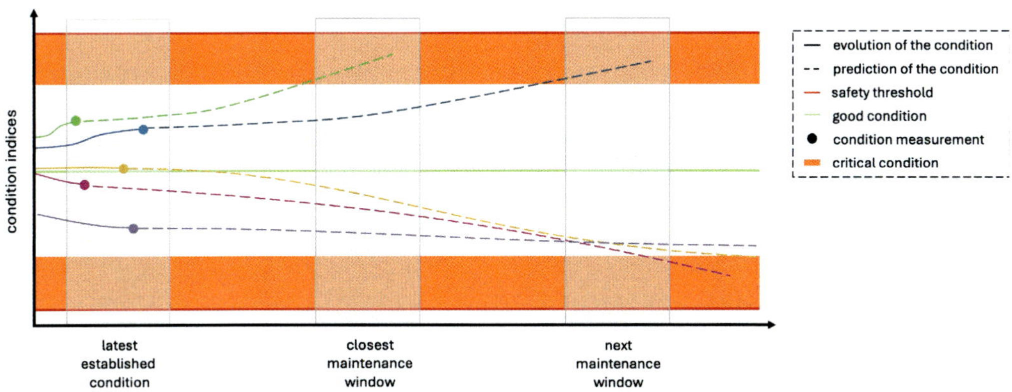

FIGURE 14.7 Window-based maintenance process and relation to condition forecasting of multiple components.

prescribed for the maintenance instance. An obvious approach not requiring a disruption of the current way of working is to align the condition-based maintenance actions with the pre-defined windows. Understanding the future condition does enable a maintenance scheme that considers multiple components.

Moreover, this approach enables the user to implement a condition-based maintenance scheme in an iterative way where the most prominent issues are addressed first.

14.5.4 Practical Considerations

For the successful implementation of an AI-based predictive maintenance (PM) scheme, there are thus several practical aspects that need to be considered:

14.5.4.1 Component Level Maintenance Record and/or Inspection Information

To assess and quantify the performance of a PM scheme it is essential to have either historic information on the maintenance actions that are done on a component or at least the inspection information on the component's condition. Using this information together with condition monitoring and forecasting, it is possible for the user to build trust in the actionable insights or prescriptions that are generated by the AI. Moreover, a business case for the PM scheme can be derived from such an assessment.

14.5.4.2 Data Quality of Records and Condition Information

Any assessment or business case depends on the quality of the underlying data. Understanding shortcomings and quantifying their impact improves the reliability of drawn conclusions. Statistical properties on the data quality and the impact on the outcomes should be derived. Error propagation methods are helpful in determining the effects on the outcomes.

14.5.4.3 Continuous Performance Monitoring

Instead of assessing the performance once or only from time to time, it can be beneficial to automate and implement the performance assessment to enable continuous performance monitoring. This is usually not the first step when implementing a PM scheme, but when a PM scheme is in operation, the monitoring of its performance reinforces the trust of the user and, early on, provides insights on when the scheme is losing performance. In that way countermeasures and remedies can be implemented before the user loses trust in the AI.

14.5.4.4 Model Quality over Time

It is well known that all models are wrong, but some are useful, as stated by George Box [13]. Model quality is a measure and insight into how well a model reflects the real-life system in relation to the observation context. Here, the context is the degradation of components. Changes in the real-life system, e.g., changing component type or greasing lubricant, may require an update to the model to reflect the real-life system behavior correctly and have the appropriate forecasting capability for a PM scheme. From a practical point of view, it means that a change in the real-life system would need to be reported and recorded with a subsequent assessment of the model's adequacy. It is a pragmatic approach to implement model quality checks together with continuous performance monitoring.

14.5.4.5 Forecasting Horizons

Longer horizons in forecasting require higher-accuracy needs for the AI. Any forecast horizon should therefore be well related to the operational needs, which are commonly quite low and relieving the accuracy requirements on the AI. Thus, it is advised to choose appropriate horizons in relation to the operational requirements and consider the confidence bounds of the forecasts.

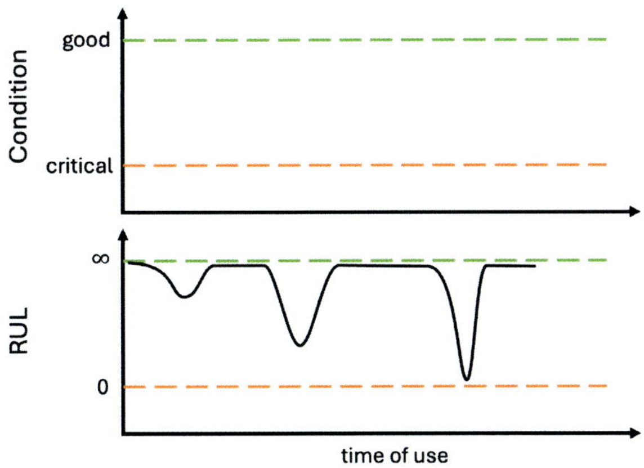

FIGURE 14.8 Conceptual sketch of the behavior of the remaining useful life (RUL) over time in relation to the condition reflected by a condition indicator.

14.5.4.6 Remaining Useful Life

An often-desired measure within PM schemes is the remaining useful life. It ideally quantifies the time of use that remains until a component fails, which is difficult to relate to from condition indicators. Usually, thresholds on condition indicators relating to safety or maintenance thresholds are used.

A confusing behavior of the remaining useful life over time is its tendency to fluctuate between low values and infinity at times. As the RUL depends on the rate of degradation or change of the condition indicator, it tends to infinity as a degradation or change in the condition settles to a locally stationary value, as can be seen in Figure 14.8. This means that no growth toward an anticipated threshold can be observed. This behavior should not be confused with the curing of the component, as it only means that the degradation has temporarily come to a stop. Interpretation of RUL over time can therefore be complicated and hard to communicate.

14.5.4.7 Iterative Implementation and Roll-Out

Introducing PM schemes and also condition-based maintenance has a transformative impact on the organization and operations. To show concrete value creation, an iterative approach to implement and roll-out within an organization is advised. It is also important to lower the adoption threshold and first involve staff with a positive attitude to change that can act as champions and ambassadors with the organization. Then targeting the most obvious and high-priority issues first, which is a result of a ranking and scoping exercise at the beginning of the work.

14.6 AI FOR TRACTION DRIVE SYSTEMS CONTROL AND AUTOMATION

Train driving control is a complex area where infrastructure, vehicles, operation, and human factors intersect, making AI applications a good enabler for optimizing the drive system of the vehicle and its control at a system level. In theory, it is also an enabler for train driving automation, but the required safety levels discourage practical applications of AI-based self-driving trains. The traction system can become even more complex when introducing several possible energy sources such as pantograph and batteries, where the system-optimal working point depends also on the energy availability, needs, and limitations of onboard energy storage systems.

The more common areas where AI techniques are used are *efficient driving strategies*, where acceleration and braking patterns of individual trains are the variables to optimize while infrastructure and/or timetable are used as constraints; and *power allocation*, where the power usage from different sources (such as catenary power, batteries, or fuel cells) is optimized for an overall energy consumption minimization.

Most authors target efficient driving strategies using either ML or Evolutionary Algorithms. In [52] the authors try to reduce the energy consumption of the run while maintaining the travel time and comfort levels; they compare two train control algorithms – expert system and reinforcement learning (RL) – and their target function for the operation is how an experienced driver would drive the train. Similarly, authors in [53] perform energy optimization for a driverless subway, in this case applying genetic algorithms (GAs).

Energy is the common proxy for efficient driving, but there are other variables that can also be targeted. In [54] costs of energy and time are determined by applying approximate dynamic programming, while in [55] safe and efficient control of heavy haul rail systems is the aim using SVMs.

When it comes to power allocation, the methods focus on bridging the complexity of drive cycle optimization analyses when different power sources are available, which are also entwined with the operational and infrastructure characteristics. Examples of this are the optimization of Hydrogen EMU's energy management using the Double Q-Learning Algorithm [56] and coming up with strategies for dynamic power allocation in hybrid storage systems using Iterative learning control [57].

All the AI applications in this section are related to minimizing the error between variables or optimizing a system. It is not totally clear why different authors prefer different methods, but they all have their modeling needs and can be grouped into three clusters: ML, Evolutionary Algorithms, Dynamic Programming.

Double Q-Learning (and its basic algorithm, Q-Learning) is a specific type of RL algorithm. RL represents the problem as intelligent agents within a dynamic environment, and agents take actions to maximize the final reward (Figure 14.9). To use RL one needs to be able to represent the problem as a Markov decision process (MDP), as the RL models include agents and dynamic decision-making. Another ML example in these cases is the use of SVMs, a supervised learning model typically used for analyzing and classifying data by maximizing the distance between a hyperplane within the dataset and the closest element from the different groups. Finally, iterative learning control is a control strategy for systems that have repetitive working modes. It is rooted in control systems design and is arguably an early example of ML.

Another cluster is EA, represented by GAs. GAs generate sets of solutions to optimization problems using stochastic combinations of the variables to optimize, in a process inspired by how natural evolution works. It's a powerful technique that needs more computational power than other optimization methods but is also able to tackle much more complex problems. The details of the implementation are not fully clear in the applied research works; it is, however, a method so widespread that application examples can be found in virtually any research area.

Dynamic programming typically simplifies decision-making by breaking the system down into consecutive decision steps, which makes it invaluable for AI applications. Here problems are typically represented as Network Models. Approximate dynamic programming further incorporates uncertainty modeling from stochastic programming into the DP framework.

All in all, the representation of vehicle operation and drive systems as agent-based or network models is a common and useful feature of many of these examples, while the specific AI technique used varies depending on the preferences or previous experience of the authors.

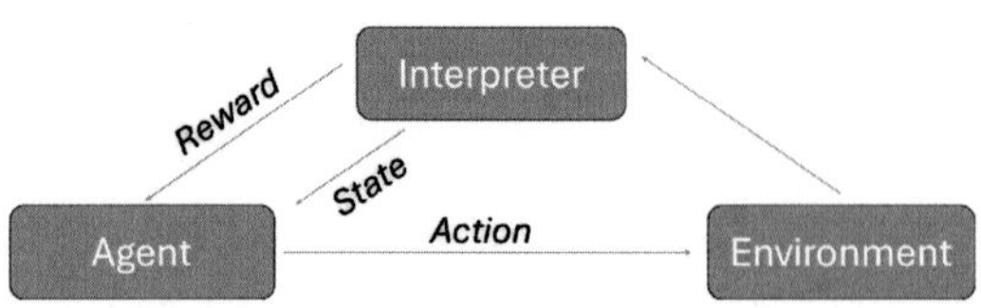

FIGURE 14.9 Reinforcement learning (RL) feedback loop representation.

14.7 AI IN THE DESIGN OF RAIL VEHICLES

For quite a long-time mathematical optimization with the help of goal functions and boundary conditions has been applied to the design of rail vehicles, like in any other application design. The more evident parameters were optimized early on, and more complex parameters and design geometries have been introduced in the optimization process, needing more complex optimization algorithms, eventually evolving into AI-based techniques. One of the most prolific areas that use AI tools is the *aerodynamic shape optimization* of the front and tail of rail vehicles. The main reason is that the design space for these complex three-dimensional geometries is extremely wide, and AI methods allow to simplify and automate the exploration of this otherwise daunting design space. Another common area is *wheel and rail profile optimization*, where AI methods can be used to avoid expert-defined objective functions in the wheel-rail contact. GAs have also been tested for the design of other components such as disc brakes (cf. Section 14.4.5).

When it comes to *aerodynamic shape optimization*, GAs are one of the most commonly used AI techniques. One application is to use evolutionary algorithms to perform an initial swipe in the variable space as a first step toward a global optimization, as in [58] where a non-dominated sorting GA (NSGA-II) is used to obtain the reference framework of the train head with an optimized global shape, and use more efficient methods for the fine-tuning of the design optimization. NSGA-II is an efficient implementation of a GA algorithm that can handle multiple objective functions separately. GAs have also been used for the optimization of slipstream effects, as in [59] where authors have merged cross-validation and GAs in a combined Kriging model, but these have also been tackled with other techniques such as ϵ-TSVR surrogate modeling [60], an implementation of SVM (see Section 14.6). Other authors [61,62] approach the minimization of the drag coefficient of a high-speed train under headwind using conventional GAs over radial basis functions networks (RBFN) surrogate models of the aerodynamic simulations to speed up the optimization process, while using an ANOVA test and Self-Organizing Maps (SOM) for analyzing the influence of each design variable.

As GAs are typically expansive in their approach, having a computationally efficient model helps speed up the optimization process. This is typically tackled by using surrogate models of the original one, similar to Kriging, for example. The best example of the use of AI in vehicle design though is [61], where three stages of the optimization methodology are performed with AI techniques: first, global optimization uses classic GAs; also, the complex aerodynamic model is made more manageable by using RBFN, a neural network ML application; and finally the analysis and visualization of the resulting designs and variables use SOM, an unsupervised ML technique that allows visualizing a high-dimensional space into a low-dimensional representation while preserving the data structure. Additionally, other examples demonstrate the use of AI as a first step for reducing the size of the design space, showcasing the breadth and flexibility of AI algorithms that can be applied in vehicle design optimization processes.

In the field of vehicle dynamics, *wheel and rail profile design optimization* has been a long-standing research topic where, as for many other optimization fields, AI has shown promising applications. Profiles are typically represented as a concatenation of curves, which creates a very broad design space. They also have geometric limits that cannot be directly inferred from the piecewise curves themselves, while also influencing comfort and safety parameters in the trainset. Classic profile design was based on vehicle dynamics expertise and iterative processes, advancing to representing the objective functions with less subjective variables to be optimized. Later, AI was incorporated with the use of Evolutionary Algorithms, like Genetic Algorithms for designing wheel profiles [63,64] and increasing the complexity even more, globally optimizing profiles that run on two different rail gauges [65]. More advanced versions of GAs like the multi-objective NSGA-II (Non-dominated Sorting Genetic Algorithm-II) have also been used for wheel profile design [66], cf. Figure 14.10, and other evolutionary algorithms are also being tested for wheel profile design in the form of Particle Swarm Optimization (PSO) [67,68], where a set of solution candidates (particles)

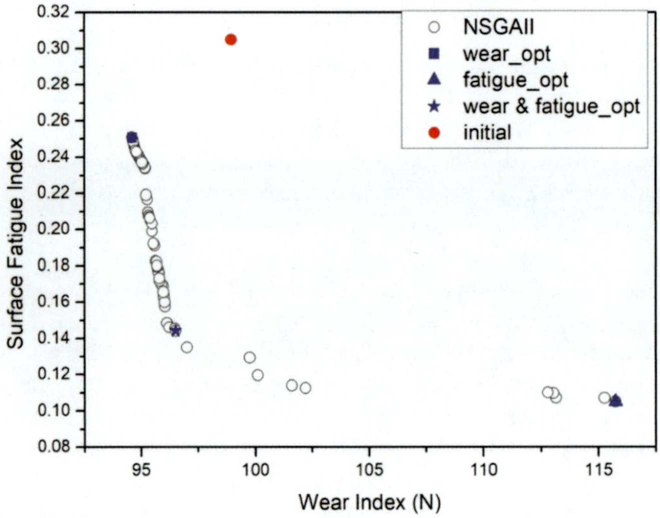

FIGURE 14.10 Multi-variable optimization Pareto front example where different local optima are displayed. (From C.-Y. Song and H.-Y. Choi, 'Multi-objective profile design optimization to minimize wear damage and surface fatigue of city train wheel', *Appl. Sci.*, vol. 12, no. 8, Art. no. 8, 2022, doi:10.3390/app12083940, CC BY 4.0.)

are moved within the design space following simple displacement criteria based on both the particle best position and the swarm's best particle position. This study also makes use of radial basis functions (RBF) to reduce the complexity of the problem to be solved by creating a surrogate model of the objective functions, as other studies presented before.

The disc brake is another very important and safety-critical component in a rail vehicle. During the design phase the disc and brake pads with the friction blocks are usually modeled with finite element models and the frictional heat developed as a critical measure is calculated for the braking process. In [69] the usage of artificial neural networks is described to optimize the friction block location on the brake pads for high-speed railway vehicles. The optimized model displays a larger hot band on the brake disc compared to the original model, leading to a more even distribution of thermal stress across the brake disc.

As a wrap-up, when it comes to rolling stock design, the uses are typically the reduction of the design space for a more manageable problem, the reduction of complexity for specific submodels, or the optimization of the desired performance attribute. The AI techniques are clustered around evolutionary methods, support vector machines, and artificial neural networks, with no clear benefit or drawback for the chosen technique.

14.8 SUMMARY

Artificial intelligence tools are a broad portfolio of mathematical algorithms and tools that act in an automated way and support railway engineers and workers in their daily tasks of designing, operating, and maintaining rail vehicles at a lower cost and with higher efficiency. The sheer number of publications on topics that apply AI techniques in rolling stock applications is vast and shows the validity of the opportunity, but it also makes it difficult for engineers to navigate that landscape and build or use their own solutions. The aim of this chapter was therefore to highlight important application areas, provide guidance on avoiding typical pitfalls, and rule out common misconceptions about AI.

AI techniques are such a general tool set, that they can be used to augment the perception, collection, and detection of data and latent environmental factors, and in that way complement human perception and intuition, and the established body of common knowledge for rolling stock systems. While AI can provide a data-driven result, the tools and algorithms are still developed by humans for a given purpose, therefore the applicability of the tool to the specific problem should be considered prior to usage. Likewise, AI tools are highly dependent on the quality and quantity of the input data, therefore it is important that the user understands the limitations of the dataset and its effect on the AI-generated insights. As the rapid development of AI tools outgrows the systematic validation of the results, the validation methodologies of AI results are not keeping up. A word of caution is that testing, validation, consistency, and gaining trust in a developed solution is essential, even though it requires substantial resources and time.

In the literature, validation is often based in comparison to other AI tools and in a rather narrow scope. Although this may indicate the efficiency and efficacy of one AI tool over several others, validation of conventional analysis methods should be considered to ensure the accuracy of the results, the explainability to the user, and that the results are supported by the common knowledge on the rolling stock system.

Control and automation aspects of rolling stock have not been the focus of this chapter, as solutions in that context are well established and adopted by the community. It is the belief of the authors that the largest impact of AI techniques for rolling stock lies in the context of condition monitoring and diagnostics, which is supported by the large number of research publications found in that area. The amount of work available on predictive maintenance is less but increasing. In the design of vehicles, AI does not seem to be used very systematically yet, besides in a few areas like aerodynamic optimization. Most suppliers of rail vehicles and services surrounding them all use AI to some extent, e.g., providing condition monitoring solutions as add-ons for the purchased rolling stock. Our opinion is that there is still a long way to go until AI can be regarded as a mature engineering technology to be applied in railway rolling stock design and maintenance. But the more it is used and exposed to real-life use cases, the better its use and benefits are understood and proven.

In summary, the successful application of AI in rolling stock is enabled by the following important aspects:

- Systematically storing data in a structured and well-documented way, including information on the data quality. It is important to note that this data storage can be decentralized, as long as data has accurate asset-, operation-, and time-stamps.
- Storing meta data on the assets, including the actions that have been undertaken on each asset or component, on what grounds an action was decided, and the prior and posterior state of the asset, all of this held digitally, or in a machine-readable format. This would reduce labeling efforts and provide insights into the true state of assets needed for the validation of AI solutions. Current asset management systems do support this, but it requires the human in the loop to provide the outcomes.
- Benchmarking conventional or classical methods with AI solutions to assess the performance and benefits to build a body of knowledge that supports the trust-building for these new approaches. For example, use publicly available data sets to evaluate specific AI solutions against each other, as there is typically no right or wrong choice of an AI solution. One of the issues with using AI tools or even conventional tools is that, prior to trying, there is no proof that it will work or not. Any solution must be tested and assessed in the application context. Benchmark studies are therefore a good source of information for evaluating how different promising solutions compare against each other and against previous conventional tools and methods.
- Start small and iterate. Success on a large scale is often hard to prove as the number of disturbing factors increases and may mask the benefits. While the benefits are there, it is simply hard to derive them with sufficient confidence. Iterating from the most pressing and solvable issue to the next allows for a better understanding of the solutions and their applicability.

Adapt the AI solution to the organization and not the organization to the solution. It facilitates the performance assessment and trust-building into the solution, and most importantly, reduces the time to get started and create a successful outcome.

REFERENCES

1. European Commission, 'AI watch: defining Artificial Intelligence : Towards an operational definition and taxonomy of artificial intelligence'. LU: Publications Office, 2020. Accessed: June 20, 2024. [Online]. Available: https://data.europa.eu/doi/10.2760/382730.
2. N. Bešinović, L. De Donato, F. Flammini, R.M. P. Foverde, Z. Lin, R. Liu, et al., 'Artificial intelligence in railway transport: Taxonomy, regulations, and applications', *IEEE Trans. Intell. Transp. Syst.*, vol. 23, no. 9, pp. 14011–14024, 2021.
3. S. A. Alstom, 'White paper: Artificial Intelligence across Alstom'. 2022. [Online]. Available: https://www.alstom.com/sites/alstom.com/files/2023/04/24/Alstom_Whitepaper_Artificial_Intelligence_across_Alstom_EN.pdf.
4. CAF SA, 'CAF digital mobility'. Accessed: June 19, 2024. [Online]. Available: https://www.cafmobility.com/en/solutions/digital-mobility/.
5. C. Bahlmann, 'Putting the A.I. on rails'. Siemens Mobility, 2024. [Online]. Available: https://www.mobility.siemens.com/global/en/company/thought-leadership-stories/putting-the-ai-on-rails.html.
6. Siemens Mobility, 'Assisted and driverless train operation'. Siemens Mobility, 2024. [Online]. Available: https://www.mobility.siemens.com/global/en/portfolio/digital-solutions-software/digital-train-solutions/assisted-and-driverless-train-operation.html.
7. S. F. Dietlmeier, R. J. Floetgen, J. Bock, and F. Urmetzer, 'IoT for rail transportation: The case of railigent', in *2022 IEEE International Conference on Big Data (Big Data)*, Osaka, Japan IEEE, 2022, pp. 3806–3813.
8. Hitachi Rail, 'What's ahead for studies into AI and collision avoidance?' Accessed: June 19, 2024. [Online]. Available: https://www.hitachirail.com/blog/2023/what-s-ahead-for-studies-into-ai-and-collision-avoidance/
9. G. Culot, G. Nassimbeni, G. Orzes, and M. Sartor, 'Behind the definition of Industry 4.0: Analysis and open questions', *Int. J. Prod. Econ.*, vol. 226, p. 107617, 2020, doi:10.1016/j.ijpe.2020.107617.
10. J. Leng, W. Sha, B. Wang, P. Zheng, C. Zhuang, Q. Liu, et al., 'Industry 5.0: Prospect and retrospect', *J. Manuf. Syst.*, vol. 65, pp. 279–295, 2022, doi:10.1016/j.jmsy.2022.09.017.
11. N. Besinovic, L. De Donato, F. Flammini, R. M. P. Foverde, Z. Lin, R. Liu, et al., 'Artificial intelligence in railway transport: Taxonomy, regulations, and applications', *IEEE Trans. Intell. Transp. Syst.*, vol. 23, no. 9, pp. 14011–14024, 2022, doi:10.1109/TITS.2021.3131637.
12. D. G. Pascual, *Artificial intelligence tools: Decision support systems in condition monitoring and diagnosis.* CRC Press, 2015.
13. G. E. Box, G. M. Jenkins, G. C. Reinsel, and G. M. Ljung, *Time series analysis: Forecasting and control.* John Wiley & Sons, 2015.
14. S. V. Stehman, 'Selecting and interpreting measures of thematic classification accuracy', *Remote Sens. Environ.*, vol. 62, no. 1, pp. 77–89, 1997.
15. D. Powers, 'Evaluation: from precision, recall and F-measure to ROC, informedness, markedness & correlation', *J. Mach. Learn. Technol.*, vol. 2, p. 37, 2011.
16. J. Opitz, 'A closer look at classification evaluation metrics and a critical reflection of common evaluation practice', *Trans. Assoc. Comput. Linguist.*, vol. 12, pp. 820–836, 2024.
17. R. Karim, W. Birk, and P.-O. Larsson-Kråik, 'Cloud-based emaintenance solutions for condition-based maintenance of wheels in heavy haul operation', in *11th International Heavy Haul Conference, (IHHA 2015)*, 21–24 June 2015, Perth, Australia. International Heavy Haul Association, Perth, 2015.
18. W. Birk, T. Dittman, R. Karim, and J. Westerberg, 'Experiences from the detection and prediction of wheel damages on railway vehicles in operation', in *Proceedings of the International Heavy Haul Association: The 13th International Heavy Haul STS Conference*, Narvik, NO, 2019.
19. Y. Ye, D. Shi, P. Krause, and M. Hecht, 'A data-driven method for estimating wheel flat length', *Veh. Syst. Dyn.*, vol. 58, pp. 1329–1347, 2020.
20. M. Z. Shaikh, Z. Ahmed, B. S. Chowdhry, E. N. Baro, T. Hussain, M. A. Uqaili, et al., 'State-of-the-art wayside condition monitoring systems for railway wheels: A comprehensive review', *IEEE Access*, vol. 11, pp. 13257–13279, 2023, doi:10.1109/ACCESS.2023.3240167.

21. J. Sresakoolchai and S. Kaewunruen, 'Wheel flat detection and severity classification using deep learning techniques', *Insight* - Non-Destr. *Test. Cond. Monit.*, vol. 63, no. 7, pp. 393–402, 2021, doi:10.1784/insi.2021.63.7.393.
22. A. Mosleh, A. Meixedo, D. Ribeiro, P. A. Montenegro, and R. Calçada, 'Condition monitoring of rolling stock supported by artificial intelligence technique', in *Digital Railway Infrastructure*, D. Ribeiro, P. A. Montenegro, A. Andersson, and M. D. Martínez-Rodrigo, Eds., in *Digital Innovations in Architecture, Engineering and Construction*. Cham: Springer Nature Switzerland, 2024, pp. 117–133, doi:10.1007/978-3-031-49589-2_6.
23. A. Mosleh, A. Meixedo, D. Ribeiro, P. Montenegro, and R. Calçada, 'Automatic clustering-based approach for train wheels condition monitoring', *Int. J. Rail Transp.*, vol. 11, no. 5, pp. 639–664, 2023, doi:10.1080/23248378.2022.2096132.
24. W. Birk and J. Westerberg, 'Novel detection and prediction tool for bearing damages on heavy haul vehicles using way-side detectors', in *Presented at the AREMA 2020 Virtual Conference & Expo, Online Conference*, Online Conference, 2020. https://www.diva-portal.org/smash/record.jsf?pid=diva2%3A1455170&dswid=1438
25. W. Hu, G. Xin, J. Wu, G. An, Y. Li, K. Feng, et al., 'Vibration-based bearing fault diagnosis of high-speed trains: A literature review', *High-Speed Railw.*, vol. 1, no. 4, pp. 219–223, 2023, doi:10.1016/j.hspr.2023.11.001.
26. X. Yan and M. Jia, 'A novel optimized SVM classification algorithm with multi-domain feature and its application to fault diagnosis of rolling bearing', *Neurocomputing*, vol. 313, pp. 47–64, 2018, doi:10.1016/j.neucom.2018.05.002.
27. A. Sharma, R. Jigyasu, L. Mathew, and S. Chatterji, 'Bearing fault diagnosis using weighted K-nearest neighbor', in *2018 2nd International Conference on Trends in Electronics and Informatics (ICOEI)*, Tirunelveli, India, May 2018, pp. 1132–1137, doi:10.1109/ICOEI.2018.8553800.
28. M. Sun, H. Wang, P. Liu, S. Huang, and P. Fan, 'A sparse stacked denoising autoencoder with optimized transfer learning applied to the fault diagnosis of rolling bearings', *Measurement*, vol. 146, pp. 305–314, 2019, doi:10.1016/j.measurement.2019.06.029.
29. J. Liu, C. Zhang, and X. Jiang, 'Imbalanced fault diagnosis of rolling bearing using improved MsR-GAN and feature enhancement-driven CapsNet', *Mech. Syst. Signal Process.*, vol. 168, p. 108664, 2022, doi:10.1016/j.ymssp.2021.108664.
30. Y. Lei, B. Yang, X. Jiang, F. Jia, N. Li, and A. K. Nandi, 'Applications of machine learning to machine fault diagnosis: A review and roadmap', *Mech. Syst. Signal Process.*, vol. 138, p. 106587, 2020, doi:10.1016/j.ymssp.2019.106587.
31. H. Wang, Z. Liu, D. Peng, and M. J. Zuo, 'Interpretable convolutional neural network with multilayer wavelet for Noise-Robust Machinery fault diagnosis', *Mech. Syst. Signal Process.*, vol. 195, p. 110314, 2023, doi:10.1016/j.ymssp.2023.110314.
32. A. Kumar and S. Harsha, 'A systematic literature review of defect detection in railways using machine vision-based inspection methods', *Int. J. Transp. Sci. Technol.*, 2024.
33. M. Aimar and A. Somà, 'Study and results of an onboard brake monitoring system for freight wagons', *Proc. Inst. Mech. Eng. Part F J. Rail Rapid Transit*, vol. 232, no. 5, pp. 1277–1294, 2018.
34. B. W. Schlake, S. Todorovic, J. R. Edwards, J. M. Hart, N. Ahuja, and C. P. L. Barkan, 'Machine vision condition monitoring of heavy-axle load railcar structural underframe components', *Proc. Inst. Mech. Eng. Part F J. Rail Rapid Transit*, vol. 224, no. 5, pp. 499–511, 2010, doi:10.1243/09544097JRRT376.
35. B. Freid, C. P. L. Barkan, N. Ahuja, J. M. Hart, S. Todorovic, and N. Kocher, 'Multispectral machine vision for improved undercarriage inspection of railroad rolling stock', in *Proceedings of the Ninth International Heavy Haul Conference*, Shanghai, CN, 2007.
36. J. M. Hart, Resendiz, E, B. Freid, Sawadisavi, S, C. P. L. Barkan, and N. Ahuja, 'Machine vision using multi-spectral imaging for undercarriage inspection of railroad equipment', in *Proceedings of the 8th World Congress on Railway Research*, Seoul, South Korea, 2008.
37. P. V. V. Kishore and Ch. R. Prasad, 'Computer vision based train rolling stock examination', *Optik*, vol. 132, pp. 427–444, 2017, doi:10.1016/j.ijleo.2016.12.060.
38. T. Schmitz, H. Fabian, N. Martin-Christopher, and C. Schindler, 'Zuverlässigkeitsanalyse einer KI-gestützten Wartungsanalge für Straßenbahnräder', in *Presented at the 4th International Railway Symposium*, Aachen, Germany, 2023.
39. L. Liu, F. Zhou, and Y. He, 'Automated status inspection of fastening bolts on freight trains using a machine vision approach', *Proc. Inst. Mech. Eng. Part F J. Rail Rapid Transit*, vol. 230, no. 7, pp. 1629–1641, 2016, doi:10.1177/0954409715619603.

40. K. Niu, Y. Dietmar, and M. Breemersch, 'Bogie health monitoring using acoustic data', in *Presented at the 4th International Railway Symposium*, Aachen, Germany, 2023.
41. J. Ning, W. Cui, C. Chong, H. Ouyang, C. Chen, and B. Zhang, 'Feature recognition of small amplitude hunting signals based on the MPE-LTSA in high-speed trains', *Measurement*, vol. 131, pp. 452–460, 2019, doi:10.1016/j.measurement.2018.08.035.
42. J. Ning, M. Fang, W. Ran, C. Chen, and Y. Li, 'Rapid multi-sensor feature fusion based on non-stationary kernel JADE for the small-amplitude hunting monitoring of high-speed trains', *Sensors*, vol. 20, no. 12, Art. no. 12, 2020, doi:10.3390/s20123457.
43. J. Guo, G. Zhang, H. Shi, and J. Zeng, 'Small-amplitude bogie hunting identification method for high-speed trains based on machine learning', *Veh. Syst. Dyn.*, vol. 62, no. 5, pp. 1–15, 2023, doi:10.1080/00423114.2023.2224906.
44. J. Sun, E. Meli, X. Song, M. Chi, W. Jiao, and Y. Jiang, 'A novel measuring system for high-speed railway vehicles hunting monitoring able to predict wheelset motion and wheel/rail contact characteristics', *Veh. Syst. Dyn.*, vol. 61, no. 6, pp. 1621–1643, 2023, doi:10.1080/00423114.2022.2086145.
45. R. Kulkarni, A. Qazizadeh, and M. Berg, 'Unsupervised rail vehicle running instability detection algorithm for passenger trains (*iVRIDA*)', *Measurement*, vol. 216, p. 112894, 2023, doi:10.1016/j.measurement.2023.112894.
46. B. Girstmair, A. Haigermoser, and P. Dietmaier, 'Advantages of using statistical models for detecting faulty components in railway bogies against using simple criteria as defined in standards', *Veh. Syst. Dyn.*, vol. 59, no. 1, pp. 56–69, 2021, doi:10.1080/00423114.2019.1662925.
47. Y. Chen, G. Niu, Y. Li, and Y. Li, 'A modified bidirectional long short-term memory neural network for rail vehicle suspension fault detection', *Veh. Syst. Dyn.*, vol. 61, no. 12, pp. 3136–3160, 2023, doi:10.1080/00423114.2022.2158879.
48. Y. Ye, P. Huang, and Y. Zhang, 'Deep learning-based fault diagnostic network of high-speed train secondary suspension systems for immunity to track irregularities and wheel wear', *Railw. Eng. Sci.*, vol. 30, no. 1, pp. 96–116, 2022, doi:10.1007/s40534-021-00252-z.
49. Y. Ye, Y. Zhang, Q. Wang, Z. Wang, Z. Teng, and H. Zhang, 'Fault diagnosis of high-speed train suspension systems using multiscale permutation entropy and linear local tangent space alignment', *Mech. Syst. Signal Process.*, vol. 138, p. 106565, 2020, doi:10.1016/j.ymssp.2019.106565.
50. D. Galar, K. Goebel, P. Sandborn, and U. Kumar, *Prognostics and remaining useful life (RUL) estimation: Predicting with confidence*. CRC Press, 2021.
51. GCU Bureau, 'General contract of use for wagons'. GCU Bureau sprl, 2024. [Online]. Available: https://gcubureau.org/.
52. J. Yin, D. Chen, and L. Li, 'Intelligent train operation algorithms for subway by expert system and reinforcement learning', *IEEE Trans. Intell. Transp. Syst.*, vol. 15, no. 6, pp. 2561–2571, 2014, doi:10.1109/TITS.2014.2320757.
53. M. Brenna, F. Foiadelli, and M. Longo, 'Application of genetic algorithms for driverless subway train energy optimization', *Int. J. Veh. Technol.*, vol. 2016, p. e8073523, 2016, doi:10.1155/2016/8073523.
54. P. Wang, A. Trivella, R. M. P. Goverde, and F. Corman, 'Train trajectory optimization for improved on-time arrival under parametric uncertainty', *Transp. Res. Part C Emerg. Technol.*, vol. 119, p. 102680, 2020, doi:10.1016/j.trc.2020.102680.
55. G. Hua, L. Zhu, J. Wu, C. Shen, L. Zhou, and Q. Lin, 'Blockchain-based federated learning for intelligent control in heavy haul railway', *IEEE Access*, vol. 8, pp. 176830–176839, 2020, doi:10.1109/ACCESS.2020.3021253.
56. Q. Li, X. Meng, F. Gao, G. Zhang, and W. Chen, 'Approximate cost-optimal energy management of hydrogen electric multiple unit trains using double Q-learning algorithm', *IEEE Trans. Ind. Electron.*, vol. 69, no. 9, pp. 9099–9110, 2022, doi:10.1109/TIE.2021.3113021.
57. X. Wang, Y. Luo, B. Qin, and L. Guo, 'Power dynamic allocation strategy for urban rail hybrid energy storage system based on iterative learning control', *Energy*, vol. 245, p. 123263, 2022, doi:10.1016/j.energy.2022.123263.
58. S. Wang, R. Wang, Y. Xia, Z. Sun, L. You, and J. Zhang, 'Multi-objective aerodynamic optimization of high-speed train heads based on the PDE parametric modeling', *Struct. Multidiscip. Optim.*, vol. 64, no. 3, pp. 1285–1304, 2021, doi:10.1007/s00158-021-02916-0.
59. Y. Shuanbao, G. Dilong, S. Zhenxu, Y. Guowei, and C. Dawei, 'Optimization design for aerodynamic elements of high speed trains', *Comput. Fluids*, vol. 95, pp. 56–73, 2014, doi:10.1016/j.compfluid.2014.02.018.

60. Z. Sun, S. Yao, and G. Yang, 'Research on aerodynamic optimization of high-speed train's slipstream', *Eng. Appl. Comput. Fluid Mech.*, vol. 14, no. 1, pp. 1106–1127, 2020, doi:10.1080/19942060.2020.1810128.
61. J. Muñoz-Paniagua and J. García, 'Aerodynamic drag optimization of a high-speed train', *J. Wind Eng. Ind. Aerodyn.*, vol. 204, p. 104215, 2020, doi:10.1016/j.jweia.2020.104215.
62. J. Muñoz Paniagua, 'Aerodynamic optimization of the nose shape of a high-speed train', PhD Thesis, Universidad Politécnica de Madrid, 2014. doi:10.20868/UPM.thesis.23321.
63. I. Persson and S. D. Iwnicki, 'Optimisation of railway wheel profiles using a genetic algorithm', *Veh. Syst. Dyn.*, vol. 41, pp. 517–526, 2004.
64. M. Novales, A. Orro, and M. R. Bugarín, 'Use of a genetic algorithm to optimize wheel profile geometry', *Proc. Inst. Mech. Eng. Part F J. Rail Rapid Transit*, vol. 221, no. 4, pp. 467–476, 2007, doi:10.1243/09544097JRRT150.
65. J. Santamaria, J. Herreros, E. G. Vadillo, and N. Correa, 'Design of an optimised wheel profile for rail vehicles operating on two-track gauges', *Veh. Syst. Dyn.*, vol. 51, no. 1, pp. 54–73, 2013, doi:10.1080/00423114.2012.711478.
66. C.-Y. Song and H.-Y. Choi, 'Multi-objective profile design optimization to minimize wear damage and surface fatigue of city train wheel', *Appl. Sci.*, vol. 12, no. 8, Art. no. 8, 2022, doi:10.3390/app12083940.
67. F. Lin, X. Dong, Y. Wang, and C. Ni, 'Multiobjective optimization of CRH3 EMU wheel profile', *Adv. Mech. Eng.*, vol. 7, no. 1, p. 284043, 2015, doi:10.1155/2014/284043.
68. Y. Qi, H. Dai, P. Wu, F. Gan, and Y. Ye, 'RSFT-RBF-PSO: A railway wheel profile optimisation procedure and its application to a metro vehicle', *Veh. Syst. Dyn.*, vol. 60, no. 10, pp. 3398–3418, 2022, doi:10.1080/00423114.2021.1955135.
69. H. R. Hong and C.-W. Ha, 'Optimizing friction block location on brake pads for high-speed railway vehicles using artificial neural networks', *Appl. Sci.*, vol. 13, no. 17, Art. no. 17, 2023, doi:10.3390/app13179634.

15 AI for Rail Infrastructure and Maintenance

Stefano Bruni, Marco Carnevale, Gabriele Cazzulani, Egidio Di Gialleonardo, and Ivano La Paglia

15.1 INTRODUCTION

Railway infrastructure is a complex asset that must be maintained to appropriate standards to ensure the safe and efficient operation of railway transportation. Therefore, infrastructure maintenance is a prominent cost issue for railway administrations, and strong efforts are made to improve maintenance efficiency by switching from traditional predetermined/preventive maintenance to more efficient condition-based and predictive maintenance action [1].

However, there are significant issues and challenges with devising and implementing modern maintenance strategies for components of railway infrastructure. First, a large variety of faults may occur in the infrastructure, each of which is in need of being addressed by specific maintenance actions. The fault types most relevant to this chapter are:

- track geometric irregularities;
- railhead defects (squats, corrugation);
- damage to railway bridges and viaducts;
- defects and damage in components of the overhead contact line (OCL).

Another big challenge is the extension of modern railway networks, which may amount to several thousand kilometers of railway lines. This means a very large effort is required to monitor the condition of the entire network, to manage the enormous amount of data gathered, and to extract valuable information about infrastructure conditions. Finally, assuming accurate information on the infrastructure condition is available, efficient strategies need to be defined to optimize the allocation of limited resources, thus meeting the required maintenance needs while keeping costs within acceptable limits.

Artificial intelligence (AI) represents a key instrument for addressing all the above-mentioned challenges since it allows the efficient, unmanned processing of large datasets: a conceptual scheme of AI-based monitoring and maintenance of the infrastructure is shown in Figure 15.1: data from multiple sources including sensor sets fitted on in-service vehicles [2] and track recording vehicles (TRVs) [3] are combined in a single database of track condition together with information from traditional visual inspection of the track and OCL or from new smart inspection methods, e.g., drones and artificial vision. The large amount of data collected from multiple sources is processed using AI techniques to identify features relevant to the state of track geometry, enhancing fault detection robustness compared to traditional methods and even predicting the evolution of the monitored asset state over time. AI-based techniques can also be used by monitoring systems installed on vehicles from the commercial fleet, in which case, the amount of information transferred to the track condition database can be reduced drastically. The information resulting from this process is synthesized into maintenance alerts, which are used to define maintenance decisions in the form of a list of prioritized maintenance interventions [4]. This last step can be done by a human operator but, more and more frequently, is supported by an AI-based decision support system (DSS).

 DOI: 10.1201/9781003492146-15

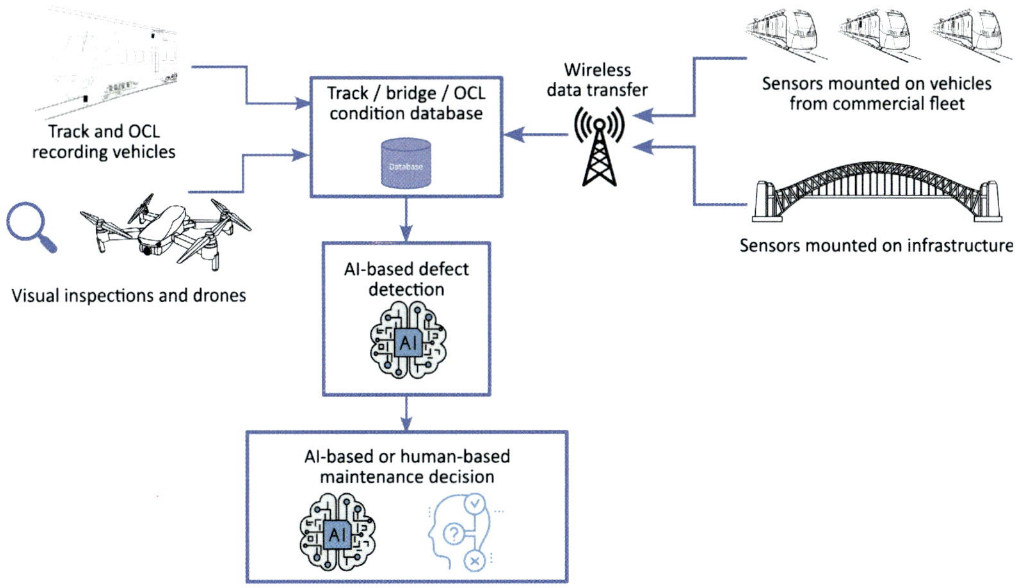

FIGURE 15.1 Conceptual scheme of AI-based monitoring and maintenance of the railway infrastructure.

15.2 AI-BASED MONITORING OF TRACK GEOMETRY

15.2.1 Track Geometry and Current Maintenance Strategy

At the infrastructure construction stage, the design geometry should be achieved with tight tolerances. However, track characteristics tend to deviate from the design values with the passage of railway vehicles. This is nowadays even more true considering the trend to increase vehicle speed and axle load, driven by the need for more sustainable and cost-effective means of transport. To determine and monitor the status of the infrastructure, a number of measurable parameters have been defined by railways standards (EN 13848-1 [5]), referred to as *track geometry parameters*: track gauge, longitudinal level (vertical alignment), cross-level, alignment, and twist (see Figure 15.2).

These parameters strongly affect the dynamic response of the vehicle, ultimately having a significant influence on running safety and ride comfort. Therefore, they can be regarded as indicators

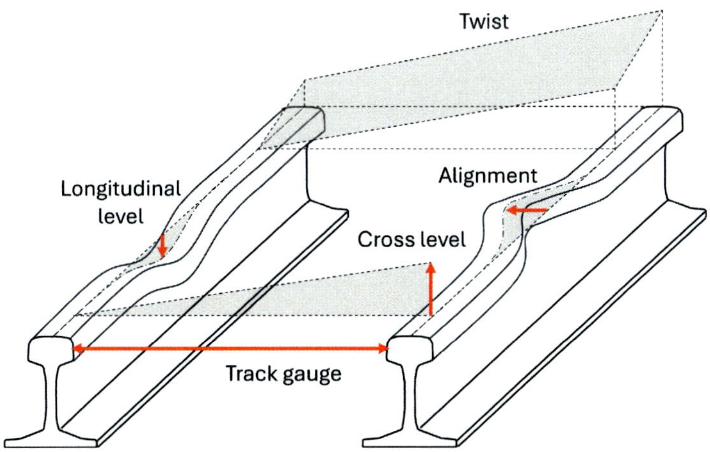

FIGURE 15.2 Track geometry defects.

of the quality of the track, facilitating the monitoring of track geometry, which is of fundamental importance for infrastructure managers to guarantee the highest level of safety. To this end, TRVs are usually employed. They are typically equipped with optical and inertial sensors to measure the track geometry and the corresponding vehicle dynamic responses. This maintenance strategy helps keep the safety of the network under control, scheduling the necessary maintenance operations to correct defects in the track geometry in a timely fashion. Ideally, frequent inspection of the entire network should be performed, but this is often not possible due to the high cost and limited availability of TRVs. Additionally, the amount of data gathered by the TRVs is huge, and a large portion of it remains unexploited. Therefore, new approaches for track geometry monitoring are being developed as presented below.

15.2.2 Modern Trends

In the past decade, the attention of both industry and academia has shifted to track monitoring through in-service vehicles, proving the suitability of this approach and taking advantage of the higher frequency of train passages along the line, progressively moving toward the possibility of implementing continuous monitoring and predictive maintenance strategies. In this respect, inertial sensors are the ones commonly installed on in-service vehicles [6] due to their versatility, robustness, and relatively low cost; applications including visual, magnetic, fiber-optic, and temperature sensors and strain gauges can be found as well [7,8]. In all the considered cases, the possibility of relying on sensor fusion techniques is also explored.

In this regard, different methodologies for track condition monitoring have been developed, which can be grouped into model-based and signal-based techniques [9]. Both techniques can address different tasks depending on the degree of detail needed and on the acknowledged effort and complexity. At first, the existence of defects can be identified by well-distinguished signatures in the measured signals. At a finer level of detail, the estimation of the defect amplitude can be targeted by adopting synthetic indicators of that/each specific defect. Finally, even more sophisticated techniques could be designed to reconstruct the track geometry profile, typically solving an inverse problem.

Given the significant amount of data stored, in addition to the well-known statistical methods, big data and machine learning (ML) techniques are nowadays often used. It can be here anticipated that the ones commonly adopted rely on artificial neural networks (ANNs) and convolutional neural networks (CNNs), genetic algorithms (GAs), Bayesian models, support vector machines (SVM), and random forest regression (RFR) [2,10].

15.2.3 AI for Track Geometry Estimation

Out of the track geometry defects introduced in Section 15.2.1, great attention has been paid to the longitudinal level, as it is the one generally determining the need for track maintenance due to its faster growth rate compared to other defects. Given the geometrical coupling between wheel and rail profiles, the vertical movement of the vehicle follows the vertical position of the rail. Thus, the sensors measuring the vertical dynamics of the vehicle can be effectively employed to estimate longitudinal-level defects. Depending on the wavelength range of interest (ranging from short 3–25 m wavelength up to 70–150 m long ones), sensors can be installed on the in-service vehicle's axle box, bogie frame, and/or carbody.

In [11], the track longitudinal level is predicted by adopting carbody accelerations from an in-service vehicle using an extended auto-encoder. The proposed neural network (NN) relies on the combined use of an encoder, a decoder, and an estimator. Specifically, at the training phase, the encoder is used to compress the information of the track longitudinal level into a latent representation, later reconstructed by the decoder. The NN estimator uses the carbody measurements to estimate this latent representation learned by the encoder. The methodology can accurately estimate the longitudinal level for wavelengths in the 5–200 m range, while the estimation of components with shorter wavelengths suffers from the filtering action of the vehicle suspensions.

An NN model, based on the combined adoption of attention mechanism, CNN, and gated recurrent unit, is adopted in [12] to estimate the longitudinal level along the Chinese high-speed railway network. The model is trained using real data from an inspection vehicle, in terms of carbody vertical and lateral acceleration, together with the vehicle speed. The comparison between measured and reconstructed data proves satisfactory, except for wavelengths shorter than 3 m, which still suffer from the filtering effect caused by vehicle suspensions.

A CNN is proposed in [13] to separate longitudinal-level defects from localized defects at the rail joints. The NN is trained considering vertical carbody acceleration measured along a main line, processed with the continuous wavelet transform (CWT) to generate frequency-milestone images. The CNN allows automatically classifying the different types of defects and their degradation rate.

Other track defect components of interest are alignment and cross level (see Figure 15.2). Compared to longitudinal-level ones, the estimation of these defects generally requires a larger set of data coming from different sensors, given the complex nonlinear relationship between rail deviations and vehicle movements in the lateral direction. In this regard, a data fusion process between lateral and vertical accelerations and pitch, roll, and yaw rate gyroscopes is often adopted. The possibility of relying on carbody vibrations to estimate alignment and cross-level defects (in addition to the longitudinal-level ones) has been investigated [14]. The proposed method relies on an SVM algorithm to classify track defects. At first, the methodology is developed based on simulation data: a baseline of defects with increasing amplitude is defined, together with the corresponding vertical and lateral accelerations and carbody roll rate. Then, the methodology is also tested considering field data from a regional train. Satisfactory results are achieved, although the possibility to further increase the effectiveness of the methodology by increasing the amount of training data is recognized.

Different classification algorithms, namely decision tree (DT), linear SVM, and Gaussian SVM, are adopted and compared in [15], to monitor alignment and cross-level irregularities from lateral and roll bogie acceleration measurement. The models are first trained with simulation data, allowing the exploration of a wide range of defect amplitudes, and later validated against data recorded along a high-speed line. SVM algorithms are proven to be more accurate than DT ones in classifying defects.

15.2.4 Challenges and Future Developments

Different AI-based strategies for track geometry estimation have been developed, generally relying on data-driven approaches based on inertial sensors installed on in-service vehicles. Conversely, very few examples of monitoring systems using video and optical systems have been observed [16] (TRVs representing a notable exception). The reason behind this is twofold: on the one hand, these latter sensors are more expensive and not particularly suited for a harsh environment like a rail vehicle; on the other, they provide a huge volume of data when recording along an entire railway line, which can be difficult to manage. This is even more true if we consider that railway lines are kept (for obvious reasons) in a healthy state by infrastructure managers, thus resulting in an unbalanced dataset. In this context, the continuous development and enhancement of big data and ML techniques can be regarded as promising for the adoption of vision and optical devices for track geometry monitoring purposes, which are expected to become more and more relevant in the years to come.

15.3 AI-BASED MONITORING OF SHORT WAVELENGTH RAIL DEFECTS

15.3.1 Classification of Short-Wavelength Defects

Before addressing the monitoring of short-wavelength rail defects, it is important to define the relevant wavelength range of these defects. For instance, EN 13848-1 [5] specifies four different wavelength ranges for track geometry. The range with the shortest wavelengths is labeled D0 and covers defects with wavelengths between 1 and 5 m. Hereinafter, the algorithms for monitoring defects with wavelengths shorter than this range will be analyzed, within the wavelength range commonly known as "rail roughness".

A rough classification of these rail defects can be given into two clusters: distributed defects and isolated defects.

Distributed defects occur over a larger area of the rail and are generally quasi-periodic in nature, creating a consistent pattern along the rail track. The main example of a distributed defect is rail corrugation: it is a common distributed defect characterized by quasi-periodic waves on the rail surface with wavelengths ranging from a few millimetres to several centimetres. A comprehensive outlook on the characteristics of rail corrugation, the root causes and the mitigation actions are presented by Grassie in [17].

Despite their small amplitude, these defects can significantly impact the performance of rail vehicles, leading to increased noise, vibrations, and accelerated wear of both the rails and the rolling stock.

On the contrary, isolated defects are localized irregularities occurring at specific points along the rails. These defects can be equally, if not more, critical in terms of their impact on rail performance. Among the different examples, the following can be mentioned:

a. Rail squats [18] are localized depressions of the rail surface, typically found at the rail head.
b. Weld defects are associated with the usage of continuously welded rails, where two rail sections are joined together using different technologies. From a macroscopic perspective, the weld guarantees a continuous rail; however, at a finer observational scale, a geometrical irregularity along the wheel–rail interface is typically found.
c. Insulated rail joints (IRJs) are used to electrically isolate sections of the rail network for signaling purposes. The joint is an assembly of two rail pieces and an insulating nylon shim, connected to each other using glue, fishplates, and fishbolts. These joints are more prone to damage than standard rails and need to be properly monitored and maintained.

15.3.2 Standard Monitoring and Inspection Techniques

Monitoring of short wavelength rail defects is a critical aspect of maintaining the safety and efficiency of rail infrastructure. Standard monitoring and inspection techniques include:

- visual inspections to detect rail surface defects (squats, weld defects, and IRJ defects);
- portable rail measuring trolleys for the measurement of short wavelength rail irregularities;
- TRVs equipped with systems for the optical measurement of the rail surface profile; these systems record the railhead geometry with cameras and lasers using the versine transfer function [5].

The above monitoring methods provide reliable and accurate results, but they have some drawbacks. Manual inspections are labor-intensive and may miss fine details. Measurements using trolleys are limited by the measurement speed, which is typically comparable to the walking speed of the operator (3–4 km/h), and measurements using optical systems on track diagnostic cars are limited by the availability of the vehicle itself, which is typically employed on many different railway lines. All that said, it is clear that standard monitoring techniques are not suitable for a continuous monitoring approach, which is important especially for short-wavelength defects.

In fact, the large dynamic loads associated with these defects may cause their rapid evolution and propagation, necessitating frequent and thorough inspection.

15.3.3 AI for Monitoring Short-Wavelength Defects

In recent years, intense research has been devoted to sensor-based detection methods, owing to the recent progress in sensor technology that allows for cost reductions, making them attractive for large-scale usage, such as the application on commercial trains.

In this regard, two main categories of sensors may be considered. The first option relies on the so-called ABA (axlebox accelerometers) systems, which are used extensively for monitoring short-wavelength defects, based on the idea that, for high-frequency phenomena, the sensor should be located as close as possible to the wheel–rail interface [19].

Classical signal processing techniques are employed, showing good correlation results between the root mean square (RMS) of the axlebox acceleration and the defect amplitude [20], but, at the same time, the dependence on the vehicle speed and track substructure is found. Recent advances [21,22] propose a simplified frequency domain approach that may take into account different factors: the vehicle speed and the coupled dynamics of the track and the wheelset. Promising results are obtained in a curved section where corrugation is found on the low rail, even if further improvements are needed to provide a better estimate of the corrugation on the high rail. A similar model-based approach is proposed by Pieringer and Kropp [23], permitting to separate wheel and rail roughness by cycle averaging of the contact force. However, the effectiveness of the algorithm is assessed only by means of numerical simulations. Approaches based on CWT and experimental mode decomposition (EMD) are proposed in [24] to detect frequency contents for corrugation defects and squats.

As far as squats are concerned, a procedure based on the use of scale-averaged wavelet power is proposed for their automatic detection [25]. The results are found to be sensitive to thresholds that were empirically tuned.

Monitoring of rail welds and IRJs using ABA is less frequently found in the literature. Preliminary analyses prove their applicability for estimating weld geometry [26] and for monitoring defects in IRJs [27]. In both cases, wavelet analysis is applied. An index associated with the intrinsic mode functions computed by the EMD of ABA on rail welds is proposed to guide track maintenance in [28]. However, the evolution over time of this index is not assessed.

From the perspective of sensor technology, the alternative relies on vision-based systems.

The feasibility of a vision-based method for rail corrugation detection is proved for the first time in [29] based on texture analysis and Gabor filtering of digital images of the rail surface.

Gazafrudi et al. [30] present a computer-vision-based measurement system for corrugation inspection. Their approach is based on a laser triangulation method avoiding chord-based methods and the consequent need of using the versine transfer function. Experimental results in the laboratory demonstrate the effectiveness of the proposed method.

As far as isolated defects are concerned, Li and Ren [31] developed a real-time visual inspection system for discrete rail surface defects. After the extraction of the rail from the acquired image, the contrast enhancement enables defect localization.

Yang et al. [32] propose a multilevel, end-to-end approach for detecting rail-isolated defects, but limitations are found in terms of the capability of classifying the different defects and the impact of installation errors and camera vibrations on the performance of the extraction algorithm.

In summary, systems based on axlebox accelerations and vision sensors have been validated in applications devoted to the detection of short-wavelength defects. However, at the present stage, they have not yet been adopted for large-scale use by rail infrastructure owners, mainly because of inadequate quantitative detection accuracy.

Many of the above-described methods rely on thresholds and algorithm parameters that have been manually tuned, which is unsuitable for large-scale deployment. In this regard, artificial intelligence enables the possibility for the various methods to be self-adaptive and overcome the limitations inherent to traditional monitoring approaches. Additionally, AI is able to deal with big data and, by its nature, can handle information in different forms. In particular, AI techniques enable the combined use of ABA and vision measurements. This helps to overcome the limitations of the two systems when considered individually and improves the accuracy of fault detection and classification, supporting maintenance management.

Some examples of AI techniques applied to the problem under analysis have already been presented in recent years; however, there is still work to be done to increase the technology readiness level and make it available to railway infrastructure owners to shift the maintenance paradigm from corrective to predictive.

Regarding corrugation monitoring, ABA systems are successfully employed by Hassanieh [33] in combination with an ML model, specifically a tuned RFR. This model is capable of reproducing the moving RMS of rail corrugation within a 100–300 mm bandwidth, particularly in highly corrugated sections.

Similarly, a data-driven, convolution-based regression model, named RCNet, is proposed by Xie [34]. In a specific scenario, the trained RCNet demonstrates the ability not only to accurately locate rail corrugation but also to identify the dominant wavelength and predict its roughness level. Another scheme based on a CNN is studied and tested using ABA on a scaled vehicle running on a 1/10-scale track [35].

Another application of CNNs, combined with a Kriging surrogate model and particle swarm optimization, developed by Xie [36], integrates the use of ABA with numerical simulations to identify the state of rail corrugation, classify its wavelength, and obtain quantitative information about its depth.

Several attempts have been made to apply AI techniques for monitoring and diagnosing discrete rail defects, primarily using image data. Two studies [37,38] demonstrate the feasibility of this approach, though they do not address defect classification.

In contrast, Faghih-Roohi et al. in [39] present an approach using a CNN capable of distinguishing between welds, joints, and squats. Additionally, the approach provides a rough classification of squats into three categories: small, medium, and large.

A successful attempt using data from a laser profilometer, processed with a supervised ML approach, is reported in [40]. However, the robustness of the detection methodology against external disturbances, such as vehicle movement when the profilometer is mounted on board, still needs to be investigated.

15.4 AI-BASED MONITORING OF RAILWAY BRIDGES

15.4.1 Introduction and Context

Bridges and viaducts represent a critical asset in railway networks. The progressive aging of the infrastructure and the need to prioritize intervention have pushed asset managers toward the use of technology as an additional tool with respect to the classical visual inspection performed by human operators. This involves both structural health monitoring (SHM), based on measurements from a number of sensors installed on the bridge, and automated inspection, based on the acquisition of images over time to identify possible defects and progressive damage.

In the past, most of the examples of bridge monitoring were related to specific case studies, allowing for a direct analysis of the dataset with strong involvement of human intelligence.

In recent years, mainly driven by some catastrophic events, many infrastructure managers have focused their attention on large-scale monitoring programs, involving the installation of hundreds of SHM systems in their networks. This also changes the paradigm from the data analysis point of view, since the large amount of information, together with the presence of multiple monitoring systems on bridges (or spans of the same bridge) having the same structural archetype, opens the door to automatic data-driven approaches instead of traditional model-based ones.

The main challenges related to bridge SHM can be summarized as follows:

- Large datasets, especially when dynamic monitoring is considered
- Strongly unbalanced datasets, where very few measurements corresponding to severe real damage scenarios have been reported in the literature
- Large environmental and operational effects, compared to measurement variations due to damages
- Correlation between damage, which is often a local phenomenon, and measurements, which are global indicators not necessarily related to damage in a direct way, in contexts where the direct measurement of the damage is not feasible

15.4.2 AI for the Removal of Environmental and Operational Effects

Data measured from a monitoring system are affected by exogenous effects. Generally speaking, this involves both environmental variables (mainly temperature and solar radiation but also wind in the case of slender structures) and operational (traffic) effects, although when considering railway bridges the second aspect can be easily addressed since the traffic is generally known. Many studies demonstrate a strong dependency of both static and modal parameters from environmental conditions [41,42], causing variations that are usually greater than those caused by damage. For this reason, there is a need to remove these effects or, in other words, to identify some damage-sensitive features which are not affected by the environment.

Traditional approaches are based on linear regression or principal component analysis [41]. A training period, where the structure is assumed to be healthy, is considered to identify the baseline. After that, residuals can be computed through a regression model, to obtain an "environment-free" parameter that can be used as a damage index. However, a complete depuration of ambient effects is difficult, since, for example, a perfect knowledge of temperature distribution along the structure is not available and the relationship with the parameters of interest may be nonlinear. In this framework, many approaches based on artificial intelligence are proposed. An SVM technique is applied in [43] to formulate the relationship between temperature and modal parameters, outperforming linear regression. Reference [44] proposes an unsupervised learning strategy based on multiple steps, which address the problem of missing data in the dataset. The strategy is tested on data coming from both steel and concrete bridges. Sarmadi [45] proposes a review and comparison of different ML techniques and an automated method for hyperparameter selection for parametric unsupervised methods. The different methods are compared for two benchmark datasets belonging to two different bridges, experiencing strong and weak environment correlation, respectively. ANNs are also widely considered an efficient way to decouple anomalies due to damage from fluctuations due to normal bridge behavior under varying external conditions [46].

15.4.3 AI for Damage Identification and Classification in Bridges

Once a reliable set of damage-sensitive (and environment-insensitive) features is established, different algorithms can be implemented to identify possible damages on bridges. Unlike other sectors in which SHM is well-established as part of the industrial best practice, in the case of bridges, there is almost no documentation in the literature of monitored bridges subjected to real damage scenarios. Moreover, when training a new monitoring system, only data corresponding to the current bridge condition, which is necessarily considered as the baseline, are available for training any algorithm. An exception is represented by a few cases, involving bridges at the end of their service life, on which some damage scenarios were deliberately created to support the development and assessment of monitoring systems. This is the case, for example, of the Z24 bridge in Switzerland [47]. However, these scenarios refer only to that particular structure, and the damages are artificially induced and do not necessarily represent what could happen in the lifecycle of a similar structure.

This means that, if relying only on data-driven methods, the approach for data analysis and anomaly detection is mainly unsupervised, with the aim of identifying the normal behavior of the structure and then tracking the occurrence of anomalies with respect to the baseline condition. In this approach, what is commonly addressed in the literature as "damage identification" or "damage detection" corresponds to detecting an anomaly and not necessarily determining the kind of damage that occurred. The work in [48] provides an extensive review of different methods, distinguishing between global techniques (referring to SHM systems aimed at characterizing the global behavior of the structure, for example, through modal parameters and typically poorly sensitive to local damage) and local techniques (referring to sensors placed locally to detect a specific damage scenario on a critical component but unable to detect the behavior of the structure as a whole). The paper also addresses specific issues and research trends related to different structural archetypes. Reference [46] provides a review of ML techniques employed for civil engineering, including bridges, classifying them based on both the algorithm type and field of application.

Focusing instead on damage classification, a supervised approach, based on the presence of labeled sets of data for normal and damaged conditions, can be implemented only by introducing a numerical model of the structure. Once calibrated, the model can be used to simulate different damage scenarios and understand the corresponding readings of the sensors or damage-sensitive features. This approach is addressed in the literature, although it is less common than the previous one. AI is employed to integrate data coming from the real structure with simulated scenarios. In [49] the authors propose a physics-guided ML approach, where a physics-based NN is integrated with a finite element model to improve damage localization and evaluation of damage severity. In [50] numerical data with and without damage are generated using a probabilistic approach, to take into account model uncertainties. In this last case, data coming from the artificial damage on the Z24 bridge [47] are used, together with simulated data, as labeled inputs to train ML algorithms for damage classification.

15.4.4 AI for Automated Inspection of Bridges

Despite the development of new data analysis methods for SHM, visual inspections of bridges still cannot be completely replaced. However, the development of AI opens the door to new visual inspection techniques that could be (at least partially) based on automatic systems for image collection combined with intelligent systems for the analysis and comparison of large image sets.

The current standard for the safety assessment of bridges is based on visual inspections performed periodically at the bridge by human operators. As an example, according to Italian guidelines, inspections are performed with a given periodicity and level of detail depending on the risk class of the bridge, but similar approaches are implemented in other countries. Given the high number of bridges in the network, these inspections require an intense manpower effort and, in some cases, the use of costly vehicles such as those equipped with the so-called by-bridge platforms. Moreover, inspections must be arranged to minimize their impact on railway service and, in the case of bridges crossing the road infrastructure, also avoid interference with road traffic, which would cause additional costs. Moreover, there is always a certain margin of error or uncertainty in the inspection results, due to the individuals performing it. For this reason, the potential of AI is fundamental to bridging two gaps:

1. Making the gathering of inspection information more efficient and (at least partially) independent from the human factor
2. Standardizing the inspection procedure and the evaluation and classification of defect severity, trying to make objective something that is currently largely subjective

The development, in recent years, of unmanned aerial vehicles has opened the possibility of retrieving such information, although some limitations are still present and related to both technical issues and flight regulation [51]. The use of aerial vehicles allows the relatively easy collection of images of places that are usually inaccessible, thus allowing a complete 3D reconstruction of the structure and the tracking of its evolution over time.

One of the most addressed aspects of automated inspection is the recognition of crack patterns, for both concrete [52] and steel [53] structures. The relevance of this specific aspect is due to their criticality toward the integrity of the structure and to the fact that checking manually crack growth on large structures is very demanding, while automated systems could provide a significant advantage with limited effort. Different AI techniques are employed for damage analysis, from simple edge detectors to CNNs. Some methods also try to extend the approach to a wider set of defects [54] through a hierarchical classification approach. The limitation of this approach is mainly represented by the under-representation of some defects in the labeled dataset for training, but this gap can be progressively reduced while the dataset is enriched with data coming from new inspections.

15.4.5 Future Perspectives

It is doubtless that SHM of railway bridges has made significant progress in recent years thanks to the development of new sensing, data management, and data processing techniques. However, its implementation on a large scale is still quite limited, although it is expected to increase in the near future.

Presently there are critical aspects limiting the widespread use of such technologies for extensive monitoring of the infrastructure, in both railway and road networks. Some of these aspects are economic, mainly the cost of installation and maintenance of the monitoring systems over time. Other aspects are of a legal nature, since there is no clear regulation about the aims and scope of the monitoring systems, thus leaving to the infrastructure managers the decision about if and how to use them.

Finally, there are aspects related to the use of the huge amount of information coming from SHM systems, which is often not easy to use for those in charge of bridge inspection and maintenance.

One of the main challenges for the development of SHM systems at the industrial level on a large scale is to make clear to the infrastructure managers what is their potential (and limitations) and how information coming from SHM could be integrated systematically into the management system of the infrastructure, providing an added value to inspections and maintenance programs.

15.5 AI FOR MONITORING OF THE OVERHEAD LINE AND PANTOGRAPH–CATENARY DYNAMIC INTERACTION

15.5.1 The Pantograph/Catenary System as a Whole

Activities for the monitoring of the OCL include geometrical and structural monitoring of constituent elements like contact wire, stagger arms, droppers, suspension, and insulators, as well as the assessment of the dynamic interaction with the pantograph. The force in the sliding contact between the strips and the wire is the main quantity commonly considered for assessing the status of the infrastructure: a large variability in the contact force, leading to the occurrence of anomalous electrical arcs, is commonly used as an indirect way to detect the presence of anomalies either in the OCL or in the pantograph.

15.5.2 Consolidated Approaches in the Railway Industry

The railway industry has been dealing with the monitoring of the pantograph/catenary system mostly by relying on traditional approaches. Pantograph monitoring is normally carried out through visual inspection by operators in a railway depot: a common procedure consists of periodically replacing the contact strips as a form of preventive maintenance, but this procedure is rather uneconomical since the strips are often replaced before the end of their useful life.

Dedicated diagnostic trains, equipped with instrumented pantographs for contact force measurement and phototubes for arc detection, periodically inspect the contact line. These trains are also equipped with an optical system for measuring the geometrical parameters of the overhead equipment such as contact wire height, gradient, thickness, and stagger. In some countries, the catenary supports are also periodically captured by cameras mounted on inspection vehicles, but fault detection still mostly relies on human interpretation of the images taken. In the presence of an anomaly, professional workers usually need to inspect the catenary and repair the abnormal parts if the defect is confirmed.

The monitoring of the overhead line from in-service trains would be desirable to reduce the costs involved and maximize the availability of the lines, but the above-mentioned quantities are hardly measurable on commercial trains, requiring complex and invasive measurement systems, not compatible with the installation costs and normal maintenance operations of vehicles in service. Therefore, the present tendency is to develop new fault detection methods, making use of cheaper and more robust sensors. In this regard, AI techniques play a relevant role, as outlined below.

15.5.3 Compact Measurement Systems for Monitoring Pantograph–Catenary Interaction

Accelerometers placed on the pantograph head are being considered as a feasible and simplified option to monitor the dynamic behavior of the system in commercial operations and to obtain, through data processing, a signature of failures in both the pantograph and catenary system. This measurement system is non-invasive and does not interfere with contact strip replacement. If repeated anomalies in the measurements are detected at specific positions from several trains of the commercial fleet, the defectiveness can be attributed to the contact line. On the contrary, a global degradation of the measurements of a single pantograph can be ascribed to a deterioration of the pantograph itself.

15.5.4 Artificial Intelligence Exploitation

In recent years, data-driven methods have improved, thanks to the introduction of ML techniques and then deep learning (DL). The number of relevant publications has expanded significantly after the year 2017 [55]. The exploitation of AI in OCL monitoring is generally more recent than the application of these techniques to other railway infrastructure components and can be considered still at a research stage.

The use of AI for monitoring the OCL pursues the same objectives described for consolidated monitoring approaches: identification of structural and geometrical defects, assessment of dynamic interaction, and detection of electrical arcs. The techniques introduced exploit both contact and non-contact measurements. Most of them rely on NNs [55].

Non-contact methods mainly use image-based methods and CNNs (more frequently deep convolutional neural networks, DCNNs) aimed at detecting defects in the OCL components like fasteners [56], insulators [57], and split pins [58]. Multi-purpose DCNNs have also been presented, aimed at identifying several components in catenary supports [59]. These techniques are limited by the necessity of having a recording camera positioned on top of the train. Intensive research has focused on improving the image quality obtained from cameras mounted onboard a train.

Contact methods rely on dynamic measurements taken during train runs on the operating pantographs, the most promising being pan-head accelerations. A first approach assesses the lateral geometry of the contact wire (i.e., stagger), using vertical accelerations of the contact strips of a pantograph [60]. The prediction of the stagger value can then be used to perform a diagnosis of the steady-arms. The research work uses data obtained from numerical simulation of pantograph–catenary interaction to generate training and validation datasets. Acceleration data from the simulation are pre-processed through time–frequency wavelet analysis and then used to train the ANN.

As an alternative to the traditional method for contact force measurement, a procedure to predict the contact force from pan-head vertical acceleration has also been proposed [61]. ANNs are trained through properly pre-processed acceleration signals to predict the standard deviation of the contact force. Both classification and regression problems are considered: the former to detect dropper faults and severe contact wire wear and to classify each input sample by the standard deviation of the interaction force, and the latter to directly predict the values of contact force standard deviations. This technique requires the measurements of the contact force to train the NN, together with pan-head acceleration as an input, and it is in an early stage of development, since it presently relies on data from numerical simulations for the training of the network. Tests of the trained network with experimental contact force and acceleration have been carried out only through data acquired from a pantograph test bench, in which the pantograph collectors are excited by applying, at the midpoint of each contact strip, the vertical displacement of the pantograph collector-head obtained from purely numerical simulations. There is, therefore, still ample space for the development of training the NN with data gathered from full-scale experimental test runs. Considering the recent boost in the application of these technologies, it is likely that this gap can be filled in a very short time. Pantograph acceleration data correlated to the measurement of the contact force are indeed available from many railway organizations worldwide, since they are gathered and used in the approval of high-speed lines and pantographs. Numerical simulations will continue to be useful to represent detailed scenarios like dropper defectiveness and contact wire wear, which are difficult to implement in full-scale track conditions.

ANNs currently show well-known limits in generalization: they provide accurate results only when fed with input samples having the same features as those used in the training stage. As an example, the analyses of other potential scenarios, such as different train operating speed ranges and the use of displacements instead of acceleration as input, show poor generalization capability.

As a hybrid approach between the two previously mentioned, research works are ongoing to enable the exploitation of image processing and CNNs to detect the contact point behavior under dynamic interaction with the pantograph. In this case, a non-contact measurement is adopted for tracking the contact point. In [62] defects in the contact wire stagger are searched through a DCNN. In [63] kernel-based object tracking is exploited for identifying the interaction of pantograph–catenary systems. The contact point between the pantograph and the catenary is tracked, and the obtained positions, properly processed through S-transform, are used to detect the irregularity in the positioning of the contact wire. Moreover, arc detection is carried out by analyzing the region of the contact point.

A method to detect the occurrence of electric arcs based on AI instead of using phototube measurements is also proposed, based on features extracted from the electric current and voltage signals collected through an instrumented pantograph. Three SVM-based classifiers are trained separately to detect the arcs, and a fuzzy integral technique is used to synthesize their results [64].

15.5.5 Automatic Vision-Based Inspection System for Pantographs

Structural failures of the pantograph are one of the main causes of damage to the OCL. Therefore, the occurrence of faults in the overhead equipment can be reduced not only through the monitoring of the infrastructure but also by extending and improving the monitoring of pantographs. Some of the most significant faults are related to contact strips with mechanical or electrical-related wear, pan-head suspensions, and collectors with broken lateral horns. State-of-art solutions for pantographs [65–68], focused on the capability to monitor pantograph head components during commercial operation, so as not to withdraw trains from commercial service. Vision and lighting systems installed on poles or portals, gathering images during train passage at speeds up to 350 km/h, enable the recognition of structural defects by image processing, ML, and cloud technologies, allowing the monitoring of key features like carbon strip thickness, pantograph uplift, damaged lateral horns, and pan-head orientations.

At the research stage, DL and image processing technologies are investigated for online condition monitoring of the pantograph contact strip [69]. Four typical surface defects are considered, and a CNN model is trained for defect recognition. Criteria for qualifying the wear condition are proposed, and the results of wear depth estimation are compared with on-site measurement data with an error generally below 1 mm.

15.6 PREDICTION OF REMAINING USEFUL LIFE (RUL) AND MAINTENANCE PLANNING

In recent years, a large research effort has been spent in several areas of engineering to move from preventive or predetermined maintenance to a predictive maintenance approach [70], in which maintenance actions are performed on demand, based on diagnostic information about the condition of the asset being maintained. Two factors play a pivotal role toward enabling this paradigm shift in maintenance strategies: prognostics and health monitoring (PHM) tools estimating the residual useful life of the assets and their components based on condition monitoring data and a maintenance planning process allowing the prioritization of maintenance actions in view of meeting a cost vs. benefit trade-off.

The implementation of predictive maintenance for the railway infrastructure has been the object of intensive research over recent years and is particularly challenging due to some specific features of railway transportation. First, specific PHM strategies need to be defined for each one of the many failure modes that can occur in the railway infrastructure: track geometry degradation, rail damage such as rolling contact fatigue, squats, corrugation, ballast damage and void sleepers, damage to bridges viaducts or tunnels, and damage of the OCL. Furthermore, the search for an optimized cost–benefit balance must factor in the effect of unavailability costs and penalties resulting from delays and speed reductions. Finally, any viable maintenance planning strategy shall incorporate safety aspects, to ensure that railway safety standards are met.

AI methods are particularly well suited to define a DSS for infrastructure maintenance, as they allow the automated and unsupervised analysis of the large amount of diagnostic data available from various condition monitoring devices. As far as track geometry is concerned, maintenance decisions are usually based on the evaluation of track quality indices (TQIs) whose values are used to trace the evolution over time of track geometric defects and prioritize maintenance intervention [71,72]. This is often managed based on simple classifications such as the limits for alert, intervention, or immediate action defined by standard EN 13848-5 [73] or using yellow and red tags [1]. In this context, AI methods such as SVM [74] and adaptive NN-based fuzzy inference systems [75] can be used to define a fully automated DSS. Other examples of AI-based techniques for the prediction of the condition of track components and classification of maintenance intervention needs can be found in [76] in which the occurrence of different types of sleeper support defects in a ballasted track are predicted using a trained deep residual NN and in [77] where the use of ML in the context of DSS is exemplified with reference to earthworks asset management and planning of track circuits' maintenance.

The AI-based DSS can then be used to feed a maintenance planning process (MPP) for infrastructure maintenance scheduling. The MPP can be defined in the form of a multi-objective optimization scheme [78,79] or as a Markov decision process [80,81] or as an ad-hoc designed algorithm [4]. Alternatively, GAs [82] and ANNs [83] can be applied to minimize the number of operations and related costs, targeting a specific TQI value.

Future developments in this area can be expected in terms of better integration of methods for RUL estimation into DSS and MPP tools, consideration of specific issues related to the use of big data (see [84] for a discussion of the context and requirements) and with the definition of digital twins of the railway infrastructure to support the detection of anomaly patterns and predict the progress of degradation.

15.7 FUTURE PERSPECTIVES: AI AND VIRTUAL ENVIRONMENT FOR TRAINING MAINTENANCE OPERATORS

Before concluding this chapter, a final potential application of AI for rail infrastructure maintenance is briefly presented. The application of AI is one of the constitutive technologies in the metaverse virtual world, where users can interact with each other and with computer-generated environments in real time, which can be exploited to enhance and radically change the training of the personnel in charge of a maintenance operation [85,86]. Thanks to the exploitation of virtual, augmented, extended reality, robotics, and haptic interfaces, it is possible to build virtual environments in which maintenance operators can be trained and carry out virtual inspections, identify issues, and explore and analyze the virtual infrastructure [87]. This has a high potential in the railway industry. Data gathered from the actual infrastructure and rolling stock monitoring can be exploited to create virtual environments through AI [55], thus transcending physical limitations, seasonality (e.g., enabling the possibility of studying ice-related problems all year round), and the need for personnel displacement for training. Moreover, the possibility to train operators in a virtual and thus controlled environment allows workers to practice in safer conditions and helps them foresee possible hazards on site.

REFERENCES

1. Higgins, C. and Liu, X. Modeling of track geometry degradation and decisions on safety and maintenance: A literature review and possible future research directions. *Proceeding of the IMechE Part F: Journal of Rail and Rapid Transit* 2018, 232(5), 1385–1397. https://doi.org/10.1177/0954409717721870.
2. Bragança, C., Souza, E. F., Ribeiro, D., Meixedo, A., Bittencourt, T. N., and Carvalho, H. Drive-by methodologies applied to railway infrastructure subsystems: A literature review—Part II: Track and vehicle. *Applied Sciences* 2023, 13, 6982. https://doi.org/10.3390/app13126982.
3. Fernandez Bobadilla, H. A. and Ullrich, M. Modern tendencies in vehicle-based condition monitoring of the railway track. *IEEE Transactions on Instrumentation and Measurement* 2023, 72. https://doi.org/10.1109/TIM.2023.3243673.
4. Consilvio, A. et al. A data-driven prioritisation framework to mitigate maintenance impact on passengers during metro line operation. *European Transport Research Review* 2024, 16, 6. https://doi.org/10.1186/s12544-023-00631-z.
5. EN 13848-1:2019. Railway applications – Track – Track geometry quality. Part 1: Characterisation of track geometry, 2019.
6. Weston, P., Roberts, C., Yeo, G., and Stewart, E. Perspectives on railway track geometry condition monitoring from in-service railway vehicles. *Vehicle System Dynamics* 2015, 53 (7), 1063–1091.
7. Papaelias, M. P., Roberts, C., and Davis, C. L., A review on nondestructive evaluation of rails: State-of-the-art and future development. *Proceedings of the Institution of Mechanical Engineers, Part F: Journal of Rail and Rapid Transit* 2008, 222 (4), 367–384.
8. Hodge, V. J., O'Keefe, S., Weeks, M., and Moulds, A., Wireless sensor networks for condition monitoring in the railway industry: A survey. *IEEE Transactions on Intelligent Transportation Systems* 2015, 16 (3), 1088–1106.
9. Ward, C. P., Weston, P. F., Stewart, E. J. C., Li, H., Goodall, R. M., Roberts, C., Mei, T. X., Charles, G., Dixon, R. Condition monitoring opportunities using vehicle-based sensors. *Proceedings of the Institution of Mechanical Engineers, Part F: Journal of Rail and Rapid Transit* 2011, 225 (2), 202–218.
10. Peinado Gonzalo, A., Horridge, R., Steele, H., Stewart, E., and Entezami, M. Review of data analytics for condition monitoring of railway track geometry. *IEEE Transactions on Intelligent Transportation Systems* 2022, 23 (12), 22737–22754, https://doi.org/10.1109/TITS.2022.3214121.

11. Li, C., He, Q., Wang, P. Estimation of railway track longitudinal irregularity using vehicle response with information compression and Bayesian deep learning. *Computer-Aided Civil and Infrastructure Engineering* 2022, 37, 1260–1276.
12. Hao, X., Yang, J., Yang, F., Sun, X., Hou, Y., and Wang, J. Track geometry estimation from vehicle–body acceleration for high-speed railway using deep learning technique. *Vehicle System Dynamics* 2023, 61, 239–259.
13. Tsunashima, H. and Takikawa, M. Monitoring the Condition of Railway Tracks Using a Convolutional Neural Network. In: Francisco Bulnes (Ed.), *Recent Advances in Wavelet Transform and Their Applications*. IntechOpen: London, UK, 2022, p. 13.
14. Tsunashima, H. Condition monitoring of railway tracks from car-body vibration using a machine learning technique. *Applied Sciences* 2019, 9, 2734.
15. De Rosa, A., Kulkarni, R., Qazizadeh, A., Berg, M., Di Gialleonardo, E., Facchinetti, A., and Bruni, S. Monitoring of lateral and cross level track geometry irregularities through onboard vehicle dynamics measurements using machine learning classification algorithms. *Proceedings of the Institution of Mechanical Engineers, Part F: Journal of Rail and Rapid Transit* 2021, 235, 107–120.
16. Ji, A., Woo, W. L., Wong, E. W. L., Quek, Y. T. Rail track condition monitoring: A review on deep learning approaches. *Intelligence & Robotics* 2021, 1 (2), 151–75. https://doi.org/10.20517/ir.2021.14.
17. Grassie, S. L., Rail corrugation: Characteristics, causes, and treatments. *Proceedings of the Institution of Mechanical Engineers, Part F: Journal of Rail and Rapid Transit* 2009, 223 (6), 581–596. https://doi.org/10.1243/09544097JRRT264.
18. Li, Z., Squats on railway rails, in *Wheel-Rail Interface Handbook*, 2009. https://doi.org/10.1533/9781845696788.1.409.
19. Grassie, S. L., Measurement of longitudinal irregularities on rails using an axlebox accelerometer system, in *Notes on Numerical Fluid Mechanics and Multidisciplinary Design*, vol. 150, 2021. https://doi.org/10.1007/978-3-030-70289-2_33.
20. Bocciolone, M., Caprioli, A., Cigada, A., and Collina, A. A measurement system for quick rail inspection and effective track maintenance strategy. *Mechanical Systems and Signal Processing* 2007, 21 (3). https://doi.org/10.1016/j.ymssp.2006.02.007.
21. Karaki, J., Faccini, L., Di Gialleonardo, E., Somaschini, C., Bocciolone, M., and Collina, A. Continuous monitoring of rail corrugation growth using an in-service vehicle, in *Lecture Notes in Mechanical Engineering*, 2022. https://doi.org/10.1007/978-3-031-07305-2_17.
22. Faccini, L., Karaki, J., Di Gialleonardo, E., Somaschini, C., Bocciolone, M., and Collina, A. A methodology for continuous monitoring of rail corrugation on subway lines based on axlebox acceleration measurements. *Applied Sciences (Switzerland)* 2023, 13 (6). https://doi.org/10.3390/app13063773.
23. Pieringer, A. and Kropp, W. Model-based estimation of rail roughness from axle box acceleration. *Applied Acoustics* 2022, 193, 108760. https://doi.org/10.1016/J.APACOUST.2022.108760.
24. Wei, X., Yin, X., Hu, Y., He, Y., and Jia, L. Squats and corrugation detection of railway track based on time-frequency analysis by using bogie acceleration measurements. *Vehicle System Dynamics* 2020, 58 (8), 1167–1188. https://doi.org/10.1080/00423114.2019.1610181.
25. Molodova, M., Li, Z., Nunez, A., and Dollevoet, R., Automatic detection of squats in railway infrastructure. *IEEE Transactions on Intelligent Transportation Systems* 2014, 15 (5), 1980–1990. https://doi.org/10.1109/TITS.2014.2307955.
26. Molodova, M., Li, Z., and Dollevoet, R. Axle box acceleration: Measurement and simulation for detection of short track defects. *Wear* 2011, 271 (1–2), 349–356. https://doi.org/10.1016/J.WEAR.2010.10.003.
27. Molodova, M., Oregui, M., Núñez, A., Li, Z., and Dollevoet, R. Health condition monitoring of insulated joints based on axle box acceleration measurements. *Engineering Structures* 2016, 123, 225–235. https://doi.org/10.1016/J.ENGSTRUCT.2016.05.018.
28. Nunez, A., Jamshidi, A., and Wang, H. Pareto-based maintenance decisions for regional railways with uncertain weld conditions using the hilbert spectrum of axle box acceleration. *IEEE Transactions on Industrial Informatcs* 2019, 15 (3), 1496–1507. https://doi.org/10.1109/TII.2018.2847736.
29. Mandriota, C., Nitti, M., Ancona, N., Stella, E., and Distante, A. Filter-based feature selection for rail defect detection. *Machine Vision and Applications* 2004, 15 (4), 179–185. https://doi.org/10.1007/s00138-004-0148-3.
30. Gazafrudi, S. M. M., Younesian, D., and Torabi, M. A high accuracy and high speed imaging and measurement system for rail corrugation inspection. *IEEE Transactions on Industrial Electronics* 2021, 68 (9), 8894–8903. https://doi.org/10.1109/TIE.2020.3013748.

31. Li, Q. and Ren, S. A real-time visual inspection system for discrete surface defects of rail heads. *IEEE Transactions on Instrumentation and Measurement* 2012, 61 (8), 2189–2199. https://doi.org/10.1109/TIM.2012.2184959.
32. Yang, H., Liu, J., Mei, G., Yang, D., Deng, X., and Duan, C. Research on real-time detection method of rail corrugation based on improved ShuffleNet V2. *Engineering Applications of Artificial Intelligence* 2023, 126, 106825. https://doi.org/10.1016/J.ENGAPPAI.2023.106825.
33. Hassanieh, W., Chehade, A., Facchinetti, A., Carman, M., Bocciolone, M., and Somaschini, C. Leveraging machine learning to predict rail corrugation level from axle-box acceleration measurements on commercial vehicles. *International Journal of Rail Transportation* 2023. https://doi.org/10.1080/23248378.2023.2220112.
34. Xie, Q., Tao, G., Lo, S. M., Yang, X., and Wen, Z. A data-driven convolutional regression scheme for on-board and quantitative detection of rail corrugation roughness. *Wear* 2023, 524–525, 204770. https://doi.org/10.1016/J.WEAR.2023.204770.
35. Haghbin, M., Chiachío, J., Muñoz, S., Escalona Franco, J. L., Guillén, A. J., Crespo Marquez, A., and Cantero-Chinchilla, S. Predicting rail corrugation based on convolutional neural networks using vehicle's acceleration measurements. *Sensors* 2024, 24, 4627. https://doi.org/10.3390/s24144627.
36. Xie, Q., Tao, G., He, B., and Wen, Z. Rail corrugation detection using one-dimensional convolution neural network and data-driven method. *Measurement* 2022, 200, 111624. https://doi.org/10.1016/J.MEASUREMENT.2022.111624.
37. Shang, L., Yang, Q., Wang, J., Li, S., and Lei, W. Detection of rail surface defects based on CNN image recognition and classification. *International Conference on Advanced Communication Technology, ICACT*, Chungcheon, South Korea, February 2018, pp. 45–51. https://doi.org/10.23919/ICACT.2018.8323642.
38. Yang, H., Wang, Y., Hu, J., He, J., Yao, Z., and Bi, Q. Deep learning and machine vision-based inspection of rail surface defects. *IEEE Transactions on Instrumentation and Measurement* 2022, 71. https://doi.org/10.1109/TIM.2021.3138498.
39. Faghih-Roohi, S., Hajizadeh, S., Nunez, A., Babuska, R., and de Schutter, B. Deep convolutional neural networks for detection of rail surface defects. *Proceedings of the International Joint Conference on Neural Networks*, Vancouver, Canada, October 2016, pp. 2584–2589. https://doi.org/10.1109/IJCNN.2016.7727522.
40. Mauz, F., Wigger, R., Gota, A. E., and Kuffa, M. Automatic detection of the running surface of railway tracks based on laser profilometer data and supervised machine learning. *Sensors* 2024, 24 (8), 2638. https://doi.org/10.3390/S24082638.
41. Deraemaeker, A. and Worden, K. A comparison of linear approaches to filter out environmental effects in structural health monitoring. *Mechanical Systems and Signal Processing* 2018, 105, 1–15.
42. Li, H., Li, S., Ou, J., and Li, H. Modal identification of bridges under varying environmental conditions: Temperature and wind effects. *Structural Control and Health Monitoring* 2010, 17 (5), 495–512.
43. Ni, Y. Q., Hua, X. G., Fan, K. Q., and Ko, J. M. Correlating modal properties with temperature using long-term monitoring data and support vector machine technique. *Engineering Structures* 2005, 27 (12 SPEC. ISS.), 1762–1773.
44. Entezami, A., Sarmadi, H., and Behkamal, B. Long-term health monitoring of concrete and steel bridges under large and missing data by unsupervised meta learning. *Engineering Structures* 2023, 279, 115616.
45. Sarmadi, H. Investigation of machine learning methods for structural safety assessment under variability in data: Comparative studies and new approaches. *Journal of Performance of Constructed Facilities* 2021, 35 (6), 04021090.
46. Flah, M., Nunez, I., Ben Chaabene, W., and Nehdi, M. L. Machine learning algorithms in civil structural health monitoring: A systematic review. *Archives of Computational Methods in Engineering* 2021, 28 (4), 2621–2643.
47. Peeters, B. and De Roeck, G. One-year monitoring of the Z24-bridge: Environmental effects versus damage events. *Earthquake Engineering and Structural Dynamics* 2001, 30, 149–171.
48. An, Y., Chatzi, E., Sim, S.-H., Laflamme, S., Blachowski, B., and Ou, J. Recent progress and future trends on damage identification methods for bridge structures. *Structural Control and Health Monitoring* 2019, 26 (10), 1–30.
49. Zhang, Z. and Sun, C. Structural damage identification via physics-guided machine learning: A methodology integrating pattern recognition with finite element model updating. *Structural Health Monitoring* 2020, 20 (4). https://doi.org/10.1177/1475921720927488.
50. Bud, M. A., Nedelcu, M., Moldovan, I., and Figueiredo, E. Hybrid approach for supervised machine learning algorithms to identify damage in bridges. *Journal of Bridge Engineering* 2024, 29 (8), 1–14.

51. Dorafshan, S. and Maguire, M. Bridge inspection: Human performance, unmanned aerial systems and automation. *Journal of Civil Structural Health Monitoring* 2018, 8 (3), 443–476.
52. Flah, M., Suleiman, A. R., and Nehdi, M. L. Classification and quantification of cracks in concrete structures using deep learning image-based techniques. *Cement and Concrete Composites* 2020, 114, 103781.
53. Munawar, H. S., Hammad, A. W. A., Haddad, A., Soares, C. A. P., and Waller, S. T. Image-based crack detection methods: A review. *Infrastructures* 2021, 6 (8), 115.
54. Hüthwohl, P., Lu, R., and Brilakis, I. Multi-classifier for reinforced concrete bridge defects. *Automation in Construction* 2019, 105, 102824.
55. Phusakulkajorn, W., Núñez, A., Wang, H., Jamshidi, A., Zoeteman, A., Ripke, B., Dollevoet, R., De Schutter, B., and Li, Z., Artificial intelligence in railway infrastructure: Current research, challenges, and future opportunities. *Intelligent Transportation Infrastructure* 2019, 2, liad016. https://doi.org/10.1093/iti/liad016.
56. Chen, J., Liu, Z., Wang, H., Nez, A., and Han, Z. Automatic defect detection of fasteners on the catenary support device using deep convolutional neural network. *IEEE Transactions on Instrumentation and Measurement* 2018, 67 (2), 257–269.
57. Kang, G., Gao, S., Yu, L., and Zhang, D. Deep architecture for high-speed railway insulator surface defect detection: Denoising autoencoder with multitask learning. *IEEE Transactions on Instrumentation and Measurement* 2019, 68 (8), 2679–2690.
58. Zhong, J., Liu, Z., Han, Z., Han, Y., and Zhang, W. A CNN-based defect inspection method for catenary split pins in high-speed railway. *IEEE Transactions on Instrumentation and Measurement* 2019, 68 (8), 2849–2860. https://doi.org/10.1109/TIM.2018.2871353.
59. Liu, W., Liu, Z., Núñez, A., Wang, L., Liu, K., Lyu, Y., and Wang, H. Multi-objective performance evaluation of the detection of catenary support components using DCNNs. *IFAC-PapersOnLine* 2018, 51 (9), 98–105.
60. Blanco, B., Errandonea, I., Beltrán, S., Arrizabalaga, S., and Alvarado, U. Panhead accelerations-based methodology for monitoring the stagger in overhead contact line systems. *Mechanism and Machine Theory* 2022, 171, art. no. 104742. https://doi.org/10.1016/j.mechmachtheory.2022.104742.
61. Gregori, S., Tur, M., Gil, J., and Fuenmayor, F. J. Assessment of catenary condition monitoring by means of pantograph head acceleration and Artificial Neural Networks. *Mechanical Systems and Signal Processing* 2023, 202, art. no. 110697. https://doi.org/10.1016/j.ymssp.2023.110697.
62. Zhang, D., Gao, S., Yu, L., Kang, G., Zhan, D., and Wei, X. A robust pantograph-catenary interaction condition monitoring method based on deep convolutional network. *IEEE Transactions on Instrumentation and Measurement* 2020, 69 (5), art. no. 8746826, pp. 1920–1929. https://doi.org/10.1109/TIM.2019.2920721.
63. Aydin, I., Karakose, M., and Akin, E. Anomaly detection using a modified kernel-based tracking in the pantograph–catenary system. *Expert Systems with Applications* 2015, 42 (2), 938–948. https://doi.org/10.1016/j.eswa.2014.08.026.
64. Aydin, I., Celebi, S. B., Barmada, S., and Tucci, M. Fuzzy integral-based multi-sensor fusion for arc detection in the pantograph-catenary system. *Proceedings of the Institution of Mechanical Engineers, Part F: Journal of Rail and Rapid Transit* 2018, 232 (1), 159–170. https://doi.org/10.1177/0954409716662090.
65. Capece, P., Chiesi, L., Pinazzi, A., Sabbatelli, M., Sacchi, M., Venturi, L., Kelsey, T., and Wilson, C. PANTOBOT-3D: An automatic vision-based inspection system for locomotive pantographs. *Paper presented at the IET Conference Publications*, Birmingham, UK, 2016.
66. Camlin Rail. Automated Pantograph Monitoring. https://camlingroup.com/rail/pantograph-monitoring (web site last viewed on July 2, 2024).
67. Pantoinspect A/S. PantoSystem - Complete System For Automated and Real-time 3D Pantograph Inspection. https://www.pantoinspect.com/ (web site last viewed on July 2, 2024).
68. ESIM srl. 3D Pantograph Inspection System. https://www.esimgroup.com/servizi-e-soluzioni/diagnostica-ferroviaria/video-ispezione-pantografi-3d/ (web site last viewed on July 12, 2024).
69. Wei, X., Jiang, S., Li, Y., Li, C., Jia, L., and Li, Y. Defect detection of pantograph slide based on deep learning and image processing technology. *IEEE Transactions on Intelligent Transportation Systems* 2020, 21 (3), art. no. 8667891, pp. 947–958. https://doi.org/10.1109/TITS.2019.2900385.
70. EN 13306:2017 Maintenance – Maintenance terminology, 2017.
71. La Paglia, I., Carnevale, M., Corradi, R., et al. Condition monitoring of vertical track alignment by bogie acceleration measurements on commercial high-speed vehicles. *Mechanical Systems and Signal Processing* 2023, 186. https://doi.org/10.1016/j.ymssp.2022.109869.

72. La Paglia, I., Di Gialleonardo, E., Facchinetti, A., Carnevale, M., and Corradi, R. Acceleration-based condition monitoring of track longitudinal level using multiple regression models. *Proceedings of the Institution of Mechanical Engineers, Part F: Journal of Rail and Rapid Transit* 2024, 238 (5), 479–488. https://doi.org/10.1177/09544097231201513.
73. EN 13848-5:2017. Railway applications – Track – Track geometry quality – Part 5: Geometric quality. Accessed: July 16, 2024. [Online]. Available: https://standards.iteh.ai/catalog/standards/cen/6b03fdaf-2d25-4f0f-96d7-39039fbeb9be/en-13848-5-2017.
74. Cárdenas-Gallo, I. et al. An ensemble classifier to predict track geometry degradation. *Reliability Engineering and System Safety* 2017, 161, 53–60.
75. Dell'Orco, M. et al. New decision support system for optimization of rail track maintenance planning based on adaptive neurofuzzy inference system. *Transportation Research Record* 2008, 2043 (1), 49–54.
76. Feng, B. et al. Data-driven railway crosstie support condition prediction using deep residual neural network: Algorithm and application. *Transportation Research Record* 2022, 2676 (3), 160–171. https://doi.org/10.1177/03611981211049423.
77. Consilvio, A. et al. On applying machine learning and simulative approaches to railway asset management: The earthworks and track circuits case studies. *Sustainability* 2020, 12 (6), 2544. https://doi.org/10.3390/su12062544.
78. Khalouli, S., Benmansour, R., and Hanafi, S. An ant colony algorithm based on opportunities for scheduling the preventive railway maintenance. *International Conference on Control, Decision and Information Technologies (CoDIT)*, St. Paul's Bay, Malta, 2016, pp. 594–599. https://doi.org/10.1109/CoDIT.2016.7593629.
79. Consilvio, A., Febbraro, A. D., and Sacco, N. A rolling-horizon approach for predictive maintenance planning to reduce the risk of rail service disruptions. *IEEE Transactions on Reliability* 2020. https://doi.org/10.1109/TR.2020.3007504.
80. Sharma, S. et al. Data-driven optimization of railway maintenance for track geometry. *Transportation Research Part C* 2018, 90, 34–58.
81. Lopes Gerum, P. C. et al. Data-driven predictive maintenance scheduling policies for railways. *Transportation Research Part C* 2019, 107, 137–154.
82. Khouzani, A. H. E., Golroo, A., and Bagheri, M., Railway maintenance management using a stochastic geometrical degradation model. *Journal of Transportation Engineering, Part A: Systems* 2017, 143 (1), Art. no. 04016002.
83. Bressi, S., Santos, J., and Losa, M. Optimization of maintenance strategies for railway track-bed considering probabilistic degradation models and different reliability levels. *Reliability Engineering & System Safety* 2021, 207, Art. no. 107359.
84. McMahon, P., Zhang, T., and Dwight, R. Requirements for big data adoption for railway asset management. *IEEE Access* 2020, 8, 15543–15564. https://doi.org/10.1109/ACCESS.2020.2967436.
85. Radhakrishnan, U., Koumaditis, K., and Chinello, F. A systematic review of immersive virtual reality for industrial skills training. *Behaviour & Information Technology* 2021. https://doi.org/10.1080/0144929X.2021.1954693.
86. Nikitin, A., Reshetnikova, N., Sitnikov, I., and Karelova, O. VR training for railway wagons maintenance: Architecture and implementation. *Procedia Computer Science* 2020, 176, 622–631. https://doi.org/10.1016/j.procs.2020.08.064.
87. Wong, E. Y. C. and Lee, P. T. Y. Virtual reality in transportation and logistics: A clustering analysis of studies from 2010 to 2023 and future directions. *Computers in Human Behavior* 2024, 153, art. no. 108082. https://doi.org/10.1016/j.chb.2023.108082.

16 Digital Twin and Artificial Intelligence for Railway Hubs Design and Management

Pierluigi Coppola, Luigi Castagna, and Fulvio Silvestri

16.1 INTRODUCTION: RAILWAY STATIONS AS MULTIMODAL AND MULTISERVICE HUBS

The diversion of travel demand toward public transport and active modes is central in the urban agendas of EU cities to reduce the negative externalities produced by private (automobile) transport and to allow also a more livable and sustainable urban development (see [1,2]). People are encouraged to use trains, buses, shared mobility services, walking, and cycling and, in so doing, to undertake multimodal trip chains incorporating different modes [3]. Despite numerous studies in the literature showing that passengers perceive significant inconvenience when interchanging between transport modes [4], railway stations play a pivotal role as primary nodes for integrating different modes and services (individual and collective, public, and private) supplied at different scales (urban, suburban, regional, national, and international); they are fundamental in ensuring seamless trips and providing a travel experience that competes effectively with that offered by private cars.

Moreover, railway stations represent not only multimodal hubs but also urban "centralities" for new development, offering opportunities for both passengers and residents in the surrounding areas [5].

Zemp et al. [6] pointed out that railway stations serve multiple functions by connecting their catchment area with the available transport services, promoting intermodality, while also facilitating commercial opportunities. As also highlighted by Wicki et al. [7] stations are increasingly perceived as public spaces and fundamental components of the identity of surrounding neighborhoods. In fact, railway stations attract not only travelers accessing transport services but also other users seeking services and amenities that are increasingly located in station buildings and forecourts, such as retail stores, shopping malls, and co-working spaces [8].

Revamping a station as a multimodal and multiservice hub represents a challenge, particularly in relation to safety and security issues [9] compounded by existing congestion in urban public transport networks [10]. Additionally, users with different objectives and needs (e.g., commuters, business travelers, tourists, station visitors) may create overcrowding and conflicting flows [11]. A comprehensive understanding of how visitors/travelers access and use (both indoor and outdoor) a station [12,13] is therefore needed to appropriately design and size transport facilities and services, to avoid crowded spaces and conflicts among pedestrians [14] and to ensure a comfortable and pleasant travel experience.

This chapter discusses the role of artificial intelligence (AI) in designing and managing railway stations as multimodal and multiservice hubs, delving into the concept of the digital twin as a combination of integrated micro- and macro-simulation models, real-time data collection, and policy implementation. This chapter is organized as follows: following this introduction, Section 16.2 presents macroscopic models for the design of mobility services in a railway station and microscopic

DOI: 10.1201/9781003492146-16

models of pedestrian flows, focusing on their applications for service planning and design. Section 16.3 examines the concept of digital twins applied to railway stations, highlighting the role of transport models and real-time data requirements, and describing an architecture for application to multimodal hubs. The concluding Section 16.4 discusses the regulatory and technological challenges related to real-world implementation.

16.2 SIMULATION MODELS FOR RAILWAY HUB DESIGN

The functional design of a railway hub typically includes the distribution and sizing of indoor and outdoor services (e.g., platforms, parking) and functions (e.g., ticket offices, shops) to facilitate travelers' movements, avoid overcrowded spaces, and prevent conflicts among pedestrian flows. The typical "what-if" approach (Figure 16.1) involves simulating different assumptions (scenarios) to assess, in advance, the impacts and levels of service available for travelers, operators, and the environment overall (i.e., so-called external impacts).

Simulated scenarios may include infrastructural interventions, such as a new tramway line or bike lane, new mobility services (e.g., bike sharing), and new urban development in the station or catchment area (e.g., new residential neighborhoods). All these changes result in modified passenger volumes using the station, a redistribution of flows among the modes of access and egress (A/E), and changes in the use of spaces within the station premises.

Simulation models play a central role in assessing the performance of the transport facilities (supply) and forecasting how passengers behave and move (demand). Depending on the scale of analysis and the theoretical assumptions behind it, they can be distinguished as follows:

- *Macroscopic models*, referring to aggregate passenger flows and simulate the performances of the components of the transportation networks (i.e., links or nodes) using aggregate functions.
- *Microscopic models*, which simulate the behavior of individual users in the network and estimate the performance of the transportation system based on the interactions among travelers.

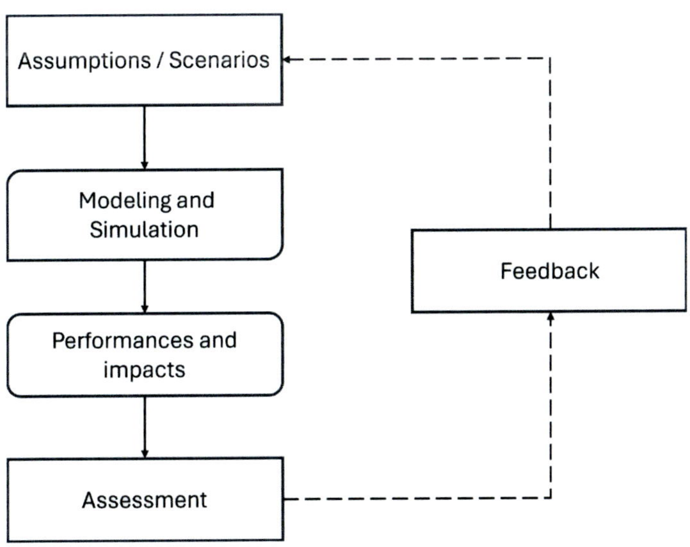

FIGURE 16.1 *What if* approach in transportation network design.

16.2.1 Macroscopic Models

In the simulation framework of a railway hub, macroscopic models allow us to estimate the A/E, and interchange flows as a function of rail services supplied in the station and the transport modes available to access or leave the station. In so doing, they support the sizing of mobility infrastructure and services (e.g., parking lots, station entrances) and help to identify necessary upgrades to improve station performance (e.g., charging station, cycling infrastructure).

The first step in building macroscopic models is the estimation of the origin–destination (OD) matrix of trips of the integrated public transport system in the study area (Figure 16.2). Typically, OD matrices are estimated by means of travel surveys, which are both highly expensive and time-consuming, and therefore not frequently updated [15]. Today, big data sources are available from transport smart cards, GPS sensors, Bluetooth, Wi-Fi hotspots, mobile network data, and mobile apps that track the movement of users or vehicles throughout their trip. Coupled with machine learning and deep learning techniques [16], mobile network data and smartphone app data can be analyzed to extract door-to-door trips and, consequently, OD matrices. Many mobile network operators and app aggregators already offer those matrices, often segmented by estimated transport mode and trip purpose. The estimation process typically involves unsupervised learning techniques, such as clustering algorithms [17].

Despite the benefits and new opportunities enabled by the use of passive data, these sources lack the behavioral accuracy of survey data, as they do not provide information on travelers' socio-economic characteristics and trip purposes, which can be collected through travel surveys. Therefore, data fusion can be a viable compromise to address this issue by integrating phone data with survey data, leveraging the strengths of both techniques [15,18,19].

Once the OD matrix is estimated, the A/E modal shares of the station can be estimated following two different approaches: discrete choice or network assignment models.

The discrete choice approach is based on the random utility models [20], allowing for the estimation of A/E mode shares using probabilistic functions of factors such as the level of service (e.g., in-vehicle time, monetary costs, comfort, number of transfers), station attributes (e.g., available

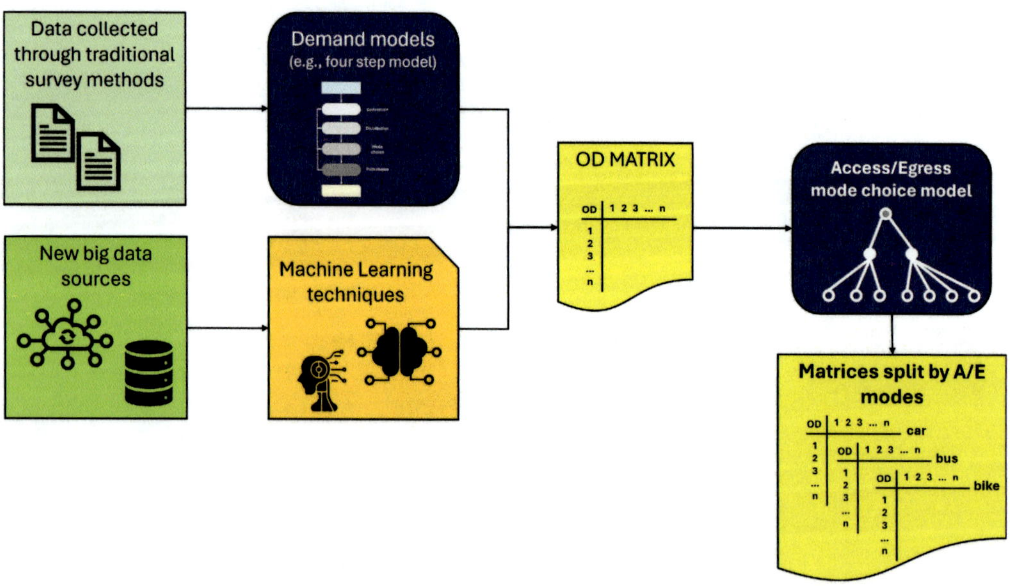

FIGURE 16.2 Discrete choice approach for the estimation of mode shares to access and to egress (A/E) from the railway hubs.

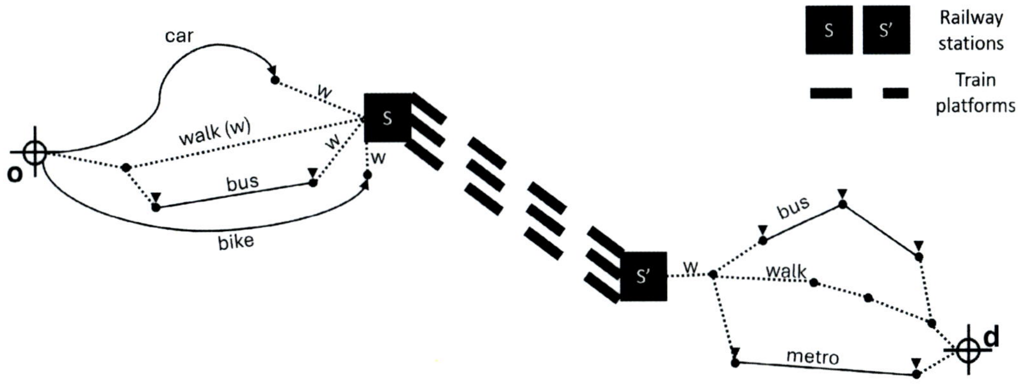

FIGURE 16.3 Example of transportation network with multimodal access to railway nodes (car, public transport, bike, walk).

parking slots, distance from the station), and socio-economic characteristics of travelers (e.g., average income, possession of a driving license).

On the contrary, the network assignment approach involves assigning OD flows to the shortest paths in a multimodal transport network, where different A/E modes are explicitly represented by links with associated generalized travel costs. This approach identifies the routes passengers use to access the station and, in doing so, helps delineate the station's catchment areas by mode (e.g., car, walk, bike, local public transport) (Figure 16.3).

16.2.2 Microscopic Models

In the context of a railway hub simulation, microscopic models simulate individual movements and their interactions within the stations; each individual regulates his own behavior and dynamics, enabling the recreation of specific local interactions, such as lane formation, collisions with other pedestrians or obstacles, overtaking, and queuing [21]. Pedestrian flows can be modeled through the representation of movements from the station's forecourt to the platforms and vice versa. From the supply side, this requires a dynamic characterization of the performance of station facilities, such as walking areas, stairs, escalators, elevators, obstacles, and gates [13] and the characteristics of the surrounding area. However, the most critical part is the modeling of pedestrian dynamics as the result of the hypotheses underlying pedestrian behavior [22] and the physical and social characteristics that may influence their ability to navigate within the station building [23].

One of the most popular models for simulating pedestrian dynamics in railway hubs is the social force model introduced by [24]. This model assumes that the interactions of each pedestrian (α,β) with the environment are described by a series of forces [23] that drive them toward their desired destination (Figure 16.4). The movement of the pedestrian results from a driving force F_α, which is the superposition of:

- the pedestrian's own desires, F_α^0
- the repulsive and attractive effects of other pedestrians, $F_{\alpha\beta}$
- the repulsive and attractive effects of obstacles, $F_{\alpha B}$

Demand data pertains to bidirectional passenger volumes from entrances to platforms (station OD matrices) and their distribution over time intervals. This data can be obtained from macroscopic models or through direct counts and video surveys. Examples of demand data useful for building the railway station model include ridership data related to passenger boarding and alighting from each transit, passengers interchanging (i.e., moving from one platform to another), and the distribution

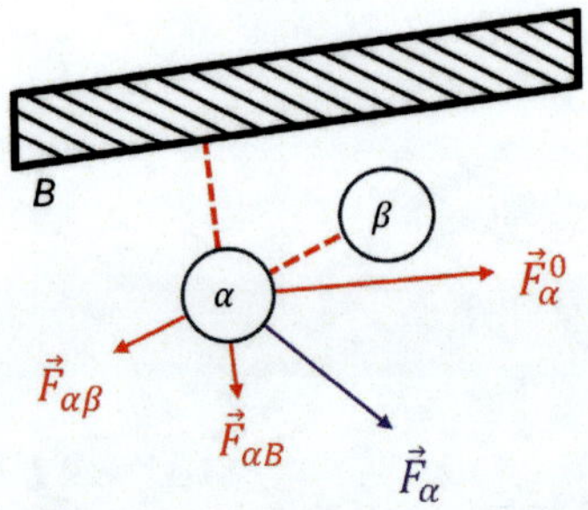

FIGURE 16.4 Forces acting on a pedestrian (α) in the social force model.

of A/E modes. Examples of outputs from pedestrian microsimulation include identifying congested areas and analyzing queue phenomena, as well as estimating the level of service of the station based on space utilization over time (e.g., density heatmaps) or journey times between different locations.

Simulations can be conducted to assess various scenarios, including crowd management strategies (e.g., access gates, escalators), layout reconfigurations (e.g., location and number of services), infrastructure upgrades (e.g., new platforms, new car parks), changes in railway or local public transport services (e.g., service timetables, platform allocations), service perturbations and disruptions, special events, and emergency evacuations. One of the key features of microscopic models is that they enable analysts to visualize dynamic processes, such as the formation of queues, bottlenecks, and congestion phenomena, through 2D and 3D graphical interfaces. Additionally, this capability improves the presentation of results to non-technical audiences [25].

A comprehensive review of microsimulation models is included in [21,26,27], while Yang et al. [28] examined both traditional and more recent methods for crowd simulation and modeling.

16.3 DIGITAL TWIN FOR RAILWAY HUB (RT) MANAGEMENT

In recent years, the availability of data in both time and space related to service supply and passenger demand flows has greatly improved the accuracy of transportation models [19]. In addition, the proliferation of information technology, notably the availability of real-time data, establishes a basis for augmenting traditional (offline) simulation techniques with actual data to construct a detailed real-time digital model of the transportation system [29,30].

Depending on the level of synchronization, the degree of automation in communication between the physical object (i.e., the real world) and the virtual environment and whether the decision-making processes for policy implementation are governed by a human or AI, the categories of digital model, digital shadow, digital twin, and intelligent digital twin (IDT) can be distinguished by adapting the classification proposed by [31–33], as shown in Figure 16.5.

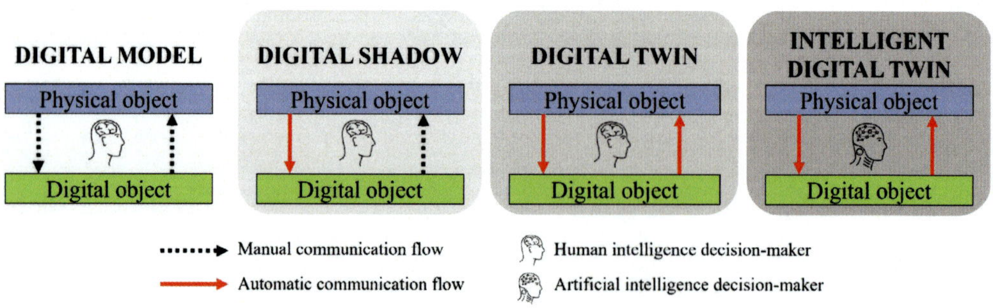

FIGURE 16.5 The evolution from digital model to intelligent digital twin. (Adapted from W. Kritzinger, M. Karner, G. Traar, J. Henjes, and W. Sihn, 'Digital Twin in manufacturing: A categorical literature review and classification', *IFAC-PapersOnLine*, vol. 51, no. 11, pp. 1016–1022, 2018, doi:10.1016/j.ifacol.2018.08.474.)

The digital model allows for the simulation of current or future scenarios in a virtual environment without automatic data communication or exchange with the real world. The digital shadow establishes a one-way real-time communication flow of data from the physical object to its digital counterpart, enabling real-time synchronization of the digital object and allowing for effective monitoring and supervision of the physical object's activities within the virtual environment. The digital twin extends the capabilities of the digital shadow by introducing an additional communication flow from the digital object to the physical object, thereby creating a closed-loop system. This bidirectional communication allows not only for supervision but also for real-time control of the transportation system.

The communication flow is the key factor distinguishing a digital model, digital shadow, and digital twin. The need for real-time synchronization between the physical and digital objects is both a key enabler and a significant constraint for the digital twin, due to the intricate technological challenges inherent in the real-time fusion and integration of large amounts of data from different sources.

In fact, the digital twin of a railway hub is based on three levels [34]:

- The *physical level* is the real-world station including the station building and equipment, facilities, vehicles, and users.
- The *digital level* is the digital replica of the physical objects.
- The *communication level* enables real-time communication, data exchange, and policies' actuation between the physical object and the digital object.

Data are sampled and transmitted from the physical object to the digital object, where they are stored, integrated, and elaborated to update the digital objects in real time. These data are processed through various stages in the digital domain, including storage, data fusion, modeling, simulation, and prediction. The resulting output, as an action to be implemented, is transmitted to the physical object through the communication level, where it is applied by real-world actuators. This establishes a closed loop between the physical and digital objects via the communication level, allowing continuous synchronization between the physical and digital worlds (Figure 16.6).

The digital twin combines transportation models with real-time data to provide an effective tool for prediction, optimization, and control. It enables a virtual environment for modeling, real-time monitoring, and predicting issues. If a potential issue is detected, the digital twin allows for the real-time simulation of alternative corrective actions, assessing their outcomes and deciding which ones to implement in the real world. Moreover, thanks to continuous synchronization with the real world, it is possible to monitor and assess the effectiveness of the proposed actions, measure the gaps between planned and actual outcomes, and learn from real-world feedback.

In the context of a railway hub, real-time data such as arrival/departure times of trains (supply) or passengers alighting/boarding (demand) are sampled from the field and transmitted to the control

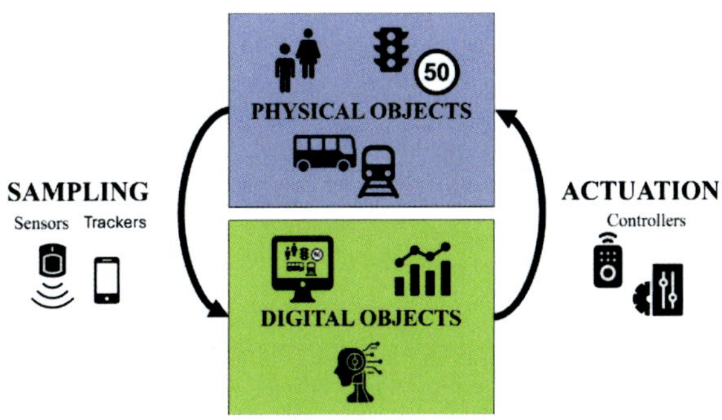

FIGURE 16.6 Closed loop between physical and digital objects. (Adapted from C. Schwarz and Z. Wang, 'The role of digital twins in connected and automated vehicles', *IEEE Intelligent Transportation Systems Magazine*, vol. 14, no. 6, pp. 41–51, 2022, doi:10.1109/MITS.2021.3129524.)

center. Here, the digital model of the station operates alongside algorithms and machine learning techniques. The sampled data enabling real-time monitoring of the station are used as input for the prediction algorithms implemented in the digital object. If some issues are detected by these algorithms (e.g., a line disruption), the digital twin starts running parallel simulations of corrective actions and policies that may reduce undesired impacts (e.g., train delays). The different outputs of the simulations are then compared by means of key performance indicators (KPIs) that allow the selection of the most appropriate policy action in relation to the expected future developments (Figure 16.7).

The strength of a digital twin lies in its ability to monitor the railway hub in real time, suggesting actions and policies to optimize performance and proactively prevent critical issues by anticipating and providing alternatives to reduce their impacts. The digital twin enables data-driven decision-making by providing operators with intervention and policy alternatives based on the measured or predicted situation. The operator (i.e., the station manager) can evaluate the KPI measures obtained from parallel simulations and choose which strategy to implement, thereby supporting human intelligence with analytics.

In a more advanced implementation, referred to in the literature as an IDT or digital twin intelligent system, the digital twin is integrated with AI to automatically predict potential issues before they manifest and implement the most appropriate interventions to prevent or mitigate their impact on the performance of the station and the transport system [35,36]. This represents a significant shift from traditional digital twin models, which rely on human analysts to interpret the outcomes of simulations and make decisions on the policies to implement. The IDT takes a proactive role in managing and optimizing the performance of complex transport systems.

In contrast to traditional digital twins, which operate in a *goal-given* manner – where specific objectives are predefined by human analysts and interventions in the physical world are automatically implemented through a *rule-based* approach – the IDTs are *goal-seeking* and *anticipatory* [33],

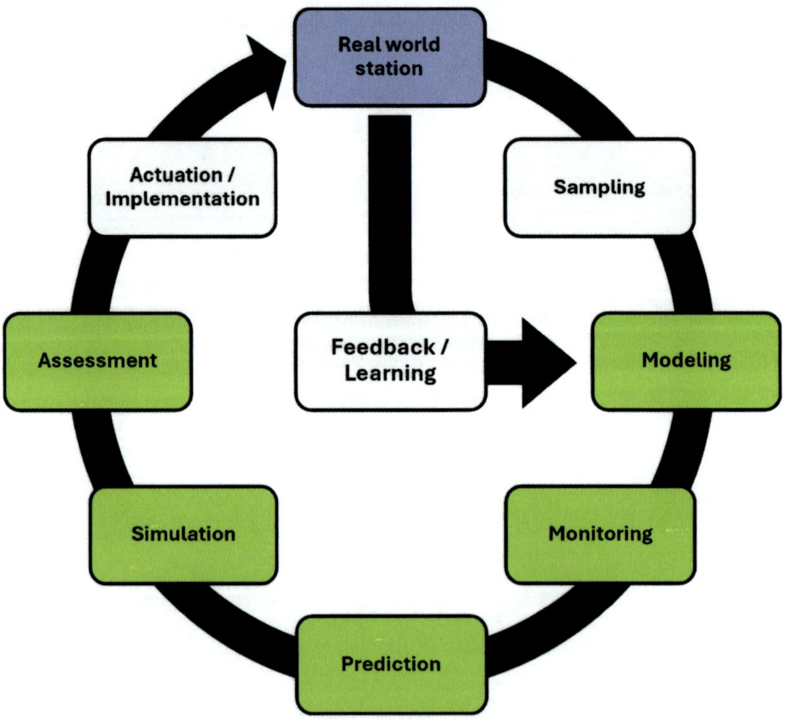

FIGURE 16.7 Example of functions for digital twinning of railway hubs.

meaning they actively pursue optimal solutions in real time to dynamically adapt to evolving objectives, constantly optimizing outcomes while minimizing resource use. By leveraging AI-driven decision-making, IDTs shift the responsibility for determining the best course of action from humans to machines. Consequently, the role of the human analysts transforms from decision-maker to supervisor, overseeing the choices made by AI. This interaction between humans and AI represents a paradigm of *augmentation* rather than *substitution*. The AI acts as a powerful tool that amplifies the human analyst's decision-making ability by providing timely and unbiased findings. Finally, human judgment remains critical for interpreting AI-driven outcomes, ensuring that the intervention decisions are aligned with ethical considerations and long-term strategies.

In both cases, whether decisions are made by human intelligence or AI, digital twinning encompasses and expands upon traditional transport simulation models [37]. Both create a digital representation of the real transportation system (e.g., the railway station) using historical data, transportation system theory, and models within a digital space. However, the digital model operates in an *open loop* with the real world, functioning offline. Once calibrated, it is able to reproduce what happens under given circumstances and to perform what-if analyses, but the model does not update itself in real-time (*offline model*). In contrast, the input data needs to be available and fed to the digital twin in *near-real time* [29] to enable the digital object to evolve over time with the physical object [38]. In this way, a *closed-loop* system is created that allows for direct interactions and real-time feedback with the real world (*online model*).

The functional architecture of the digital twin of a railway hub is schematically represented in Figure 16.8. The architecture builds on the digital twin framework proposed by Bao et al. [37] and Padovano et al. [39], extending its application to encompass the entire station system.

The input data can be static, such as BIM and CAD files used for constructing the static model of the station and its surroundings, GTFS files for detailed descriptions of transit arrival and departure schedules, and other historical data used for the construction, calibration, and validation of models and algorithms. To maintain the digital twin's accuracy and enable near-real-time synchronization with events occurring at the station, online data must be sampled and transmitted to the digital model.

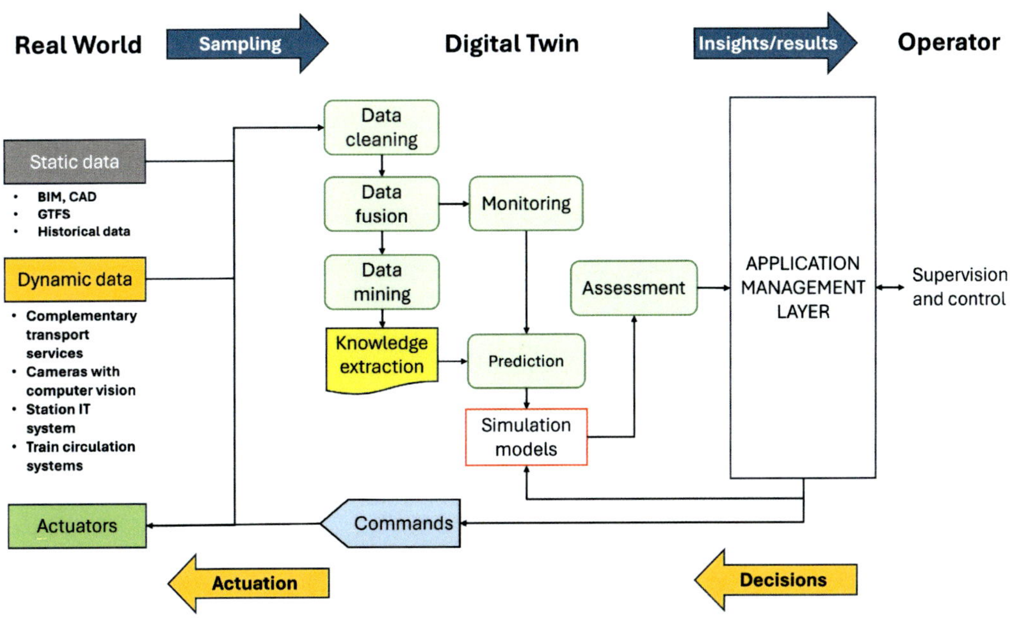

FIGURE 16.8 Functional architecture of a digital twin for a railway hub.

A comprehensive and up-to-date overview of activities within the hub requires data collection from various sources. These include interchange infrastructures with A/E transport modes (e.g., parking lots, local public transport stops, bike-sharing stations), inner station areas, and the platform level. Complementary transport infrastructures that provide real-time data include smart parking systems, which offer information on parking occupancy and vehicle stay duration, and electric vehicle charging stations (for both private and shared vehicles), which provide data on stall occupancy and estimated charging completion times. Additional real-time data can be collected regarding the availability of last-mile transportation options in station areas, such as shared scooters and bicycles (both station-based and free-flow), as well as the occupancy levels of bike station stalls.

For public transport, intelligent bus stop signs can be installed to provide real-time crowding estimates. Automated vehicle monitoring and automatic passenger counting systems offer real-time data on vehicle locations within the station's area of interest and estimates of the number of passengers on board each vehicle. Additionally, the development of vehicle-to-everything systems will enhance data acquisition regarding vehicular traffic around the station.

Equipping the multimodal hub with cameras that use computer vision technology is also beneficial. These cameras can detect, identify, track, and classify vehicle and pedestrian flows, providing real-time data and insights on infrastructure congestion.

Further data for the digital twin can be integrated from the station's infrastructure supervision and management systems (e.g., SCADA) and railway traffic control systems. These systems provide information on train arrivals, scheduled and actual arrival tracks, delays, and estimated passenger counts. Additionally, the station's gates can provide estimates of the number of passengers accessing the platform level who will soon board a train, as well as those moving from the platform level to the station premises and complementary transport facilities after alighting from a train.

In the *digital object*, the above-mentioned data will undergo processes of data cleaning, fusion, and mining to extract useful information (*knowledge extraction*) through prediction algorithms, machine learning techniques, and cloud computing. This enables the real-time monitoring of the station and its surrounding areas, allowing for the *prediction* of potential problems and the identification of alternative solutions that can be tested proactively in real time by parallel simulation (*anticipatory control*).

16.4 CONCLUSIONS AND PERSPECTIVES

AI and digital twinning are in the early stages of adoption within the railway sector and have been primarily focusing on maintenance, monitoring, and inspection of equipment [40,41]. The application of these technologies in railway hub management remains notably underexplored, with limited studies and practical implementations documented in the literature [39].

Big data generated from smart card transactions and passenger transit through station gates have been utilized to infer passenger routes and train choices [42], which can then be applied in simulation models. These data have also been used to develop machine learning and deep learning algorithms for estimating passenger flows in railway systems [41]. While most current applications focus on long-term predictions, a promising research avenue could be integrating real-time data with historical data to achieve short-term forecasts [41,43] and actuated action policy management. To this aim, the digital twinning of railway hubs represents a promising direction. However, several technological, legal, and security challenges must be addressed for its full realization.

Legally, the primary challenges involve individual privacy and the integration of data from systems managed by different operators, such as railway infrastructure managers, railway and public transport operators, and service providers. Technologically, challenges include the availability and integration of data, as well as the construction and validation of models. Synchronizing the physical and digital object levels presents an additional challenge, requiring data to be sampled and provided to the digital model in *near-real time*, while decisions made by the digital model must have direct connections to actuators for execution and effectiveness monitoring.

Finally, significant security issues must be addressed. The digital twin of the railway hub processes a vast amount of information, centralizing control over the entire railway node and station system. Therefore, it is crucial to implement high levels of redundancy and cybersecurity to prevent potential system failures and protect against cyber-attacks.

REFERENCES

1. European Commission. Directorate general for mobility and transport, Sustainable transport infrastructure charging and internalisation of transport externalities: Executive summary. LU: Publications Office, 2019. Accessed: June 21, 2024. [Online]. Available: https://data.europa.eu/doi/10.2832/246834.
2. N. Van Oort and M. Yap, 'Innovations in the appraisal of public transport projects', in *Advances in Transport Policy and Planning*, vol. 7. Elsevier, pp. 127–164, 2021. doi:10.1016/bs.atpp.2020.07.008.
3. R. Schakenbos, L. L. Paix, S. Nijenstein, and K. T. Geurs, 'Valuation of a transfer in a multimodal public transport trip', *Transport Policy*, vol. 46, pp. 72–81, 2016, doi:10.1016/j.tranpol.2015.11.008.
4. M. Yap, H. Wong, and O. Cats, 'Passenger valuation of interchanges in urban public transport', *Journal of Public Transportation*, vol. 26, p. 100089, 2024, doi:10.1016/j.jpubtr.2024.100089.
5. A. Monzón, S. Hernández, and F. Di Ciommo, 'Efficient urban interchanges: The City-HUB Model', *Transportation Research Procedia*, vol. 14, pp. 1124–1133, 2016, doi:10.1016/j.trpro.2016.05.183.
6. S. Zemp, M. Stauffacher, D. J. Lang, and R. W. Scholz, 'Classifying railway stations for strategic transport and land use planning: Context matters!', *Journal of Transport Geography*, vol. 19, no. 4, pp. 670–679, 2011, doi:10.1016/j.jtrangeo.2010.08.008.
7. M. Wicki, S. Hauller, T. Bernauer, and D. Kaufmann, 'Beyond a transport node? What residents want from transforming railway stations', *European Planning Studies*, pp. 1–23, 2023, doi:10.1080/09654313.2023.2236145.
8. M. Triggianese, R. Cavallo, N. Baron, and J. Kuijper, Eds., *Stations as Nodes: Exploring the Role of Stations in Future Metropolitan Areas from a French and Dutch perspective.* Delft: TU DElft Open, 2019.
9. P. Coppola and F. Silvestri, 'Assessing travelers' safety and security perception in railway stations', *Case Studies on Transport Policy*, vol. 8, no. 4, pp. 1127–1136, 2020, doi:10.1016/j.cstp.2020.05.006.
10. N. Coulombel, E. Munch, and C. Pivano, 'Travel demand management: The solution to public transit congestion? An ex-ante evaluation of staggered work hours schemes for the Paris region', *Transport Policy*, vol. 137, pp. 48–66, 2023, doi:10.1016/j.tranpol.2023.04.004.
11. D. Li, M. Schwartz, S. S. Sohn, S. Yoon, V. Pavlovic, and M. Kapadia, 'Microscopic modeling of attention-based movement behaviors', *Transportation Research Part C: Emerging Technologies*, vol. 162, p. 104583, 2024, doi:10.1016/j.trc.2024.104583.
12. K. Halldórsdóttir, O. A. Nielsen, and C. G. Prato, 'Home-end and activity-end preferences for access to and egress from train stations in the Copenhagen region', *International Journal of Sustainable Transportation*, vol. 11, no. 10, pp. 776–786, 2017, doi:10.1080/15568318.2017.1317888.
13. F. S. Hänseler, M. Bierlaire, and R. Scarinci, 'Assessing the usage and level-of-service of pedestrian facilities in train stations: A Swiss case study', *Transportation Research Part A: Policy and Practice*, vol. 89, pp. 106–123, 2016, doi:10.1016/j.tra.2016.05.010.
14. N. Molyneaux and M. Bierlaire, 'Controlling pedestrian flows with moving walkways', *Transportation Research Part C: Emerging Technologies*, vol. 141, p. 103672, 2022, doi:10.1016/j.trc.2022.103672.
15. R. J. Lee, I. N. Sener, and J. A. Mullins, 'An evaluation of emerging data collection technologies for travel demand modeling: From research to practice', *Transportation Letters*, vol. 8, no. 4, pp. 181–193, 2016, doi:10.1080/19427867.2015.1106787.
16. M. Luca, G. Barlacchi, B. Lepri, and L. Pappalardo, 'A survey on deep learning for human mobility', *ACM Computing Surveys*, vol. 55, no. 1, pp. 1–44, 2023, doi:10.1145/3485125.
17. P. Coppola, F. Silvestri, F. De Fabiis, and L. Barbierato, 'Disaggregate travel demand analysis using big data sources: Unsupervised learning methods for data-driven trip purpose estimation', *Transportation Research Procedia*, vol. 82, pp. 1824–1838, 2024.
18. T. F. Welch and A. Widita, 'Big data in public transportation: A review of sources and methods', *Transport Reviews*, vol. 39, no. 6, pp. 795–818, 2019, doi:10.1080/01441647.2019.1616849.
19. L. Willumsen, *Use of Big Data in Transport Modelling.* Paris: OECD Publishing, 2021.
20. M. E. Ben-Akiva and S. R. Lerman, 'Discrete choice analysis: Theory and application to travel demand', in *MIT Press Series in Transportation Studies*, vol. 9. Marvin Manheim, Editor, MIT Press, Cambridge, MA, 1985.

21. F. Martinez-Gil, M. Lozano, I. García-Fernández, and F. Fernández, 'Modeling, evaluation, and scale on artificial pedestrians: A literature review', *ACM Computing Surveys*, vol. 50, no. 5, pp. 1–35, 2018, doi:10.1145/3117808.
22. M. Dubroca-Voisin, B. Kabalan, and F. Leurent, 'On pedestrian traffic management in railway stations: Simulation needs and model assessment', *Transportation Research Procedia*, vol. 37, pp. 3–10, 2019, doi:10.1016/j.trpro.2018.12.159.
23. X. Chen, M. Treiber, V. Kanagaraj, and H. Li, 'Social force models for pedestrian traffic – State of the art', *Transport Reviews*, vol. 38, no. 5, pp. 625–653, 2018, doi:10.1080/01441647.2017.1396265.
24. D. Helbing and P. Molnár, 'Social force model for pedestrian dynamics', *Physical Review E*, vol. 51, no. 5, pp. 4282–4286, 1995, doi:10.1103/PhysRevE.51.4282.
25. I. Kim, R. Galiza, and L. Ferreira, 'Modeling pedestrian queuing using micro-simulation', *Transportation Research Part A: Policy and Practice*, vol. 49, pp. 232–240, 2013, doi:10.1016/j.tra.2013.01.018.
26. C. Caramuta, G. Collodel, C. Giacomini, C. Gruden, G. Longo, and P. Piccolotto, 'Survey of detection techniques, mathematical models and simulation software in pedestrian dynamics', *Transportation Research Procedia*, vol. 25, pp. 551–567, 2017, doi:10.1016/j.trpro.2017.05.438.
27. D. C. Duives, W. Daamen, and S. P. Hoogendoorn, 'State-of-the-art crowd motion simulation models', *Transportation Research Part C: Emerging Technologies*, vol. 37, pp. 193–209, 2013, doi:10.1016/j.trc.2013.02.005.
28. S. Yang, T. Li, X. Gong, B. Peng, and J. Hu, 'A review on crowd simulation and modeling', *Graphical Models*, vol. 111, p. 101081, 2020, doi:10.1016/j.gmod.2020.101081.
29. K. Kušić, R. Schumann, and E. Ivanjko, 'A digital twin in transportation: Real-time synergy of traffic data streams and simulation for virtualizing motorway dynamics', *Advanced Engineering Informatics*, vol. 55, p. 101858, 2023, doi:10.1016/j.aei.2022.101858.
30. I. C. Semanjski, *Smart Urban Mobility: Transport Planning in the Age of Big Data and Digital Twins*. Elsevier, Amsterdam, Netherlands; Cambridge, MA, 2023.
31. W. Kritzinger, M. Karner, G. Traar, J. Henjes, and W. Sihn, 'Digital Twin in manufacturing: A categorical literature review and classification', *IFAC-PapersOnLine*, vol. 51, no. 11, pp. 1016–1022, 2018, doi:10.1016/j.ifacol.2018.08.474.
32. C. Schwarz and Z. Wang, 'The role of digital twins in connected and automated vehicles', *IEEE Intelligent Transportation Systems Magazine*, vol. 14, no. 6, pp. 41–51, 2022, doi:10.1109/MITS.2021.3129524.
33. M. Grieves, 'Intelligent digital twins and the development and management of complex systems [version 1; peer review: 4 approved]', *Digital Twin*, vol. 2, no. 8, 2022, doi:10.12688/digitaltwin.17574.1.
34. Z. Wang, R. Gupta, K. Han, H. Wang, A. Ganlath,N. Ammar, et al., 'Mobility digital twin: Concept, architecture, case study, and future challenges', *IEEE Internet Things J.*, vol. 9, no. 18, pp. 17452–17467, 2022, doi:10.1109/JIOT.2022.3156028.
35. F. Emmert-Streib, 'What is the role of AI for digital twins?', *AI*, vol. 4, no. 3, Art. no. 3, 2023, doi:10.3390/ai4030038.
36. T. Kreuzer, P. Papapetrou, and J. Zdravkovic, 'Artificial intelligence in digital twins—A systematic literature review', *Data & Knowledge Engineering*, vol. 151, p. 102304, 2024, doi:10.1016/j.datak.2024.102304.
37. L. Bao, Q. Wang, and Y. Jiang, 'Review of digital twin for intelligent transportation system', in *2021 International Conference on Information Control, Electrical Engineering and Rail Transit (ICEERT)*, Lanzhou, China: IEEE, 2021, pp. 309–315, doi:10.1109/ICEERT53919.2021.00064.
38. L. Wright and S. Davidson, 'How to tell the difference between a model and a digital twin', *Advanced Modeling and Simulation in Engineering Sciences*, vol. 7, no. 1, p. 13, 2020, doi:10.1186/s40323-020-00147-4.
39. A. Padovano, F. Longo, L. Manca, and R. Grugni, 'Improving safety management in railway stations through a simulation-based digital twin approach', *Computers & Industrial Engineering*, vol. 187, p. 109839, 2024, doi:10.1016/j.cie.2023.109839.
40. S. Ghaboura, R. Ferdousi, F. Laamarti, C. Yang, and A. E. Saddik, 'Digital twin for railway: A comprehensive survey', *IEEE Access*, vol. 11, pp. 120237–120257, 2023, doi:10.1109/ACCESS.2023.3327042.
41. R. Tang, L. De Donato, N. Bešinović, F. Flammini, R. M. P. Goverde, Z. Lin, et al., 'A literature review of Artificial Intelligence applications in railway systems', *Transportation Research Part C: Emerging Technologies*, vol. 140, p. 103679, 2022, doi:10.1016/j.trc.2022.103679.
42. F. Ghofrani, Q. He, R. M. P. Goverde, and X. Liu, 'Recent applications of big data analytics in railway transportation systems: A survey', *Transportation Research Part C: Emerging Technologies*, vol. 90, pp. 226–246, 2018, doi:10.1016/j.trc.2018.03.010.
43. M. Gallo, G. De Luca, L. D'Acierno, and M. Botte, 'Artificial neural networks for forecasting passenger flows on metro lines', *Sensors*, vol. 19, no. 15, p. 3424, 2019, doi:10.3390/s19153424.

17 Ethical Considerations for the Development and Deployment of AI in Rail

Yit Hong Choo, Douglas Creighton, Fariza Sabrina, Hong Shen, and MD Mamunur Rashid

17.1 INTRODUCTION

Artificial intelligence (AI) holds the promise of significant advancements in the rail industry, from enhancing operational efficiency to improving safety, equity, and fairness. The rapid advancement and integration of AI systems into everyday life, however, brings forth significant ethical considerations that must be addressed to ensure responsible use and societal benefit. This chapter provides a detailed overview of these ethical considerations, specifically focusing on the development and deployment of AI in the rail industry. It is informed by established ethical AI frameworks and guidelines substantiated with current and relevant references.

Frameworks for ethical AI, such as the Ethical Use of Artificial Intelligence Framework by NASA [1] and the Ethics Guidelines for Trustworthy AI by the European Commission (EC) [2], are essential for providing guidance on how to develop and implement AI in the rail industry. In essence, these guidelines offer extensive direction for maneuvering through the intricate realm of AI ethics, stressing the requirement for AI systems to adhere to legality, ethics, and resilience. It highlights the significance of fairness, transparency, accountability, privacy, and safety principles, all of which are essential in preventing AI systems from perpetuating prejudices and prejudice. Making AI operations and decision-making processes understandable to users fosters trust and accountability, while mechanisms to ensure that developers and users of AI systems are responsible for the outcomes of their technologies provide a means for redress in cases of harm or misuse.

Ethical AI principles can enhance operational efficiency, safety, and fairness, ensuring that AI-driven systems for scheduling, maintenance, and providing timely customer service are equitable and transparent [2–4]. The commitment to ethical AI is not a task for a single entity but a collective responsibility involving stakeholders from industry, academia, civil society, and policymakers. The responsible use of AI relies on the important contributions of every stakeholder, and by following these ethical principles, we can leverage the possibility of AI for positive change while upholding societal values, defending individual rights, and advancing the greater good.

Fairness is crucial in the rail industry to prevent biases and discrimination [2–4]. Regular audits of AI systems are essential to detect and mitigate biases, ensuring equitable service for all passengers, including designing accessible ticketing systems for passengers with disabilities. Additionally, AI can optimize train schedules and crew assignments to enhance efficiency, but ensuring these systems are accountable and fair will prevent biases and ensure equitable working conditions for employees. Transparency involves providing clear explanations for AI decisions and openly communicating the workings of AI systems to build trust among passengers and employees [2–4]. Accountability requires clear lines of responsibility for AI system outcomes and the implementation of regular audits to ensure compliance with ethical standards and industry regulations [2–4].

Privacy and data governance are highlighted as essential considerations, emphasizing the need for stringent data protection measures and ensuring that passengers and employees have control

DOI: 10.1201/9781003492146-17

over their data [2,4]. Technical robustness and safety are also critical to ensure that AI systems can handle failures gracefully without causing service disruptions and are extensively tested to operate safely in various scenarios [2,4]. Finally, the importance of human control is underlined, emphasizing the need for mechanisms for human oversight and intervention in AI-driven processes and the training of rail industry employees on the capabilities and limitations of AI systems [3,4].

Social responsibility is a crucial aspect of AI development and deployment [2–4]. Ensuring that AI systems treat all individuals and groups without bias is fundamental. This involves designing AI with fairness in mind, conducting regular audits to detect and prevent discriminatory outcomes, and considering the diverse needs of different populations, including marginalized and vulnerable groups. This approach helps create an inclusive society where everyone can benefit from AI advancements.

AI can also contribute to solving societal challenges, such as improving healthcare, promoting environmental sustainability, and advancing education [5]. Therefore, the development and deployment of AI systems should take into account their environmental impact, promoting sustainable practices and reducing carbon footprints. This comprehensive approach addresses both the technical and broader social and environmental implications of AI.

Ethical governance is another essential aspect of responsible AI development and deployment. Organizations must establish and adhere to ethical guidelines for AI development and deployment, informed by interdisciplinary perspectives and stakeholder input [3,4]. These guidelines should prioritize transparency, accountability, and respect for human rights. Regular checks and reviews of AI systems are vital to safeguard that they uphold ethical standards and societal values.

Public engagement and education are critical components of responsible AI development and deployment [1,3,4,6]. Efforts should be made to inform the general public about the advantages and possible drawbacks of AI technologies. This includes transparent communication and accessible information that is understandable by all. Additionally, engaging various stakeholders, including the public, in discussions about AI ethics and governance can foster greater trust and ensure diverse perspectives are considered. By working together, we can create a more responsible and accountable AI ecosystem that prioritizes social responsibility and promotes positive societal outcomes.

The ethical considerations for the development and deployment of AI in the rail industry are complex and require a comprehensive approach. By adhering to established ethical principles and frameworks, and through the collective efforts of stakeholders, we can harness the transformative potential of AI while safeguarding societal values, protecting individual rights, and promoting the common good. Ensuring fairness, transparency, accountability, privacy, and safety in AI systems will not only enhance operational efficiency and service quality in the rail industry but also build public trust and contribute to a more equitable and sustainable future.

17.2 ETHICAL AI CONSIDERATIONS IN RAIL SYSTEMS

Railway systems are composed of infrastructure, rolling stock, and operations, which collectively facilitate the transportation of passengers and freight along a network of tracks [7]. The infrastructure includes physical rails, guideways, bridges, tunnels, catenary systems, and other structures essential for train movement. Rolling stock refers to self-propelled vehicles such as locomotives, passenger coaches, and freight cars, maintained at facilities like depots or yards [8]. Common issues affecting these assets encompass component failures, human error, and external factors like weather conditions or third-party interference.

Traditionally, the railway industry has relied on conventional methods such as regular maintenance scheduling, incident investigation, weather monitoring, security measures, communication protocols, and so on to address common issues [8]. Nevertheless, there are emerging advanced AI applications such as infrastructure status tracking systems powered by AI, passenger train line-scan sensors, and cameras, fiber optic acoustic sensors for detecting rail and wheel issues, trespassers, and level crossings, and AI-based algorithms for identifying damage in pantographs, wheels, and brake blocks [8].

The potential of AI in robust rail logistic planning is also being explored [3,9]. The railway infrastructure is a highly complex system with dynamic characteristics that change over time due to usage, degradation processes, and maintenance [8,10]. The dynamics vary across different locations such as bridges, tunnels, stations, and curves. The infrastructure is also impacted by various stochastic factors, including weather conditions, making it a dynamic, continuous, distributed, and stochastic system [8,10]. This complexity requires the development of new intelligent methods for practical solutions.

Furthermore, monitoring and maintaining health conditions is essential for ensuring service safety, availability, and reliability while prolonging the lifecycle of assets and infrastructure. Standardization and governance for big data management and analytics are necessary for advanced railway networks to effectively monitor infrastructure conditions and manage lifecycle costs [11,12]. Proper data management allows asset managers to shift from reactive to predictive maintenance, as seen in the literature [13,14]. The growing use of AI in various fields enables the automation of decision support tools by linking data with decisions and facilitating analysis of both asset-specific and whole-system behavior.

Railway systems are also characterized as complex systems, and require holistic management approaches [3,8]. These systems encompass infrastructure, rolling stock, and operations, including scheduling and dispatching, traffic management, infrastructure maintenance, and technology integration [3,8,15,16]. They serve multiple purposes, such as passenger transportation, freight transportation, goods movement, and energy transmission [9]. Passenger transportation involves moving people over short or long distances for work, leisure, or education. Freight transportation focuses on delivering goods across the country or region, supporting industries and businesses by ensuring efficient product movement [17]. Goods movement includes transporting raw materials, finished products, and other cargo between industries, businesses, and consumers. Energy transmission involves delivering resources like coal and oil from production sites to power plants, refineries, or storage facilities, ensuring a reliable energy supply.

Efficient and effective management of rail systems involves employing various tools and strategies [16]. These include scheduling and dispatching to coordinate train movements, traffic management to regulate train flow and maintain safety, infrastructure maintenance to ensure the proper functioning of critical components, and technology integration to implement advanced systems that enhance safety, efficiency, and capacity [18,19]. The modern transportation network benefits from the presence of railway systems, as they help reduce congestion, decrease emissions, and enhance capacity in comparison to other modes of transportation. Their contribution to the transportation industry is evident in their ability to support both passenger and freight movement while providing a sustainable and efficient means of transportation.

17.2.1 Application Aspects of AI Technologies in Rail Domains

The railway sector comprises various sub-domains, each with distinct characteristics and challenges. Tang et al. [9] identified seven critical sub-domains of AI applications in railways, including maintenance and inspection, safety and security, autonomous driving and control, traffic planning and management, revenue management, transport policy, and passenger mobility [4]. These sub-domains are essential for understanding the overall functioning and performance of railway systems. Maintenance and inspection are crucial for ensuring the proper operating condition of railway systems through preventive and corrective measures. Safety and security are fundamental concerns, with the aim of minimizing the likelihood of accidents and hazards to ensure the safety of passengers and assets.

Autonomous driving and control involve developing systems for trains to operate automatically, enhancing efficiency and competitiveness [20]. Traffic planning and management focus on capacity management, timetabling, and resource allocation to increase efficiency. Revenue management optimizes service availability and price to maximize revenue growth. Transport policy involves developing strategies to attain specific objectives relating to the functioning and performance of the transport system. In addition, passenger mobility denotes the transfer of individuals utilizing transportation methods, highlighting cost-effectiveness, availability, and security.

It is crucial to comprehend AI applications in various areas to improve railway operations and services. Therefore, the following section will delve into different AI approaches, specific AI techniques and applications, and the associated data used in railway sub-domains. These include maintenance and inspection, safety and security, traffic planning and management, and passenger mobility.

17.2.1.1 Maintenance and Inspection

The maintenance and inspection subdomain within the realm of railway infrastructure management encompasses a range of activities focused on evaluating the condition and operational status of complex mechanical and electrical systems. In this context, AI has been successfully applied to address various challenges, including:

i. *Defect detection*: Identifying physical imperfections such as faults, missing components, and scratches within railway infrastructure
ii. *Fault detection*: Identifying anomalies in electrical and mechanical components, followed by diagnosis to determine the root cause of the fault
iii. *Defect prediction*: Forecasting the progression of defects, such as track deterioration
iv. *Failure prediction*: Predicting when a component's behavior deviates from its nominal operating characteristics
v. *Maintenance planning*: Incorporating decision support systems that facilitate dynamic scheduling of maintenance and inspection activities
vi. *Autonomous maintenance systems*: Enabling intelligent and automated inspection procedures

For example, it explores various AI/machine learning (ML) methods for detecting defects in railway tracks, rolling stocks, and catenaries, such as AdaBoost [21], Latent Dirichlet Allocation [22], deep neural networks (DNNs) [23,24], support vector machines (SVMs) [25], fully convolutional neural network (CNN) [26], and multitask learning framework [27], for identifying track-related defects like fastening system issues, track geometry anomalies, and rail defects. Additionally, it investigates AI/ML approaches for detecting defects in track geometry using neural networks [28], SVMs [29], and big data [30].

Furthermore, the study examines the application of AI/ML techniques for identifying surface and structural defects in rails, employing CNN-based architectures [31–35], transfer learning approaches [36], traditional ML methods such as linear SVM [37], random forest 38], unsupervised approaches in combination with principal component analysis (PCA) [10], and semi-supervised approaches [39].

In addition, AI/ML is applied in detecting defects in rolling stock components like bogies, wheels, and underframe components using SVMs, AdaBoost, and CNNs [17,40,41]. Moreover, studies such as Zhou et al. [42] used traditional ML techniques (SVMs, PCA, LDA, ANNs), and Lee et al. [43] proposed deep learning approaches (RNNs, deep belief networks, capsule neural networks, deep residual networks) to address fault detection and diagnosis. Predictive maintenance and autonomous maintenance are also explored, showcasing how AI can predict defects and estimate the remaining useful life of components, as well as how autonomous systems can inspect and maintain railway components without human intervention [44,45].

17.2.1.2 Safety and Security

To mitigate safety and security concerns associated with individual trains or entire railway systems, researchers have explored the potential of AI techniques in addressing complex practical scenarios.

Researchers have investigated AI-based approaches to risk management, failure/error estimation, and railway accidents.

Various AI-based methods were proposed for risk assessment, including fuzzy reasoning approaches, fuzzy analytical hierarchy decision-making processes, and particle swarm optimization algorithms. For instance, studies by An et al. [46], Gul and Celik [47], and Zhang et al. [48] aimed to qualitatively identify risk clusters and corresponding control measures. In addition, studies have focused on the maintenance and failure estimation of track circuits, a crucial device for train detection. For example, Zilko et al. [49] and Li et al. [50] employed Bayesian networks and knowledge rules-based methods to predict disruption lengths and simulate reasoning processes.

Furthermore, decision tree algorithms have been used to identify patterns and predict behavior in hazardous events at railway stations, while adaptive neural fuzzy inference systems have been proposed for identifying qualified candidates among level crossings. For example, studies by Alawad et al. [51] and Ćirović and Pamučar [52] aimed to optimize safety measures. Researchers have employed various data sources, including expert scoring, historical accident records, and vision captures of track defects or surroundings, to inform their AI-based approaches.

17.2.1.3 Autonomous Train Driving and Train Control

The use of AI in autonomous train driving and vehicle control has attracted significant attention in recent years due to its potential to improve safety and optimize efficiency. Its powerful learning and computation capabilities have led to the development of various approaches to address complex problems, such as energy-efficient driving and verification and validation.

Researchers have applied a range of AI-based methods to optimize energy-efficient driving, including genetic algorithms [53], approximate dynamic programming (ADP) [54], reinforcement learning (RL) [55], and SVMs [56]. For example, Wang et al. [54] used ADP to learn the costs of energy and time over iterations, while Yin et al. [55] introduced two train control algorithms that operated similarly to an experienced driver, reducing energy consumption while maintaining comfort levels and punctuality. Furthermore, Hua et al. [56] proposed an approach that incorporated blockchain-based Federated Learning and SVM-based intelligent control models to achieve safe and efficient control of heavy haul rail systems.

AI can also be used in the verification and validation of railway control by learning from expert knowledge without the presence of an expert [57]. In addition, big data analysis has been employed to assess the performance of algorithms. For instance, a model based on fuzzy Resource Description Framework and uncertainty was employed by Zhang [58] for different high-speed train control systems. Moreover, researchers have mainly utilized information on real life and theoretical characteristics of infrastructures (e.g., road conditions) and vehicle components (e.g., engines) to evaluate the performance of algorithms. Existing studies have also gathered data from prior knowledge from the expert (e.g., experienced driver) to develop and fine-tune ML models.

17.2.1.4 Traffic Planning and Management

The integration of AI techniques is increasingly recognized as a promising approach to tackling the challenges posed by the growing complexity and unpredictability of traffic patterns in rail transportation systems. AI methods are applied to various aspects of traffic planning and management, including timetabling, routing, shunting, capacity management, traffic analysis and prediction, and disruptions management, with the overarching goals of enhancing efficiency, safety, and reliability [3,8,9]. Recent research has demonstrated the effectiveness of AI in analyzing delay patterns and predicting conflicts in train scheduling. For instance, Wang and Zhang [59] utilized a gradient-boosted regression tree model to study the impact of weather conditions and scheduled timetables on train delays. Similarly, Cerreto et al. [60] employed k-means clustering to identify significant delay patterns and classify the primary causes of delay occurrences.

Evolutionary-based techniques, such as genetic algorithms and particle swarm optimization, have been utilized to optimize train scheduling and rescheduling processes. For example, Wang and Zhang [59] proposed a genetic algorithm-based particle swarm optimization method to minimize secondary delays and reduce the number of trains exceeding a predefined delay threshold. Additionally, RL approaches have been leveraged to develop decision-making systems for traffic planning and management. Obara et al. [18] introduced a deep Q-network RL algorithm to optimize train scheduling and minimize average total delay, while Roost et al. [19] applied a model-free asynchronous advantage actor-critic RL algorithm to reschedule trains during disruptions.

Deep learning techniques, including fully connected neural networks (FCNN) and long short-term memory (LSTM) layers, have been employed to predict train delays and optimize traffic management strategies. Huang et al. [61] developed a train delay prediction model, FCLL-Net model integrating a fully connected neural network with two LSTM layers to analyze operational interactions between trains and forecast delays. Furthermore, expert systems and knowledge-based decision support systems have been deployed to devise cost-effective solutions for traffic planning and management. Deng et al. [62] proposed an expert system utilizing cost functions based on the total sum of train delays to optimize train scheduling.

Moreover, effective AI-based studies for traffic planning and management necessitate the utilization of various historical data sources, such as realized traffic movements, infrastructure occupation data, weather records, existing train scheduled timetables, the topology of the rail network, and accident event data. In conclusion, the application of AI techniques holds substantial promise in revolutionizing traffic planning and management, offering more precise predictions, optimizing scheduling decisions, and enhancing overall system performance. Nonetheless, further research is required to fully explore the potential of AI in this domain and develop practical solutions for real-world implementation.

17.2.1.5 Passenger Mobility

In the realm of passenger mobility in the rail industry, the goal is to provide satisfactory transportation for passengers. There is a significant focus on analyzing passenger demand to predict traffic, including both mainline and urban railway systems. Hence, accurate prediction of passenger traffic is essential for efficient and effective service delivery. For this reason, AI has become an important technology in this context, offering various approaches to improve the accuracy of traffic prediction.

Deep learning architectures have been particularly prominent in this domain, with applications ranging from feed-forward artificial neural networks (ANNs) to long short-term memory recurrent neural networks (LSTM-RNN). For instance, Gallo et al. [63] employed ANNs to predict metro onboard passenger flows, while Liu et al. [64] developed an end-to-end passenger flow prediction architecture that integrated domain knowledge and deep learning. Other studies have explored the use of graph convolutional networks and three-dimensional CNNs for passenger forecasting at railway stations [65].

In addition to these AI-based approaches, traditional supervised learning methods have also been applied to predict and classify passenger load categories [66]. The choice of data source has been critical in these studies, with historical ridership data being commonly used. This data can be derived from smart card transactions or tickets, although the long collection time of smart card data may hinder short-term passenger flow prediction. In contrast, Liu et al. [64] forecasted passenger flow onboard based on passenger counts at station turnstiles, which can be retrieved in near real-time. These findings highlight the potential of AI to improve the accuracy and efficiency of passenger mobility services in the rail industry.

17.2.1.6 Freight and Logistics

AI has significantly transformed freight and logistics in rail systems, enhancing efficiency, safety, and reliability. The primary applications of AI in this domain include predictive maintenance, autonomous operations, demand forecasting, and real-time tracking and monitoring. AI systems provide real-time tracking and monitoring of freight cars and cargo [67]. These systems use Internet of Things devices and AI algorithms to monitor the condition and location of cargo, ensuring timely deliveries and reducing the risk of loss or damage. AI can also optimize routing and scheduling, taking into account real-time traffic and weather conditions.

AI facilitates the development of autonomous trains, which can operate with minimal human intervention. These trains use AI algorithms for navigation, obstacle detection, and decision-making, ensuring safe and efficient operations. Autonomous operations are particularly beneficial for long-haul freight trains, as they can reduce operational costs and improve scheduling accuracy. AI-powered demand forecasting tools analyze historical data and external factors (such as economic indicators and weather conditions) to predict future demand for rail freight services. This allows rail operators to optimize resource allocation, improve load planning, and enhance overall service efficiency [68].

17.2.2 Ethics Issues

- Transparency
- Diversity and fairness
- Safety and reliability
- Privacy and security
- Societal and environmental well-being
- Education and awareness
- Accountability and explicability

The application of AI in rail systems raises several ethical considerations that must be taken into account. On the one hand, AI has the potential to revolutionize the way we operate and manage rail networks, improving safety, efficiency, and passenger experience. On the other hand, as AI becomes increasingly integrated into rail operations, it is essential that we consider the potential ethical implications of its use.

Safety is one of the primary concerns when it comes to AI. The effectiveness of AI relies heavily on the integrity of its training data. If this data is flawed or incomplete, the consequences could be severe, endangering passengers, crew, and the environment [3,4]. AI applications that directly impact passenger safety such as autonomous driving and control mechanisms present significant ethical dilemmas [69]. These systems must balance ensuring maximum safety with maintaining passenger comfort. For example, an AI's decision to initiate braking when detecting an obstacle must be precisely calibrated. Additionally, the risk of false positives in obstacle detection requires careful consideration. The AI's response to unexpected track intrusions, whether by animals or vehicles, involves ethical complexities that need thorough examination and resolution [3]. By addressing these ethical challenges, we can harness AI's potential to enhance rail system safety without compromising the ethical standards that protect human life and the environment.

Another ethical consideration is privacy. As AI processes large amounts of data, including passenger and employee information, ensuring that this information remains confidential is vital [3,4,6,70]. The potential for AI-powered surveillance systems to infringe on individuals' privacy rights is a serious concern, especially in the context of rail networks where sensitive information may be shared among multiple stakeholders. Rigorous data protection measures and transparent policies are essential to prevent misuse and unauthorized access to personal data, thereby safeguarding individual privacy rights.

The risk of cyber security breaches is also a significant ethical consideration. As AI systems become increasingly connected and dependent on cloud-based services, they are more vulnerable to cyber-attacks, data breaches, and unauthorized access [3,6,70]. These vulnerabilities can compromise the safety and functionality of rail systems, potentially leading to operational disruptions, financial losses, and threats to passenger safety. To mitigate these risks, it is imperative to implement robust cybersecurity measures, including encryption, regular security audits, and real-time threat detection systems. Additionally, developing comprehensive incident response plans and ensuring continuous monitoring can help protect AI-powered rail systems from malicious actors and safeguard sensitive data against potential breaches.

Furthermore, the application of AI in rail systems must also consider social impact. The development and deployment of AI technologies can have significant social implications, such as job displacement, changes in work dynamics, and impacts on local communities [71]. As AI replaces human workers in certain roles, it is essential that we prioritize retraining and upskilling programs to ensure a smooth transition for affected employees. This not only helps mitigate the negative effects of job displacement but also empowers workers with new skills that are relevant in an increasingly automated industry. Additionally, it is crucial to engage with local communities to address their concerns and ensure that AI implementations do not disproportionately affect certain groups. Communication about the benefits and challenges of AI in rail systems can help build trust and acceptance among stakeholders.

In addition, the use of AI in rail systems must also consider environmental impact. While AI-powered systems may reduce energy consumption and emissions in some cases, their development and deployment can still have negative environmental implications, such as resource extraction and processing, which can contribute to greenhouse gas emissions and climate change [71]. To mitigate these risks, it is crucial that we prioritize sustainable practices throughout the entire AI lifecycle, from design to disposal. Ultimately, the responsible development and deployment of AI in rail systems require a thoughtful consideration of these ethical concerns. By acknowledging and addressing these potential issues, we can create safer, more efficient, and environmentally sustainable rail networks that benefit both passengers and the environment.

17.3 CHALLENGES IN IMPLEMENTING ETHICAL AI IN RAILWAYS

The implementation of AI in railways promises significant advancements in efficiency, safety, and customer satisfaction. It also presents ethical and practical challenges that must be addressed to ensure responsible deployment, including privacy, safety, cybersecurity, and integrity. Below, we discuss these challenges, identify specific issues arising from AI deployment in rail applications, analyze real-world applications, and draw parallels with case studies from other sectors.

AI systems in railways will increasingly rely on large volumes of data collected from systems, passengers, employees, and the environment. This data may include personal information, travel patterns, and behavioral analytics. The challenge lies in ensuring the privacy of individuals by implementing stringent data protection measures and complying with relevant regulations such as the General Data Protection Regulation (GDPR) [4]. AI applications, such as autonomous trains and predictive maintenance, directly impact the safety of operations. Ensuring that AI systems are reliable and capable of handling unexpected situations without causing harm is critical.

Furthermore, AI systems in railways are susceptible to cyber-attacks, which could lead to significant disruptions or safety hazards. Implementing robust cybersecurity measures to protect AI systems from malicious attacks and ensuring data integrity is a significant challenge. AI systems must operate with high levels of accuracy and integrity to ensure reliable decision-making. This involves ensuring data quality, avoiding biases in AI algorithms, and maintaining the integrity of AI-driven decisions. Determining who is responsible for decisions made by AI systems, particularly in autonomous or semi-autonomous operations, is also challenging. Establishing clear accountability frameworks and legal standards to address liability issues is essential.

Real-world applications of AI in railways highlight several ethical issues. For instance, predictive maintenance uses AI systems to analyze sensor data from trains and infrastructure to predict failures and schedule maintenance proactively. Ethical concerns include ensuring that data collected from sensors does not infringe on individual privacy, verifying the accuracy of predictions to prevent accidents, and maintaining data quality to avoid biases in maintenance scheduling algorithms. Autonomous trains operate using AI for navigation, speed control, and obstacle detection. Ethical issues here include assuring that autonomous systems can handle all operational scenarios safely, determining responsibility for incidents involving autonomous trains, and protecting autonomous systems from hacking and unauthorized access.

AI-powered applications provide real-time updates, personalized services, and enhanced security measures for passengers. Ethical issues include protecting personal data collected from passengers to offer bespoke services, ensuring that AI-driven services do not discriminate against any group of passengers, and clearly communicating how AI systems use passenger data. AI optimizes train schedules, manages rail traffic, and minimizes delays through real-time analysis and decision-making. Ensuring the accuracy and reliability of data used in traffic management systems, clarifying responsibility for decisions made by AI in managing rail traffic, and making the decision-making process of AI systems understandable to operators and stakeholders are crucial.

17.4 AI IMPLEMENTATION CASE STUDIES FROM OTHER SECTORS

In this subsection, we will discuss the ethical challenges of using AI in other applications such as healthcare, agriculture, and smart cities.

17.4.1 Ethical Challenges in Healthcare

Although using AI in clinical healthcare practices has enormous promise to improve patient care, it also presents ethical issues. However, what is the relationship between the informed consent standards and the use of AI to support patient care? Even though obtaining informed consent will be one of the biggest obstacles to implementing AI in clinical practice right away, this is an urgent issue that has not received sufficient attention in the ethical community. It is important to consider when (if ever) it is appropriate to apply the principles of informed consent in the field of clinical AI. How responsible are clinicians for informing patients about the intricacies of AI, such as the types of ML utilized, the data inputs involved, and the potential biases or limitations in the data being used? These questions can be particularly difficult to address when the AI functions through "black-box" algorithms, stemming from complex ML methods that may be challenging for clinicians to fully comprehend [72]. Ensuring safety is a major obstacle for AI in the healthcare industry. It emphasizes the critical need for AIs to be both safe and efficient. The question arises as to how we can guarantee that AIs deliver on their commitments. To fully harness the potential of AI, it is essential for stakeholders, especially AI developers, to prioritize two important factors: the accuracy and credibility of the data sets and transparency [73].

17.4.2 Ethical Challenges in Agriculture

AI technology in agriculture holds promise for enhancing farming methods, yet it also presents privacy worries. These systems depend on gathering and utilizing sensitive data, such as farm management data and the personal information of farmers. AI technologies could significantly enhance livestock monitoring and automated farming practices in agriculture. Nevertheless, it is crucial to consider the possible impact that these innovations could have on animal well-being. Specifically, there are ethical issues to consider regarding the application of AI in areas like livestock monitoring and automated farming operations. The integration of AI technology in agriculture could result in major shifts in workforce dynamics and socio-economic consequences for farming communities. With the automation of tasks, there is a possibility of workers being replaced by AI systems, potentially affecting small-scale farmers and rural communities and creating obstacles related to livelihoods and financial security. The importance of ensuring fairness and transparency in AI algorithms extends beyond simply adhering to regulations – it fundamentally involves fostering trust and ensuring equitable treatment for all individuals in the agricultural industry. When AI technology is utilized in critical decision-making processes such as forecasting crop yields or determining loan recipients, any instance of bias can lead to disparities in how farmers are treated, ultimately impacting their livelihoods and competitiveness within the industry. Government policies and regulations are essential in steering the progress and implementation of AI in agriculture. To promote the ethical adoption of AI in farming, it is important to create regulatory structures that tackle the main obstacles and promote responsible and sustainable use of these technologies [74].

17.4.3 Ethical Challenges in Smart City Developments

The advantages of incorporating AI in smart cities revolve around the idea of making things more effective and safer, and promoting better economic results for residents. AI-powered ML systems in "smart cities" are equipped with the transformative abilities of AI, improving efficiency and reducing environmental impact. The primary concern when using AI in urban areas is the ethical dilemmas that arise. Utilizing big data, digital tools, and AI can lead to concerns about surveillance and privacy. The potential threats to civil liberties and personal privacy are significant due to these technologies. A major moral issue is striking a proper balance between safeguarding human rights and utilizing AI for city optimization. Considering the EU's leading stance with the GDPR, it is logical for the union to take strong actions to guarantee the ethical use of AI. Nevertheless, being cautious in this way can cause a delay in moving from coming up with ideas to putting them into action. It is crucial to have responsible AI governance frameworks and practices in place to address these issues. Nonetheless, ensuring the ethical and compliant use of AI is a complex and time-consuming endeavor [75].

The integration of AI into the rail industry introduces a range of complex ethical challenges, akin to those elucidated in the EC Report on the Ethics of Connected and Automated Vehicles [5]. The report highlights the crucial need to incorporate ethical considerations into the development of AI technologies, with a specific focus on safety, stakeholder engagement, and responsible innovation. These principles are equally applicable to the rail industry, which must navigate both technical and ethical complexities to ensure the safe and equitable deployment of AI systems.

The responsible innovation (RI) approach, advocated by the report, accentuates inclusive and responsive dialogue among stakeholders to integrate ethical principles into technological advancements [76]. This methodology can be instrumental for the rail industry, which is similarly complex and has broad societal impacts [76]. Given the socio-technical complexity of transportation systems, ethical AI development in rail must prioritize stakeholder engagement, integrate ethical guidelines with legislative measures, and ensure technologies reflect societal values and needs. This approach not only addresses current challenges but also lays a foundation for RI in future transportation technologies.

By adopting an inclusive, iterative, and values-driven approach, the rail industry can develop AI systems that enhance safety, efficiency, and equity, while maintaining public trust and accountability. This holistic strategy ensures that technological advancements align with ethical principles and contribute positively to societal welfare. However, comprehensive and continuous stakeholder engagement poses challenges when implementing ethical AI due to the logistics of organizing stakeholder meetings and obtaining iterative feedback can be resource-intensive, and there is a risk of dominant voices overshadowing minority perspectives.

Additionally, adapting ethical guidelines and legislative frameworks to the dynamic nature of AI technologies presents a significant challenge. The need for ongoing dialogue and revisions based on new data and experiences can be difficult to commit within rigid legislative frameworks. Another challenge is embedding societal values into AI systems, which is complex and may face resistance from those advocating for purely technical solutions [5,76]. Achieving a balance between technical efficiency and ethical responsiveness requires interdisciplinary collaboration and a willingness to embrace complexity. Translating abstract ethical principles into concrete design specifications is also a non-trivial task, necessitating iterative, context-specific deliberations that can be time-consuming and require continuous stakeholder involvement. As AI systems become more autonomous, outlining the boundaries of human responsibility can become blurred, necessitating clear guidelines and accountability frameworks to ensure that responsibility is appropriately assigned and understood.

Furthermore, balancing transparency with privacy concerns in data sharing and implementing mechanisms to ensure that all voices are heard and valued in inclusive deliberation processes presents its own set of challenges. Overall, addressing these challenges requires an inclusive, iterative, and values-driven approach that places importance on stakeholder engagement, ethical guidelines, societal values, interdisciplinary collaboration, and clear accountability frameworks. Moreover, the intersection of technological and political challenges in the deployment of AI systems presents complex issues that require careful consideration and strategic resolution. Two key challenges that demand attention are bias and accountability in AI systems and narratives and ideological bias.

The bias, explainability, and accountability in AI systems are of paramount importance in ensuring the ethical and equitable operation of such systems. The recommendation for algorithmic auditing, as outlined in the EC Report by Kroll [77], underscores the need for transparent and accountable AI systems. However, developing and implementing robust auditing frameworks for AI in the rail industry is a multifaceted task that necessitates the collaboration of technologists, ethicists, and policymakers. Moreover, ensuring transparency and accountability involves the establishment of mechanisms for public and stakeholder scrutiny, adding a layer of complexity to the process.

It is often necessary to explain the decision-making process of a system, especially when it comes to significant or unexpected decisions. This becomes challenging when dealing with AI systems, as they often consist of complex black boxes. This has raised concerns within the industry. For example, using a deep learning system to automate court sentencing led to uninterpretable predictions. This is unacceptable in high-stress situations, such as in railway operations (Global Railway Review). AI systems also bring about a new level of accountability, as multiple decision-makers contribute to building the system. Who is responsible if something goes wrong? Is it the person labelling the data, the engineer who developed it, or someone else, like the delivery manager? Given the complexity of AI systems, these questions become even more difficult to answer.

Similarly, the influence of ideological narratives on public and policy discourse about technology, as highlighted in the EC Report [5,78], raises concerns about the potential impact of ideological biases in the implementation of AI. Overcoming such entrenched biases calls for a transformative approach to stakeholder dialogue that empowers relegated voices and encourages critical examination of prevailing assumptions. However, this endeavor may prove politically and socially challenging, particularly when powerful stakeholders have vested interests in specific technological trajectories.

The rail industry faces significant challenges in deploying AI, requiring collaboration across various stakeholders such as technologists, ethicists, policymakers, and marginalized communities. It is crucial to develop comprehensive strategies that address bias, ensure accountability, and challenge ideological narratives. A holistic, interdisciplinary approach is essential to navigate the complex landscape of AI in the rail industry. Ethical challenges must be carefully managed, prioritizing safety, stakeholder inclusivity, individual rights, public safety, and RI principles. Continuous monitoring, algorithmic auditing, and critical analysis of dominant narratives are necessary to support these efforts. By proactively addressing these challenges, the rail industry can harness AI to improve safety and efficiency while upholding ethical standards and public trust.

17.5 MITIGATING ETHICAL RISK IN AI DEVELOPMENT AND DEPLOYMENT

The advent of AI in the railway industry promises significant advancements in efficiency, safety, and operational management. However, it also raises substantial ethical concerns that necessitate comprehensive strategies to mitigate potential risks. To address these ethical challenges, establishing a robust framework that ensures AI systems are lawful, ethical, and robust is imperative. This framework should encompass safety and product liability regulations tailored to the railway industry's unique requirements, particularly on autonomous decision-making. Ensuring that AI systems adhere to stringent safety standards and legal accountability is essential. Explainable AI (XAI), which aims to make AI decision-making processes transparent and understandable, is crucial in this context [4]. In safety-critical railway subsystems, the ability to trace and understand AI decisions will enhance trust and facilitate broader acceptance among stakeholders [79].

Cybersecurity is another significant concern in the railway industry [8]. Comprehensive cybersecurity measures are essential to protect data and information, fostering confidence among stakeholders. Effective cybersecurity will ensure the integrity and availability of data necessary for AI training and development. Investing in and researching cybersecurity technologies and practices is imperative to safeguard sensitive information and prevent malicious attacks. Additionally, the lack of standard definitions and terminologies for AI technologies in the railway industry hampers the development of cohesive policies and regulations [6]. Establishing transparent and standardized terms will facilitate better communication among stakeholders, leading to more effective governance and smoother integration of AI technologies.

Several strategies can ensure the ethical development and deployment of AI in the railway sector. Security-by-design and privacy-by-design involve embedding security and privacy features from the initial design stage of AI systems to pre-emptively address potential risks [70]. This initiative-taking approach ensures responsible data handling practices throughout the AI lifecycle, enhancing the trust and safety of AI applications. Developing and implementing regulated risk assessment guidelines tailored for railway systems will help identify potential threats, assess vulnerabilities, and determine corrective actions [70]. These guidelines should be systematic and comprehensive to cover various risk scenarios.

Transparent data governance frameworks are also crucial. Implementing clear and accessible data policies that balance innovation with privacy-preserving mechanisms ensures that AI systems handle data responsibly and transparently [6]. Such frameworks should prioritize informed user consent, data security, and ethical data usage. Collaboration among railway operators, researchers, and government representatives can facilitate knowledge sharing and the development of best practices [69]. This collaborative approach will help in addressing ethical concerns and promoting the ethical development of AI technologies.

Addressing the skills gap between AI principles and railway engineering practices is essential [15]. Providing training and education programs for railway engineers on AI techniques, ethics, and safety implications will enhance their understanding and ability to implement AI solutions effectively. Developing AI solutions that are transparent and interpretable, particularly for safety-critical applications, is also crucial [8]. Employing XAI techniques will make AI decision-making processes clear to human operators, fostering trust and acceptance. Finally, establishing specific certification procedures for AI systems in railways, similar to existing safety certifications, will ensure that AI systems meet safety, reliability, and ethical compliance standards before deployment [71]. This involves developing clear standards and evaluation criteria for AI technologies.

Mitigating ethical risks in the development and deployment of AI in the railway industry requires an intricate approach. Establishing a robust framework, enhancing cybersecurity, ensuring explainability, and promoting standardization are critical steps. By implementing strategies such as security-by-design, robust risk assessments, transparent data governance, collaboration, skills development, and standardized certification, the railway industry can ensure the ethical and responsible integration of AI systems. These measures will not only enhance safety and reliability but also foster trust and acceptance among stakeholders, ultimately leading to the successful adoption of AI technologies in the railway sector.

17.6 PRACTICAL TASKS IN THE INTEGRATION OF ETHICAL AI

The integration of ethical AI in the rail industry presents both opportunities and challenges. To realize value from such implementations, the industry must address several critical aspects systematically.

This discussion focuses on transparency, diversity, non-discrimination, fairness, safety, security, privacy, societal and environmental well-being, accountability, education, and awareness.

17.6.1 Transparency

Transparency is fundamental to building trust in AI systems within the rail industry. It involves clear communication of an AI system's capabilities and limitations to all stakeholders, setting realistic expectations. XAI is not yet widely used in the rail industry, but it is expected to become very important, especially for safety-critical systems. XAI can help us understand how AI makes decisions, such as the reasons for train delays, which could improve safety and speed up the approval process by regulators. This transparency ensures that everyone involved understands why AI makes certain decisions, which helps build trust and makes it easier to implement and operate AI systems.

17.6.2 Diversity and Fairness

Ethical AI systems should be inclusive and avoid biases that could disregard specific groups or perpetuate discrimination. Designing AI with these principles ensures that systems are accessible to all, including individuals with disabilities, and considers the input of all relevant stakeholders throughout the AI lifecycle. In the context of staff scheduling, AI should not only optimize operational efficiency but also consider staff well-being, ensuring fair work patterns and adequate breaks. Such inclusive design enhances workforce satisfaction and promotes a fairer working environment.

17.6.3 Safety and Reliability

Safety and reliability are paramount in rail technology, especially with AI applications in autonomous driving, signaling, and control. AI systems must adhere to stringent legislative standards to ensure operational safety. Ethical considerations come into play in scenarios like obstacle detection, where AI must decide on actions that balance passenger safety with comfort. For instance, detecting an animal or vehicle on the track requires a swift and appropriate response to mitigate risks. Legislative frameworks and rigorous testing are essential to validate these systems, ensuring they perform reliably under various conditions.

17.6.4 Privacy and Security

Data privacy and security are significant concerns in implementing AI systems. These systems must respect individuals' privacy rights, obtain consent for data collection and usage, and also have the ability to detect various types of poisoning to training data and model parameters. This is especially critical considering the large span of rail systems to governmental, monitorial databases and passengers personal information, and the distributed way of data collection and AI model training. Effective methods for privacy and security assurance including data privacy protection, privacy-preserving computing, and secure ML are essential for protecting data from breaches, utilizing privacy-preserved data in the process of learning, ensuring the robustness of both data and AI models against poisoning attacks, to safeguard the against any compromise of individuals' privacy and the operation security of AI-embedded rail systems. By prioritizing privacy and security assurance, rail companies can maintain public trust and comply with data protection regulations, such as GDPR, enhancing the ethical standing of their AI implementations.

17.6.5 Societal and Environmental Well-Being

Ethical AI should also consider the broader impact on society and the environment. Sustainable practices and ecological responsibility should be at the forefront of AI development. For instance,

AI can improve maintenance schedules to reduce waste from consumable component replacements. By promoting sustainability, the rail industry can contribute to environmental conservation while achieving operational efficiencies, thus aligning with global sustainability goals.

17.6.6 Education and Awareness

Raising awareness and educating all stakeholders about AI's potential impacts is vital. Continuous education and training programs can inform stakeholders about ethical considerations and equip them to engage meaningfully in AI implementation processes. Training a new generation of experts in AI ethics and disseminating research findings to the public ensures an informed community that can contribute to responsible AI development.

17.6.7 Accountability and Explicability

Accountability mechanisms are crucial for the ethical deployment of AI. Companies should identify potential impacts of their AI systems early and ensure compliance with relevant norms through "by-design" concepts like privacy-by-design and security-by-design. Establishing internal review boards and providing ethical training can help address dilemmas and emerging ethical issues. These mechanisms, coupled with legal oversight, ensure that AI systems operate within ethical boundaries and are held accountable for their actions.

Explicability in the rail industry is a crucial principle that ensures AI systems are both understandable and accountable. This dual focus is essential for fostering trust among stakeholders and for ethical deployment [80]. Intelligibility of AI technologies. The first aspect of explicability, intelligibility, involves providing clear explanations of how AI systems operate. It is imperative that the decision-making processes of AI, such as those involved in train scheduling or delay predictions so that stakeholders are transparent and comprehensible to all relevant parties, including non-experts, can comprehend and scrutinize AI decisions. This transparency reduces the opacity often associated with AI technologies and mitigates the opacity that often surrounds AI technologies, enabling stakeholders to scrutinize and understand the operational logic behind AI decisions. By making the inner workings of AI systems intelligible, the rail industry can enhance the credibility and reliability of these technologies, facilitating smoother implementation and acceptance.

The second aspect, accountability, is equally important. It addresses the need to identify who is responsible for AI-generated outcomes. In case the event of a negative outcome like an accident outcome, such as an accident or system failure, clear accountability ensures failure and understanding who holds responsibility is crucial for taking appropriate corrective actions are taken. This aspect establishes explicability and ensures that there is a clear chain of responsibility, essential for and that those involved in the deployment and operation of AI systems are held accountable for their performance. By establishing accountability, the rail industry can address and rectify issues effectively, thereby maintaining safety and reliability standards.

Together, intelligibility and accountability are essential components that form the cornerstone of explicability in the rail industry. By integrating explicability, rail companies guarantee this principle not only enhances transparency but also ensures that their AI systems are efficient, ethical, and in alignment with societal values and legal standards. This, in turn, enhances trust and acceptance among stakeholders. By incorporating explicability into their AI deployments, rail companies can ensure that these systems are not only effective and efficient but also ethical and trustworthy. This comprehensive approach to AI governance is essential for realizing the full potential of AI in improving the rail industry's operations while safeguarding the interests of all stakeholders involved.

As a result, implementing ethical AI in the rail industry requires a comprehensive approach that addresses transparency, diversity, safety, privacy, societal well-being, accountability, education, and explicability. Continuous stakeholder engagement and a proactive stance on ethical considerations can lead to safer, more efficient, and sustainable railway systems. By adhering to these principles, the rail industry can harness the benefits of AI while respecting fundamental rights and promoting public trust, ultimately achieving value for both the industry and society.

17.7 RELEVANT RAILWAY STANDARDS AND REGULATORY CONSIDERATIONS

Railway systems are safety-critical, requiring stringent compliance with standards such as CENELEC EN 50126, EN 50128, and EN 50129 to ensure software, including AI algorithms, is safe and reliable [81–83]. These standards cover system safety, software development, and certification. For AI in railways, these need adaptation to handle dynamic, data-driven systems.

AI models must be explainable for regulatory and safety reasons. Standards like the Ethics Guidelines for Trustworthy AI demand transparency and traceability of decisions, a challenging feat for complex models like deep learning [2]. Railway AI developers must ensure both the reliability and comprehensibility of their systems.

Standards like ISO/IEC 23894:2023 highlight the need for robust data governance in AI, ensuring accurate, current data that complies with privacy and security laws [4]. The EC's Ethics Guidelines for Trustworthy AI and NASA's ethical AI framework emphasize fairness, transparency, accountability, and safety in AI systems [1,2].

These frameworks advocate for human oversight, safety, and fairness, which are crucial for lawful, ethical AI deployment in railways. They also stress making AI operations understandable to build trust and accountability, ensuring developers and users are responsible for outcomes.

The AI in the rail industry should adhere to ethical and regulatory standards to ensure fairness, transparency, accountability, privacy, and safety. Responsible AI use in railways should promote socially beneficial outcomes. These guidelines provide a solid foundation for safely integrating AI into railway systems, enhancing efficiency, safety, and societal goals like sustainability and fairness.

17.8 CONCLUSION, FUTURE DIRECTIONS, AND CHALLENGES

17.8.1 Conclusion: Integrating Ethics into AI for a High-Performing, Safe, and Sustainable Rail Sector

The integration of AI into the railway sector presents a significant opportunity to transform operations, enhancing efficiency, safety, and sustainability [3,4,8,84]. However, this potential can only be fully realized if ethical considerations are prioritized throughout all aspects of AI development and deployment [3,4,8,84]. Neglecting these considerations can result in unintended consequences that undermine public trust and hinder technological progress [3,85]. Therefore, it is imperative to embed ethical principles at the core of AI initiatives in the rail sector to ensure that technological advancements align with societal values and expectations.

17.8.2 Future Directions: Toward Ethically Aligned AI in Rail

Moving forward, the rail industry must transcend a purely technical focus and embed ethical considerations into every stage of the AI lifecycle [86]. This necessitates fostering a culture of responsibility where developers and operators are held accountable for the ethical implications of their work [71,87]. By doing so, the industry can create AI systems that are not only effective but also align with human values and societal norms. Emphasizing the creation of AI systems that respect human dignity, privacy, and autonomy is essential [1,3,85]. This involves considering the social and ethical impacts of AI applications, from data collection to decision-making processes.

Given the safety-critical nature of railway operations, it is crucial that AI systems are explainable [3,4,85]. XAI ensures that the decision-making processes are understandable and justifiable to human operators and regulators, fostering trust and facilitating oversight [3]. Enhancing transparency in AI systems aids in identifying and mitigating biases and errors, ensuring that AI-driven decisions are fair and reliable.

- *Prioritizing ethical D&D:* The rail industry must move beyond a purely technical perspective and embed ethical considerations into every stage of the AI lifecycle. This includes promoting a culture of responsibility, where developers and operators are accountable for the ethical implications of their work. The focus should be on creating AI systems that not only are effective but also align with human values and societal expectations.
- *Explainable, transparent, and trustworthy AI:* The safety-critical nature of many railway operations demands a high level of transparency in AI systems. XAI will be crucial to ensure that the decision-making processes of AI systems are understandable and justifiable to human operators and regulators.
- *Standardization and regulation:* A robust framework of standards and regulatory frameworks regulations is essential to guide the ethical development and deployment of AI in the rail sector [4,79]. This includes setting and developing clear guidelines for data privacy, algorithmic bias mitigation, and safety certification for AI systems [6,15,79]. As AI technology evolves, regulatory frameworks must remain flexible and responsive to new challenges, ensuring continuous alignment with the latest advancements and ethical standards [3,4,15,70].

17.8.3 Challenges: Navigating the Evolving Landscape of AI Ethics in Rail

As AI systems gain sophistication and autonomy, and become more sophisticated and autonomous, new ethical challenges are likely to emerge. This includes navigating complex dilemmas related to liability in the event of AI system failures becomes complex [9,79,88]. Clear protocols for assigning responsibility are needed to address this issue. Ensuring safe and effective, managing human–machine interaction in safety-critical situations is paramount [3,85]. This involves designing interfaces and protocols that facilitate seamless collaboration between human operators and AI systems. AI applications should promote equity by ensuring the fair distribution of benefits across society and preventing the exacerbation of existing inequalities caused by AI.

The rapid pace of AI development necessitates the ongoing evolution of ethical frameworks, standards, and regulations [4,15]. Existing guidelines may not fully address the specific challenges posed by AI in the rail sector, highlighting the need for continuous research and interdisciplinary collaboration [15]. Effective ethical frameworks require input from various disciplines, including ethics, law, engineering, and social sciences, to address the multifaceted challenges posed by AI [71,86].

In conclusion, integrating ethical considerations into AI development and deployment in the rail sector is not just a technical necessity but a moral imperative. By focusing on ethical design and deployment, promoting transparency and trust, and developing robust regulatory frameworks, the rail industry can ensure that AI contributes to a high-performing, safe, and sustainable future. Continuous adaptation and interdisciplinary collaboration are essential to navigate the evolving ethical landscape and address the emerging challenges posed by AI in rail systems.

REFERENCES

1. McLarney E, Gawdiak Y, Oza N, et al. *NASA Framework for the Ethical Use of Artificial Intelligence (AI)*, National Aeronautics and Space Administration, 2021.
2. Smuha NA. The EU approach to ethics guidelines for trustworthy artificial intelligence. *Computer Law Review International.* 2019;20(4):97–106.

3. Besinovic N, De Donato L, Flammini F, et al. Artificial intelligence in railway transport: Taxonomy, regulations, and applications. *IEEE Transactions on Intelligent Transportation Systems.* 2022;23(9):14011–14024.
4. Flammini F, Tang T, Liu R, et al. *RAILS Deliverable D 1.1: Definition of a Reference Taxonomy of AI in Railways*, Shift2Rail, 2020.
5. Santoni de Sio F. The European Commission report on ethics of connected and automated vehicles and the future of ethics of transportation. *Ethics and Information Technology.* 2021;23(4):713–726.
6. Singh P, Elmi Z, Lau Y-y, et al. Blockchain and AI technology convergence: Applications in transportation systems. *Vehicular Communications.* 2022;38:100521.
7. López-Aguilar P, Batista E, Martínez-Ballesté A, et al. Information security and privacy in railway transportation: A systematic review. *Sensors.* 2022;22(20):7698.
8. Phusakulkajorn W, Núñez A, Wang H, et al. Artificial intelligence in railway infrastructure: Current research, challenges, and future opportunities. *Intelligent Transportation Infrastructure.* 2023;2:liad016.
9. Tang R, De Donato L, Bešinović N, et al. A literature review of Artificial Intelligence applications in railway systems. *Transportation Research Part C: Emerging Technologies.* 2022;140:103679.
10. Famurewa SM, Zhang L, Asplund M. Maintenance analytics for railway infrastructure decision support. *Journal of Quality in Maintenance Engineering.* 2017;23(3):310–325.
11. Ghofrani F, He Q, Goverde RMP, et al. Recent applications of big data analytics in railway transportation systems: A survey. *Transportation Research Part C: Emerging Technologies.* 2018;90:226–246.
12. McMahon P, Zhang T, Dwight RA. Approaches to dealing with missing data in railway asset management. *IEEE Access.* 2020;8:48177–48194.
13. Allah Bukhsh Z, Saeed A, Stipanovic I, et al. Predictive maintenance using tree-based classification techniques: A case of railway switches. *Transportation Research Part C: Emerging Technologies.* 2019;101:35–54.
14. Wang Q, Bu S, He Z. Achieving predictive and proactive maintenance for high-speed railway power equipment with LSTM-RNN. *IEEE Transactions on Industrial Informatics.* 2020;16(10):6509–6517.
15. Seisenberger M, ter Beek MH, Fan X, et al. Safe and secure future AI-driven railway technologies: Challenges for formal methods in railway. In *Leveraging Applications of Formal Methods, Verification and Validation. Practice. Lecture Notes in Computer Science*, Springer, Rhodes, Greece, 2022, pp. 246–268.
16. Guler H. Decision support system for railway track maintenance and renewal management. *Journal of Computing in Civil Engineering.* 2013;27(3):292–306.
17. Liu L, Zhou F, He Y. Automated visual inspection system for bogie block key under complex freight train environment. *IEEE Transactions on Instrumentation and Measurement.* 2016;65(1):2–14.
18. Obara M, Kashiyama T, Sekimoto Y, editors. Deep reinforcement learning approach for train rescheduling utilizing graph theory. *2018 IEEE International Conference on Big Data (Big Data)*, Seattle, WA, USA, 10–13 December 2018, 2018.
19. Roost D, Meier R, Huschauer S, et al., editors. Improving sample efficiency and multi-agent communication in RL-based train rescheduling. *2020 7th Swiss Conference on Data Science (SDS)*, Luzern, Switzerland, 26–26 June 2020, 2020.
20. Yin J, Tang T, Yang L, et al. Research and development of automatic train operation for railway transportation systems: A survey. *Transportation Research Part C: Emerging Technologies.* 2017;85:548–572.
21. Trinh H, Haas N, Li Y, et al., editors. Enhanced rail component detection and consolidation for rail track inspection. *2012 IEEE Workshop on the Applications of Computer Vision (WACV)*, Breckenridge, CO, USA, 9–11 January 2012, 2012.
22. Feng H, Jiang Z, Xie F, et al. Automatic fastener classification and defect detection in vision-based railway inspection systems. *IEEE Transactions on Instrumentation and Measurement.* 2014;63(4):877–888.
23. Wang S, Dai P, Du X, et al., editors. Rail fastener automatic recognition method in complex background. *Tenth International Conference on Digital Image Processing (ICDIP 2018)*, Shanghai, China, 2018. SPIE.
24. Wei X, Yang Z, Liu Y, et al. Railway track fastener defect detection based on image processing and deep learning techniques: A comparative study. *Engineering Applications of Artificial Intelligence.* 2019;80:66–81.
25. Gibert X, Patel VM, Chellappa R, editors. Robust fastener detection for autonomous visual railway track inspection. *2015 IEEE Winter Conference on Applications of Computer Vision*, Waikoloa, HI, USA, 5–9 January 2015, 2015.

26. Giben X, Patel VM, Chellappa R, editors. Material classification and semantic segmentation of railway track images with deep convolutional neural networks. *2015 IEEE International Conference on Image Processing (ICIP)*, Quebec City, QC, Canada, 27–30 September 2015, 2015.
27. Gibert X, Patel VM, Chellappa R. Deep multitask learning for railway track inspection. *IEEE Transactions on Intelligent Transportation Systems*. 2017;18(1):153–164.
28. Sadeghi J, Askarinejad H. Application of neural networks in evaluation of railway track quality condition. *Journal of Mechanical Science and Technology*. 2012;26(1):113–122.
29. Tsunashima H. Condition monitoring of railway tracks from car-body vibration using a machine learning technique. *Applied Sciences*. 2019;9(13):2734.
30. Zarembski AM, editor Some examples of big data in railroad engineering. *2014 IEEE International Conference on Big Data (Big Data)*, Washington, DC, USA, 27–30 October 2014, 2014.
31. Soukup D, Huber-Mörk R, editors. Convolutional neural networks for steel surface defect detection from photometric stereo images. *International Symposium on Visual Computing*, Las Vegas, NV, USA, 2014. Springer.
32. Faghih-Roohi S, Hajizadeh S, Núñez A, et al., editors. Deep convolutional neural networks for detection of rail surface defects. *2016 International Joint Conference on Neural Networks (IJCNN)*, Vancouver, BC, Canada, 24–29 July 2016, 2016.
33. Jamshidi A, Faghih-Roohi S, Hajizadeh S, et al. A big data analysis approach for rail failure risk assessment. *Risk Analysis*. 2017;37(8):1495–1507.
34. Santur Y, Karaköse M, Akin E, editors. A new rail inspection method based on deep learning using laser cameras. *2017 International Artificial Intelligence and Data Processing Symposium (IDAP)*, Malatya, Turkey, 16–17 September 2017, 2017.
35. Santur Y, Karaköse M, Akin E. An adaptive fault diagnosis approach using pipeline implementation for railway inspection. *Turkish Journal of Electrical Engineering and Computer Sciences*. 2018;26(2):987–998.
36. Shang L, Yang Q, Wang J, et al., editors. Detection of rail surface defects based on CNN image recognition and classification. *2018 20th International Conference on Advanced Communication Technology (ICACT)*, Chuncheon, Korea (South), 11–14 February 2018, 2018.
37. Gao S, Szugs T, Ahlbrink R, editors. Use of combined railway inspection data sources for characterization of rolling contact fatigue. *Proceedings of the 12th European Conference on Non-Destructive Testing (ECNDT)*, Gothenburg, Sweden, 2018.
38. Santur Y, Karaköse M, Akın E. Learning based experimental approach for condition monitoring using laser cameras in railway tracks. *International Journal of Applied Mathematics Electronics and Computers*. 2016;4(Special Issue-1):1–5.
39. Hajizadeh S, Núñez A, Tax DMJ. Semi-supervised rail defect detection from imbalanced image data. *IFAC-PapersOnLine*. 2016;49(3):78–83.
40. Krummenacher G, Ong CS, Koller S, et al. Wheel defect detection with machine learning. *IEEE Transactions on Intelligent Transportation Systems*. 2018;19(4):1176–1187.
41. Yao Z, He D, Chen Y, et al. Inspection of exterior substance on high-speed train bottom based on improved deep learning method. *Measurement*. 2020;163:108013.
42. Zhou F, Xia L, Dong W, et al., editors. Fault diagnosis of high-speed railway turnout based on support vector machine. *2016 IEEE International Conference on Industrial Technology (ICIT)*, Taipei, Taiwan, 14–17 March 2016, 2016.
43. Lee J, Choi H, Park D, et al. Fault detection and diagnosis of railway point machines by sound analysis. *Sensors (Basel)*. 2016;16(4):549.
44. Eker OF, Camci F, Guclu A, et al. A simple state-based prognostic model for railway turnout systems. *IEEE Transactions on Industrial Electronics*. 2011;58(5):1718–1726.
45. Hu L-Q, He C-F, Cai Z-Q, et al. Track circuit fault prediction method based on grey theory and expert system. *Journal of Visual Communication and Image Representation*. 2019;58:37–45.
46. An M, Chen Y, Baker CJ. A fuzzy reasoning and fuzzy-analytical hierarchy process based approach to the process of railway risk information: A railway risk management system. *Information Sciences*. 2011;181(18):3946–3966.
47. Gul M, Celik E. Fuzzy rule-based Fine–Kinney risk assessment approach for rail transportation systems. *Human and Ecological Risk Assessment: An International Journal*. 2018;24(7):1786–1812.
48. Zhang H, Yuan M, Liang Y, et al. A risk assessment based optimization method for route selection of hazardous liquid railway network. *Safety Science*. 2018;110:217–229.
49. Zilko AA, Kurowicka D, Goverde RMP. Modeling railway disruption lengths with Copula Bayesian Networks. *Transportation Research Part C: Emerging Technologies*. 2016;68:350–368.

50. Li Y, Chen D, Yang X, editors. Expert system applied to high-speed railway track circuit coding and its simulation. In *Artificial Intelligence and Computational Intelligence*, Berlin, Heidelberg: Springer Berlin Heidelberg, 2011, pp. 517–524.
51. Alawad H, Kaewunruen S, An M. Learning from accidents: Machine learning for safety at railway stations. *IEEE Access*. 2020;8:633–648.
52. Ćirović G, Pamučar D. Decision support model for prioritizing railway level crossings for safety improvements: Application of the adaptive neuro-fuzzy system. *Expert Systems with Applications*. 2013;40(6):2208–2223.
53. Brenna M, Foiadelli F, Longo M. Application of genetic algorithms for driverless subway train energy optimization. *International Journal of Vehicular Technology*. 2016;2016:1–14.
54. Wang P, Trivella A, Goverde RMP, et al. Train trajectory optimization for improved on-time arrival under parametric uncertainty. *Transportation Research Part C: Emerging Technologies*. 2020;119:102680.
55. Yin J, Chen D, Li L. Intelligent train operation algorithms for subway by expert system and reinforcement learning. *IEEE Transactions on Intelligent Transportation Systems*. 2014;15(6):2561–2571.
56. Hua G, Zhu L, Wu J, et al. Blockchain-based federated learning for intelligent control in heavy haul railway. *IEEE Access*. 2020;8:176830–176839.
57. Nowakowski W, Bojarczak P, Łukasik Z, editors. Verification and validation of railway control systems using an expert system. *Intelligent Transport Systems–From Research and Development to the Market Uptake: First International Conference, INTSYS 2017*, Hyvinkää, Finland, November 29–30, 2017, Proceedings 1, 2018. Springer.
58. Zhang D. High-speed train control system big data analysis based on the fuzzy RDF model and uncertain reasoning. *International Journal of Computers Communications & Control*. 2017;12(4):577–591.
59. Wang P, Zhang Q-p. Train delay analysis and prediction based on big data fusion. *Transportation Safety and Environment*. 2019;1(1):79–88.
60. Cerreto F, Nielsen BF, Nielsen OA, et al. Application of data clustering to railway delay pattern recognition. *Journal of Advanced Transportation*. 2018;2018:1–18.
61. Huang P, Wen C, Fu L, et al. Modeling train operation as sequences: A study of delay prediction with operation and weather data. *Transportation Research Part E: Logistics and Transportation Review*. 2020;141:102022.
62. Deng W-t, Zhao H-b, He H. Research on heavy haul railway dispatching system based on fuzzy expert system. *DEStech Transactions on Computer Science and Engineering*. 2018. https://dpi-journals.com/index.php/dtcse/article/view/24719.
63. Gallo M, De Luca G, D'Acierno L, et al. Artificial neural networks for forecasting passenger flows on metro lines. *Sensors (Basel)*. 2019;19(15):3424.
64. Liu Y, Liu Z, Jia R. DeepPF: A deep learning based architecture for metro passenger flow prediction. *Transportation Research Part C: Emerging Technologies*. 2019;101:18–34.
65. Zhang J, Chen F, Guo Y, et al. Multi-graph convolutional network for short-term passenger flow forecasting in urban rail transit. *IET Intelligent Transport Systems*. 2020;14(10):1210–1217.
66. Heydenrijk-Ottens L, Degeler V, Luo D, et al., editors. Supervised learning: Predicting passenger load in public transport. *CASPT Conference on Advanced Systems in Public Transport and TransitData*, Brisbane, Australia, 2018.
67. Brezulianu A, Geman O, Popa IV. Artificial intelligence component of the FERODATA AI engine to optimize the assignment of rail freight locomotive drivers. *Applied Sciences*. 2023;13(20):11516.
68. Tsolaki K, Vafeiadis T, Nizamis A, et al. Utilizing machine learning on freight transportation and logistics applications: A review. *ICT Express*. 2023;9(3):284–295.
69. Singh P, Dulebenets MA, Pasha J, et al. Deployment of autonomous trains in rail transportation: Current trends and existing challenges. *IEEE Access*. 2021;9:91427–91461.
70. Lopez-Aguilar P, Batista E, Martinez-Balleste A, et al. Information security and privacy in railway transportation: A systematic review. *Sensors (Basel)*. 2022;22(20):7698.
71. Shahriari K, Shahriari M, editors. IEEE standard review—Ethically aligned design: A vision for prioritizing human wellbeing with artificial intelligence and autonomous systems. *2017 IEEE Canada International Humanitarian Technology Conference (IHTC)*, Toronto, ON, Canada, 2017. IEEE.
72. Yu K-H, Beam AL, Kohane IS. Artificial intelligence in healthcare. *Nature Biomedical Engineering*. 2018;2(10):719–731.
73. Gerke S, Minssen T, Cohen G. Ethical and legal challenges of artificial intelligence-driven healthcare. In *Artificial Intelligence in Healthcare*, Adam Bohr, Kaveh Memarzadeh, Editors, Academic Press, United Kingdom, pp. 295–336, 2020.

74. Ryan M. The social and ethical impacts of artificial intelligence in agriculture: Mapping the agricultural AI literature. *AI & Society*. 2023;38(6):2473–2485.
75. Diaz M. AI in smart cities: The benefits and challenges Gemmo AI2024 [cited 2024]. Available from: https://gemmo.ai/ai-in-smart-cities/
76. Bonnefon JF, Shariff A, Rahwan I. The social dilemma of autonomous vehicles. *Science*. 2016;352(6293):1573–1576.
77. Kroll JA. *Accountable Algorithms*. Princeton University, Princeton, NJ, 2015.
78. Blok V. Look who's talking: responsible innovation, the paradox of dialogue and the voice of the other in communication and negotiation processes. *Journal of Responsible Innovation*. 2014;1(2):171–190.
79. Perez-Cerrolaza J, Abella J, Borg M, et al. Artificial intelligence for safety-critical systems in industrial and transportation domains: A survey. *ACM Computing Surveys*. 2024;56(7):1–40.
80. Floridi L, Cowls J. A unified framework of five principles for AI in society. *Harvard Data Science Review*. 2019;1. https://hdsr.mitpress.mit.edu/pub/l0jsh9d1/release/8.
81. CENELEC. CENELEC – EN 50126-1: Railway applications – The specification and demonstration of reliability, availability, maintainability and safety (RAMS) – Part 1: Generic RAMS process, 2017.
82. CENELEC. CENELEC - EN 50129: Railway applications – Communication, signalling and processing systems – Safety related electronic systems for signalling, 2019.
83. CENELEC. CENELEC – EN 50128: Railway applications – Communication, signaling and processing systems, Software for railway control and protection systems, 2020.
84. Singh P, Elmi Z, Krishna Meriga V, et al. Internet of Things for sustainable railway transportation: Past, present, and future. *Cleaner Logistics and Supply Chain*. 2022;4:100065.
85. Pflanzer M, Traylor Z, Lyons JB, et al. Ethics in human–AI teaming: principles and perspectives. *AI and Ethics*. 2023;3(3):917–935.
86. Trentesaux D, Karnouskos S. Ethical behaviour aspects of autonomous intelligent cyber-physical systems. In *Service Oriented, Holonic and Multi-agent Manufacturing Systems for Industry of the Future. Studies in Computational Intelligence*, Borangiu, T., Trentesaux, D., Leitão, P., Giret Boggino, A., Botti, V., Editors, Springer, Valencia, Spain, pp. 55–71, 2020.
87. Dignum V. Ethics in artificial intelligence: Introduction to the special issue. *Ethics and Information Technology*. 2018;20(1):1–3.
88. Plissonneau A, Trentesaux D, Ben-Messaoud W, et al. AI-based speed control models for the autonomous train: a literature review. *2021 Third International Conference on Transportation and Smart Technologies* (TST), Tangier, Morocco, pp. 9–15, 2021.

18 Integration of AI in Digital Twin Technologies

Esteban Bernal Arango

18.1 INTRODUCTION

Artificial intelligence (AI) and digital twins (DT) are two data-based emerging technologies whose combination appears to be a natural link promising synergy to improve various aspects of railway systems across a wide spectrum. Starting from the first principles, a DT is a "digital representation of a target entity with data connections that enable convergence between the physical and digital states at an appropriate rate of synchronization" [1]. Contextualizing this definition to an engineering system such as a railway, a DT is a system that generates new knowledge, through interactive data exchange between multiphysics and data-based models and a physical asset, and such knowledge is generated according to requirements that directly enable the addition of progressive improvements on a daily basis across a railway system lifecycle during planning, operation, maintenance, and management phases [2]. AI, on the contrary, although not a new technology, has considerably advanced in recent years due to developments in computational processing power, availability of large amounts of data, and a variety of new use cases across different sectors [3]. In general, AI can be defined as "a range of technologies that can learn from data or experiences that would otherwise require human intelligence" [4].

AI systems, developed using various tools like neural networks, are trained and tested to meet specific performance expectations. These systems are applied in areas such as computer vision and natural language processing and may evolve as new data is integrated. AI offers advantages over traditional methods by processing larger data volumes, improving efficiency, reducing risks in hazardous environments, and minimizing human error. Its flexibility to learn from new information enables ongoing improvements, fostering opportunities to enhance safety, security, and operational efficiency. It becomes evident that there is a latent synergy between DTs and AI technologies, and their natural interaction can leverage the functionalities they provide. See, for example, the DT for rail surface damage prediction in Figure 18.1 [5] and the framework for a high-speed train DT simulation platform in Figure 18.2.

It becomes evident that there are numerous data connections between the physical and digital entities of the DT. Each DT interface component processes data through a series of transformations, converting raw data into actionable information for operational and maintenance decisions. In a typical DT service or function, this might involve signal transformation, filtering, and cleaning, followed by indicator generation that enables other components to recommend or execute operational and maintenance decisions. Traditional algorithms rely on predefined rules derived from logical processes and expert knowledge built upon experience to generate outputs from inputs, meaning they function best in systems without uncertainty. As DTs are large and complex systems, maintaining them with traditional algorithms can become unsustainable due to the need for increasingly sophisticated and very specific rules. In contrast, AI algorithms address this challenge by allowing the machine to define relationships between inputs and outputs. Machine learning (ML) algorithms adapt and evolve with new data, eliminating the need for manual coding adjustments for each new dataset [6,7]. As shown in Figure 18.3, AI enables a paradigm shift, where the data transformation rules are not designed by a human but are deducted by a computational algorithm.

DOI: 10.1201/9781003492146-18

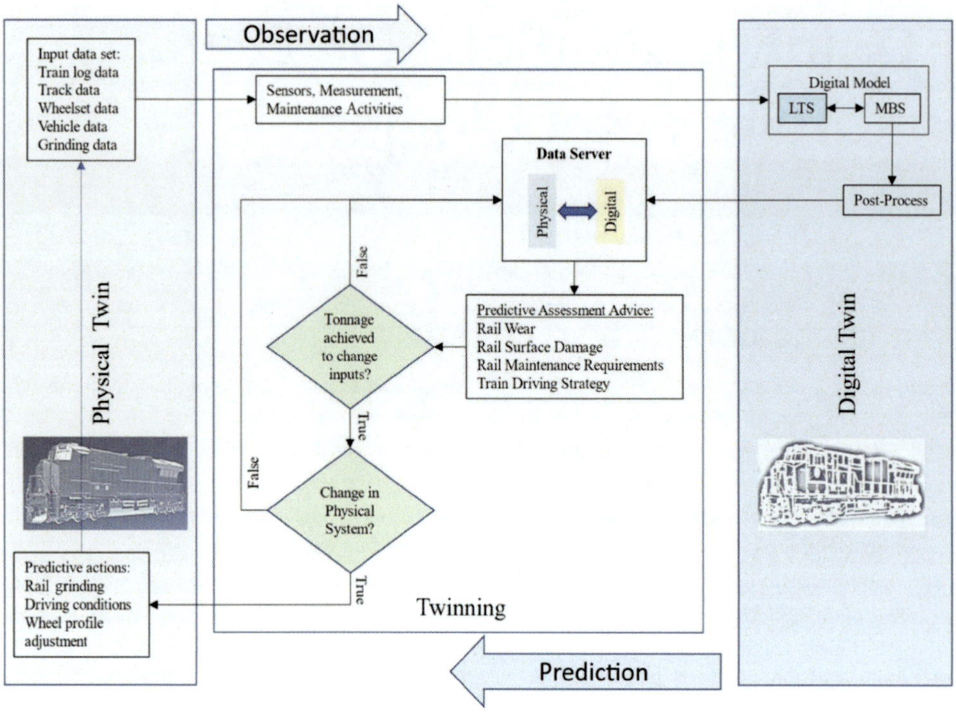

FIGURE 18.1 Methodology to develop a DT for prediction of rail surface damage. (From S. Ahmad, M. Spiryagin, Q. Wu, E. Bernal, Y. Sun, C. Cole, et al., "Development of a Digital Twin for prediction of rail surface damage in heavy haul railway operations," *Veh Syst Dyn, vol.* 62, no. 1, pp. 41–66, 2024, with permission.)

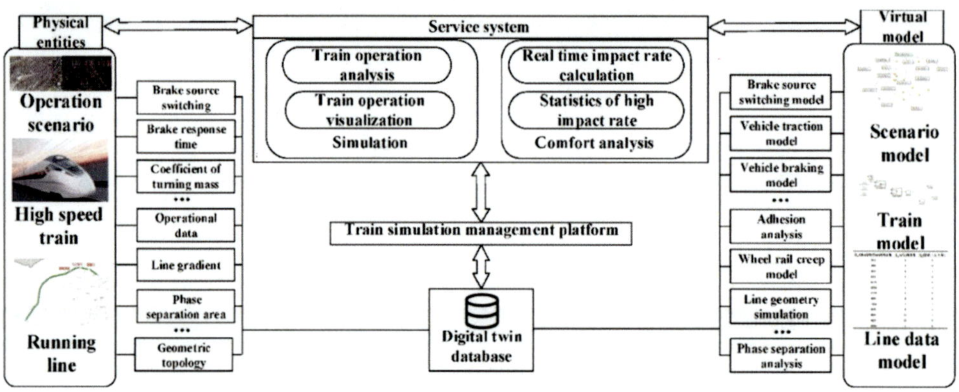

FIGURE 18.2 Framework for a high-speed train digital twin simulation platform. (From M. Spiryagin, J. Edelmann, F. Klinger, and C. Cole, "Vehicle system dynamics in digital twin studies in rail and road domains," *Veh Syst Dyn*, vol. 61, no. 7, pp. 1735–1784, 2023, doi:10.1080/00423114.2023.2188228, with permission.)

This paradigm shift requires careful consideration when integrating AI into DTs that involve safety-critical systems, particularly regarding system architectural principles to first make the DT viable in effectively adding value to railway systems, and second to ensure safety in contexts where AI and DT will access data and, in some cases, have the autonomy to make operational decisions that may impact system integrity and safety. In this context, in addition to the typical DT system architecture requirements [8], the AI component of the DT system should consider the principles of boundaries, monitoring, and modularization—principles originally developed for nuclear applications but relevant and contextual to railway systems [4]:

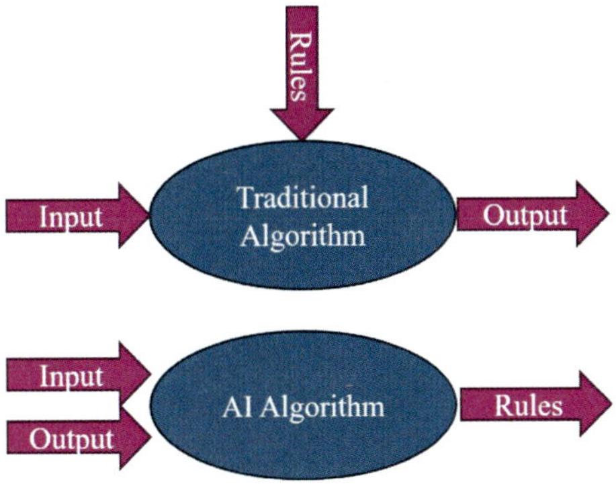

FIGURE 18.3 AI paradigm shift. (Redrawn from F. Balali, J. Nouri, A. Nasiri, and T. Zhao, *Data Intensive Industrial Asset Management: IoT-based Algorithms and Implementation*, Springer Nature, 2020.)

- *System boundaries:* Establishing clear system boundaries is essential for maintaining the integrity and security of AI-enabled DTs for railway systems. These boundaries define the AI DT subcomponent scope and interaction with its environment, ensuring secure, reliable data exchange and compatibility with existing assets. Properly designed boundaries help prevent errors, unsafe actions, and disruptions in safety-critical processes. Key considerations include optimizing data availability to avoid inefficiency, constraining software input and output to reduce computer processing power requirements, and using isolation principles to limit AI functionality and reduce risks in railway operations.
- *Monitoring:* Continuous monitoring of AI modules inside DTs is crucial for detecting anomalies, preventing failures, and ensuring system integrity, particularly in AI-enabled DT submodules that execute vehicle-track operational decisions. This involves tracking AI performance, accuracy, and error rates and detecting potential security threats, such as adversarial attacks. Monitoring strategies also include the use of anomaly detection and access control measures. Effective monitoring, integrated with system boundaries, ensures accurate AI outputs and promotes trust in the system's reliability and usability inside the organization.
- *Modularization:* As a relevant element for general DT system architectures, modularization in AI components of a DT facilitates flexibility, maintainability, and explainability by breaking down the AI into smaller, independent modules with specific functions. This approach contrasts with monolithic AI, which handles data from multiple sources as a single entity. Modular AI DT subcomponents allow for easier adaptation and validation in different operational environments, such as transferring models from one domain to another. Careful consideration must be given to risks, operational performance, and training data limitations when reusing AI models in new contexts. For example, a modular AI bearing anomaly detection model may work well for a certain type of vehicle but fail to identify faults for the same bearing installed on a different vehicle type or when the same vehicle operates on a different type of track. The modular design also allows for easier fault detection and isolation, improving system reliability and transparency, which is harder to achieve with monolithic AI models.

In addition to these general principles for the integration of AI technologies in safety-critical systems, railway systems pose challenges when implementing DTs and AI. Modern railway systems feature mobile and fixed assets with various subsystem components equipped with sensors that acquire and store information for component-based operational decisions. However, this information is typically limited in quantity, quality, and accessibility, often restricted to component manufacturers or specific departments, which limits opportunities for collaboration and new knowledge generation. While today's systems gather substantial data, the challenge lies in effectively utilizing this information.

DTs aim to generate new knowledge through interactive data exchange between data-based and physics-based models and physical entities, ultimately adding value across railway operations, maintenance, and management. General AI applications, and especially the ones embedded inside DTs, require infrastructure that enables experimentation and continuous improvement, particularly in developing predictive maintenance algorithms [9].

Given that railways are complex multisystem entities involving several engineering disciplines, the integration of AI technologies in DTs must be systematized and standardized to effectively and positively impact railway systems. Having established the context of AI's role in DTs, this chapter covers machine learning operations (MLOps) as a key approach to integrating AI into DTs, provides examples of AI applications in railway DTs, and discusses the challenges and opportunities in their implementation.

18.2 MLOps AND RAILWAY DIGITAL TWINS

Unlike advanced simulations that recreate the behavior of a system under static operational conditions, DTs are dynamic and evolving entities that mirror the behavior of the physical systems along their lifecycle [10]. DTs perform complex simulations to generate data-driven and physics-driven decisions to optimize the planning, monitoring, and control of physical assets [11]. In this context, it has been argued that AI is the "central puzzle piece" required to overcome the challenges of implementing and enabling DTs to effectively use real-time data streams in the process of adding value to industrial processes [12]. For a DT to effectively add value to a railway system, it must integrate AI frameworks, concepts, theories, and day-to-day railway operations maintenance and management. However, automating and integrating AI algorithms (and DTs) is highly challenging, as most AI and ML workflows are not automated. The majority of ML projects fail or fall short of expectations, as many ML proof-of-concept ideas never reach production [13,14]. Current applications of AI technologies in railway systems are at very early stages and mostly limited to maintenance and inspection of particular components, with few examples in the subdomains of passenger mobility, transport policy, management, security, control, or autonomous operations [15]. Hence, it is expected that deploying an AI application inside a DT will prove even more challenging. In this context, the concept of MLOps becomes a critical technology in the integration of AI into DTs [14]. MLOps are a multidisciplinary approach that brings together experts from various domains to work collaboratively in an iterative manner. It builds upon the well-established practices of DevOps (Development and Operations), which have been widely adopted in software development, such as continuous integration (CI) and continuous delivery (CD). These principles are integrated with data engineering techniques and ML technologies to create a holistic framework for developing, deploying, and maintaining AI-powered applications [11]. For this purpose, Kreuzberger et al. [14] identified principles, technical components, and roles for an end-to-end MLOps architecture, which can be applied for general AI deployments, including the integration of AI into DTs for railway systems. A combined MLOps and DT architecture is shown in Figure 18.4, developed using the principles presented by Kreuzberger et al. [14] and van Bruggen et al. [16]. The following subsections describe the stage of each MLOps and elaborate on how MLOps and DTs can interact inside the framework. The end-to-end process spans from initial ideation through to model deployment and monitoring in production. Key stages include:

- *MLOps product initiation*: The process starts with a business stakeholder identifying a business problem or opportunity that could be potentially addressed using ML inside the DT environment. A solution architect evaluates technology options and defines the high-level system architecture. Data scientists and data engineers work to specify the ML problem (i.e., classification, regression), according to the business goal. Finally, the raw

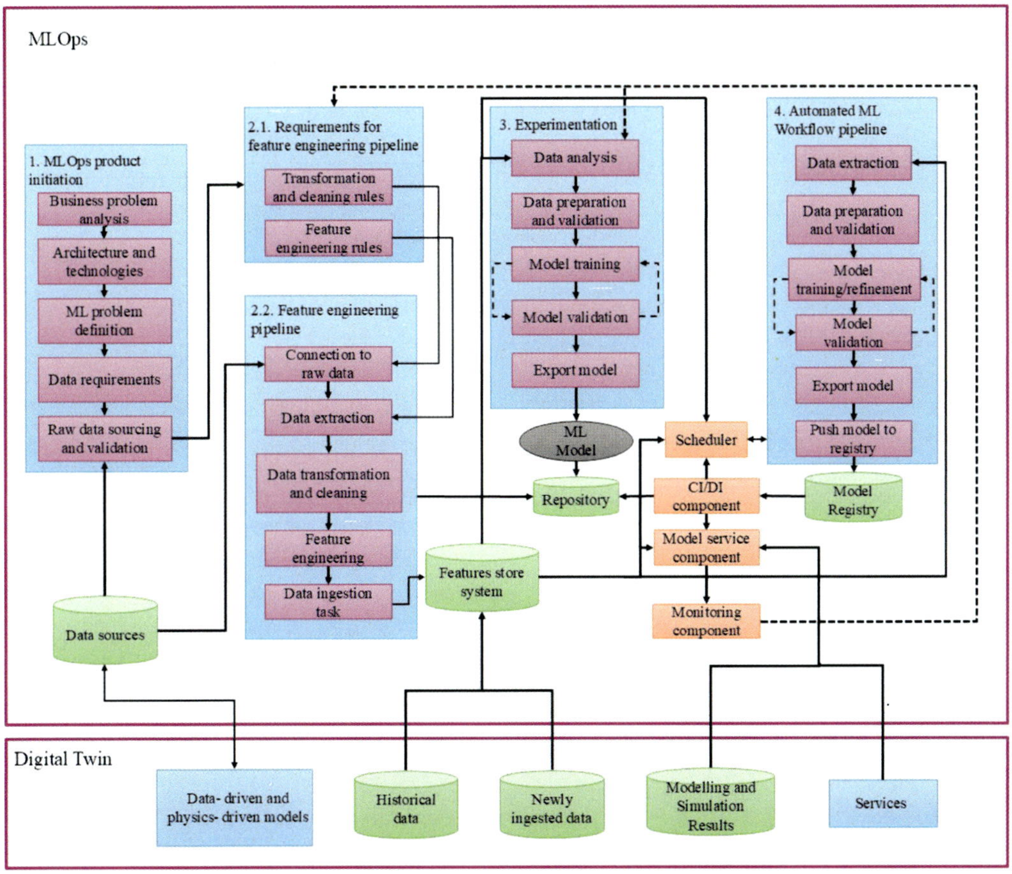

FIGURE 18.4 End-to-end MLOps architecture for integrating AI into DTs.

data is sourced for the initial validation of the ML problem business case, considering that the data must be appropriately labeled to validate the feasibility of the ML algorithm [14].

As an example, consider a railway operator that wants to develop a predictive maintenance regime for wheelset bearings as part of their DT. The product owner works with data scientists to formulate this as a supervised ML problem using, for example, data from on-board and wayside sensors, inspection logs, and maintenance records to generate a bearing predictive maintenance indicator, for example, an anomaly detection qualitative indicator or a remaining-useful-life numeric indicator [17,18] to optimize maintenance schedules.

The team identifies vibration and temperature data from wireless sensors, along with maintenance history data from asset management systems, as key data inputs. Inside the data sources, the team must locate temperature and vibration data correctly labeled for healthy and defective bearings. In the absence of relevant data sources or to augment the capabilities of the ML algorithm, the DT can generate additional data from physics-based and data-based models to complement the initial dataset for the MLOps product initiation.

- *Feature engineering pipeline*: With the use case and data sources clarified, data engineers build a feature engineering pipeline to extract, transform, and store relevant features for ML model development. This involves defining data filtering, transformation, and cleaning logic; calculating data features; and loading feature data into the features store system. ML features will be the inputs to the AI model. Feature engineering rules are initially set at this stage, but these will be iteratively adjusted in the future, based on feedback from experimental modeling stages or from monitoring components that evaluate the ML solution performance [14]. At the pipeline requirements stage, this process is performed manually to verify its viability. Then, the final feature engineering pipeline is automated and located inside the MLOps architecture.

Typical features for bearing acceleration data include time-domain and frequency-domain information such as mean value, peak-to-peak, variance, standard deviation, peak factor, and skewness factor [19]. Temperature features could include the temperature differences between bearings on the same vehicle [20]. Static features like manufacturer, installation date, and maintenance history are joined from enterprise asset management systems. The DT services ensure the feature pipelines are constantly updated with information from vehicle systems and maintenance records.

- *Experimentation*: Data scientists then use the feature data to iteratively train and evaluate candidate ML models. This involves connecting to the features store system, preparing training/test datasets, and experimenting with different model architectures, algorithms, and hyperparameters. Software engineers assist in automating model training code. The best-performing model is saved to a model registry [14]. In this stage, the DT plays a crucial role as it can generate synthetic features that are not measurable with existing sensors or recreate "what if" scenarios based on the vehicle-track multisystem models to increase the robustness of the specific business solution and the safety of the system in the presence of a fault. Furthermore, the DT can provide new features on the bearing prognosis of ML algorithms; these can be, for example, environmental indicators, passenger flow, track condition, or other information that may influence the remaining useful life of a bearing.
- *Automated ML workflow pipeline*: The model training pipeline is automated into a reproducible workflow inside the MLOps architecture. DevOps engineers and ML engineers configure the pipeline to automatically schedule retraining when new feature data is available, use versioned features and models, execute training with the predefined algorithm and hyperparameters, evaluate model performance, and push the results to the model registry if successful. ML metadata stores track training iterations and results. When the model is promoted to production, CI/CD pipelines package and deploy the model to target environments for real-time or batch inference. The served model is continuously monitored for input data drift, service levels, and prediction quality. If issues are detected, the MLOps system can automatically initiate a retraining process [14]. For the DT predictive maintenance application, the model is deployed to an online prediction service that uses real-time sensor data streamed from the train. Predictions are used to optimize maintenance schedules and trigger maintenance actions when bearings near the end of life. Model performance is monitored, and the model is retrained if bearing failure rates deviate from predictions.

In summary, MLOps are a key technology toward the integration of AI into railway DTs. MLOps provide a robust framework for automating the lifecycle of ML models, from data ingestion and transformation, model training and validation, to deployment and monitoring. By adopting MLOps best practices, railway organizations can accelerate the use of AI to enhance DTs for a variety of use cases.

18.3 EXAMPLES OF AI IN DIGITAL TWINS

The use of AI in railway DTs is still in an emerging state, with most reported applications in the scientific literature reaching only the *features engineering requirements* stage of the previously presented MLOps architecture. The following paragraphs present some examples of AI-powered DTs for railway applications reported in the literature.

Padovano et al. [10] demonstrate the successful application of simulation-based DTs to enhance passenger safety and crowd flow management in railway stations. The authors developed a proof of concept of a simulation-based DT for a railway station, able to recreate crowd dynamics, with capabilities for real-time monitoring, predictive analytics, and decision support. The DT integrates data from various sources, such as train schedules, passenger information, and station equipment status, to provide station managers with actionable insights for proactive crowd management such as intelligent gate management and staff allocation. The study validates the effectiveness of the proposed methods and technical solutions, addressing challenges related to data integration, synchronization between the virtual and physical environments, and human–system interaction. The authors argue that the developed DT empowers users, operations managers, station managers, customer experience managers, crowd control staff, and security managers, to make informed decisions and optimize the station performance in terms of safety, efficiency, and passenger experience.

Augmented DTs (ADTs) are implementations that use AI algorithms in the form of surrogate models to replace explicit physics-based models. DTs are especially useful for multiphysics, multi-model problems that require real-time detailed modeling. In [21] an ADT for railway systems is developed by leveraging multibody simulations, high-performance computing, and ML to estimate the derailment risk of individual wagons in real time, considering various operational conditions such as train configuration, speed, curve radius, and wagon loading variations. The ADT utilizes lateral coupler forces estimated from the locomotive longitudinal coupler forces measured onboard, along with other vehicle-track information, to predict the wagon derailment index using a surrogate model. The surrogate model was developed by training simple ML regression models on the results of a massive numerical program consisting of 2,100 vehicle multibody simulations. The proposed ADT enables automated train control systems to consider the derailment risk in real time, which would otherwise be unfeasible due to current onboard computing power limitations. The study demonstrates the effective use of parallel computing and multibody dynamics techniques to generate large datasets for ML applications in DTs for railway systems.

A comparable approach was developed by Bosso et al. [22]. The authors propose a novel simulation technique that enhances the capabilities of longitudinal train dynamics (LTD) simulators by integrating surrogate models of wagons, which act as DTs of multibody simulations. The surrogate models are developed using kernel-based regressions, specifically, support vector machine and least-squares support vector machine, and are trained with the outputs of computationally expensive multibody (MB) simulations of the wagon. This approach significantly reduces the computational time required to calculate safety indexes for all wagons in the train, as it eliminates the need for time-consuming MB simulations once the surrogate models are available. The results demonstrate the feasibility of the proposed approach, with the surrogate models accurately replacing the detailed MB model for all considered simulation scenarios. The authors argue that this methodology paves the way for LTD simulators that can evaluate not only in-train forces but also the main quantities describing the safety and running behavior of each wagon in the train consist.

Liu et al. [23] developed a DT approach for rapid performance evaluation of railway bogie suspension systems. The method integrates surrogate models to represent the suspension system's physical mechanisms and its interaction with the physical world. A consistency metric function allows quick estimation of the bogie operation state, and an LHS-Adam algorithm adaptively updates the model structure. Experiments on a bogie suspension system were performed to validate the method's effectiveness. The DT accurately predicts key performance indicators, such as the maximum derailment index and the root mean square (RMS) of car body lateral and vertical accelerations. The DT features progressively change the wheel–rail equivalent conicity due to wheel wear, maintaining prediction accuracy along the bogie component's lifecycle. The DT also efficiently evaluates different suspension parameter combinations, supporting component selection. According to the authors, compared to traditional simulation-based DTs, the proposed approach significantly reduces computation time for performance evaluations.

18.4 CHALLENGES FOR AI IN DIGITAL TWINS

Development and implementation of AI and DT technologies is an intricate and multidisciplinary endeavor involving specific subject domains in addition to knowledge of software development, data analytics, ML, and project management [23]. While DTs specifically developed for railway systems offer significant potential, the organizational distribution between vehicle operators and infrastructure owners in certain regions poses even more challenges. Successful deployment requires ample data for proper calibration of the AI models. Additionally, a shift in mindset is required among rail operators, infrastructure owners and authorities, promoting cooperation, data sharing, and the consolidation of business resources. Changes in business models, along with financial investments and a comprehensive strategy to address cyber threats, are also essential for successful adoption [24].

In essence, an effective AI-powered DT requires significant organizational changes, interdisciplinary and interdepartmental collaboration inside an organization, and the development of new links between internal stakeholders. Stohr et al. [9] identified five generative mechanisms in the implementation of AI for predictive maintenance in different types of organizations, as shown in Table 18.1, which are also relevant for the implementation of AI technologies in DTs for railway applications. These generative mechanisms summarize the stages an organization must go through to successfully implement AI and DT technologies as part of normal operations.

TABLE 18.1
Generative Mechanisms Encountered in Implementation of AI Technologies Inside Organizations [9]

Mechanism	Definition
Experimentation	Organizations aiming to effectively implement AI technologies must invest in resources for experimentation. These experiments play a crucial role in overcoming limitations and readying the technology for efficient application
Knowledge building and integration	Investing in experimentation resources through training and recruitment promotes collaboration across departments and disciplines. This knowledge-building process supports experimentation with AI-based solutions and facilitates ongoing knowledge creation. However, as the necessary knowledge for AI-enabled DTs becomes more established inside the company, the need for continued investments in training and hiring new personnel will diminish
Data	Engaging with previously gathered data and implementing quality-control measures leads to new insights and stimulates further data collection. However, this cycle of data enrichment is mitigated through a saturation process, where the need for additional data requirements decreases as the amount of usable data grows
Anxiety	The opaque nature of AI can gradually raise user insecurity, potentially leading to resistance toward both ongoing and future projects
Inspiration	Giving employees the freedom to innovate enables them to achieve positive outcomes through experimentation, which, when effectively communicated, inspires others to do the same

REFERENCES

1. ISO/IEC, *BS ISO/IEC 30173:2023 – Digital Twin – Concepts and Terminology*, International Organization for Standardization (ISO) and International Electrotechnical Commission (IEC): ISO/IEC, 2023.
2. M. Spiryagin, J. Edelmann, F. Klinger, and C. Cole, "Vehicle system dynamics in digital twin studies in rail and road domains," *Veh Syst Dyn*, vol. 61, no. 7, pp. 1735–1784, 2023, doi:10.1080/00423114.2023.2188228.
3. P. Stone, R. Brooks, E. Brynjolfsson, R. Calo, O. Etzioni, G. Hager, et al., "Artificial intelligence and life in 2030: The one hundred year study on artificial intelligence," arXiv preprint arXiv:2211.06318, 2022.
4. Canadian Nuclear Safety Commission, UK Office for Nuclear Regulation, and US Nuclear Regulatory Commission, *Considerations for Developing Artificial Intelligence Systems in Nuclear Applications*, Canadian Nuclear Safety Commission, United Kingdom Office for Nuclear Regulation and The United States Nuclear Regulatory Commission, 2024.
5. S. Ahmad, M. Spiryagin, Q. Wu, E. Bernal, Y. Sun, C. Cole, et al., "Development of a Digital Twin for prediction of rail surface damage in heavy haul railway operations," *Veh Syst Dyn*, vol. 62, no. 1, pp. 41–66, 2024.
6. F. Balali, J. Nouri, A. Nasiri, and T. Zhao, *Data Intensive Industrial Asset Management: IoT-based Algorithms and Implementation*, Springer Nature, 2020.
7. L. Bottou, "Large-scale machine learning with stochastic gradient descent," in *Proceedings of COMPSTAT'2010: 19th International Conference on Computational Statistics*, Paris France, August 22–27, 2010, Keynote, Invited and Contributed Papers, Springer, 2010, pp. 177–186.
8. E. Bernal, M. Spiryagin, J. F. Santa-Marín, A. Toro, Q. Wu, and C. Cole, "Real-time digital twin for railway systems," in *IAVSD*, Ottawa, 2023.
9. A. Stohr, P. Ollig, R. Keller, and A. Rieger, "Generative mechanisms of AI implementation: A critical realist perspective on predictive maintenance," *Inf Organ*, vol. 34, no. 2, p. 100503, 2024, doi:10.1016/j.infoandorg.2024.100503.
10. A. Padovano, F. Longo, L. Manca, and R. Grugni, "Improving safety management in railway stations through a simulation-based digital twin approach," *Comput Ind Eng*, vol. 187, p. 109839, 2024.
11. F. Theusch, L. Seemann, A. Guldner, S. Naumann, and R. Bergmann, "Towards machine learning-based digital twins in cyber-physical systems," 2023 [Online]. Available: https://www.dfki.de/en/web/about-us/employee/person/feth02.
12. M. Groshev, C. Guimarães, J. Martín-Pérez, and A. de la Oliva, "Toward intelligent cyber-physical systems: Digital twin meets artificial intelligence," *IEEE Commun Mag*, vol. 59, no. 8, pp. 14–20, 2021.
13. R. Kocielnik, S. Amershi, and P. N. Bennett, "Will you accept an imperfect AI? Exploring designs for adjusting end-user expectations of ai systems," in *Proceedings of the 2019 CHI Conference on Human Factors in Computing Systems*, Glasgow, UK, 2019, pp. 1–14.
14. D. Kreuzberger, N. Kühl, and S. Hirschl, "Machine learning operations (mlops): Overview, definition, and architecture," *IEEE Access*, vol. 11, pp. 31866–31879, 2023.
15. R. Tang, L. De Donato, N. Bešinović, F. Flammini, R.M.P. Goverde, Z. Lin, et al., "A literature review of Artificial Intelligence applications in railway systems," *Transp ResPart C Emerg Technol*, vol. 140, p. 103679, 2022, doi:10.1016/j.trc.2022.103679.
16. A. H. van Bruggen, K. Kruger, A. H. Basson, and J. Grobler, "An architecture to integrate digital twins and machine learning operations," in *International Workshop on Service Orientation in Holonic and Multi-Agent Manufacturing*, Springer, Annecy, France, 2023, pp. 3–14.
17. I. Germoglio Barbosa, A. de O. Lima, J. R. Edwards, and M. S. Dersch, "Development of track component health indices using image-based railway track inspection data," *Proc Inst Mech Eng F J Rail Rapid Transit*, 2024, doi:10.1177/09544097231224825.
18. Y. Zang, W. Shangguan, B. Cai, H. Wang, and M. G. Pecht, "Hybrid remaining useful life prediction method. A case study on railway D-cables," *Reliab Eng Syst Saf*, vol. 213, p. 107746, 2021.
19. C. Yu, Y. Ning, Y. Qin, W. Su, and X. Zhao, "Multi-label fault diagnosis of rolling bearing based on meta-learning," *Neural Comput Appl*, vol. 33, pp. 5393–5407, 2021.
20. E. Bernal, M. Spiryagin, and C. Cole, "Onboard condition monitoring sensors, systems and techniques for freight railway vehicles: A review," *IEEE Sens J*, vol. 19, no. 1, pp. 4–24, 2019, doi:10.1109/JSEN.2018.2875160.
21. E. Bernal, Q. Wu, M. Spiryagin, and C. Cole, "Augmented digital twin for railway systems," *Veh Syst Dyn*, 2023, doi:10.1080/00423114.2023.2194543.

22. N. Bosso, M. Magelli, R. Trinchero, and N. Zampieri, "Application of machine learning techniques to build digital twins for long train dynamics simulations," *Veh Syst Dyn*, vol. 62, no. 1, pp. 21–40, 2024.
23. X. Liu, X. Han, H. Wang, and G. Liu, "A modelling and updating approach of digital twin based on surrogate model to rapidly evaluate product performance," *Int J Adv Manuf Technol*, vol. 129, no. 11–12, pp. 5059–5074, 2023, doi:10.1007/s00170-023-12646-w.
24. W. Phusakulkajorn, A. Núñez, H. Wang, A. Jamshidi, A. Zoeteman, B. Ripke, et al., "Artificial intelligence in railway infrastructure: Current research, challenges, and future opportunities," *Intell Transp Inf*, vol. 2, p. liad016, 2023.

19 Future of Digital Twins in the Rail Industry

Maksym Spiryagin, Stefano Bruni, and Colin Cole

19.1 TRENDS AND DIRECTIONS IN DIGITAL TWIN TECHNOLOGIES

Building on the foundation of smart factories and human-centered manufacturing (Industry 4.0 and Industry 5.0, respectively), the speculative Industry 6.0 concept envisions a future with even more advanced human–machine collaboration, automation, and adaptability. Understanding these advancements requires a proper grounding in the theoretical foundations discussed in the next sections.

19.1.1 System of System vs. Dynamic Stem of System

One possible direction for Industry 6.0 proposes the development of dynamic systems that can continuously adapt over time. It leads to an understanding that a standard definition of 'system of systems' (SoSs) considering non-stoppable digitalization efforts in industries will be further transformed to dynamic SoSs (dynaSoSs), which are characterized by special dynamics (e.g., more frequent context changes) with regard to their continuous adaptation, as mentioned earlier [1,2]. The definition of SoSs is provided in the ISO/IEC/IEEE 21841 standard [3] as a 'set of systems or system elements that interact to provide a unique capability that none of the constituent systems (CS) can accomplish on its own' where CS is defined as an 'independent system that forms part of a system of systems'. In dynaSoS, there are some distinguishing features in comparison with the traditional SoS as defined in [1]: 'more frequent and faster changes of CS in the SoS; more frequent and faster context changes; more frequent and faster changes of the goals of the SoS'. To clarify it as in [1], if we consider the transportation system of a city, it will include the following individual CSs for infrastructure, pedestrians, railway vehicles, trams, vehicles, bicycles, and scooters. And the behavior of each CS will change because the behaviors of other CSs change, and it is also subject to changes in the entire system due to weather changes, pollution, etc. Considering a growth in the number of individual CSs, this assumes that the dynaSoS systems would utilize digital twins and possess cognitive abilities exceeding those of Industry 4.0 and 5.0. This would involve advanced artificial intelligence (AI) that can learn and adapt to both known and unforeseen changes in real time. By analyzing data from a wide range of sources, these AI systems could make decisions and take action even in situations with uncertainties and inconsistencies. This corresponds well with six characteristic aspects of dynaSoSs that are classified as [2]:

1. Heterogeneous Open Systems (Managerial Independence and Operational Independence).
2. Continuously Improved and Innovation-Driven (Evolutionary Development).
3. Complex Systems (Emergent Behavior).
4. Distributed Systems (Geographic Distribution).
5. Data-Intensive Systems.
6. AI-based Autonomy of Constituent Systems.

The five characteristics listed within four parentheses are also applicable to SoS. However, the last two, which include AI, autonomous systems, big data, and data science, are unique defining features

DOI: 10.1201/9781003492146-19

of dynaSoSs. At this stage, the dynamics of a digital twin evolves, necessitating versioning to ensure its instantiation alongside an individual system. This versioning also helps maintain consistency with changes to the physical asset, such as repairs or upgrades. Therefore, a digital twin can be classified as a small dynaSoS with a complexity level ranging from low to high. It may or may not have multiple constituent systems. This explains the progress achieved in Industry 4.0 and 5.0 concepts, where digital twins are major drivers of digital transformation for existing SoSs, as well as their existing limitations. As a result, an incomplete understanding of the interrelationships within a dynaSoS, as well as the computational and data demands, hinders further progress.

19.1.2 Metaverse

The current applications of the metaverse in industrial settings are quite weak, but that trend is rising steadily and will rely on the implementation of the digital twin technologies. The term is defined in the IEEE standard 3812.1:2023 [4] as:

> Metaverse refers to a kind of experience in which the outside world is perceived by the users (human or non-human) as being a universe that is built upon digital technologies as a different universe ('Virtual Reality'), a digital extension of our current universe ('Augmented Reality'), or a digital counterpart of our current universe ('Digital Twin').

There is also some more clarification on the inclusion provided: 'the metaverse includes both extended reality and digital twins, defined by various systems including Internet of Things, Information Technology, and the Internet'. In this case, it is possible to consider the metaverse as an extension of digital twin scopes. As defined in [5], the metaverse represents 'the layer between you and reality', and it is a practical realization of a '3D virtual shared world where all activities can be carried out with the help of augmented and virtual reality services'. From a commonsense perspective, considering use cases of the metaverse as analyzed in [6], the metaverse scenarios are mirror worlds (reflection of the physical world), virtual worlds (augment the economic and social life of the physical world), lifelogging (record the intimate states and life histories of users), and augmented reality (enhance the external physical world for the individual). As stated in [7], the goal of the industrial metaverse is to enable a more collaborative workforce through an immersive, intuitive, and real-time environment. AI would handle highly repetitive tasks, freeing up humans to focus on their strengths: driving creativity, innovation, and inspiration. Key enabling technologies specified in [7] are AI, digital twin, decarbonization technologies, automation and additive manufacturing, and edge and connectivity. The potential applications of the industrial metaverse extend further, including [7]:

- Simulating and testing different scenarios in a virtual world.
- Collaborating virtually on real-world projects, in real-time, and from any location.
- Reducing costs, emissions, and resource consumption.

This raises the question: how can the industrial metaverse be applied to the rail industry, or even more broadly, to transportation systems? Reference [7] provides some initial directions for exploration:

- Decentralized energy production with the deployment of renewable energy, energy storage, and green hydrogen solutions to decarbonize transportation systems
- Circular economy design with the utilization of new digital technologies to design products that support circular economies and improve resource efficiency within the transportation sector
- Integrated mobility solutions with a focus on the integration of regional and high-speed trains with local electric mobility infrastructure for cars, buses, trucks, metros, etc.

Reference [8] offers evaluations of the metaverse's potential impact on transportation systems, highlighting four key areas:

- Immersive services with the delivery of a new level of user experience through immersive technologies
- Healthier transportation systems where novel design approaches in transportation systems promote well-being and reduce negative environmental impacts
- Intelligent transport systems that develop more sophisticated and interconnected transportation networks
- Optimal and flexible transportation systems that create adaptable and efficient transportation systems that meet evolving needs

There are certainly a number of challenges associated with integrating digital twins into the metaverse environment. These challenges include [9]:

- Virtual–physical integration, which also encompasses the level of data integration between the real world, digital twins, and the metaverse.
- User experience and interaction that focuses on ensuring a smooth and intuitive way for users to interact with digital twins in the metaverse.
- Resource management for effective handling of the computational resources needed to run complex digital twins within the metaverse.
- Interoperability that enables seamless communication and data exchange between different digital twins and metaverse platforms.
- Data privacy and security that focuses on protecting sensitive data collected from both the physical world and user interactions within the metaverse.
- Scalability for building systems that can accommodate a growing number of digital twins and users within the metaverse.
- Real-time synchronization that aims to maintain consistent and up-to-date data flow between the physical world, digital twins, and the metaverse.

Addressing these aspects requires the development of relevant implementation concepts, involving digital and integration techniques. As highlighted in [10], digital twins can significantly enhance the realism and functionality of the metaverse. This creates a more immersive experience, potentially transforming interacting techniques and approaches within both virtual and physical spaces, opening up new ways to develop collaborative, exploratory, and problem-solving approaches across numerous application fields. However, these considerations should be specifically tailored at the development stages of digital twins and the metaverse to the use cases, operational needs, and scenarios within the rail domain.

19.1.3 Prospective Directions in Digital Twin Developments

As more and more digital twins are created, use cases will expand significantly alongside a multitude of challenges and opportunities. As analyzed in [11], there will be critical trade-offs to consider in terms of the long-term perspectives on digital twin developments:

- *Balancing model complexity and cost*: Simpler models are cheaper but less accurate, while complex models offer higher accuracy but can be prohibitively expensive. Each case requires careful evaluation to find the optimal balance.
- *Assessing AI opportunities and risks*: The potential of AI in this field needs a thorough assessment to understand both its benefits and potential drawbacks.

- *Standardization for a maturing field*: To ensure the field reaches maturity, validation benchmarks and international standards are urgently needed.

These considerations point toward the need for major technological advancements in software engineering, data science, and quantum computing. By analyzing current trends and revising long-term initiatives outlined in the 2020 NASA Technology Taxonomy [12] for Technology Area TX11 'Software, Modeling, Simulation, and Information Processing', the possible technical areas for digital twin technological developments can be identified and formulated as shown in Table 19.1.

The computing area shows strong interest in four major areas: edge computing, high-performance computing, cloud computing, and quantum computing. The first three areas leverage the achievements and design principles of classical computer technology. They are rapidly evolving, and descriptions of their developments and potential applications are provided in previous chapters of this handbook. Quantum computing, on the contrary, is still an experimental technology in its early stages of development. It harnesses the power of quantum mechanics to tackle problems that are too complex for classical computers. A study published in [13] suggests that quantum computing holds promise for solving complex data-intensive and optimization problems.

It is anticipated that the modeling area should focus further development on integrated hardware and software models; human–system performance modeling; physical and multidisciplinary modeling for online and real-time tasks; and frameworks, platforms, and standards. These items are still quite challenging, and their future directions are quite similar to those stated in [12]:

- Integrated hardware and software modeling should be focused on the development of the following technologies: intelligent hardware and software interface reasoning framework, automated design specification knowledge capture system, dynamic behavior modeling, etc.
- Human–system performance modeling should concentrate on the integration of new and relevant human-related technologies into digital twins.
- Physical and multidisciplinary modeling should aim toward the development of mathematical models to quantify the physical processes. These models should be integrated within a single application domain while considering their multidisciplinary nature and their application for online and real-time digital twin platforms.
- Frameworks, platforms, and standards are the fields that should be further investigated, analyzed, and conceptually developed in digital twin technologies considering the lack of industry-approved documentation in rail and cross-dependent industry domains.

TABLE 19.1
Technological Development Areas from Digital Twin Prospectives

Computing	Modeling	Simulation	Information Processing
Edge computing	Integrated hardware and software models	Distributed simulations	Big data concept and application
High-performance computing	Human–system performance modeling	Integrated system lifecycle simulations	Intelligent data understanding
Cloud computing	Physical and multidisciplinary modeling for online and real-time tasks	Simulation of dynamic system of systems	Data integration, iterability and processing in dynamic system of systems and metaverse platforms
Quantum computing	Frameworks, platforms, and standards	Simulation-based training and decision support systems	Data security and privacy

The simulation area is quite ambitious because it requires generating data by application of a digital twin that can be used along the lifecycle of a physical system. While simulation concepts are continuously developing, there are some technologies that require specific knowledge and the resolution of implementation and deployment difficulties. It is anticipated that the following directions require further development in comparison to what is discussed in [12]:

- Distributed simulations encompass immersive computational environments, infrastructure, and the network integration of heterogeneous computer systems for the relevant simulation tasks.
- Integrated system lifecycle simulations should be established based on the multi-decadal data-driven concept for data storage, processing, and simulation framework to comprehensively analyze the entire lifecycle of a physical system.
- Simulation of dynamic systems of systems should focus on tighter dynamic integration between models, data, and decision-making within each system and across systems. The primary emphasis should be on bidirectional interactions, not only between the virtual world systems within the SoSs but also between the physical system and its corresponding virtual counterpart.
- Simulation-based training and decision support systems should be focused on the development of digital-human-in-the-loop simulation systems for digital twins. This direction still requires significant developments not only in the rail domain but also across other industries such as automotive and aerospace.

The area of information processing is experiencing significantly faster development due to its strong alignment with research and development topics in other industries. These developments are well-defined in terms of both their general directions and their specific applications in the rail industry. However, there are some specific new areas that should be further developed in terms of specific rail-related application tasks: big data concept and application; intelligent data understanding; data integration and iterability and processing in dynamic systems of systems and metaverse platforms; and data security and privacy. It's important to note that research in this area can be interconnected. For example, big data technology plays a central role in unlocking the full potential of the metaverse. Of course, ethical considerations around data privacy and security will be dictated by critical requirements as the metaverse takes shape. However, when forming big data clusters and developing processing algorithms using AI techniques, the operational environment and specific requirements of the rail domain must also be considered and further investigated.

The list of other prospective directions in digital twin developments can be further extended considering particular and specific application aspects; however, base core topics and areas will stay the same not only for rail digital twin systems but also for other intelligent systems used in other domains.

Though specific operational tasks and applications will influence the finer points of digital twin development and suggest other prospects or even new directions, the core technological areas as outlined in Table 19.1 hold immense promise across various domains, including but not limited to rail systems. These areas could encompass advancements in real-time data integration, the incorporation of AI for predictive maintenance, the leveraging of building information models in digital twins, and the development of robust interoperability standards that enable seamless communication between digital twins across different sectors. Therefore, the scope of technological developments is theoretically limitless, and any rapid progress or advancements in computational capabilities could significantly expand the scope in the near future.

19.2 EMERGING APPLICATIONS

Emerging applications often share common trends across various industry sectors. In the railway sector, these trends allow us to identify key emerging areas for digital twins:

- *Asset management and predictive maintenance*: Digital twins in this area are designated for real-time monitoring and prediction of lifecycle and maintenance procedures for physical assets and their components to reduce downtime and maintenance costs. In the case of AI prospects in such applications, AI-powered digital twins can monitor the condition of railway physical assets, aiding in asset management tasks. AI algorithms analyze real-time monitoring and computed data from trains, tracks, infrastructure, and other physical assets or third-party systems. This analysis allows for the detection of anomalies and the prediction of potential failures before they occur. Additionally, AI can automatically detect changes, defects, or non-compliance with standards in assets, reducing reliance on manual processes. By integrating AI models into digital twins, human operators benefit from decision support, recommendations, and insights. This integration leverages the strengths of both artificial and human intelligence.
- *Design and modernization*: Digital twins are focused on the virtual approbation of new and existing design concepts and their further optimization for physical assets to understand performance, life cycle, associated costs, and risks before actual physical implementation. By implementing AI in the design and modernization process, rolling stock and railway operators can simulate system expansions, upgrades, new line constructions, or passenger flow or car movements in problematic locations. These virtual capabilities facilitate the assessment of potential impacts and plan optimization before any physical changes occur.
- *Operational efficiency and safety*: Digital twins assist in replicating train operations, infrastructure conditions, and their potential failures, along with passenger flow variability to optimize operations, improve safety and performance, and reduce operational delays. By implementing AI in this application area, different scenarios such as optimization of train schedules, routing, energy usage, and other operational parameters can be achieved. AI can also analyze data on crew scheduling, training, and performance to optimize human resource allocation inside railway company operational practices. In addition, it can be applicable to simulate potential hazards, accidents, or failures, helping identify risks and develop mitigation strategies to improve railway safety.
- *Data integration and cross-system collaboration*: Digital twins are used as a platform for data integration and data sharing concepts among different systems operated by stakeholders including manufacturers, operators, and infrastructure managers, fostering collaboration and system-level optimization. The main challenge here is that different systems (engineering design tools, sensor networks, maintenance platforms) use different data formats and languages. This heterogeneity creates hurdles in seamlessly integrating data into a digital twin, ultimately hindering its effectiveness. To overcome this, the data structures and terminologies of various systems should be translated into a common language understandable by digital twins. In this domain, AI applications can play a crucial role by compensating for data uncertainties and inconsistencies encountered during digital twin integration.

19.3 ROADMAP FOR RAIL DIGITAL TWIN DEVELOPMENTS

The analysis performed in reference [14] presented a roadmap which indicated that the development of digital twin simulations and architectures are both making steady progress. This trend is likely to continue if advancements in computing, simulation, and information technology continue apace. These advancements will accelerate big data processing performance, ultimately enabling the creation of more precise and accurate digital twins within their operational environments. In this case, the digital twin strategic roadmap that covers applications of vehicle system dynamics as presented in Figure 19.1 can be considered a valid roadmap for simulation-based digital twins.

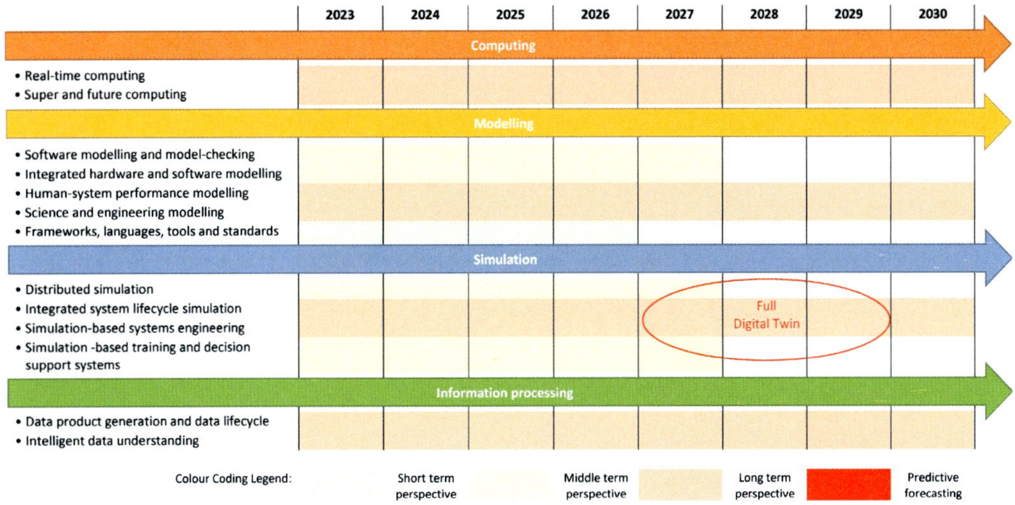

FIGURE 19.1 Digital twin strategic roadmap established on simulation-based technologies. (From Spiryagin M, Edelmann J, Klinger F, et al. Vehicle system dynamics in digital twin studies in rail and road domains. *Vehicle System Dynamics* 2023, vol. 61(7), 1737–1786, with permission.)

However, this roadmap excludes predictions for timelines in AI advancements and quantum computing development. These areas are inherently difficult to predict due to their reliance on ongoing research and breakthroughs across various scientific disciplines and industries. The IBM quantum roadmap [15] outlines two key milestones: 'Scale quantum computing' by 2027 and 'Deliver a fully error-corrected system' by 2029. Consequently, this also indirectly aligns with Figure 19.1, which projects the completion of a full digital twin by 2029 using classical computing technologies. However, any significant progress in dynamic systems of systems and metaverse platforms likely hinges on the successful development of a fully error-corrected quantum computer. This scenario, while conceptually interesting, might be more philosophical than realistic in the near term.

REFERENCES

1. Heinrich J, Balduf F, Becker M, et al. Industry voices on software engineering challenges in dynamic systems of systems. In *2023 IEEE/ACM 11th International Workshop on Software Engineering for Systems-of-Systems and Software Ecosystems (SESoS)*, Melbourne, Australia, 2023, pp. 58–64.
2. Adler R, Elberzhager F, Falcão R, et al. Defining and researching "Dynamic systems of systems". *Software* 2024, vol. 3(2), 183–205.
3. ISO/IEC/IEEE 21841:2019. *ISO/IEC/IEEE International Standard – Systems and Software Engineering – Taxonomy of Systems of Systems*, 19 July 2019, pp. 1–20, doi:10.1109/IEEESTD.2019.8766998.
4. IEEE Std 3812.1-2023. *IEEE Standard for General Requirements for Identity Framework for Metaverse*, IEEE Consumer Technology Society, 21 September 2023. ISBN 979-8-8557-0312-2.
5. Damar M. Metaverse shape of your life for future: A bibliometric snapshot. *Journal of Metaverse* 2021, vol. 1(1), 1–8.
6. Dwivedi YK, Hughes L, Baabdullah AM, et al. Metaverse beyond the hype: Multidisciplinary perspectives on emerging challenges, opportunities, and agenda for research, practice and policy. *International Journal of Information Management* 2022, vol. 66, 102542.
7. Siemens. Managing Megatrends, Siemens AG, Munich, Germany, 2023. Available at: https://www.siemens.com/global/en/company/about/strategy/siemens-megatrends.html. Accessed on 28 May 2024.
8. Njoku JN, Nwakanma CI, Kim DS. Building a metaverse for transportation systems: A brief review and demonstration. In *The Proceedings of 2023 Fourteenth International Conference on Ubiquitous and Future Networks (ICUFN)*, Paris, France, 2023, pp. 544–549.

9. Alryalat M, Rajan R, Dhir S, et al.. Assessing the challenges to digital twins in the metaverse environment: Strategic management decisions perspective. *Kurdish Studies* 2023, vol. 11(2), 3037–3053.
10. Cruz M, Oliveira A. Where are we now? - Exploring the metaverse representations to find digital twins. *Electronics* 2024, vol. 13(10), 1984.
11. The increasing potential and challenges of digital twins. Editorial. *Natural Computational Science* 2024, vol. 4, 145–146.
12. 2020 NASA Technology Taxonomy. National Aeronautics and Space Administration, Washington, DC, 2020. Available at https://www3.nasa.gov/sites/default/files/atoms/files/2020_nasa_technology_taxonomy.pdf. Accessed on 28 May 2024.
13. Rasheed A, San O, Kvamsdal T. Digital twin: Values, challenges and enablers from a modeling perspective. *IEEE Access* 2020, vol. 8, 21980–22012.
14. Spiryagin M, Edelmann J, Klinger F, et al. Vehicle system dynamics in digital twin studies in rail and road domains. *Vehicle System Dynamics* 2023, vol. 61(7), 1737–1786.
15. IBM Technology Atlas. Quantum Roadmap. Available at https://www.ibm.com/roadmaps/quantum/. Accessed on 29 May 2024.

20 Conclusion

Maksym Spiryagin, Stefano Bruni, and Colin Cole

The editors' intent in preparing and publishing this handbook was to systemize and make available the existing knowledge in digital twin (DT) and artificial intelligence (AI) technologies with their specific applications to rail domain tasks. A railway network is commonly considered an integrated system of systems performing highly complex technical tasks. It is, therefore, necessary that the constructive study presented in this handbook should be comprehensively detailed with regard to the main system, subsystems, and components for the development of DT and AI technologies and their proper integration in the desired computerized final solution. As a result, the chapters of the handbook have been structured starting from the basic theoretical background and terminology and progressing to more detailed explanations of DT and AI theories, approaches, and analysis techniques applicable to railway applications. Some reasonable ideas on frameworks and methodologies built based on DT and AI technologies are analyzed and presented for those interested in actual existing concepts applicable to the rail domain. The editors and contributors hope that readers have found the information flow properly systemized to allow easy following and understanding of topics presented in each chapter. In addition, each chapter contains many references, which should allow readers to further investigate the existing theories and knowledge base in the case of a particular field of personal interest or even to further develop any of the presented materials to address a future possible problem to achieve a final solution related to rail domain-specific DT- and/or AI-related tasks.

Readers with enquiries regarding any materials presented in this handbook are free to directly contact the editors or the contributors of each chapter. Constructive review and feedback on this handbook are always welcome.

DOI: 10.1201/9781003492146-20

Index

Made in the USA
Las Vegas, NV
31 July 2026

f04e5199-63d8-4569-9890-51b6fda80c51R01